Ion Spectroscopies for Surface Analysis

METHODS OF SURFACE CHARACTERIZATION

Volume 1 VIBRATIONAL SPECTROSCOPY OF MOLECULES ON SURFACES
Edited by John T. Yates, Jr., and Theodore E. Madey

Volume 2 ION SPECTROSCOPIES FOR SURFACE ANALYSIS
Edited by A. W. Czanderna and David M. Hercules

Ion Spectroscopies for Surface Analysis

Edited by
A. W. Czanderna
Solar Energy Research Institute
Golden, Colorado

and

David M. Hercules
University of Pittsburgh
Pittsburgh, Pennsylvania

PLENUM PRESS • NEW YORK AND LONDON

Library of Congress Cataloging-in-Publication Data

Ion spectroscopies for surface analysis / edited by A.W. Czanderna and
David M. Hercules.
p. cm. -- (Methods of surface characterization ; v. 2)
Includes bibliographical references and index.
ISBN 0-306-43792-9
1. Solids--Surfaces--Analysis. 2. Secondary ion mass
spectrometry. 3. Surface chemistry. 4. Surface chemistry.
I. Czanderna, Alvin Warren, 1930- . II. Hercules, David M.
III. Series.
QC176.8.S8I66 1991
541.3'3--dc20 91-19636
CIP

ISBN 0-306-43792-9

A Division of Plenum Publishing Corporation
233 Spring Street, New York, N.Y. 10013

Printed in the United States of America

Contributors

Christopher H. Becker, Molecular Physics Laboratory, SRI International, Menlo Park, California 94025

Kenneth L. Busch, School of Chemistry and Biochemistry, Georgia Institute of Technology, Atlanta, Georgia 30332

A. W. Czanderna, Applied Sciences Branch, Solar Energy Research Institute, Golden, Colorado 80401

Leonard C. Feldman, AT&T Bell Laboratories, Murray Hill, New Jersey 07974

Barbara J. Garrison, Department of Chemistry, Pennsylvania State University, State College, Pennsylvania 16801

David M. Hercules, Department of Chemistry, University of Pittsburgh, Pittsburgh, Pennsylvania 15260

C. J. Powell, Surface Science Division, National Institute of Standards and Technology, Gaithersburg, Maryland 20899

E. Taglauer, Surface Physics Department, Max-Planck-Institut für Plasmaphysik, EURATOM-Association, D-8046 Garching bei München, Germany

Nicholas Winograd, Department of Chemistry, Pennsylvania State University, University Park, Pennsylvania 16802

About the Series

A large variety of techniques are now being used to characterize many different surface properties. While many of these techniques are relatively simple in concept, their successful utilization involves employing rather complex instrumentation, avoiding many problems, discerning artifacts, and carefully analyzing the data. Different methods are required for handling, preparing, and processing different types of specimen materials. Many scientists develop surface characterization methods, and there are extensive developments in techniques reported each year.

We have designed this series to assist newcomers to the field of surface characterization, although we hope that the series will also be of value to more experienced workers. The approach is pedagogical or tutorial. Our main objective is to describe the principles, techniques, and methods that are considered important for surface characterization, with emphasis on how important surface characterization measurements are made and how to ensure that measurements and interpretations are satisfactory, to the greatest extent possible. At this time, we have planned four volumes, but others may follow.

The first volume brought together a description of methods for vibrational spectroscopy of molecules on surfaces. Most of these techniques are still under active development; commercial instrumentation is not yet available for some techniques, but this situation could change in the next few years. The current state of the art of each technique was described as were their relative capabilities. An important component of the first volume was the summary of the relevant theory.

This book is the first of two volumes that contain descriptions of the techniques and methods of electron and ion spectroscopies which are in widespread use for surface analysis. These two volumes are and will be largely concerned with techniques for which commercial instrumentation is available. The books are intended to fill the gap between a manufacturer's handbook, and review articles that highlight the latest scientific developments.

A fourth volume will deal with techniques for specimen handling, beam artifacts, and depth profiling. It will provide a compilation of methods that have proven useful for specimen handling and treatment, and it will also address the common artifacts and problems associated with the bombardment of solid surfaces by photons, electrons, and ions. A description will be given of methods for depth profiling.

Surface characterization measurements are being used increasingly in diverse areas of science and technology. We hope that this series will be useful in ensuring that these measurements can be made as efficiently and reliably as possible. Comments on the series are welcomed, as are suggestions for volumes on additional topics.

C. J. Powell
Gaithersburg, Maryland
A. W. Czanderna
Golden, Colorado
D. M. Hercules
Pittsburgh, Pennsylvania
T. E. Madey
New Brunswick, New Jersey
J. T. Yates, Jr.
Pittsburgh, Pennsylvania

Preface

Determining the elemental composition of surfaces is an essential measurement in characterizing solid surfaces. At present, many approaches may be applied for measuring the elemental and molecular composition of a surface. Each method has particular strengths and limitations that often are directly connected to the physical processes involved. Typically, atoms and molecules on the surface and in the near surface region may be excited by photons, electrons, ions, or neutrals, and the detected particles are emitted, ejected, or scattered ions or electrons.

The purpose of this book is to bring together a discussion of the surface compositional analysis that depends on *detecting* scattered or sputtered *ions,* and the methods emphasized are those where instruments are commercially available for carrying out the analysis. For each topic treated, the physical principles, instrumentation, qualitative analysis, artifacts, quantitative analysis, applications, opportunities, and limitations are discussed.

The first chapter provides an overview of the role of elemental composition in surface science; compositional depth profiling; stimulation by an electric field, electrons, neutrals, or photons and detection of ions; and then stimulation by ions, and detection of ions, electrons, photons, or neutrals.

The second chapter deals with the molecular dynamics involved for sputtered particles, which are important in secondary ion mass spectrometry (SIMS). The third chapter deals with particle-induced desorption, particularly, as SIMS is applied to studying organic materials. The fourth chapter describes a SIMS method for surface analysis by postionizing sputtered neutrals with a laser. Thus, in the first four chapters various aspects of SIMS are considered that complement the book by Benninghoven, Rüdenauer, and Werner. The fifth chapter is on Rutherford backscattering, nuclear reaction analysis, and hydrogen forward scattering as a high-energy (~1 MeV) technique. The sixth chapter describes ion scattering as a low-energy (~1 keV) method for surface compositional and adsorbate-structural analysis. The final chapter presents comparisons of the major techniques described in this volume and in a volume to follow on Auger and X-ray photoelectron spectroscopies. Several ASTM standards relevant to ion spectroscopies for surface compositional analysis are reprinted in the Appendix; these were originated by the E42 Committee on Surface Analysis.

In view of the recent treatise on SIMS[1] we have focused on several rapidly developing aspects of SIMS. We recognize that there are other important topics in this area, such as electron-impact postionization of sputtered neutrals, that have not been included. These will have to be addressed in the future. The contributions in this volume also complement the book by Wilson, Stevie, and Magee.[2]

The editors are deeply gratefully to the authors whose work made this book possible, and for taking the time from their active research programs to prepare their contributions.

A. W. Czanderna
Golden, Colorado
D. M. Hercules
Pittsburgh, Pennsylvania

[1] A. Benninghoven, F. G. Rüdenauer, and H. W. Werner, *Secondary Ion Mass Spectrometry, Basic Concepts, Instrumental Aspects, Applications and Trends,* Wiley, New York (1987).

[2] R. G. Wilson, F. A. Stevie, and C. W. Magee, *Secondary Ion Mass Spectrometry, A Practical Handbook for Depth Profiling and Bulk Impurity Analysis,* Wiley, New York (1989).

Contents

1

Overview of Ion Spectroscopies for Surface Compositional Analysis

A. W. Czanderna

Glossary of Acronyms

AEM	Analytical electron microscopy
AES	Auger electron spectroscopy
APFIM	Atom probe field ion microscopy
ARUPS	Angle-resolved ultraviolet photoelectron spectroscopy
ATR	Attenuated total reflection
BET	Brunauer, Emmett, and Teller
BLE	Bombardment-induced light emission
CDP	Compositional depth profiling
CVD	Chemical vapor deposition
DIET	Desorption induced by electronic transitions
EDX	Energy dispersive x-ray analysis
EELS	Electron energy loss spectroscopy
EIID	Electron impact ion desorption
ELEED	Elastic low-energy electron diffraction
EM	Electron microscopy

A. W. Czanderna • Applied Sciences Branch, Solar Energy Research Institute, Golden, Colorado 80401.

Ion Spectroscopies for Surface Analysis (Methods of Surface Characterization series, Volume 2), edited by Alvin W. Czanderna and David M. Hercules. Plenum Press, New York, 1991.

EMP	Electron microprobe (analyzer)
EPMA	Electron probe microanalysis (analyzer)
ER	Electromagnetic radiation
ESCA	Electron spectroscopy for chemical analysis (see also XPS)
ESD	Electron-stimulated desorption
ESDIAD	Electron–stimulated desorption ion angular distribution
EXAFS	Extended x-ray fine structure spectroscopy
FAB–SIMS	Fast atom bombardment–secondary ion mass spectrometry
FEM	Field emission microscopy
FIM	Field ion microscopy
FIMS	Field ion mass spectrometry
FT-IR	Fourier transform infrared (spectroscopy)
FWHM	Full width at half-maximum
GDMS	Glow discharge mass spectrometry
GDOS	Glow discharge optical spectrometry
G/S	Gas/solid interface
HEED	High-energy electron diffraction
HFS	Hydrogen forward scattering spectrometry
HV	High vacuum
HVEM	High-voltage electron microscopy
IIR	Ion-induced radiation
ILEED	Inelastic low-energy electron diffraction
IMMA	Ion microprobe mass analyzer
INS	Ion neutralization spectroscopy
IR	Infrared (spectroscopy)
ISS	Ion scattering spectroscopy
LAMMA	Laser microprobe mass analysis
LEED	Low-energy electron diffraction
L/G	Liquid/gas interface
LIMS	Laser ionization mass spectrometry
L/L	Liquid/liquid interface
LMP	Laser microprobe
LPE	Liquid phase epitaxy
LRS	Laser Raman spectroscopy
L/S	Liquid/solid interface
MBE	Molecular beam epitaxy
MRIS	Multiple reflectance infrared spectroscopy
NRA	Nuclear reaction analysis
NSS	Neutral scattering spectrometry
PAI	Post-ablation ionization
PIXE	Particle-induced x-ray emission
PV	Photovoltaic

PVD	Physical vapor deposition
QMS	Quadrupole mass spectrometer
RAIR	Reflection absorption infrared
RBS	Rutherford backscattering spectroscopy
RGA	Residual gas analysis
RHEED	Reflection high-energy electron diffraction
SALI	Surface analysis by laser ionization (of sputtered neutrals)
SAM	Scanning Auger microscopy
SARISA	Surface analysis by resonance ionization of sputtered atoms
SBEC	Single binary elastic collision
SCANIIR	Surface compositional analysis by neutral and ion impact radiation
SECS	Solar energy conversion systems
SEM	Scanning electron microscopy
SEXAFS	Surface extended x-ray fine structure
S/G	Solid/gas interface
SIMA	Secondary ion mass analyzer
SIMS	Secondary ion mass spectrometry
S/L	Solid/liquid interface
SNMS	Secondary neutral mass spectrometry
S/S	Solid/solid interface
STEM	Scanning transmission electron microscopy
STM	Scanning tunneling microscopy
S/V	Solid/vacuum interface
TED	Transmission electron diffraction
TEM	Transmission electron microscopy
TOF–AP	Time-of-flight atom probe
TOF–MS	Time-of-flight mass spectrometry
TOF–SIMS	Time-of-flight secondary ion mass spectrometry
UHV	Ultra-high vacuum
UPS	Ultraviolet photoelectron spectroscopy
UV	Ultraviolet
VHV	Very high vacuum
XPS	X-ray photoelectron spectroscopy (see also ESCA)
XRD	X-ray diffraction
XRF	X-ray fluorescence

1. Purposes

The purposes of this chapter are (1) to provide a brief overview of the role of surface compositional analysis (i.e., surface analysis) in surface characterization, (2) to provide a brief overview of the methods of surface analysis involving ion bombardment of a solid or ion detection, and (3) to discuss briefly each of the methods of surface analysis using the ion spectroscopies that are not treated by the other chapters in this volume. Comparisons and contrasts of the most widely used methods of surface analysis are discussed in Chapter 7 of this volume. Extensive literature citations are provided, so that further details about surface analysis methods can be obtained. Both the surface (i.e., the outer monolayer of atoms on a solid) and the interface (i.e., the boundary between two compositionally different solids) are included in the term "surface" as it is used in this chapter.

2. Introduction

2.1. Role of Surface Analysis in Surface Characterization

An overview of the most important phenomena in surface science related to studying materials is presented first in this section. The three phases of interest are the solid (S), liquid (L), and gas (G) phases, none of which is infinite. The boundary region between these phases, i.e., the surface phase, has fundamentally different properties from those of the bulk. The possible boundaries are the S/S, S/L, S/G, L/L, and L/G surfaces. These boundaries are studied in surface science to develop an understanding of phenomena and to develop theories that will permit predicting future events. Some of the broad *topical areas* of study at the S/G, S/L, and S/S surfaces are listed in Table 1. (See also Tables 11 and 12, Chapter 7 in this volume.) An understanding of these topics is enhanced by applying the methods of surface analysis. A list of acronyms

Table 1. Topical Study Areas at Different Interfaces Between the Solid (S), Liquid (L), and Gas (G) Phases

Interface		Topical area of study
Solid A / Solid B	S/S	Corrosion, grain boundary passivation, adhesion, delamination, epitaxial growth, nucleation and growth, abrasion, wear, friction, diffusion, boundary structure, thin films, solid state devices, mechanical stability, creep.
Liquid / Solid	S/L	Wetting, spreading, lubrication, friction, surface tension, capillarity, electrochemistry, galvanic effects, corrosion, adsorption, nucleation and growth, ion electromigration, optical properties, cleaning techniques.
Gas / Solid	S/G	Adsorption, catalysis, corrosion, oxidation, diffusion, surface states, thin films, condensation and nucleation, permeation, energy transfer.

and abbreviations used in this chapter has been given.

The S/S, S/G, and S/L surfaces can be studied using surface analysis. As recent publications show,[1–16] the experimental work conducted to study these surfaces is now very extensive; the theoretical treatments are difficult. In science, it is customary to adopt a model based on an ideal situation and to compare the behavior of real systems with the ideal model. What is a realistic view of the boundary at a solid surface? It is not the ideal plane of infinite dimensions, but on an atomic scale it consists of different crystal planes with composition, structure, orientation, and extent that are fixed by the pretreatment of the solid. Imperfections such as an isolated atom, a hole, an edge, a step, a crevice, a corner, and a screw dislocation may also coexist on the surface. Wide variations in the microscopic topography may also adversely influence the stability of the surface.

Solid surfaces are frequently treated in a variety of ways (e.g., outgassing, chemical reduction, flashing a filament, ion bombardment, cleavage, field desorption, or depositing a thin film[17]) to minimize uncertainty about the initial composition of the surface. However, impurities may accumulate during some of these treatments at or in the boundary, in trace or larger quantities, and drastically alter the behavior of the boundary. Following a controlled use of the solid, a reexamination of the surface permits us to evaluate the influence of that use on the measured properties of the surface. Characterization of the surface before, after, and if possible, during the use of the material is clearly required.

For an overview of characterizing a solid surface, consider these questions: How much surface is there and where is it located? Is the surface real or clean? What solid form does the surface have? What is its topography and structure? What thermodynamic processes occur? How do surface species migrate? What is the equilibrium shape of surfaces? What is the depth of the surface phase? How much gas or liquid is adsorbed and where? What is the nature of the adsorbate–solid interaction? How should these phenomena be studied? The history of studying these effects in gas adsorption on solid surfaces alone leads us to recognize that careful experimentation is the primary necessity in surface science. As each of the characterization questions are briefly addressed in the following paragraphs, the experimental methods deemed most appropriate for studying materials will be indicated.

The concept of a surface is not well defined, especially for most samples of industrial interest. The structure and composition of a material often deviate from their bulk values at depths of only nanometers and in a spatially inhomogeneous manner. As is well known,[18] each method of surface spectroscopy has its own depth and areal resolution and, therefore, presents its unique view of the inhomogeneous region of the solid. The information available from surface analysis includes identifying the elements present, their lateral distribution, and their depth distribution; structural and topographical information can also be obtained.[19]

Compositional analysis, the primary subject of this chapter (and book), involves determining three quantities. The elemental identity, i.e., the atomic number, is of primary interest. However, it is also desirable to know the chemical state of the species, e.g., whether it is elemental, oxidized, or reduced. Finally, it is necessary to determine the spatial distribution of the chemical species. An important trend in compositional surface analysis is the demand for greater and greater lateral resolution; e.g., a resolution of 20 nm can now be routinely reached by using Auger spectroscopy. The study of compositional differences between grains and the grain boundary region is an extremely important area of materials science for deducing how these differences change the materials properties.[20]

Structural surface analysis also involves three levels of desired information.[19] For an ideal, atomically flat, single crystal, the structure is specified by the geometry in each cell of the surface unit mesh. Secondly, real surfaces contain defects, such as dislocations, steps, kinks, ledges and grain boundaries. Finally, a new trend is emerging for determining local atomic order for a particular chemical species. Determining surface structure is essentially limited to studies of solid–vacuum (S/V) or S/G interfaces. This subject is treated only briefly; there are numerous books on techniques such as LEED, ELEED, ILEED, RHEED, etc.

Most studies of chemical bonding at surfaces are related to the S/G and S/L interfaces. The latter can be probed with ion and electron beams but only with difficulty.[21] The adsorption of gases on solids can be considered a significant branch of surface science. Although adsorption is involved in many processes of interest to the materials scientist (e.g., corrosion, contamination, and catalysis), the scope of this volume would have to be broadened considerably to deal with the use of probe beams for studying the phenomena related to adsorption.[22] Similarly, atomistic dynamics and the electronic structure at the S/V or the S/G interfaces are not considered,[19] even though ion beams may be used as part of the surface analysis.

2.1.1. Surface Area

Following Langmuir's pioneering work about 70 years ago, in which the importance of structure, composition, and bonding to chemisorption were demonstrated, Brunauer, Emmett, and Teller (BET) provided a means for deducing the surface area from multilayer physical adsorption isotherms.[23] The BET method for analyzing adsorption isotherms is used routinely by hundreds of laboratories with commercially available equipment.[24] When the surface is located internally, hysteresis is observed between the adsorption and desorption branches of the isotherm. The hysteresis results from capillary filling of internal pores. An excellent treatise on the surface area and porosity of solids is available,[25] extensive examples from recent studies of this subject have been cataloged,[26] and the parameters used to differentiate between physical and chemical adsorption have been tabulated.[26] Visual observation of the topographical features of solids is possible using electron microscopic (EM) examination of replicas of surfaces, or a scanning electron microscope (SEM); light scattering, the stylus method, and scanning tunneling microscopy also provide information about topography. An assessment of the external surface area can be made from SEM or EM photographs, but obviously the internal area is not directly observable. Thus, qualitative and quantitative answers can be provided to questions about how much surface there is and where it is located, except for S/S surfaces.

2.1.2. Real and Clean Surfaces; Solid Forms

Broadly speaking, real surfaces are those obtained by ordinary laboratory procedures, e.g., mechanical polishing, chemical etching, industrial processes, etc. Such a surface may react with its environment and be covered with an oxide (generally), possibly with chemisorbed species and by physically adsorbed molecules from the surroundings. Real surfaces have been studied extensively because they are easily prepared, readily handled, and amenable to many types of measurements. The real surface is the one encountered in most practical applications.

Clean surfaces, which may be obtained by outgassing, chemical reduction, cleaving, field desorption, ion bombardment and annealing, preparing a thin film in ultrahigh vacuum (UHV), or flashing a filament,[17] are more difficult to prepare and keep clean. Clean single-crystal surfaces are useful for comparing theories with experimental results. They can be prepared from ordered single crystals and can then be perturbed in carefully controlled experiments. Fundamental studies on clean, well-ordered surfaces are increasing the basic understanding of S/G interactions in the modeling tradition of the kinetic theory, the dilute solution, the point particle, and the harmonic oscillator.

In surface studies, various solid forms are used, such as powders, foils, vacuum-prepared thin films, coatings, filaments, cleaved solids, field emitter tips, single crystals, and polycrystalline solids. These may be formed from metals, semiconductors, and compounds.

2.1.3. *Structure and Topography*

A clear distinction must be made between structure and topography, which are frequently used interchangeably (and incorrectly) in the literature. Topography can be illustrated with a few SEM photographs showing contouring, hills and valleys, and superatomistic surface features. Structure refers to the repetitive spacing of atoms in a surface grating. The results of surface structural determinations show that the lattice spacing in the bulk and at surfaces is the same for most metals, but that halide and many semiconductor surfaces may be relaxed or reconstructed. The most common methods of deducing surface structure are by the diffraction of low-energy electrons, either elastically or inelastically (ELEED and ILEED), by reflected high-energy electrons (RHEED), and by using the imaging techniques of the field ionization microscope (FIM).[1,8–10,12,13] The value of the LEED, RHEED, and FIM techniques to fundamental surface science is evident in the extensive results that are secured by using them; many examples can be found in the journal *Surface Science*.

2.1.4. *Surface Thermodynamics, Equilibrium Shape, and Diffusion*

Surface atoms are in a markedly different environment from that of bulk atoms. Surface atoms are surrounded by fewer neighboring atoms, which are in a surface-unique anisotropic distribution, compared with those in the bulk. The surface phase has a higher entropy, internal energy, work content (Helmholtz energy), and free energy (Gibbs energy) per atom than the bulk phase. For isotropic solids, surface free energy and surface tension are equivalent.[8–10] At equilibrium, the solid surface will develop the shape that corresponds to the minimum value of total surface free energy. The surface free energy of solids can be calculated for surfaces of different structures. The relative magnitude of surface free energy can be deduced from experiments involving thermal faceting, grain boundary grooving, etc. The low index planes are (usually) the most stable because they have the lowest free energy. All small particles and all large, flat, polycrystalline surfaces, which are characteristic of the type encountered in most applications, possess a relatively large surface excess energy and are thermodynamically unstable.

Based on calculations made for inert gases,[27] ionic halides,[28] and semiconductors,[29] the depth of the surface phase, i.e., that part of the interface with unique properties, may extend to 5 nm. In the latter two cases, substantial deviations in the actual surface structures from those characteristic of a truncated but otherwise bulk solid have been reported from theoretical analyses of LEED intensities from single crystals. The reconstruction or rearrangement of the surface involves movements up to 0.05 nm in the surface atomic plane and possibly in the first bulk layer as well.[29]

Surface atoms are restrained from interatomic motion by nearest-neighbor bonding, but these bonds are weaker than those at a corresponding position in the bulk. The potential barriers for surface diffusion are lower than those for bulk diffusion, so less activation energy is required to produce surface diffusion processes.[22] The mechanism of surface diffusion may change with temperature, because the surface population of adatoms, adions, vacancies, ledges, and the like, or other conditions such as structure and ambient atmosphere, may change.

2.1.5. Amount Adsorbed and Nature of Adsorbate/Solid Interactions

This discussion applies primarily to the G/S surface, although the general concepts are valid for the L/S surface. Adsorption is the accumulation of a surface excess of two immiscible phases. Except for multilayer physical adsorption, the forces of interaction limit the amount adsorbed to one atomic layer, i.e., up to a monolayer for chemisorption. Mass gain, volumetric, and radiotracer techniques are used to measure the amount of gas adsorbed directly. Indirect techniques require relating the amount adsorbed to the increase or decrease in intensity of some "output particle" such as desorbed gas, scattered ions, ejected Auger electrons, or quanta of radiation. The direct techniques are limited in their applicability, but as to the indirect techniques, there is some danger in assuming that the change in signal intensity varies linearly with adsorbate coverage.[22,26]

Adsorbed gases may bind to the surface nondissociatively or dissociatively, remain localized or have mobility, form two-dimensional surface compounds, or incorporate themselves into the solid, either as an absorbed entity or as part of a new compound. The bonding interaction between a chemisorbed species and the solid depends on the particular geometric configuration, fractional coverage, and electronic interaction. To understand the bonding interaction, a wide diversity of measurement techniques are used, such as infrared (IR) spectroscopy, magnetic susceptibility, electron spin and nuclear magnetic resonance, work function, conductance, LEED, impact desorption, FEM, FIM, electron energy loss, ultraviolet photoelectron spectroscopy, and so on. The most significant recent methods using vibrational spectroscopy to study molecules on surfaces were treated in the first volume of this series.[30]

2.1.6. Surface Composition or Purity

The useful properties of real surfaces often depend on the presence of chemical groups, e.g., impurities, that are extraneous to the bulk composition. These tend to prevent the self-minimization of the surface energy of the solid. They also influence the growth kinetics, topography, surface diffusion coefficients, and residence time of adsorbed species. Depending on the surface energy, impurities may concentrate at the surface or be incorporated into the bulk. These processes, which involve mass transport to, from, and along the surface, may produce significant time-dependent changes in the properties of the material. Therefore, it is extremely important to be able to identify the elemental composition of solid surfaces for both scientific and technological studies.

Until about 20 years ago, experimental techniques for identifying the elements on a surface did not exist. Since then, commercial instruments have become readily available and have been developed for the single purpose of measuring the elemental composition of surfaces. The basic concepts of surface analysis will be discussed in more detail in Section 3. Briefly, the role of surface analysis is *crucial* in the characterization of a solid surface, both in scientific and technological studies.

2.2. Surface Atom Density and Ultrahigh Vacuum

The surface atom density of a typical solid is of the order of 10 atoms/nm^2. The time it takes to form a monolayer of adsorbed atoms on a clean solid surface is given by $t_m = (N/A)(1/vB)$, where N/A is the surface atom density (atoms/cm^2), v is the arrival rate of the impinging gas, and β is the sticking coefficient. From the kinetic theory, $v = P/(2\pi mkT)^{1/2}$, where P is the pressure, m is the mass of the gas molecule, k is Boltzman's constant, and T is the temperature in K. At a pressure of 10^{-6} Torr (0.133 mPa) and 300 K, v ranges from 3 to 5 collisions/nm^2 s for the molecular masses of gases in unbaked vacuum systems. When $\beta = 1$, t_m will be 2 to 3.3 s for our typical solid. While β is much less than unity for many gas/solid encounters, especially when a monolayer of adsorbate coverage is approached, this calculation provides the sobering realization that pressures of 10^{-9} Torr or less are necessary in a surface analysis system for general applicability. Otherwise, the composition measured may be that formed by reaction with the residual gases in the vacuum (e.g., H_2O and CO in an unbaked system or H_2 and CO in a baked system). A nomograph, which Roth presented as Figure 1.1,[31] shows the interrelationships of pressure, degrees of vacuum, molecular incidence rate, mean free path, and monolayer formation time *when β is unity*.

In most surface analysis systems, the sample is introduced from air into an analysis chamber held at 10^{-9} Torr or better, via a fast entry air lock. A third intermediary chamber is frequently used because of vacuum design considerations as well as to provide a convenient location for outgassing or processing the sample, or both. All as-received samples introduced from ordinary laboratory air have surfaces contaminated with adsorbates from the atmosphere. Carbon from various adventitious carbon containing gases is the principal contaminant, but O, Cl, S, Ca, and N are also commonly detected. The contamination layer may be removed by ion bombardment, but wary analysts are careful to watch for surface diffusion of contaminants from the unbombarded sample surface into the cleaned area. Further details about sample preparation, handling, processing, and treatment, as well as on the materials used for UHV systems, are given in a later volume in this series.[32]

2.3. Compositional Depth Profiling

Many studies in science and technology require determining the depth distribution of elements. Surface sensitive probe beams can be used for this purpose in combination with the most commonly used method of ion etching, which is also known as ion erosion, ion milling, sputter etching, compositional depth profiling (CDP), depth profiling, and (atomic) layer by layer microsectioning. Ideal and experimental depth profiles are shown in Fig. 1 for a multilayer stack of partially oxidized aluminum on partially oxidized silicon. Sputtering, the process of removing material by bombarding the surface with an energetic ion beam (typically 0.5–10 keV argon ions), in principle provides a means for atomically microsectioning the solid, and the sputtering time can be related to depth. The principle of microsectioning is shown in Fig. 2, to indicate the goal of obtaining an analysis at each atomic layer into the solid. As illustrated, the available drill, i.e., an ion beam, produces a Gaussian-shaped crater, so the beam is rastered to remove an area of material that may range from 1 to 100 mm^2. The composition is obtained from analyzing the surface in the central portion of the crater.

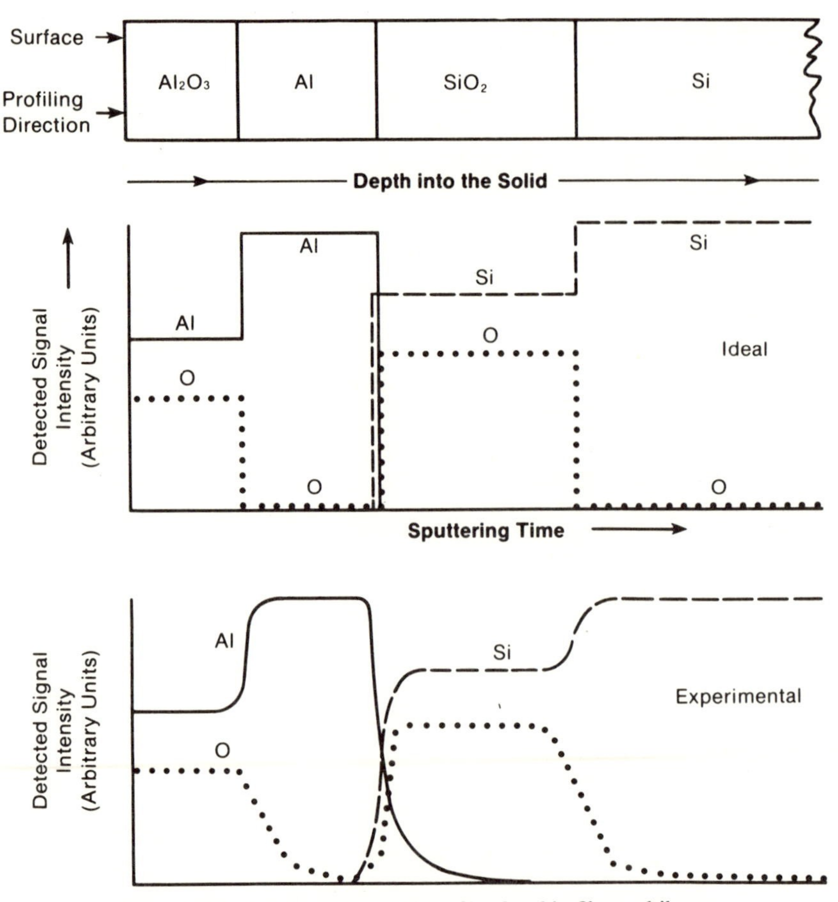

Figure 1. Ideal and experimental profiles for thin-film multilayers.

The advantages of sputtering for in-depth analysis are considerable. Physical sputtering is an atomic process, so, in principle, atomic depth resolution is possible. Depth profiling can be performed *in situ* in vacuum, which means surface compositional analysis and sputtering can generally be performed simultaneously. When pure inert gases are used, clean surfaces are produced, and reoxidation and recontamination can be avoided. Reasonable erosion rates are achieved, e.g., 0.1–100 nm/min, so microsectioning to depths of 1000–2000 nm are accomplished within reasonable periods of time.

The sputtering process itself limits these considerable advantages, however, as we can see by comparing the ideal and experimental profiles in Fig. 1. A zone of mixing is created from knock-on effects, which

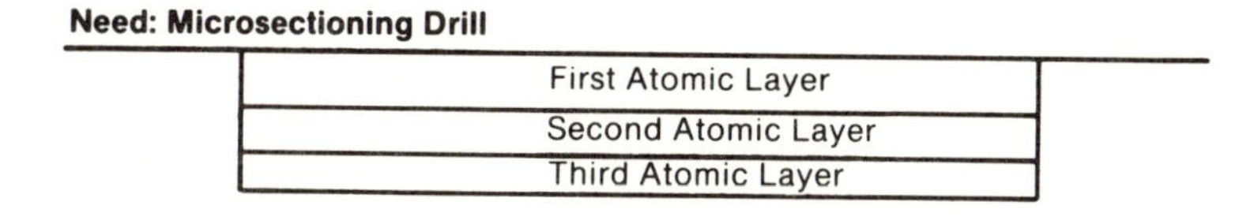

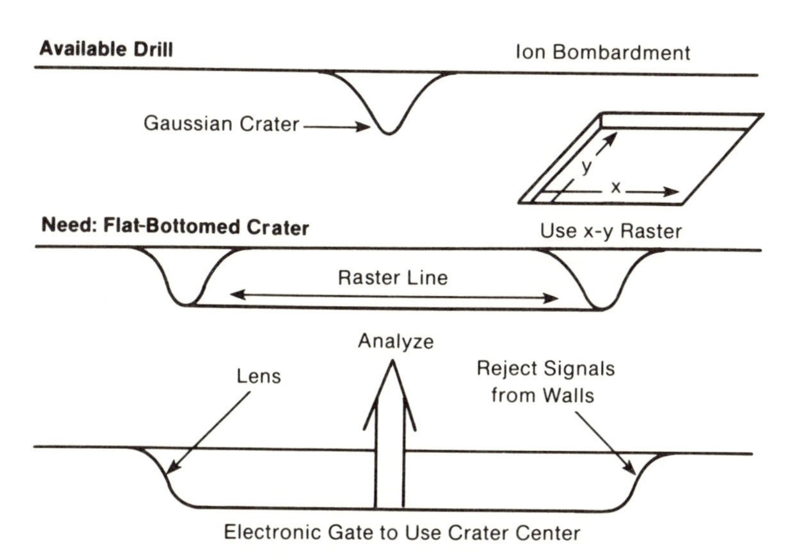

Figure 2. Objectives and practice of depth profiling. The objective of atomic layer microsectioning cannot be achieved because of the Gaussian shape of the ion intensity.

broadens the depth resolution. Preferential sputtering may occur, and surface roughness may develop. Structure may be destroyed and new chemical states may be formed. Bulk and surface diffusion of the target may be enhanced during sputtering. Matrix effects may change the rate of erosion (e.g., resulting from compositional and structural differences; see Appendix, ASTM E-673) as the process proceeds. The bombarding gas ions are implanted into the solid; this is particularly serious when reactive gases such as oxygen or nitrogen are used. Redeposition of sputtered material occurs and complicates the analysis when it occurs on the surface being analyzed. Enhanced adsorption of residual gases from the vacuum may also occur, placing more stringent requirements on the vacuum level maintained during in-depth analysis.

2.3.1. Sputtering Mechanism, Yield, and Rate

All current approaches to understanding physical sputtering are based on a model in which a sequence of binary elastic collisions occurs

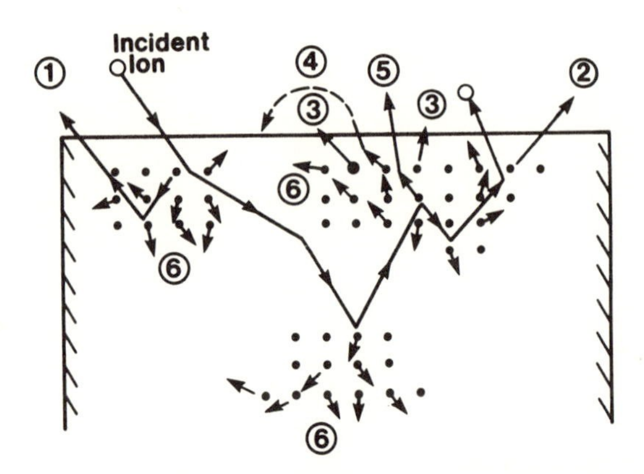

Figure 3. Several of the possible collision processes that occur during ion bombardment. Surface atoms are ejected from being hit from (1) above or (2) below; some do not receive enough energy to leave the surface (3), or they leave and fall back (4). Occasionally, bulk atoms receive enough energy to escape (5); others simply receive energy (6).

(Fig. 3), i.e., the ion interacts with one target atom at a time.[33] During the multiple collision process, the energy of the incident ion is transferred to lattice atoms until the incident ion is implanted or manages to escape from the solid. Winters[33] has compared theory and experiment regarding sputtering. As higher-energy ions are used, more energy is transferred to the lattice, and the mean escape depth of the target atoms increases.[18,34] For ideal depth profiling, the mean escape depth should be only from the outer monolayer, which can be accomplished only at low erosion rates. Consequently, to achieve high erosion rates for reaching depths of 10,000–20,000 nm, the increased amount of energy that must be transferred to the lattice per unit time results in many of the limitations mentioned in Section 2.3. Most of the incident ion energy does not result in ejection of lattice atoms but is transformed into increases in the local temperature of the solid and into the ejection of electrons and photons.

Many parameters govern the sputtering yield, S (i.e., the number of target atoms ejected per incident ion). Since the binary collision theory is based on the classical mechanics of colliding masses, it is not surprising that S depends on the energy, mass, and angle of incidence of the projectile ion[33,35] as well as the binding energy of the solid.[36] Other factors influence S as well, such as the surface roughness, the crystal orientation of the target domains, the electronic properties of the target, the ionization potential of the target, surface cleanliness, residual gases present, and the concentration of implanted ions. Wehner,[35] Werner,[18] and Fine[37] have compiled sputtering yields for commonly used bombardment energies. For 1-keV noble gas ions at normal incidence onto metal surfaces, S varies from 0.08 for Pt with He ion bombardment up to 11.2 for Cd under Ar ion bombardment. Variations in S up to 40 times have been reported for energy ranges from 0.1 to 70 keV (e.g., Ar ion bombardment of copper).

The erosion or sputtering rate during ion bombardment of solids is given by $dz/dt = 0.00624\, SJ^{+}A/\rho$, where dz/dt is in nm/min, S is in atoms per incident ion, J^{+} is the ion beam current density in $\mu A/cm^2$, A is the molecular weight of the target in g/mol, and ρ is the density of the target in g/cm^3.[35] *Erosion rate,* which is the time-dependent removal of material during ion bombardment, is a term preferred to *sputtering rate,* because it is easy to confuse sputtering *rate* and sputtering *yield.* According to Wehner,[35] when 2-keV Ar ion bombardment at $0.15\ mA/cm^2$ is used, the erosion rates for stainless steels, Ta, SiO_2, and Ta_2O_5 range from 7 to 13 nm/min; for Pt, the rate is 23 nm/min; Cu, Au, and Ag range from 30 to 40 nm/min. Werner has developed a nomogram relating dz/dt to J^{+} for various values of S at an A/ρ of 10.[18] He has also published other useful nomograms for selecting various beam parameters and erosion rates as well as for use in SIMS.

2.3.2. *Instrumentation*

Depth profiling is accomplished by using a primary ion source, i.e., an ion gun, to direct ions onto the target. A typical ion source is shown schematically in Fig. 4. A gas is introduced into an ionization chamber at pressures of 10^{-5}–10^{-1} Pa and is ionized by a discharge process. The ions generated are extracted from the ionization region by using an electric field. After mass separation—which is optional depending on the purity

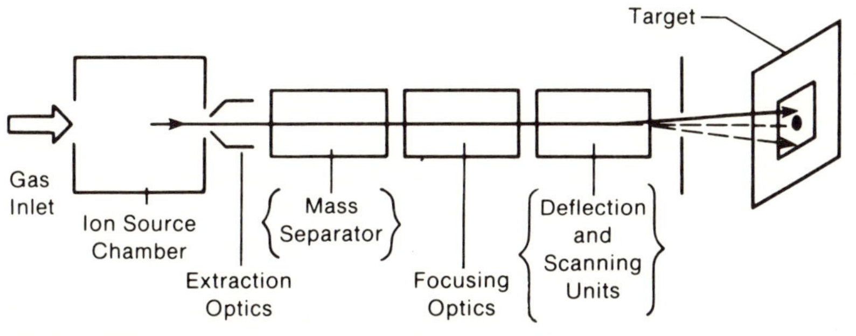

Figure 4. Schematic of an ion gun for a primary beam. The mass separator and scanning units are optional, as indicated by the brackets.

of the initial gas or the desired fidelity of the final beam—the beam is focused to the desired diameter. A Gaussian-shaped beam is typically obtained with the maximum ion intensity on the axis of the beam and the beam diameter stated as the full width at half the maximum (FWHM) intensity of the beam. The stated current density uses the FWHM diameter. It is also important to be aware that ion bombardment occurs on the sample at diminishing intensities at radial distances greater than the FWHM diameter, producing a crater that is approximately an inverted Gaussian shape. To produce a crater with a flat bottom, the ion beam may then be deflected with electric fields to raster the beam over a surface area, e.g., from 2×2 mm up to 10×10 mm.

Analysis of a part of the flat crater bottom may require two additional modifications of the depth profiling apparatus. One of these is to accept the signal electronically only when the ion beam is incident on the central part of the crater. The other is to add a lens that rejects signals from outside the gated area when detecting ions for identifying the surface elements present. The quality of the profile is improved by the progression and the permutations of using rastering, gating, and an electronic lens. Alternative techniques, such as mechanically aperturing the beam or bombarding the surface with a large beam and analyzing it with a small beam, are useful when the instrumental capabilities cannot be expanded to include all the possible ion gun improvements.

2.3.3. Data Obtained and Typical Results

A plot similar to that shown in Fig. 1 is typically obtained where the abscissa is the ion bombardment time. Bombardment time can be converted to depth by measuring the crater depth with a profilometer after erosion; e.g., a stylus instrument can be used to detect changes in surface topography down to about 2.5 nm. Those using this method usually assume that the specimen was homogeneous and eroded uniformly. In terms of multilayers of pure metals, this assumption is usually good, but greater caution must be exercised in dealing with multilayers of metals, alloys, oxides, and semiconductors. As a minimum, the erosion rate for each pure material in the multilayer stack must be calibrated.

There are many factors that influence the depth resolution of an interface because of the ion bombardment process. As an example, Fig. 5 shows the difference in interfacial width in a 300-nm-thick Ta_2O_5 sample, where the middle 100-nm-thick oxide layer was formed in oxygen-18. The 200-nm-deep interface is broadened more than the one at a depth of 100 nm. This result is generally true, but as yet no simple relationship between interfacial width and depth has been established. In fact, there is

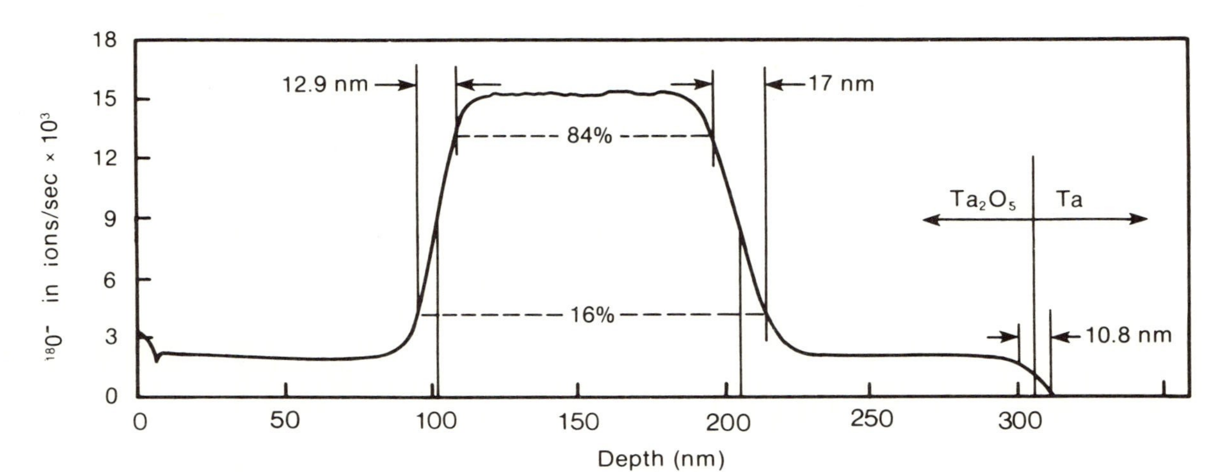

Figure 5. ^{18}O depth profile in a Ta_2O_5 sample used to evaluate depth resolution. The central 100 nm of the film was enriched with ^{18}O. The 16% and 84% points of the signal level represent ±2 standard deviations from the half maximum value and serve to evaluate depth resolution. The detected ion was $^{18}O^-$. (Adapted from McHugh, Ref. 34.)

considerable complexity associated with the possible causes.[18,37,38] The depth resolution depends not only on the quality of the depth profile, but also on factors related to the analysis method (e.g., SIMS, ISS, AES), instrumental imperfections, the sample, and the type of ion bombardment. The various factors include causes from the basic sputtering process, the original and final sample roughness, ion transport, and the incident ion energy. Examples of artifacts related to the limitations of the ion etching process have been illustrated by Wehner.[35] Artifacts related to the sample include surface roughness, crystallite orientation, lateral variation in impurities from surface diffusion, too great an angle of incidence, matrix effects, and an altered surface layer composition compared with that of the bulk. Artifacts resulting from ion bombardment include the incident ion energy, ion implantation, amorphization of the surface, enhanced adsorption, enhanced bulk diffusion, knock-in and knock-up effects, preferential sputtering, cascade mixing, redeposition, and changes in the chemical states of the target species.

Numerous examples are available of depth profiles taken by various authors using various methods of surface analysis, and depth profiling has been treated by a number of authors.[12,18,34–40] Werner and Boudewijn[32] have prepared a detailed treatment of depth profiling using sputtering methods, for a later volume in this series.

3. Overview of Compositional Surface Analysis by Ion Spectroscopies

Of the parameters for characterizing the G/S and S/S interface, the most important is probably the elemental composition.[22,41] A large number of probes have been developed to determine the composition, which is directly related to its importance to the technology of surfaces. Lichtman, who more than a decade ago identified 56 separate techniques for measuring the composition of surfaces,[42] indicates that the number is now more than 350.

3.1. Effects of Energetic Ion Impact on Surfaces

When a solid surface is bombarded by ions in the keV range, complex processes of energy transfer and interactions occur on the surface and in the near surface region. As a result of the impact, atomic and molecular particles, electrons, and photons are emitted (ejected) from the surface. The ejected particles may be in neutral, excited, or charged (+ or − secondary ions) states and originate from the surface zone. The ejected secondary electrons may result from various surface processes, such as Auger deexcitation and Auger neutralization of the impinging primary ions or emitted secondary particles. They may also result from bulk processes, such as ionization and Auger transitions. The emitted photons are generated from deexcitation and neutralization processes above the surface, on the surface, and in the bulk.[40] These processes are discussed in considerable detail by Winograd and Garrison in Chapter 2.

Energy transfer by ion bombardment also has an effect in the region of primary ion impact, in addition to the obvious sputtering processes. The primary ions will be implanted and surface and near surface atoms will be knocked into the bulk. As the primary ion and recoiling lattice atoms undergo a collision cascade,[33] atoms in the lattice will be displaced both inward and outward in a zone of mixing. As a result, changes in the surface structure, such as defect formation, amorphization of structure, bond breaking, and forming, may occur in the surface damage zone.

Energetic neutral particle beams have the same effects on emission processes and changes in the surface zone. When other probes are incident on the surface (e.g., electrons, photons, and electric fields), ions can be ejected from the surface or surface zone. These processes will be described for each generic technique in Section 5.

3.2. Stimulation and Detection in Ion Spectroscopies

Various schemes have been used to organize conceptually the methods of compositional surface analysis, but perhaps the most logical approach is to consider the mode of surface excitation as input and the mode of detection as output.[42] Since this brief overview is restricted to considering the most widely used techniques for ion spectroscopies, the several diagrams used by Lichtman can be simplified to those shown in Fig. 6. The most physically descriptive or commonly accepted acronyms and abbreviations have been used in this figure. Numerous other variations of these acronyms are listed in Table 2. The most unfortunate acronym falls under IIR; some authors had the gall to name this technique BILE.

The input particle has a controlled energy, mass, or intensity (or all of these); the appropriate mass or energy of the output particle is

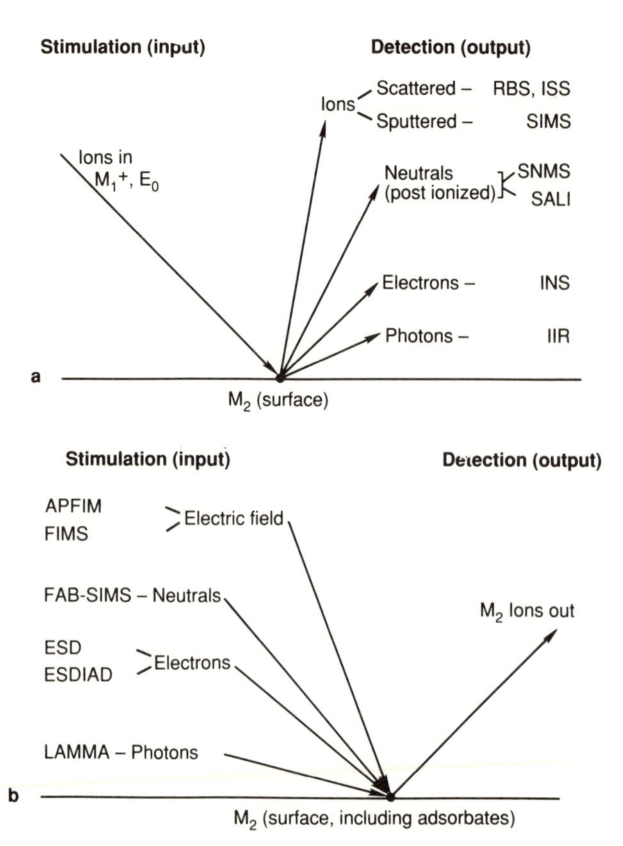

Figure 6. Generalized scheme for ion spectroscopies for surface analysis: (a) stimulation by ions; (b) detection of ions.

measured by the detector. Detectors may be concerned with the mass and charge of the particle, its energy, and spatial distribution. The interest, of course, is to correlate the information detected with the elements on the surface.

4. Ion Spectroscopies Using Ion Stimulation

In this section, brief descriptions of the ion spectroscopies that depend on ion stimulation are given. In the case of ion detection, a summary of the advantages and limitations is also provided. Since these are the principal topics of this book, Sections 4.1 and 4.4 provide a useful overview.

Table 2. Overview of Ion Spectroscopies for Surface Analysis—Acronyms and Abbreviations

Detection	Acronym or abbreviation	Definition(s)
		Stimulation or Excitation by Ions
Ions	ISS	Ion scattering spectrometry (<5–10 keV)
	LEIS	Low-energy ion scattering
	NIRMS	Noble gas ion reflection mass spectrometry
Ions	RBS	Rutherford backscattering (spectrometry)
	BS	Backscattering spectrometry
	MEIS	Medium-energy ion scattering (~20–500 keV)
	HEIS	High-energy ion scattering (~0.5–3 MeV)
Ions	HFS	Hydrogen forward scattering (H specific)
Ions	NRA	Nuclear reaction analysis (light element specific)
Ions	SIMS	Secondary ion mass spectrometry
	IPM	Ion probe microscopy
	IMMA	Ion microprobe mass analysis (analyzer)
Electrons	INS	Ion neutralization spectroscopy
	IIAES	Ion-induced Auger electron spectroscopy
Neutrals (postion-ized)	SNMS	Secondary neutral mass spectrometry
	MSSN	Mass spectroscopy of sputtered neutrals
	SALI	Surface analysis by laser ionization (neutrals)
Photons	IIR	Ion-induced radiation
	BLE, BILE	Beam-induced light emission
	IIXE	Ion-induced x-ray emission
	PIXE, PIX	Particle- (or proton-) induced x-ray emission
	SCANIR	Surface composition by analysis of ion impact radiation
	SIIP	Spectroscopy of ion-induced photons
	SFS	Spontaneous fluorescence spectroscopy
	IBSCA	Ion beam spectroscopy chemical analyzer
	GDOS	Glow discharge optical spectroscopy
	IEX	Ion-excited x-rays
	BXE	Bombardment-excited x-rays
	IILE	Ion-induced light emission
	IMXA	Ion microprobe x-ray analysis
	PRSP	Polarized radiation for scattered particles
Photons	REEP	Radiation from excited ejected particles
	ESDEF	Electron-stimulated desorption of energetic particles (that emit photons)
		Detection of Ions
Electrons	ESD	Electron-stimulated desorption
	EIID	Electron-induced ion desorption
	EPSMS	Electron probe surface mass spectrometry
	ESDIAD	Electron-stimulated desorption ion angular distribution
E	FIMS	Field ion mass spectrometry
E	APFIM	Atom probe field ion microscope
	APMS	Atom probe mass spectrometry
	TOF–AP	Time-of-flight atom probe
	FAB–SIMS	Fast atom beam–secondary ion mass spectrometry
Neutrals	LAMMA	Laser (ablatian) microprobe mass spectrometry
Photons	LIMS	Laser (post) ionization (of ablated particles) mass spectrometry

4.1. Ion Detection: SIMS, ISS, RBS, NRA, HFS

4.1.1. Secondary Ion Mass Spectrometry (SIMS)

In SIMS, a beam of primary ions is directed toward the sample surface, where most of the energy of the ions is dissipated into the near surface region of the solid by a series of binary collisions. As a result of both the increase in the energy of the solid and the multiple collisions, surface and near surface atoms and molecular clusters are ejected (sputtered) from the solid.[33] Some of the particles are sputtered as secondary ions, and these may be directed into a mass spectrometer for analysis. Although early SIMS experiments were carried out in the late 1930s, and the feasibility of SIMS was clearly demonstrated by Herzog and Viehböck in 1949,[43] this activity was not practiced widely until the late 1950s as a result of Honig's work.[44] A large body of information about secondary ion emission and the applications of SIMS to surface and thin film analysis have been documented in reviews by McHugh,[34] Blaise,[45] and Werner,[46] by those who provide emphasis on aspects related to SIMS,[33,35,47,48] and in several conferences on SIMS,[49] as well as in the treatise by Benninghoven, Rüdenauer, and Werner.[40]

In Chapter 2 in this volume, Winograd and Garrison provide new insight about molecular dynamics. In Chapter 3, Busch provides a focused review of particle-induced desorption of organic materials.

The detected signal of the ith element in SIMS requires controlled destruction of the surface and subsurface of the sample to produce a secondary ion current $I_i^{\pm}$. The current for the ith element depends on many factors that may be related by the expression

$$I_i^{\pm} = I_p f_i^{\pm} C_i S_i \eta_i \tag{1}$$

where I_p is the incident ion current (ions/s), S_i is the sputtering yield of both ions and neutrals (particles/incident ion), $f_i^{\pm}$ is the fraction of the particles sputtered as ions, C_i is the concentration of the ith element (corrected for isotopic abundance) in the sputtered volume, η_i is the collection efficiency of the SIMS instrument, and $\pm$ refers to a positive or a negative particle.

Most SIMS instruments can be operated in a mode to detect either positive or negative ions, so I_i^+ or I_i^- can be considered separately in Eq. (1). Since C_i depends on the sample selected, consideration will now be given to S_i, I_p, and f_i.

The factors affecting the sputtering yield during ion bombardment were summarized in Section 2.3.1. The most important parameters influencing the sputtering yield are the energy, mass, and incidence angle of the projectile ion. Typical bombardment energies range from 0.5 to 20 keV and argon is the gas most frequently used. For example, S increases steadily up to about 50 keV for $Ar^+ \rightarrow Cu$, and a threefold increase in S occurs between 1 and 10 keV. The combined effects of mass, energy, and angle for sputtering copper with Xe^+ at 110 keV and 60° from the normal provides a sputtering yield that is about 250 times greater than when 1-keV He^+ is used at normal incidence. Chambers and Fine of the National Institute of Standards and Technology plan to complete a compilation of evaluated sputtering yields in 1990.

For the fraction $f_i^{\pm}$ of secondary ions produced, theoretical work to date indicates that Auger deexcitation of excited species leaving the solid and surface ionization may be a major source of the secondary ion fraction.[(34)] Resonance ionization of ground state sputtered neutrals is thought to make a less important contribution. The ion-forming processes occurring near the surface are complicated and must be treated quantum mechanically. Although a theoretical understanding of these matters has not yet been attained, more insight about the ion-forming process is provided in Chapters 2 and 3 of this volume. The following facts, moreover, have been determined experimentally. Most sputtered particles leave the surface with energies of only a few eV. The secondary ion yields from nonconducting metal compounds are typically 10^2–10^3 times the yields from the corresponding metals.[(43)] The mean velocity of sputtered particles is greater for materials with high binding energies (insulators versus metals), and the higher velocity components of the distribution of sputtered particles are ionized with greater efficiency. The secondary ion species are mainly singly charged atomic and molecular ions (mainly dimers). Doubly charged atomic species and trimers are also formed, which can be detected by high sensitivity mass spectrometers, and which play an important role in producing the complex spectra. Furthermore, the presence of anions (e.g., oxides) results in the formation of M_xO_y species, where x and y are integers.[(43)] In using I^+, for example, to obtain C for a particular element, the fraction f^+ must be weighted appropriately to include the influences of clusters and multiply charged ions.

The current density and chemical nature of the primary ion also have a marked influence on I^+ or I^-. Since I_p (ions/s) $= 0.25\,\pi d^2 j$, where d is the diameter (FWHM) of a Gaussian-shaped beam and j is the current density (ions/cm^2 s), we see that I^+ or I^- is directly proportional to the number of ions incident on the target. The proportionality remains valid even for current densities as large as 10^5 μA/cm^2. A potential problem at high current densities is an increase in the sample temperature. The chemical identity of the primary ion is important, especially when reactive gases are used. Many examples may be found in which reactive gases, particularly oxygen and nitrogen, have been used to increase the secondary ion current.

The SIMS instruments have been categorized as secondary ion mass analyzers (SIMA) and ion microprobes (the acronym IMMA is frequently used for this instrument). The latter provides imaging capabilities. In conventional SIMA, the ion optics are arranged to focus all secondary ions at a point in front of the detector, which precludes the possibility of imaging. Distinctions between ion microprobes are sometimes made on the basis of the different methods used to obtain an image of the sample. Ion microprobes cost 3–10 times more than SIMA. The secondary ions may be mass analyzed by using a quadrupole mass filter or a conventional mass spectrometer. Most SIMA use a single focusing mass spectrometer or a quadrupole mass filter, whereas the ion microprobes typically use a double focusing mass spectrometer. The advantages and disadvantages of the various schemes used to detect the secondary ion current have been treated in depth in various review articles.[34,36,47] The outstanding difference is an improvement of 500–1000 times in the lateral resolution in the ion microprobes—at a cost 3 to 10 times that of SIMA with a single focusing mass spectrometer or a quadrupole.

The most obvious advantage of SIMS compared with other surface analysis techniques is that it identifies elements at a low detection limit on the surface of conductors, semiconductors, and insulators. Other advantages include the ability to detect H and He on the surface, which is physically impossible with AES and impractical with XPS. The outstanding feature of SIMS is a detection limit to as low as 10^{-6} of a monolayer, depending on the beam size and how fast the surface is sputtered away.[34] As little as 10^{-18} g of a sample species may be sufficient to provide a detectable signal. Thus, using care in the instrument and bombardment parameters, signals can be restricted to 1–2 monolayers. *Static and dynamic SIMS* are terms used to divide the erosion rate of the solid into regimes so that only surface species (static) or species from the surface and bulk (dynamic) are detected. The sensitivity of SIMS to different isotopic masses provides new possibilities for studying corrosion mechanisms, self-diffusion phenomena, and surface reactions involving any exchange of atomic species. Isotopic labeling of both cations and anions could be used for studying reaction mechanisms. Despite the apparent potential, the literature contains relatively little mention of isotopic labeling with SIMS analysis of the surface. The ability to depth-profile and to maintain a constant monitor of the composition is one of the outstanding features of a SIMS apparatus. The obvious application here is to detect, down to about a 0.1 ppm atomic fraction, impurity accumulations at S/S interfaces lying hundreds of atomic layers below the outer surface. Ion microprobes provide capability to image the surface under investigation. This, combined with a lateral resolution as low as 100 nm, provides a capability for analyzing the composition of individual grains in polycrystalline materials and for determining the distribution of elements (bulk or trace) across the surface.

The principal limitation of SIMS is that it destroys part of the sample, preventing it from being analyzed further; there is no chance for a second look at the same spot on the sample. The factors causing large variations in the production of secondary ions make routine quantifications a remote hope unless standards are used in well-studied systems. Matrix effects, e.g., the variation in the signal of the same element in different chemical environments, can alter the detectability of elements by factors up to 10^5.

4.1.2. Ion Scattering Spectrometry (ISS)

The energy regimes for ion scattering are of the order of keV for ion scattering spectrometry (ISS, also referred to as low-energy ion scattering, or LEIS), 100 keV for medium-energy ion scattering (MEIS), and MeV for Rutherford backscattering (RBS, also referred to as high-energy ion scattering, or HEIS). In ISS, a collimated monoenergetic beam of ions of known mass (M_1) is directed toward a solid surface, and the energy of the ions scattered from the surface is measured at a particular angle. The energy E_0 of the projectile ion is reduced to E during a collision with a surface atom (M_2), and the intensity of scattered ions is measured and normally presented versus an E/E_0 ratio rather than as an energy loss spectrum. From the number of scattered ions of mass M_1 appearing at particular E/E_0 ratios, information may be deduced about the mass and number density of the various surface species in the target, i.e., the elemental composition of the surface. Since Smith[(50)] first demonstrated the feasibility of ISS as an analytical tool, a number of review articles have been written by other early users of ISS.[(51–56)] Smith's noteworthy elucidation is that important aspects of the energy loss spectrum can be explained by considering the ion–surface atom collision as a simple two-body event. The chapter by Taglauer in this volume probably provides the most complete coverage now available about ISS.

Although the experimental facilities one needs vary with the goals for a particular problem, the essential needs for ion scattering experiments are a vacuum chamber, an ion gun, a target, and ion energy analyzer. Ions are produced in a source by electron bombardment of a gas, as the source is held at the desired accelerating potential (~0.3–3 keV). Ions are drawn from it by a negatively biased electrode, and the ion beam is formed by an ion focusing lens system. For most applications, only one type of singly charged ion species is desired in the beam, e.g., the ions of ^{3}He, ^{4}He, ^{20}Ne, or ^{40}Ar.

Scattered ions are detected by using an energy analyzer either of the sector cylindrical mirror or hemispherical type and an electron multiplier detector. For most instruments, the scattering angle θ is fixed, but in custom-made spectrometers it may be varied.

The main feature of ISS spectra is that only one peak appears for each element, or isotope of each element, as predicted from theory. Thus, six elements will yield only six peaks, a situation that is considerably simpler than for other surface spectroscopies. An ISS spectrum is simple, but the peaks are broad and resolution between neighboring elements in the periodic table is poor, especially for high-Z elements.

There are several major advantages to using ISS as a surface sensitive technique in addition to the obvious ability to identify elemental masses on solid surfaces of samples that are conductors, semiconductors, and insulators. Qualitatively, the current detection limits can be as low as 0.01–0.001 monolayers for the light elements and 0.001–0.0001 monolayers for heavier elements. Depth profiling is routinely accomplished by using the same ion gun for etching and supplying the projectile ions for scattering. Monolayer sensitivity is the strongest advantage of ISS because the detected signal results primarily, and possibly only, from atoms in the outer monolayer. Structural information can be deduced about the arrangement of two or more elements on single crystalline substrates. Structural information can also be deduced by using multiple scattering, particularly double scattering. Isotopes of the same element can be detected when all other factors are constant. Chemical information can be deduced about the influence of different chemical environments on the detected signal for elements engaging in quasiresonant transfer processes.[57]

The strongest limitation of ISS is that the mass resolution is poor, especially for high-Z masses. This results from broad peaks (0.02–0.03 in E/E_0) and poor energy resolution when M_1 is much less than M_2. Thus, a small concentration of an element with a mass near M_2 cannot be easily resolved. Moreover, the surface under investigation will be damaged by the bombardment from the projectile ions. Even though reduced bombardment energies (~300–700 eV) and low current densities (0.4–40 $\mu A/cm^2$) may be employed, sputtering of the outer monolayer is unavoidable.

4.1.3. Rutherford Backscattering Spectrometry (RBS), Nuclear Reaction Analysis (NRA), and Hydrogen Forward Scattering Spectrometry (HFS)

In RBS, a collimated monoenergetic beam of ions of known mass M_1 is directed toward a solid material, and the energy of the ions scattered is measured at a particular angle. The initial energy E_0 of the projectile ion is reduced to E during the passage through the solid and by the collision with a surface or bulk atom. The intensity of scattered ions is measured and presented in an energy loss spectrum. From the intensity of scattered ions of mass M_1 appearing at the reduced energy E, information may be deduced about the number density, mass, and depth distribution of atoms at the surface or in the bulk. Particle accelerators in the 1–3-MeV range became readily available in the late 1940s, and in the 1950s, RBS instruments were applied to analyze bulk composition.[58] Thin film and surface analysis with RBS, which were demonstrated in the late 1950s and early 1960s, respectively, have grown dramatically in the last decade, as evident in a book devoted to the subject that was published in 1978.[59] That book, by Chu, Mayer, and Nicolet, contains a comprehensive bibliography of review articles, books on related material, and categorized applications.

Chapter 5 in this volume, by Feldman, is an excellent starting point before one delves into the more extensive treatments.[39,59] Feldman also treats NRA and HFS, and no further comments will be made in this chapter about these specialized, and very useful, high-energy methods.

RBS for detecting ions scattered from the surface and bulk atoms of a solid is typically performed with incident energies of 1–3 MeV, rather than 0.5–3 keV, as in ISS. The energy of the elastically scattered ions is reduced during a binary elastic collision, provided that there are no nuclear reactions. The scattering cross section is greatly reduced at MeV energies, so backscattered particles reach the detector typically after traversing several hundred nanometers into the solid. The ion loses an average energy, which is termed the *stopping cross section,* primarily because of collisions with electrons in the solid. The collision encounters are inelastic (quantum mechanical), so a monoenergetic incident ion energy will emerge from the solid with an energy distribution about its initial energy as a result of statistical fluctuations, which is termed *energy straggling.* The energy of a backscattered ion is then reduced by the stopping cross section before and after the binary collision, so the mean energy of the ion provides a measure of the depth at which the collision occurred. Most significantly for RBS, the central force field is well known, which means an expression for the cross section can be derived.

About 1 in 10,000 ions will be backscattered, so the analysis is essentially nondestructive. Because very few backscattered ions will have a scattering angle near 170°, beam widths of 1–2 mm are required and the energy resolution is compromised to improve the signal-to-noise ratio.

The essential experimental needs for RBS are a particle accelerator (usually 1–3 meV helium or proton ions), a solid state detector, and a multichannel analyzer, aside from the obvious needs for vacuum and a target. The important properties of particle accelerators (e.g., Van de Graaff generators) are well documented.[59] Extensive data are available for assisting in RBS data analysis as a result of nearly 40 years of using accelerators in nuclear physics. These include relative abundances of isotopes, kinetic factors (E/E_0, where E is the energy after a binary collision from an incident energy E_0) for all atomic masses at various scattering angles for both ^{4}He and ^{1}H, stopping cross sections for the elements for various incident ion energies, scattering cross sections between helium and all elements for various backscattering angles, and the yield from the surface at various scattering angles of ^{4}He.[59] The main features of RBS spectra are that only one peak appears for each element (or isotope) present, and the peak width is directly related to the thickness of the element. The peak height can be calculated from first principles, so RBS is quantitative. The data can be gathered in a matter of minutes, so the analysis is effectively nondestructive. The dynamics of changes such as interdiffusion can be studied *in situ*.

The outstanding advantage of RBS is that it provides quantitative and nondestructive in depth compositional analysis. The time needed to acquire data is only a few minutes. The essentially nondestructive nature of RBS allows studying a single sample with materials processing variables such as temperature and time. The technique is more sensitive to high-Z elements for which atomic fractions down to 0.0001 can be determined. RBS as it is normally practiced is not surface sensitive, and the high-vacuum chambers generally used preclude a serious study of surfaces unless a system is designed with UHV capabilities. When single crystals or epitaxial layers are used, structural information, both at the surface and from the bulk, can be obtained by using channeling and blocking techniques.

As for disadvantages, different isotopes of the same material are not generally resolvable, and the sensitivity for low-Z elements is limited because the scattering cross sections are up to 1000 times smaller than those for high-Z elements. Thus, detectabilities of an atomic fraction of 0.1–0.01 are typically available for Be through F in the periodic table. Where isotopes are resolvable, there are similar advantages for RBS as stated for SIMS and ISS but at lower detection sensitivities for the low-Z

elements and with a lack of resolution between neighboring elements with large atomic numbers. The incident ions are typically in a 1-mm FWHM beam, so the lateral resolution is poor compared with that of the other surface techniques. Therefore, lateral sample uniformity is important; SEM sample analysis is also recommended to identify scratches, dust, and other defects. No chemical information can be extracted from the data, however.

4.1.4. General Comparisons of SIMS, SNMS, ISS, RBS, AES, and XPS

Comparisons of aspects of SIMS, SNMS, ISS, RBS, AES, and XPS are listed in Tables 1–4 in Chapter 7 of this volume. Summaries of the principal advantages and limitations of each of these techniques are given in Tables 5–10 in Chapter 7 of this volume.

4.2. Photon Detection of Ion-Induced Radiation

When a solid is bombarded with energetic ions or neutrals, electromagnetic radiation (ER) ranging from the x-ray region to the IR can be emitted. Hence, the general abbreviation, ion induced radiation (IIR), has been suggested. When bombardment is with MeV H^+ or He^+, characteristic radiation results from this particle-induced x-ray emission (PIXE) and is used for analysis of solids. When bombardment is in the keV regime (e.g., SIMS, ISS), ER from the IR to the near UV is emitted and bombardment-induced light emission (BLE) is the most frequently used term for this regime.[40] The major disadvantage is a depth resolution of about 100 nm or more for both PIXE and BLE.

4.2.1. Particle-Induced X-Ray Emission (PIXE)

X-ray emission results from deexcitation of atoms that have been ionized in the near surface region by the passage of MeV H^+ or He^+. Useful rates of inner shell ionization (keV electrons) require that incident MeV H^+ ions have $1836E$, where E is the kinetic energy of the electron.[39] X-ray production cross sections for K_α radiation versus proton energy have been deduced for 0.5–10 MeV protons.[60] X-ray production is related to both the ionization cross section and the fluorescence yield. Experimentally, 1–4-MeV particles in beam diameters of 0.01 to 1 mm are used for excitation. The x rays are detected by either a wavelength dispersive spectrometer or an energy dispersive spectrometer.[61] The elemental sensitivity ranges from 0.1 to 10 ppm beginning with Na (cross sections are low for elements below Na).

Destruction of the sample is minimal, and lateral resolution ranges from 0.01 to 1 mm. Other major advantages include a considerable reduction in background radiation relative to that from an electron microprobe analyzer, where x rays from several hundred nanometers below the surface are typically detected. Since the knowledge of the inelastic cross sections is well known because of a considerable amount of work in nuclear physics, known data can be used to analyze the results. The major disadvantage of this technique is that proton penetration is hundreds of nanometers (see Chapter 5 in this volume) so x rays are obtained from the entire excitation volume; i.e., the technique is not uniquely surface sensitive. However, for analyzing the near surface region, which is of great interest in materials science, the technique provides obvious opportunities.

4.2.2. Bombardment-Induced Light Emission (BLE)

Light emission from the UV through the IR results from keV ion bombardment of solids. During the collision cascade, the impinging ion energy is transferred to the solid, and a broad distribution of electronic excitations result in both the incident and sputtered particles, as well as in the near surface region of the solid. Deexcitation processes result in both broadband (continuum) spectra and line spectra.[62]

The broadband spectra result from the intrinsic luminescence of the solid, emission from the sputtered particles, and from excited molecules on or below the surface. The luminescence emission originates as a result of the excitation of electrons or radiative recombination of electron–hole pairs. The emission from the sputtered particles results from various deexcitation processes as the particles are in transit above the surface. The radiation from excited molecules on or below the surface is thought to result from excited CN and possibly other molecules bound to the surface. Broad vibrational bands can be observed that are believed to be associated with exciton recombination.[63]

The line spectra, which are of interest for elemental identification, result from deexcitation processes of excited sputtered particles (with lifetimes of 10^{-7}–10^{-9} s) and scattered primary ions, as well as from surface atoms. The line spectrum for each element is superimposed on a background continuum for that element, as shown schematically in Fig. 7. BLE is simple in its experimental concept. An ion beam strikes a sample in a scattering chamber and the signals are detected with a monochromator equipped with a photomultiplier or a spectrograph with photographic recording. The technique is reportedly sensitive to all elements to 0.1 ppm with information depths as low as 1 nm, and a lateral

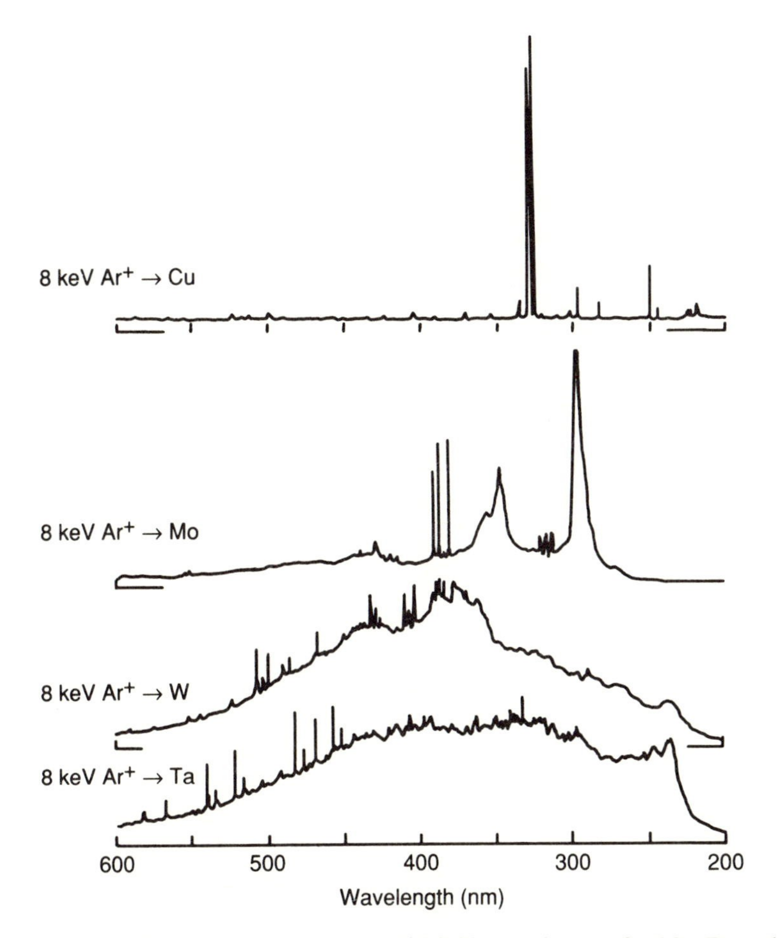

Figure 7. Optical spectra induced by impact of 8-keV argon ions on Ca, Mo, Ta, and W. (Adapted from White *et al.*, Ref. 62.)

resolution of 0.1 mm.[62] However, a depth resolution of 100 nm is typically indicated and, as with SIMS, the technique is intrinsically destructive. To understand the complications of using BLE for routine analysis of complex samples, one need only superimpose spectra from several elements to note the varying background and propensity at spectral lines because of the large number of allowed transitions for the various elements. Except for the early attempts at x-ray analysis with *ion* microprobes (IMXA), no commercially made instruments are available for BLE.

The principal value of BLE is its contribution to our understanding of the details of the collision processes and the interaction of excited particles as well as about the atomic processes that influence the ejection of positive ions, negative ions, and excited neutrals. More detail is available in the review article by White *et al.*[62] These authors named their technique SCANIR. Some significant results were obtained in about 1980 on "light versus distance" measurements to obtain information about the energy distribution of excited atoms, and more recently using Doppler-shift laser fluorescence spectroscopy.[63] A clear theoretical description and understanding of the basic principles of the various excitation processes, which result in the formation of or the deexcitation of sputtered particles, are not yet available.

4.3. Electron Detection; Ion Neutralization Spectroscopy (INS)

Although ion probes cause electrons to be emitted from surfaces, this method of surface analysis has not been pursued very much. Basically, this is because it is easier to produce secondary electrons by primary electron (AES) or photon (XPS, UPS) probes. However, the significant work by Hagstrum must be mentioned about the technique he designated ion neutralization spectroscopy.[64–67] Here, the surface is probed with low-energy (typically, a few eV) ions. When the ion approaches the surface, it is neutralized; the neutralization energy can be transmitted to an electron at the surface, giving it sufficient energy to be emitted. An analysis of the energy of these emitted electrons can provide information about the electronic nature of the surface under study. The experimental system is not simple,[67] and the technique has been exploited only in a limited number of cases. Hagstrum's early work is quite significant because it contributed to our understanding of fundamental electron transition processes at the surface. However, many of the electron-in, electron-out spectroscopies, as well as the photon-in, electron-out spectroscopies, provide similar information.

The electronic transition of greatest interest involve those shown diagrammatically in Fig. 8, i.e., resonance neutralization, resonance ionization, Auger neutralization, and Auger deexcitation. All these occur less than 1–1.5 nm from the surface and are thought to be the principal causes of neutralization of ions approaching *or* receding from a surface. Although Hagstrum usually worked with incident ions with energies from 1000 eV down to 10 eV or less, his results have been applied to keV scattering in ISS and sputtered ions in SIMS (typically, less than 10 eV). Results obtained by the fixed ion approximation Hagstrum used are considerably improved when the continuous motion of an incoming ion, perturbation theory, and other quantum mechanical considerations are used to calculate the distance from the surface where resonance

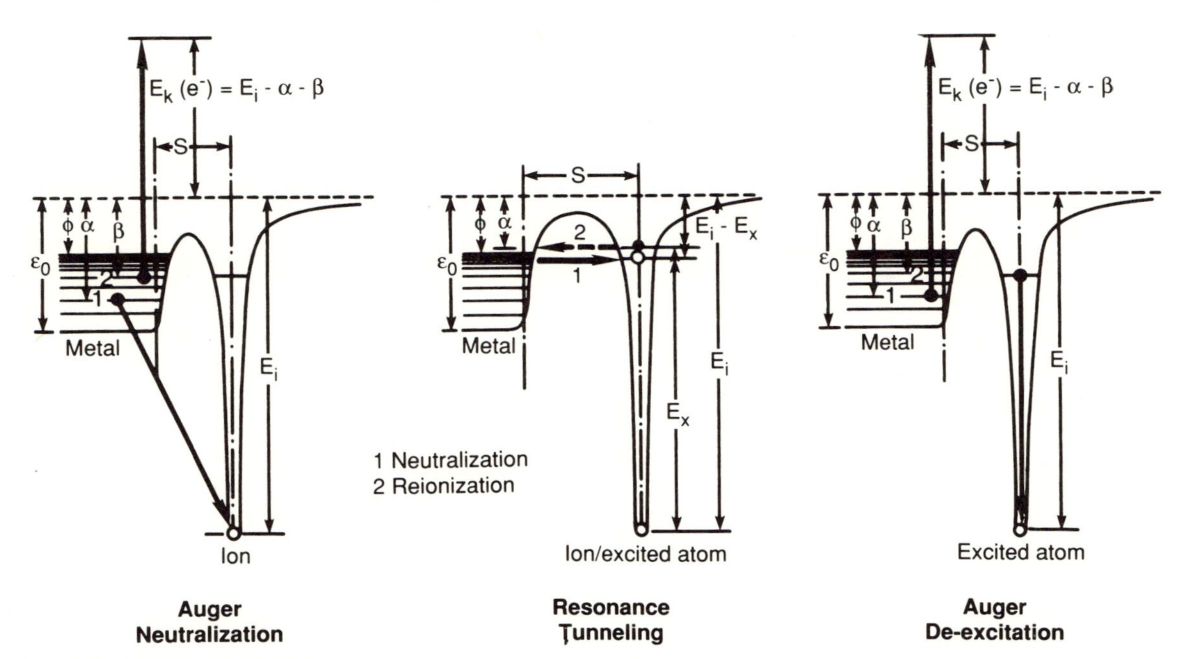

Figure 8. Energy level diagrams for a helium ion near a metal surface that illustrate electron tunneling and Auger processes for ion neutralization. (Adapted from Hagstrum, Ref. 67.) E_i is the ground state level of the ion; $E_k(e^-)$ is the kinetic energy of the ejected electron; ϕ is the work function of the metal; ε_0 is the energy of the vacuum level above the bottom of the conduction band; α and β are energies below the vacuum level; S is the distance of the ion from the metal surface.

neutralization of keV ions is probable.[68] However, Hagstrum's work contributed greatly to out understanding about electron tunneling processes for charged and neutral particles in their ground and excited states, and about the relative rates at which the processes occur. In addition, direct evidence was obtained for the variation in energy levels in the band structure of the solid resulting from the particle–solid interactions. Finally, the two-electron, Auger-type electron spectroscopy was developed (i.e., INS). As Hagstrum wrote in 1978,[67] "in order that Auger neutralization be the only electron ejection process possible for an incident ion (to give INS unique surface analytical capabilities), it is necessary that resonance tunneling processes not occur." The resonance tunneling calculation of Moyer and Orvek[68] raise serious questions about whether the requirement for INS can be met. That an ion-in, electron-out surface analysis method has not been developed is supportive of Lichtman's[38] prediction in 1975, "I do not expect that a technique of ions in, electrons out will be pursued very actively in the future." An in depth review of INS is available.[67]

4.4. Neutral (Postionized) Detection: SNMS, SALI, SARISA

For ions in, neutrals out, it is necessary to postionize neutral particles to perform mass spectrometry on them. During the last several decades, sputtering has been accomplished by using ion beams, and postionization has been done by the interaction of neutrals with excited atoms,[69] electrons,[70–73] or photons.[74–77]

In essence, sputtered neutral mass spectrometry (SNMS) is very similar to SIMS. A low-energy ion beam is used to sputter the surface, and a mass spectrometer is used to analyze secondary particles.[72] However, in SNMS, the true secondary ions are deliberately discarded and the secondary neutrals are postionized for subsequent detection. Postionization can be achieved by using a low-pressure plasma, an electron beam, or a laser. In each case, the ionization cross section is predictable and is largely independent of any matrix effects (although there are variations in the angular and energy distributions of sputtered neutrals). As a result, SNMS is a quantitative technique with a useful narrow range of sensitivities.

The use of a radiofrequency (rf) discharge plasma can result in a postionization efficiency of 10^{-2}. A significant advantage of this approach is that the plasma not only postionizes, it can also be used to sputter the target. The low energy of the impinging ions (100 ev to 1 keV) results in a minimization of atomic collision mixing (see Chapter 2); hence, the technique may be used to provide depth profiles with extremely good depth resolution. Such depth profiles may be obtained rapidly since the ion current density can exceed 1 mA/cm^2 over a 0.5-cm^2 area. Two disadvantages of plasma postionization are the background of ions generated by residual gas particles and limited lateral resolution.

A method for electron beam postionization with some useful advantages has been described by Lipinsky *et al.*[73] This method uses deflection plates to suppress the original secondary ions and a series of apertures with a high-energy bandpass to discriminate against the transmission of residual gas ions. With an electron beam postionizing current of 3 mA and sputtering ion gun with 20 μA current, bulk analysis of sub-ppm is possible. Surface monolayer analysis with a detectability of 0.01 at. % is also possible. This method is entirely compatible with a conventional SIMS analysis system, and the researcher can switch rapidly between the two modes.

In addition to the descriptions in this volume of SALI and SARISA by Becker, there are interesting summaries of applications of the technique by others.[74,76] Descriptions of using electron beams for postionization of sputtered neutrals are also available.[39,40,72,73]

5. Ion Spectroscopies Using Ion Detection

In this section, ion spectroscopies in which ions are detected following stimulation by electrons, photons, neutrals, and electric fields are briefly discussed. Techniques involving ejected and scattered ions from ion stimulation were summarized in Section 4.1.

5.1. Electron Stimulation: ESD, ESDIAD, EPMA

5.1.1. Electron Stimulated Desorption (ESD)

The term *electron-stimulated desorption* (ESD) is used in a general way to denote physical and chemical changes in the surface region caused by bombardment with low-energy electrons (<500 eV). An alternate term is *electron-induced ion desorption* (EIID).

When a chemisorbed layer on a metal is bombarded with a beam of electrons, part of their kinetic energy is expended in excitation of the adspecies. Desorption of ions results from electronic excitation to an antibonding state by incident electrons of sufficient energy (>10 eV). A transition of an electron from the electronic ground state of the adsorbate–adsorbent interaction potential to the repulsive part of the ionized adsorbate–adsorbent potential curve (or of some other excited-state potential curve) takes place, and this is followed by desorption of the adsorbate with a range of kinetic energies between 0 and 10 eV. The cross sections for desorption are typically from 10^{-17} to 10^{-22} cm^2, compared with values of 10^{-15}–10^{-16} cm^2 for gas phase dissociative ionization, and they result from an increased ionization probability of excitation of the adspecies by the nearby metal.

One of the first review articles on ESD was that of Lichtman and McQuistan[78] in 1965, in which they emphasized instrumental methods as well as the potential of ESD (or their term, EIID) as an analytical tool for examining surfaces of undefined composition. Other review articles followed and are included in the comprehensive review by Madey and Yates published in 1971.[79] ESD is now covered in a number of textbooks on chemisorption, the topical area of surface science in which it has its principal application. More recent results are contained in a series of conference proceedings on DIET, i.e., desorption induced by electronic transitions,[80] and the continuing workshops on inelastic ion surface collisions.[81]

In principle, ESD can yield information about (1) the excitation and deexcitation of the adsorbate-surface complex, (2) electron tunneling through the surface barrier, and (3) energy transfer in the formation of the chemisorption bond. As yet, only qualitative conclusions are possible. However, desorption studies have demonstrated that surface changes may frequently be induced by the impact of low-energy electrons and they highlight the difficulty of avoiding surface damage in electron spectroscopic investigations.

5.1.2. Electron Stimulated Desorption Ion Angular Distribution (ESDIAD)

In ESDIAD, a surface containing adsorbed molecules is bombarded by a focused, low-energy (10–1000 eV) electron beam. As in ESD, desorption of atomic and molecular ions results, but ESDIAD takes advantage of the discrete cones of ion emission and in directions determined by the orientation of surface molecular bonds, which are broken by the excitation. The desorbed ions are displayed as patterns using a grid–microchannel-plate–fluorescent-screen array. When the patterns are properly interpreted, the *local* bonding arrangement of the chemisorbed species can be ascertained. Ions are only a small fraction of species desorbed but are easily manipulated, and an ESDIAD apparatus is usually accompanied by a quadrupole mass spectrometer (QMS). Most ESD ions are atomic, and H^+, O^+, F^+, and Cl^+ are the most frequently studied particles; the most common molecular ions are CO^+, OH^+, and OH^-. Madey's review chapter is a good starting point for gaining more insight into ESDIAD.[82] More recent reviews are also available.[83,84]

The obvious advantage of ESDIAD is its ability to determine local order in chemisorbed layers, a significant need in surface science (see also Section 2.1.5). It also highlights a major problem of electron-induced damage in AES surface analysis, as Pantano and Madey will discuss in a later volume in this series. There are two major disadvantages, however, to ESDIAD (as well as to ESD). One is that not every adsorbed species gives rise to a detectable signal of ions. Some adsorbed components give rise to large signals (e.g., some states of adsorbed CO and O) while some yield no ions (such as adsorbed N). A second disadvantage is that no direct information is obtained about the substrate. The latter is most significant in terms of the subject of this volume.

5.2. Photon Stimulation: LAMMA, LIMS

In laser microprobe mass analysis (LAMMA), a focused laser strikes the sample. A volume of the surface region is evaporated by the laser pulse, partly as a plasma.[85,86] The ions are extracted from the plasma and analyzed by a time-of-flight (TOF) or sector-type mass spectrometer. The information depth is typically 100–1000 nm, so LAMMA (or the laser microprobe, LMP[87]) does not qualify as a surface analysis technique. Nonvolatile inorganic compounds can be analyzed from characteristic spectra similar to those obtained by SIMS.[88]

The advantages of LAMMA may be related to comparable near surface bulk analysis techniques such as x-ray fluorescence (XRF), electron probe microanalysis (EMP), energy dispersive x-ray analysis (EDX), and PIXE. For the volume analyzed, LAMMA is sensitive to all elements to the 0.1 ppm level, has a lateral resolution of 1000 nm, "profiles" materials rapidly, provides isotopic detection, analyzes thick or thin films, and can be used on organics, insulators, semiconductors, and metals. The disadvantages of LAMMA are that the technique is destructive, has limited reproducibility and hence poor precision in quantitative analysis, and requires 10–100 laser shots to yield good statistics. The lateral resolution (1000 nm) of a single shot is then usually degraded with multiple shots.

To overcome the low ion yields, about 10^{-4} in the plasma volume, Odom and Schueler[89,90] incorporated a second, focused, high-power Nd:YAG laser beam that irradiates the neutral plume produced by the primary (ablation) laser to increase the ion yields and sensitivity. They have coined another term, *PAI,* for postablation ionization. The enhanced ionization is produced via nonresonant multiphoton ionization processes in the same manner as Becker accomplished with SALI (see Section 4.4 and Chapter 4, this volume). The laser ionization mass spectrometry (LIMS)/PAI technique is claimed to be 100–1000 times more sensitive than the ordinary LAMMA or LMP.[89] Thus, this technique can now be used to obtain depth resolutions of 5–10 nm with "finesse" laser shots.[90]

5.3. Neutral Stimulation: FAB-SIMS, NSS

The neutral stimulation of surfaces can be accomplished using fast atom bombardment (FAB). The primary motivation for using a neutral beam in SIMS was to avoid major charging problems that result when ion beams are used, especially with insulating and organic samples. As interest in polymer surfaces has increased, FAB has become especially important. A FAB gun is typically an ordinary ion gun[18,39] to which a charge exchange chamber is added beyond the ion focusing optics to neutralize the ions from the ion gun. Deflection plates are then used to remove the ions from the "focused" beam just before its emergence from the exit aperture. Leys[91] provides brief descriptions of the FAB gun and recent applications. A more complete description of the gun is given by Busch in Chapter 3 of this volume; complete articles are also available.[92–94] Leys presents several excellent examples of how SIMS data are improved for inorganic insulators, polyethylene, and fluorocarbons.[91] Additional applications are presented in several other contributions in *Desorption Mass Spectrometry.*[91]

Several investigators have carried out neutral scattering spectrometry (NSS) using a FAB gun[95] or a comparable source.[96,97] The most important research issue in ISS is to understand the neutralization processes of the incident ion beam. All three papers[95–97] described attempts to secure comparable NSS and ISS spectra from the same clean targets. The NSS peak is sharply defined for helium and is 20–24 eV below that for a single binary elastic collision (SBEC). The neutral beam has to be ionized (up to 25 eV) during the collision encounter with the surface, which accounts for the downshift in the SBEC peak energy. These data were used to explain why ISS peaks may be 0–30 eV below the SBEC peak. The ion yields from ISS are 20–1000 times greater than for NSS. Thomas *et al.*[95] think that NSS could be a quantitative surface analysis technique for certain systems, but follow-on work to support this speculation has not been possible. As with aspects of some other ion spectroscopies, the NSS and ISS work by Aono and collaborators[96,97] aims to understand the fundamental physics and surface science of particle–surface collision encounters.

5.4. Electric Field Stimulation: APFIM, FIMS

The atom probe field ion microscope (APFIM) is a microanalytical tool of ultimate sensitivity for use in fundamental investigations of surface phenomena as well as in metallurgical applications. The operator may view his specimen, remove at his discretion a single surface atom or a few atoms or molecules from a selected surface area, and identify these particles by their mass. The historical development of the APFIM has been briefly described by its inventor, Erwin Müller.[98]

In using the field ion microscope (FIM), investigators first etch the specimen to form a sharp needle, usually about 10–100 nm in diameter. The FIM tip is held in vacuum a few centimeters from a channel plate intensifier and phosphor screen. When a positive voltage of 5–30 kV is applied to the tip, an electric field of more than 20 V/nm is formed perpendicular to the surface of the tip. Helium gas is backfilled into the UHV chamber and individual helium atoms are ionized near the atomic steps, kinks, etc., on the emitter surface where the field is highest. These ions are accelerated immediately to form an image of the tip on the phosphor screen. Resolution is improved if the tip is cooled to 78 K and below, since the thermal vibration of the temperature-accommodated helium ions is reduced. By raising the field further, individual surface atoms can be field evaporated to clean the tip by removing the surface atoms layer by layer, thus revealing the structure of the tip in atomic detail. By applying a voltage pulse, the mass-to-charge ratio of an individually field desorbed atom can be resolved in a TOF mass spectrometer (MS).[99] A commercial APFIM instrument is available.

Unfortunately, the APFIM is not a technique suitable for general applicability. The sample must take the form of a sharp needle or a particle where a high electric field can be formed,[100] and the material must be capable of withstanding high electrical stress during imaging and field evaporation. Recent work employing pulsed lasers to provide the evaporation pulse has extended the range of materials amenable to the technique to include not only the traditional metals (tungsten, iridium, steel, aluminum) but also semiconductors such as gallium arsenide and silicon in whisker form.[101] The pulsed laser APFIM also has a mass resolution ($\Delta M/M$) of 0.001–0.0001 and an energy resolution of 0.05–0.6 eV for analysis during field desorption (layer-by-layer micro-sectioning).[100]

The obvious applications of FIM to surface physics led to major advances in our knowledge about atomic arrangements on surfaces in the 1960s, and the APFIM was introduced in 1968.[102] It thus joined AES, ISS, SIMS, and XPS as methods in that era for identifying elements on a surface.[103] For example, an adequate demonstration of atom layer by atom layer microsectioning and compositional analysis can be found in *Methods of Surface Analysis* or in more recent reviews.[99] Although only about 15 APFIMs are now in used throughout the world, a new interest in nanometer level depth profiling and two-dimensional, nanometer level lateral resolution has also stimulated new interest in the APFIM.[104]

For the field ion mass spectrometry (FIMS) developed by Block and co-workers,[105–107] the same basic arrangement is used as that in the APFIM, i.e., a FIM to image the tip, a probe hole in the screen, and a TOF–MS for analyzing field desorbed products. Although the FIMS could function as an APFIM, the purposes of the experiments generally are to understand fast reactions on specific crystal planes of the FI tip. Adsorption–desorption phenomena and their relationship to catalytic reactions carried out in VHV or UHV have dominated much of the work in this field. As it has with APFIM, using the FIMS has led to a considerably greater understanding of fundamental physical and chemical processes on surfaces. This field of study has often been referred to as a "high-field (surface) chemistry," because the electric fields used in the experiments, ca. 10 V/nm, disturb the usual properties of atoms and molecules. Thus, the electric field becomes a new variable from a thermodynamic viewpoint; e.g., like temperature and pressure, the electric field alters the properties of ensembles. As has been shown, the high fields are not necessarily an artifact.[106] For the purposes of this book, FIMS is similar to ESDIAD; i.e., it is a superb technique for studying adsorbates on surfaces, but it is not a general purpose method for analyzing the chemical composition of surfaces.

References

1. D. O. Hayward and B. M. W. Trapnell, *Chemisorption,* Butterworths, London (1964); F. C. Tompkins, *Chemisorption of Gases on Metals,* Academic, New York (1978); T. N. Rhodin and G. Ertl, *Nature of the Surface Chemical Bond,* North-Holland, Amsterdam (1979).
2. E. A. Flood, ed., *The Solid–Gas Interface,* Vols. 1 and 2, Dekker, New York (1967).
3. R. B. Anderson, ed., *Experimental Methods in Catalytic Research,* Academic, New York (1968).
4. M. Green, ed., *Solid State Surface Science,* Dekker, New York (1969).
5. R. F. Gould, ed., *Interaction of Liquids at Solid Surfaces,* ACS Publications, Washington, D.C. (1968).

6. J. R. Anderson, ed., *Chemisorption and Reactions on Metallic Films,* Vols. I and II, Academic, New York (1969 and 1971).
7. K. L. Chopra, *Thin Film Phenomena,* Krieger, New York (1970).
8. G. A. Somorjai, *Principles of Surface Chemistry,* Prentice-Hall, Englewood Cliffs, New Jersey (1972).
9. J. M. Blakely, *Introduction to the Properties of Crystal Surfaces,* Pergamon Press, Oxford (1973).
10. A. W. Adamson, *Physical Chemistry of Surfaces,* Wiley-Interscience, New York (1982).
11. P. F. Kane and G. B. Larrabee, eds., *Characterization of Solid Surfaces,* Plenum, New York (1974).
12. A. W. Czanderna, ed., *Methods of Surface Analysis,* Elsevier, Amsterdam (1975).
13. N. B. Hannay, ed., *Treatise on Solid State Chemistry,* Vols. 6A and 6B, *Surfaces,* Plenum, New York (1976).
14. R. Vanselow and others, eds., *Chemistry and Physics of Solid Surfaces,* Vols. I, II, and III, CRC Press, Cleveland (1977, 1979, and 1982); Vols. IV, V, VI, and VII, Springer-Verlag, Berlin (1982, 1984, 1986, and 1988).
15. S. R. Morrison, *The Chemical Physics of Surfaces,* Plenum, New York (1977).
16. J. E. E. Baglin and J. E. Poate, eds., *Thin Film Phenomena—Interfaces and Interactions,* Proc. Electrochem. Soc., 78-2 (1978).
17. R. W. Roberts and T. A. Vanderslice, *Ultrahigh Vacuum and Its Applications,* Prentice-Hall, Englewood Cliffs, New Jersey (1963).
18. H. Werner, in: *Electron and Ion Spectroscopy of Solids* (L. Fiermans, J. Vennik, and W. Dekeyser, eds.), p. 419, Plenum, New York (1978).
19. C. B. Duke, in: *Industrial Applications of Surface Analysis* (L. Casper and C. Powell, eds.,) p. 12, American Chemical Society, Washington, D.C. (1982).
20. R. W. Balluffi, G. R. Woolhouse, and Y. Komen, in: *The Nature and Behavior of Grain Boundaries* (H. Hu, ed.), p. 41, Plenum, New York (1972).
21. J. O'M. Bockris, *Mat. Sci. Eng.* **53,** 47 (1982).
22. A. W. Czanderna, *J. Vac. Sci. Technol.* **17,** 72 (1980).
23. S. Brunauer, P. Emmett, and E. Teller, *J. Am. Chem. Soc.* **60,** 309 (1938).
24. E. Robens, in: *Microweighing in Vacuum and Controlled Environments* (A. W. Czanderna and S. P. Wolsky, eds.), p. 127, Elsevier, Amsterdam (1980).
25. S. J. Gregg and K. S. W. Sing, *Adsorption, Surface Area and Porosity,* Academic, New York (1982).
26. A. W. Czanderna and R. Vasofsky, *Prog. Surf. Sci.* **9,** 45 (1979).
27. R. Shuttleworth, *Proc. R. Soc. London* **62A,** 167 (1949).
28. G. C. Benson and K. S. Yun, in: *The Solid–Gas Interface* (E. A. Flood, ed.), Vol. 1, p. 203, Dekker, New York (1967).
29. C. Duke, *Crit. Rev. Solid State Matls. Sci.* **8,** 69 (1978).
30. J. T. Yates, Jr. and T. E. Madey, eds., *Vibrational Spectroscopy of Molecules on Surfaces,* Plenum, New York (1987).
31. A. Roth, *Vacuum Technology,* North-Holland, Amsterdam (1976).
32. A. W. Czanderna, C. J. Powell, and T. E. Madey, eds., *Sample Handling, Beam Artifacts, and Depth Profiling,* Plenum, New York. (in preparation).
33. H. F. Winters, in: *Radiation Effects on Solid Surfaces* (M. Kaminsky, ed.), p. 1, American Chemical Society, Washington, D.C. (1976).
34. J. A. McHugh, in: *Methods of Surface Analysis* (A. W. Czanderna, ed.), p. 223, Elsevier, Amsterdam (1975).
35. G. K. Wehner, in: *Methods of Surface Analysis* (A. W. Czanderna, ed.), p. 1, Elsevier, Amsterdam (1975).

36. R. E. Honig, *Thin Solid Films* **31,** 89 (1976).
37. G. P. Chambers and J. Fine, "Evaluated Absolute Elemental Sputtering Yields for Surface Analysis," *J. Phys. Chem. Ref. Data* (to be published).
38. S. Hofmann, in: *Practical Surface Analysis by AES and XPS* (D. Briggs and M. P. Seah, eds.), pp. 141–179, Wiley, New York (1983).
39. L. C. Feldman and J. W. Mayer, *Fundamentals of Surface and Thin Film Analysis,* pp. 69–96, North-Holland, Amsterdam (1986).
40. A. Benninghoven, F. G. Rüdenauer, and H. W. Werner, *Secondary Ion Mass Spectrometry,* pp. 761–912 (CDP) or 937–949 (SNMS), Wiley, New York (1987).
41. A. W. Czanderna, *Solar Energy Mat.* **5,** 349 (1981).
42. D. Lichtman, in: *Methods of Surface Analysis* (A. W. Czanderna, ed.), p. 39, Elsevier, Amsterdam (1975).
43. R. F. K. Herzog and R. P. Viehböck, *Phys. Rev.* **76,** 855 (1949).
44. R. E. Honig, *J. Appl. Phys.* **29,** 549 (1958).
45. G. Blaise, in: *Materials Characterization Using Ion Beams* (J. P. Thomas and A. Cachard, eds.), p. 143, Plenum, New York (1978).
46. H. Werner, in: *Electron and Ion Spectroscopy of Solids* (L. Fiermans, J. Vennik, and W. Dekeyser, eds.), p. 324, Plenum, New York (1978).
47. J. M. Morabito and R. K. Lewis, in: *Methods of Surface Analysis* (A. W. Czanderna, ed.), p. 279, Elsevier, Amsterdam (1975).
48. C. W. Magee, W. L. Harrington, and R. E. Honig, *Rev. Sci. Instrum.* **49,** 477 (1978).
49. A. Benninghoven *et al.,* eds., *Secondary Ion Mass Spectrometry, SIMS II, SIMS III, SIMS IV, and SIMS V,* Springer Series in Chemical Physics, Vols. 9, 19, 36, and 44, Springer-Verlag, Berlin (1979, 1982, 1984, and 1986, respectively); *SIMS VI,* Wiley, New York (1988).
50. D. P. Smith, *J. Appl. Phys.* **38,** 340 (1967).
51. D. P. Smith, *Surf. Sci.* **25,** 171 (1971).
52. H. H. Brongersma, *J. Vac. Sci. Technol.* **11,** 231 (1974).
53. T. M. Buck, in: *Methods of Surface Analysis,* (A. W. Czanderna, ed.), p. 75, Elsevier, Amsterdam (1975).
54. E. Taglauer and W. Heiland, *Appl. Phys.* **9,** 261 (1976).
55. E. Taglauer and W. Heiland, in: *Applied Surface Analysis* (T. L. Barr and L. E. Davis, eds.), p. 111, ASTM, Philadelphia, Pennsylvania (1980).
56. A. C. Miller, in: *Treatise on Analytical Chemistry, Pt. 1* (J. D. Winefordner, ed.), Vol. 11, 2nd Ed., p. 253, Wiley, New York (1989).
57. T. Rusch and R. L. Erickson, in: *Inelastic Ion Surface Collisions* (N. H. Tolk, J. C. Tully, W. Heiland, and C. W. White, eds.), p. 73, Academic, New York (1977).
58. S. Rubin and V. K. Rasmusen, *Phys. Rev.* **78,** 83 (1950).
59. W. K. Chu, J. W. Mayer, and M-A. Nicolet, *Backscattering Spectrometry,* Academic, New York (1978).
60. T. A. Cahill, *Ann. Rev. Nucl. Part. Sci.* **30,** 211 (1980).
61. E. T. Williams, *Nucl. Instrum. Methods Phys. Res.* **B3,** 211 (1980).
62. C. W. White, E. W. Thomas, W. F. Van der Weg, and N. H. Tolk, in: *Inelastic Ion–Surface Collisions* (N. H. Tolk, J. Tully, W. Heiland, and C. W. White, eds.), pp. 201–252, Academic, New York (1977).
63. G. Betz, *Nucl. Instrum. Methods Phys. Res.* **B27,** 104 (1987).
64. H. D. Hagstrum, *Phys. Rev.* **96,** 335 (1954).
65. H. D. Hagstrum, *Phys. Rev.* **150,** 495 (1966).
66. H. D. Hagstrum and G. E. Becker, *Phys. Rev. B* **4,** 4187 (1971).
67. H. D. Hagstrum, in: *Electron and Ion Spectroscopy of Solids* (L. Fiermans, J. Vennik, and W. Dekeyser, eds.), pp. 273–323, Plenum, New York (1978).

68. C. A. Moyer and K. Orvek, *Surf. Sci.* **114,** 295 (1982); and **121,** 138 (1983).
69. J. W. Coburn and E. Kay, *Appl. Phys. Lett.* **19,** 350 (1971).
70. R. E. Honig, in: *Advances in Mass Spectrometry* (J. D. Waldron, ed.), Vol. 1, p. 162, Pergamon Press, London (1959).
71. A. Benninghoven and F. Kirchner, *Z. Natforsch.* **18a,** 1008 (1963).
72. H. Oechsner, in: *Thin Film and Depth Profile Analysis* (H. Oechsner, ed.), pp. 63–85, Springer-Verlag, Berlin (1984).
73. D. Lipinsky, R. Jede, O. Ganschow, and A. Benninghoven, *J. Vac. Sci. Technol.* **A3,** 2007 (1985).
74. C. H. Becker and K. T. Gillen, *Anal. Chem.* **56,** 1671 (1984); *Appl. Phys. Lett.* **45,** 1063 (1984).
75. D. M. Gruen, M. J. Pellin, C. E. Young, and M. H. Mendelsohn, *Phys. Scr.* **T6,** 42 (1983).
76. C. E. Young, M. J. Pellin, W. F. Calaway, B. Jørgensen, E. L. Schweitzer, and D. M. Gruen, *Nucl. Instrum. Methods Phys. Res.* **B27,** 119 (1987).
77. J. W. Burnett, J. P. Biersack, D. M. Gruen, B. Jørgensen, A. R. Krauss, M. J. Pellin, E. L. Schweitzer, J. T. Yates, Jr., and C. E. Young, *J. Vac. Sci. Technol.* **A6,** 2064 (1988).
78. D. Lichtman and R. B. McQuistan, *Prog. Nucl. Energy, Ser. IX* **4** (pt. 2), 95 (1965).
79. T. E. Madey and J. T. Yates, Jr., *J. Vac. Sci. Technol.* **8,** 525 (1971).
80. N. H. Tolk, M. M. Traum, J. C. Tully, and T. E. Madey, eds., *Desorption Induced by Electronic Transitions, DIET I,* Springer-Verlag, Berlin (1983); W. Brenig and D. Menzel, eds., *DIET II and DIET III,* Springer-Verlag, Berlin (1985, 1987); D. Menzel, *Nucl. Instrum. Methods Phys. Res.* **B13,** 507 (1986).
81. Ph. Avouris, F. Bozso, and R. E. Walkup, *Nucl. Instrum. Methods Phys. Res.* **B27,** 136 (1987).
82. T. E. Madey, in: *Inelastic Particle-Surface Collisions* (E. Taglauer and W. Heiland, eds.), pp. 80–103, Springer-Verlag, Berlin (1981).
83. T. E. Madey, *Science* **234,** 316 (1986).
84. J. T. Yates, Jr., M. D. Alvey, K. W. Kolasinski, and M. J. Dresser, *Nucl. Instrum. Methods Phys. Res.* **B27,** 147 (1987).
85. F. Hillencamp, E. Unsoeld, R. Kaufmann, and R. Nitsche, *Appl. Phys.* **8,** 341 (1975).
86. F. Hillencamp, in: *Secondary Ion Mass Spectrometry* (*SIMS V*), (A. Benninghoven, R. J. Colton, D. S. Simons, and H. W. Werner, eds.), p. 471, Springer-Verlag, Berlin (1986).
87. J. A. J. Jansen and A. W. Witmer, *Spectrochim. Acta* **37B,** 483 (1983).
88. S. W. Graham, P. Dowd, and D. M. Hercules, *Anal. Chem.* **54,** 649 (1982).
89. B. Schueler and R. W. Odom, *J. Appl. Phys.* **61,** 4652 (1987).
90. R. W. Odom and B. Schueler, *Thin Solid Films* **153,** 1 (1987).
91. J. A. Leys, in: *Desorption Mass Spectrometry* (P. A. Lyon, ed.), pp. 145–159, ACS Symposium Series No. 291, American Chemical Society, Washington, DC (1985).
92. D. J. Surman and J. C. Vickerman, *Appl. Surf. Sci.* **9,** 108 (1981).
93. V. N. Klaus, *Vakuum Technik* **31,** 106 (1982).
94. M. Barber, R. Bordoli, R. Sedgwick, and A. Taylor, *Nature* **293,** 270 (1981).
95. T. M. Thomas, H. Neumann, A. W. Czanderna, and J. R. Pitts, *Surf. Sci.* **175,** L737 (1986).
96. R. Souda, M. Aono, C. Oshima, S. Otani, and Y. Ishizawa, *Surf. Sci.* **150,** 59 (1985).
97. M. Aono and R. Souda, *Nucl. Instrum. Methods Phys. Res.* **B27,** 55 (1987).
98. E. W. Müller, in: *Methods of Surface Analysis* (A. W. Czanderna, ed.), pp. 329–378, Elsevier, Amsterdam (1975).
99. T. T. Tsong, *Surf. Sci. Rep.* **8,** 127 (1988).

100. J. A. Panitz, A. L. Pregenzer, and R. A. Gerber, *J. Vac. Sci. Technol.* **A7,** 64 (1989).
101. T. T. Tsong, S. B. McLane, and T. J. Kinkus, *Rev. Sci. Instrum.* **53,** 1442 (1982).
102. E. W. Müller, J. A. Panitz, and S. B. McLane, *Rev. Sci. Instrum.* **39,** 83 (1968).
103. A. W. Czanderna, ed., *Methods of Surface Analysis,* Elsevier, Amsterdam (1975).
104. T. T. Tsong, *J. Vac. Sci. Technol.* **A7,** 1758 (1989).
105. J. H. Block, *Z. Phys. Chem. (NF)* **39,** 169 (1963).
106. J. H. Block and A. W. Czanderna, in: *Methods of Surface Analysis* (A. W. Czanderna, ed.), pp. 379–446, Elsevier, Amsterdam (1975).
107. J. H. Block, in: *Chemistry and Physics of Solid Surfaces VI* (R. Vanselow and R. Howe, eds.), Springer-Verlag, Berlin (1986).

2

Surface Structure and Reaction Studies by Ion–Solid Collisions

Nicholas Winograd and Barbara J. Garrison

1. Introduction

Bombardment of targets by energetic particle beams has become an important tool for the characterization and modification of solids and surfaces. With this approach, the kinetic energy of the primary particle, typically a few eV to a few thousand eV, exceeds the binding interactions normally present in chemical bonds. Because of this energy difference, a novel and intriguing chain of events is rapidly set in motion subsequent to the impact event. Atoms may be significantly displaced from their equilibrium positions. Original chemical bonds may be broken with new ones formed. Energetic collisions may give rise to electronic excitation, ionization, and desorption of atomic or molecular components of the bombarded solid.

These phenomena have led to important applications. A major impetus for research has been, of course, in the microelectronics area, where ion implantation of dopant ions[1] and reactive ion etching of semiconductors are hot topics.[2] There is also interest in evaluating surface processes which occur when energetic ions and molecules present

Nicholas Winograd and Barbara J. Garrison • Department of Chemistry, Pennsylvania State University, University Park, Pennsylvania 16802.

Ion Spectroscopies for Surface Analysis (Methods of Surface Characterization series, Volume 2), edited by Alvin W. Czanderna and David M. Hercules. Plenum Press, New York, 1991.

in the space environment impinge on a diverse array of synthetic materials ranging from light metallic composites to protective polymer overlayers deposited on nonlinear optical materials.[3] The morphology of extraterrestrial surfaces is believed to be dominated to some degree by interaction with the solar wind or other energetic ions.[4]

Other applications involve an analysis of the primary particle or of the desorbed material. Examination of the fate of the scattered primary particle forms the basis of ion scattering spectrometry (ISS) in the 10–10,000-eV range and Rutherford backscattering at still higher energies. These are important structural tools since the scattering trajectories are strongly influenced by the nuclear positions of surface atoms. Both approaches are extensively utilized as surface analytical techniques. Complete reviews of both of these ion-beam methods are presented by Taglauer in Chapter 6 and Feldman in Chapter 5.

Measurement of the intensity and mass of the ions that desorb from the surface forms the basis of secondary ion mass spectrometry (SIMS). Since the initial observation of these ions by J. J. Thomson n 1910,[5] this field has expanded almost exponentially. Currently, SIMS is widely utilized in trace analysis in combination with depth profiling of the near surface region.[6] This profiling is accomplished by eroding the surface layers of the target using the same energetic particle used to produce the secondary ions. It is also possible to produce spectacular element-specific images of surfaces with a spatial resolution below 100 nm using either a highly focused, rastered ion beam[7] or a specially designed imaging detector.[8]

Perhaps the most important aspect of SIMS for chemical analysis is that many of the desorbed ions may be molecular in nature and may reflect the composition of the bombarded surface. Since Benninghoven[9] suggested that these experiments be performed at very low ion doses—before the surface composition is significantly altered—it has been found that a vast array of normally nonvolatile molecules may be desorbed intact, without significant fragmentation. Similar observations have been made from bombarded liquid targets, even at higher doses.[10] The weight of some molecular ions has exceeded 10,000 amu.[11] These species are now detectable, largely because of the highly sensitive time-of-flight (TOF) detectors also being developed by the Benninghoven group.[12] It is an apparent paradox that fragile, heavy molecules such as insulin thermally decompose at slightly elevated temperatures, yet desorb intact when the solid is belted with a keV particle. A complete discussion of these topics by Busch is included in Chapter 3. As a bonus to all of this, the desorption occurs specifically from surface layers,[13] opening important applications in catalysis studies where it may be possible to directly identify heterogeneous reaction intermediates.

Detection of the desorbed neutral material from bombarded solids dates to the 1850s where erosion of cathodes in gas discharge sources was of concern.[14] Although we find the term most offensive, this desorption is commonly referred to as "sputtering." The study of this phenomenon was thought to be important to gain a fundamental picture of the particle/solid interaction. Early experimental results were of dubious value, however, owing to poor vacuum conditions and inefficient detection schemes. During the last 10 years, several new approaches using methods whereby the desorbed neutral species is ionized either by plasma,[15] electron,[16] or laser[17] has basically solved the detection problem. In fact, as Becker points out in Chapter 4, the most sensitive surface analyses are now performed with desorbed neutral species. High-quality measurements are now posible which finally are yielding to their promise of providing data directly comparable to theory. In addition, this development perhaps offers a new direction for explosive research growth leading to even more particle/solid interaction applications.

Although these experiments have certainly had a diverse and important impact on materials science research, it is apparent that the details of the particle-impact event are very complicated. In fact, at first glance, it might appear that it is not possible to go beyond the simple "hand-waving" stage. After one of our lectures, a doubting observer referred to our experiments as "hand-grenade science." A major goal of this chapter, then, is to convince the reader that there is, indeed, a great deal known about the nuclear motion that leads to radiation damage and desorption. Further, there are impressive attempts to quantitatively understand the basis for electronic excitation and ionization. The tools for leap-frogging beyond the hand-waving stage are at least with us, if not completely developed.

The basic idea behind our strategy of gaining an atomic-level understanding of this process is illustrated by examining the possible parameters that are measurable and calculable. These are shown schematically in Fig. 1. The incident particle is usually ionized,[18]

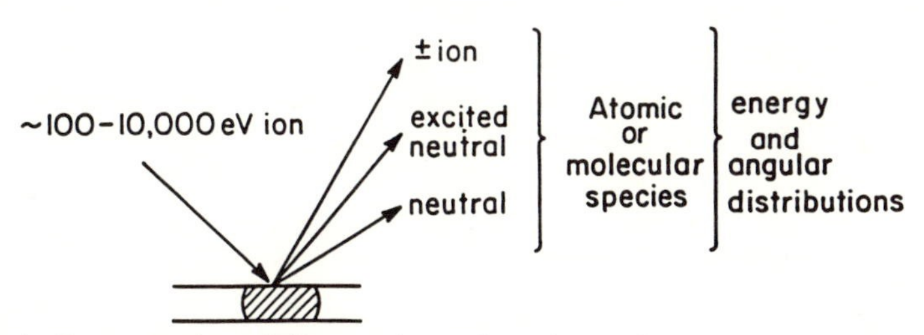

Figure 1. Representation of the experimental geometry of a typical ion–solid interaction experiment with important measurable quantities.

accelerated and directed at a specific incident angle toward the target. As we have mentioned, the kinetic energy of this particle initiates atomic motion near the point of impact. A portion of these atoms may have a momentum component that is directed away from the target and thus desorption of various species is observed to occur. Note also that radiation damage resulting in chemical modification of the surface region is an integral part of the bombardment process. From the figure, however, it is apparent that there is a wealth of experimental information potentially available to help us fully understand the dynamics of this complex event. Particles may desorb by way of a variety of channels to produce ground state and excited state neutral atoms, positive or negative ions as well as a multitude of molecular cluster species. Of particular importance is that each component possesses its own kinetic energy and angular distribution. These distributions play an essential role in validating proposed theories of either the nuclear motion or of the electronic excitation. It is our view that predictions of an angle-integrated and/or a kinetic energy-integrated quantity are not sufficient for such validation.

The focus of this chapter will be quite narrow relative to the wide scope of this field. We shall concentrate only on developing ideas and experiments that can yield quantitative comparisons between experiment and theory and shall emphasize those applications that translate directly into new insights into the chemistry or structure of surfaces. We shall try as hard as we can to avoid any hand-waving discussions and will only present the formalism behind fairly rigorous approaches. Specifically, we will be interested in presenting techniques that can perform detailed trajectory measurements for most of the reaction channels given in Fig. 1. A critical theme will be to emphasize the interrelationship of all these observables and to develop the best theoretical and experimental approaches that result in quantitative models for each process. We will begin with a discussion of the relevant experimental tools that are needed to obtain the information in Fig. 1. We will next provide a short review of the theory of sputtering, followed by an extensive development of the use of computer simulations to model the nuclear motion near the point of impact of the primary ion. A discussion of the current state of the theory of excitation and ionization will precede a section on the application of these ideas to problems of particular relevance to surface science.

2. The Experimental Approach

Within our narrow focus, the experimental goals are painfully clear. The target should be a well-characterized single crystal so that the initial atomic locations are as well known as possible. The particle beam source should be flexible with regard to incident energy, mass, and angle. The detector must be able to count desorbed particles with doses well below that of the number of surface atoms or molecules. Ideally, the device should also be able to selectively detect ionic and neutral components in ground and excited states and to resolve energy and angular distributions. It is with these measurements that theories can be carefully checked, the implications of beam methods for surface structure studies can be fully assessed, and the analytical and technological uses can be exploited to the maximum extent.

Through the early 1960s, experiments were restricted largely to weight-loss measurements on uncharacterized single-crystal surfaces.[19] After this time, a rapid series of advances greatly expanded our understanding of this process. Mass spectrometric detectors became popular for detection of secondary ions.[20] At about the same time, Thompson reported his elegant time-of-flight experiments performed on ejected neutral atoms to yield velocity measurements.[21] The legendary work of Wehner to unravel many angular distribution patterns from single crystals and his fluorescence technique to obtain velocity distributions still has an enormous influence on present day designs.[22] Yet all of these experiments were performed using poor vacuums and high incident ion doses. Electron micrographs of the experimental surfaces with the eery morphological features left one somewhat uneasy.[23]

Two critical developments from Benninghoven's laboratory sparked the development of modern experiments. First, he showed that using mass spectrometric detectors in the SIMS mode, a measurable signal could be obtained from just $\sim 10^{12}$ incident ions/cm^2.[9] This observation allowed the study of monolayers by bombarding only 1 in a thousand surface atoms—a configuration he termed "static" SIMS since the surface condition remained unchanged to the observer during the time of measurement.

His second major contribution was that he noted the appearance of organic molecular ions in these static SIMS spectra which were representative of the organic surface under investigation.[24] These early studies have, of course, given way to fast atom bombardment (FAB) measurements[10,18] with all their associated hoopla. The relevant experimental details for these studies are given in Chapter 3.

No apparatus yet exists that satisfies all of our initial measurement criteria. Advances in experimental surface science during the 1970s allowed major progress in this direction, however. A first attempt at performing detailed trajectory measurements in UHV under static-mode conditions was first initiated about 8 years ago.[25] This apparatus, shown in Fig. 2, is capable of detecting a wide variety of azimuthal and polar

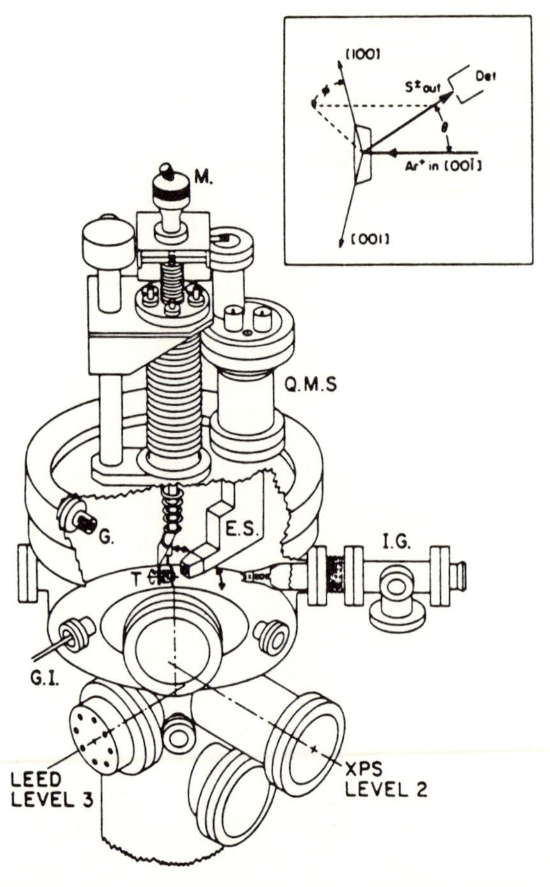

Figure 2. Schematic view of the spectrometer. The components illustrated include M, crystal manipulator; Q.M.S., quadrupole mass spectrometer; I.G., primary ion source; E.S., energy spectrometer; G, Bayard–Alpert gauge; T, crystal target; and G.I., gas inlet. Auxiliary components are omitted for graphical clarity. The SIMS experimental geometry and coordinate system are defined in the inset. (From Ref. 25.)

angles for a normally incident beam. The detector is a quadrupole mass spectrometer that could be rotated under UHV conditions with respect to the incident ion beam. Moreover, a medium resolution energy selector consisting of a 90° sector provides the capability of examining how the angular distributions change with secondary ion energy. The chamber was designed with two levels. The SIMS measurements are performed on the top level while LEED/Auger/XPS analysis could be obtained by vertically translating the sample to a lower level. This instrument has been very successful in providing ion trajectory data, as presented in the next sections, and has yielded a measure of understanding about differences in behavior between ions and the predicted behavior of the neutral flux obtained from molecular dynamics simulations. It provides no information about neutral trajectories nor does it yield state-selective information about electronically excited species.

The neutral data are potentially obtainable using efficient postionization sources to excite the atom or molecule after it has desorbed from the surface but before it has reached the detector. The first promising experiments involving lasers utilized fluorescence spectroscopy to probe these species.[26] The wavelengths of the fluorescence could be tied to the original electronic states of the desorbing atoms. As a bonus, the speed of the desorbing component could be determined by measuring its Doppler shift.[27] Unfortunately, this method was never sufficiently sensitive to be utilized in the static mode. Other schemes, including efficient electron impact ionization of the desorbed particles[16] and low-energy plasma ionization[15] offer the potential for static measurements on monolayers. As of yet, however, the necessary sensitivity does not appear to have been demonstrated.

Laser-induced ionization overcomes many of the deficiencies of current approaches. Successful experiments have been possible using both multiphoton resonance ionization (MPRI)[17] and nonresonant ionization.[28] These methods are described in detail by Becker in Chapter 4. Here, we summarize only several inherent advantages of MPRI detection specific to carrying out ion beam experiments on well-defined surfaces. First, a pulsed laser is used to create ions from the ejected neutral species. This pulse generates an ideal timing signature for time-of-flight (TOF) detection, one of the most efficient mass analysis techniques. Secondly, MPRI requires relatively low power to achieve nearly 100% ionization efficiency. This means that the laser beam may be quite large in size, permitting the spatial imaging of subsequently ionized species. Finally, MPRI is highly selective. Using tunable dye lasers, it is possible to excite only those atoms whose electronic spacings match the laser frequency. This aspect of MPRI means that it is possible to achieve extremely low background signals and to accurately count just a few particles.

A schematic diagram of an instrument designed to measure both the energy- and angle-resolved neutral (EARN) distributions[29] is shown in Fig. 3. The sample is placed in the top level of a two-level chamber similar in design to that shown in Fig. 2. This level is equipped with quartz laser ports, the EARN detector, and a differentially pumped pulsed ion gun. The bottom level is equipped in the usual fashion with Auger/LEED for auxiliary surface analysis. The tunable photon radiation

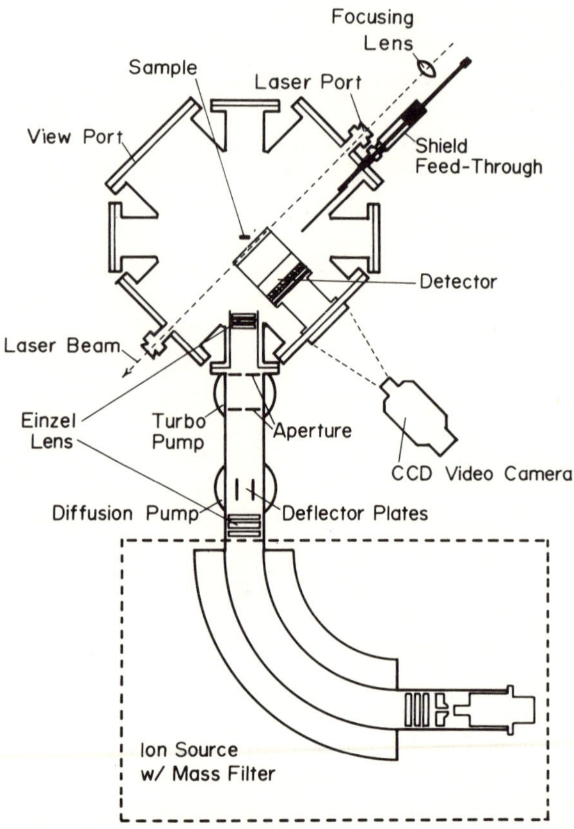

Figure 3. Schematic cross section of the EARN apparatus. The sample is shown in the position for normal-incident ($\theta_i = 0°$) ion bombardment. Not drawn to scale. (From Ref. 29.)

is generated using commercially available dye lasers pumped by a Q-switched Nd:YAG laser. This laser system is equipped with standard nonlinear optical mixing capabilities so that tunability over the range of 2–7 eV is possible. The resulting radiation takes the form of a 6-ns pulse which is repeated at a rate of up to 30 Hz. The laser beam itself is reshaped from an initial 0.5-cm circular configuration to a 0.1-cm-thick ribbon using a 225-cm local length cylindrical lens. Note that it is positioned about 1.5 cm above and parallel to the target.

The timing scheme for the EARN experiments is critical and is shown schematically in Fig. 4. In general terms, the experiment proceeds as follows: (i) A 200-ns pulse of approximately 2.5×10^6 ions is focused to a 2-mm spot on the sample; (2) upon impact of the ion pulse, an ion

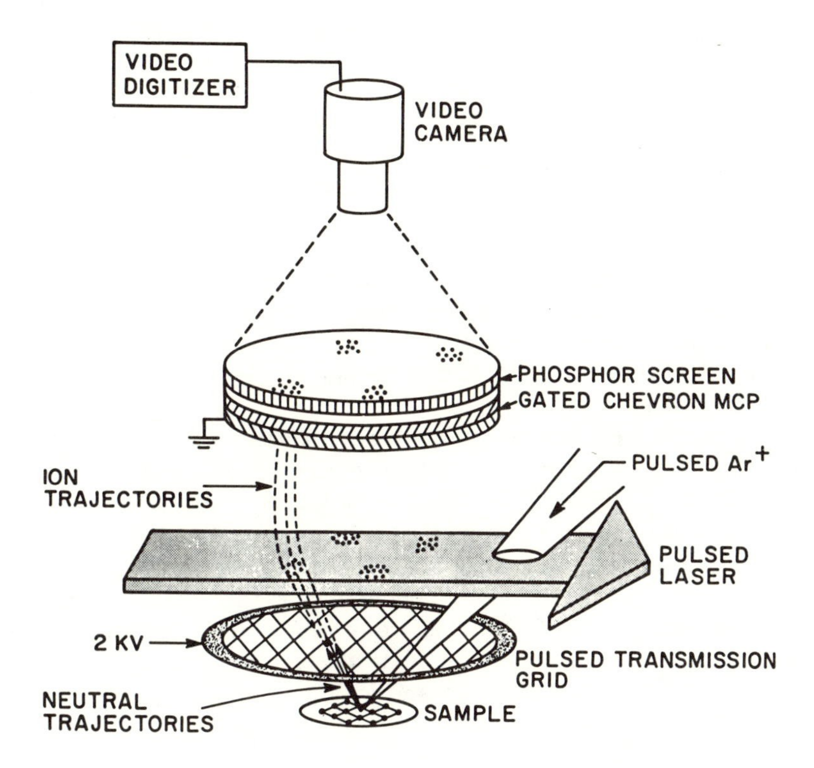

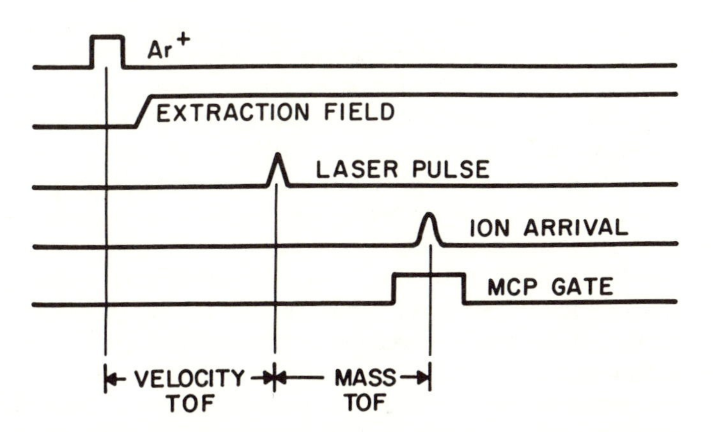

Figure 4. Schematic representation of the timing scheme for the EARN experiment. Note that the velocity of the desorbing atom is defined by the time between the ion pulse and the laser pulse. The MCP is gated to reduce unwanted signal. See text for more details.

extraction field is activated for the duration of the measure; (3) the laser pulse (~1 mJ) selectively ionizes a small volume of neutral particles at a time τ_E after the ion-pulse impact, thus defining the time of flight (TOF) of the probed particles; (4) the ionized particles are then accelerated by the ion-extraction field and arrive at the front of a microchannel plate (MCP) assembly at a time τ_m later, τ_m being governed primarily by the mass-to-charge ratio of the ionized particles. The impacting particles are detected on a phosphor screen by a CCD video camera during a time when the MCP is gated to an active state. The gate is used to discriminate against signals from scattered laser light and from ions striking the MCP at other times due to differences in either mass or ionization scheme. This image can then be accessed by a computer system for digital image data processing.

Depending on the count rate for a given experiment, 400–1000 individual images are integrated to give an average ejection pattern for one value of τ_E. During the course of a full experiment, 30–60 images, each corresponding to a different τ_E value, are recorded. The collection of τ_E images are then corrected for differences in the integration volume elements between the desired energy and angle space and the Cartesian-integrated space in which the data are recorded. After application of the appropriate transformation Jacobian, the images are sorted into an intensity map of ejection angles and kinetic energies. The geometry of the detection system allows determination of all polar angles for a given azimuthal direction with kinetic energies ranging from 0 to 50 eV. The angle resolution of about 8° is determined solely by the size of the incident ion beam, while the energy resolution of 15% for 50-eV particles and 4% for 5-eV particles is determined mainly by the incident ion beam pulse-width. This device is very important to a number of the basic ideas developed in this chapter. The data from this instrument provide a set of high-quality state-selected trajectory information which puts enormous constraints on theoretical models.

There are many other exciting new approaches to performing ion-beam/solid experiments involving liquid-metal ion sources, double-focusing mass spectrometers, fourier transform methods, triple quadrupoles, and advanced TOF analyzers, to name just a few. These schemes are beyond the scope of this work but certainly indicate that the level of sophistication of measurement is rapidly improving, boding well for the quality of data forthcoming from future experiments. In the next section, we begin to see how these measurements are intertwined with rapid theoretical advances to provide a fairly complete atomic and molecular view of the entire ion-bombardment event.

3. How To View The Process

To understand the bombardment process in detail, it is necessary to describe the series of events following the impact of the primary particle. Chronologically, the incoming particle first strikes a target atom. This atom usually attains considerable momentum directed into the bulk and collides with neighboring atoms, which produce secondary displacements. Some of these atoms have the possibility of being scattered back toward the surface with sufficient kinetic energy to escape from the sample. Many of the other atoms in the target continue moving inside the solid, slowing down only as they experience more and more collisions. The primary particle and the first target atom finally either come to rest inside the solid (implantation) or are ejected themselves. A schematic picture of some of these events, calculated using computer simulations we will describe in this section, is shown in Fig. 5. The ejection process is a complex event involving multiple atomic or molecular collisions initiated by the primary particle. Besides these atomic motions within the solid, there are concurrent electronic processes which cause species to leave the solid as secondary ions, neutrals, and electronically excited ions or neutral particles.

As shown in Figs 1 and 5 there is a multitude of events that are associated with this process including the ejection of atoms and molecules from the solid, reflection of the primary particle, implantation of the primary particle, lattice damage, and mixing within the solid, to name a few. The emphasis in this section is on the nature of the species that eject and how their properties give us insight into the characteristics of the original surface. In addition, it is our belief that models that predict detailed experimental observables are the only ones that are possible to use in a predictive mode. This emphasis thus controls which models we choose to describe most fully and which ones we only mention in passing. In addition since our goal is to ultimately describe the original surface properties, it is convenient to start with a well-defined and reproducible surface. Many of the examples described below are applicable to single-crystal targets.

The actions of interest are those in which ionic and neutral atoms and molecules eject from the surface and reach a detector. In the laboratory, it is possible to make measurements of the energy distributions, angular distributions, yields (number of particles ejected per incident particle), and cluster yields as shown in Fig. 1. The goal of the theoretical models is to predict and explain all or some subset of the experimental data. If there is good agreement between the calculated and

Figure 5. Positions of the atoms in a Rh{111} surface before and after 3-keV Ar ion bombardment. In the final positions the Ar ion appears near the center of the crystal. This trajectory was chosen at random. In some cases there is considerably more atomic motion and in some cases less.

experimental results then hopefully the model can give insight into the physical processes. Ideally, the model can also be a guiding light for new experiments. In this section we shall briefly review some of the existing approaches to predicting nuclear motion but will place perhaps undue emphasis on the model preferred by the authors. An extensive review has recently been compiled by Harrison[30] which provides excellent background information. Electronic excitations will be considered in Section 4.

There are basically two approaches to calculating the energy dissipation of the primary ion. The first utilizes a statistical model to solve the transport equations for the momentum deposition. This approach allows the yield to be expressed in terms of a simple formula which can be easily used by experimentalists. The second approach receiving considerable attention is to model the impact more explicitly using classical trajectory methods to follow the motion of the relevant particles. This method requires a large computer and generally does not provide simple formulas.

3.1. Transport Theories

There have been a number of attempts to derive analytical expressions that predict various properties of the sputtering phenomenon. The most widely used are those of Sigmund,[31–33] whose yield formulas are ubiquitous throughout the sputtering literature and of Thompson,[21] who has provided a relationship for the yield as a function of secondary particle energy. In both of these approaches, the initial assumption is that each target atom generates one secondary collision cascade in which energy is shared by a series of collisions. Next, it is assumed that the collisions themselves occur only between two atoms at a time. This assumption is referred to as the binary collision approximation (BCA). These approximations have a number of important consequences. First, any electronic energy loss processes are ignored. This is probably a realistic approximation for metals bombarded at less than 50 keV by heavy particles, although for insulators such as alkali halides,[34] coulombic interactions may contribute to sputtering. For elastic collisions, the assumption that each collision sequence can be treated separately restricts application of the theory to the region of so-called linear collision cascades.[31,35] As shown in Fig. 6a, this condition applies only when a small fraction of all atoms within a given volume are in motion. When more than one cascade is initiated by the primary ion, as shown in Fig. 6b, or when a large fraction of the atoms near the impact point are moving, the approach begins to breakdown. This region is often referred

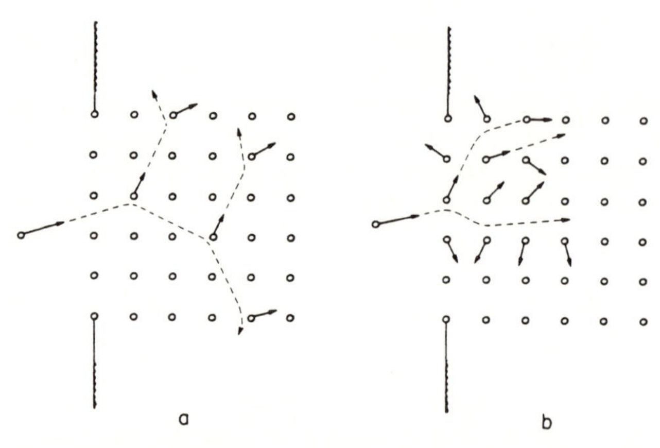

Figure 6. (a) Linear collision cascade. The structure is preserved and a small fraction of the atoms is in motion. (b) Dense cascade (spike). The structure is destroyed locally. All atoms within the spike volume are in motion. (From Ref. 35.)

to as a dense cascade or a spike. One final consequence is that the BCA should only apply to collisions which exceed several hundred eV. For example, in Fig. 6a, the sizes of the atoms are drawn such that it is impossible for more than two atoms to interact simultaneously. If the atomic radii had been drawn such that adjacent atoms nearly touched, as in the case for collision energies below ~10 eV, then binary collisions are indeed improbable. It is disturbing that most of the collisions that lead to the ejection of atoms in the solid are those whose energies are below 10 eV as illustrated by the secondary particle energy distributions shown in Fig. 7a.

In the Sigmund theory, the solid is approximated by an amorphous array of atoms that interact with differential cross sections developed for calculations aimed toward predicting the range of energetic atoms in solids.[31] The nuclear stopping cross section $S_n(E)$, where E is the energy of the ejected particle, can be evaluated from these quantities. The flow of energy through the cascade is then determined using the Boltzmann transport equation. The sputtering yield is evaluated from the flux of particles that crosses an imaginary plane at a point in the solid where the target atom is given its initial energy. To include the effect of the surface, it is assumed that the atom must surmount a potential barrier, U, acting on the perpendicular component of the velocity. This potential will act to keep slow-moving particles from escaping the barrier and will deflect the faster-moving ejecting particles away from the surface normal. This approximation is described as the planar surface binding model.[31,32] If the collision cascade is restricted to the linear regime by utilizing

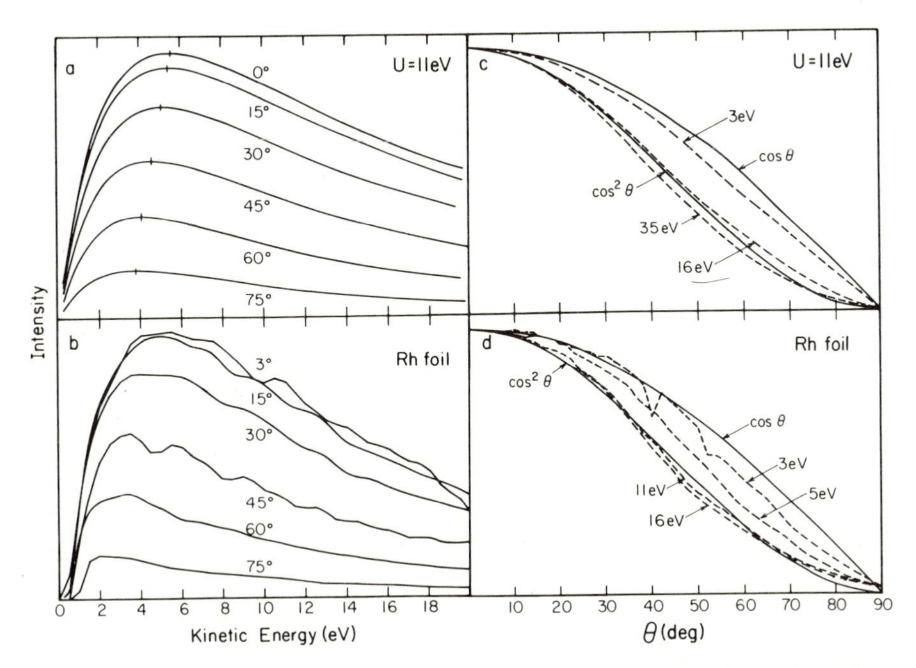

Figure 7. Energy distributions of Rh atoms ejected from a Rh foil for various polar angles, θ, as measured from the surface normal. (a) Calculated from Eq. (5). The cross marks denote the positions of the maxima. (b) Experimental results from Ref. 46. The primary ion beam was Ar^+ with a kinetic energy of 5 keV aimed perpendicular to the surface. (c) Calculated from Eq. (5). (From Ref. 47.) (d) Polar distributions of Rh atoms ejected from a Rh foil for various energies. All curves are peaked normalized at $\theta = 0°$. Experimental results from Ref. 46. The energy ranges are ± 1 eV.

bombardment energies below a few keV and by utilizing light ion and target masses, the equation for the sputtering yield becomes

$$S = \frac{CS_n(E)}{U} \tag{1}$$

where C is a constant for a given primary ion energy and target material. The simplicity of this result—that the sputtering yield is proportional to the nuclear stopping power of the solid and inversely proportional to the surface binding energy—accounts, in part, for the popularity of the Sigmund theory. In fact, it has been rather successfully applied in a number of instances.[36] The direct correlation between the stopping power and the yield is particularly striking as is illustrated in Fig. 8 for Ar^+ on Cu.[32,37] The relationship between yield and surface binding

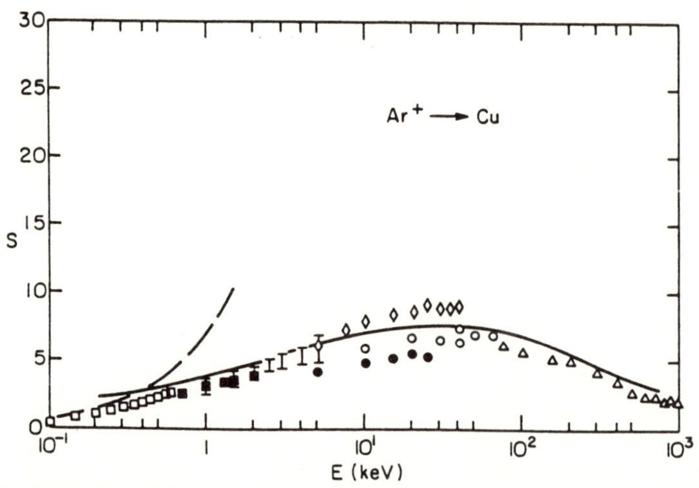

Figure 8. Sputtering yield as a function of the primary ion energy. The lines are calculated yields from $S_n(E)$ while the points are experiment curves taken from various groups. From Refs. 31 and 32.

energy (or sublimation energy of the solid) has been used to explain differential sputtering[38] and variations in sputtering yield across the periodic table.[31,39] The theory has also been applied to situations where it should not work, probably since for many years it was the only real theoretical approach available to experimentalists.

A similar theoretical model has been developed by Thompson to explain the energy distribution of ejected atoms.[21] His model grows from an expression for the flux of atoms inside the solid crossing any internal surface plane. For an ejecting particle of kinetic energy E, this flux is given as

$$\phi(E)\, dE\, d\Omega = [C'E \cos\theta/(E + U)^{n+1}](2\pi \sin\theta)\, d\theta\, dE \qquad (2)$$

where $\phi(E)$ is the flux density energy distribution with units of number per unit area per unit time per unit energy. The solid angle $d\Omega = \sin\theta\, d\theta\, d\phi = 2\pi \sin\theta\, d\theta$; $d\phi$ is the azimuthal angle about which the flux is assumed to be isotropic. According to Thompson,[21] $n = 2$ and C' is a constant for a given primary ion, energy, and target material. The appearance of U in Eq. (2) is a consequence of the incorporation of the planar surface binding model in the theoretical development.

The predictions of Eq. (2) are that (1) the peak in the energy distribution occurs at U/n, (2) the peak position is *independent* of θ, and (3) the polar distribution is *independent* of E. Over the years this relationship has fit the experimental data quite well. This agreement is remarkable since the underlying assumption in the Thompson model is that the atoms only undergo binary collisions. As stated above for particles that eject with energies $<2U$, the desorption process undoubtedly occurs via multiple collisions. In fact, an attractive interaction, e.g., surface binding energy, is inconsistent with binary collisions as attractive interactions are long ranged and binary collisions are only valid for close encounters. Generally, in any one experiment, either the energy distribution at one polar angle is measured or the polar angle distribution for a large energy bandwidth is measured. In the first case, the constants U and sometimes n of Eq. (2) are fitted to the data. In the case of the polar distributions, it has been observed[40–42] that the polar distribution is often closer to $\cos^2 \theta$ than $\cos \theta$.

Recently energy and angle resolved (EARN) distributions of neutral atoms ejected from In and Rh foils have been measured.[29,43–47] In these experiments polar distributions at several energies and energy distributions at different polar angles were obtained simultaneously.[46] In this case the primary particle was Ar^+ with 5 keV of energy directed normal to the surface. There are three interesting deviations from the predictions of the Thompson model. (1) The peak position in the energy distribution shifts to a lower value as the polar angle from normal increases (Fig. 7a). Each of the individual curves, however, if U and n are used as parameters, can be well represented by Eq. (2), as seen in Fig. 7b. (2) The polar distribution becomes narrower at higher kinetic energies (Fig. 7c), so that at high energies the distribution is approximately $\cos^2 \theta$. (3) The value of U needed to fit the data is larger than the heat of sublimation.

These discrepancies have been accounted for by analyzing the various assumptions of the Thompson model.[47,48] A reasonable agreement with the experimental distributions can be made by assuming that the velocity distribution in the solid is *not isotropic*. This assumption is completely reasonable. In a Thompson model, conceptually an infinite bulk solid is considered and then an imaginary plane is used to define the surface. In the real solid, the surface undoubtedly influences the distribution of velocities.

If one assumes that the distribution of particle energies and directions inside the solid is given by

$$f_1(E_i\theta_i) = \cos^m \theta_i / E_i^2 \tag{3}$$

where the value of n of Eq. (2) is taken to be two. In order to get the distribution outside the solid, application of the same transformations as

Thompson results in the final distribution of the yield as

$$Y(E, \theta) \propto \frac{E \cos \theta}{(E + U)^3} \frac{[(E \cos^2 \theta + U)^m]^{1/2}}{[(E + U)^m]^{1/2}} \tag{4}$$

In Thompson's case the value of m is 0, and Eq. (4) reduces to Eq. (2). The choice of $m = 2$ fits the experimental data reasonably well and simplifies the mathematics. In this case Eq. (4) becomes

$$Y(E, \theta) \propto \frac{E \cos \theta}{(E + U)^4} (E \cos^2 \theta + U) \tag{5}$$

The energy and polar distributions as predicted by Eq. (5) for Rh are shown in Figs. 7a and 7c, respectively. A value of $U = 11$ eV resulted in a reasonable fit to the experimental data. The agreement between the calculated and experimental curves is remarkable. The energy peak position shifts to lower energy as the polar angle increases. In the case of the polar distributions, at very low energies the distribution is nearly $\cos \theta$. As the energy of the particles increases, the polar distribution becomes narrower (Fig. 7c). In the limit of extremely high energies the predicted distribution becomes $\cos^3 \theta$. If all energies are averaged the polar distribution is

$$Y(\theta) \propto \cos \theta (2 \cos^2 \theta + 1)/3 \tag{6}$$

where $Y(\theta = 0°) = 1$. The deviation of Eq. (6) from $\cos^2 \theta$ is less than 0.03 for all values of θ.

Computer simulations as described in Section 3.2 for the keV particle bombardment of an amorphous indium sample were performed by Lo *et al.*[49] in order to verify the assumption of anisotropic velocity distributions. The angular velocity distributions were determined for the surface layer as a function of elapsed time after the initiation of the collision cascade. Their results as seen in Fig. 9 indicate that the presence of the free surface (neglected in the Thompson model) cause significant anisotropy in the outermost surface region. The distribution quickly became isotropic in the subsurface region.[49] Of note is that the simulations reproduced the shift in peak position in the energy distribution with polar angle and the narrowing of the polar distribution with increasing desorbed particle energy.

The final issue is the choice of a value for U. As a matter of convenience (and lacking a better choice), the value of the surface binding energy is often equated to the heat of sublimation of the solid, ΔH_s. The choice of ΔH_s for the energy parameter in the prediction of the sputtering yield gives reasonable agreement with the measured

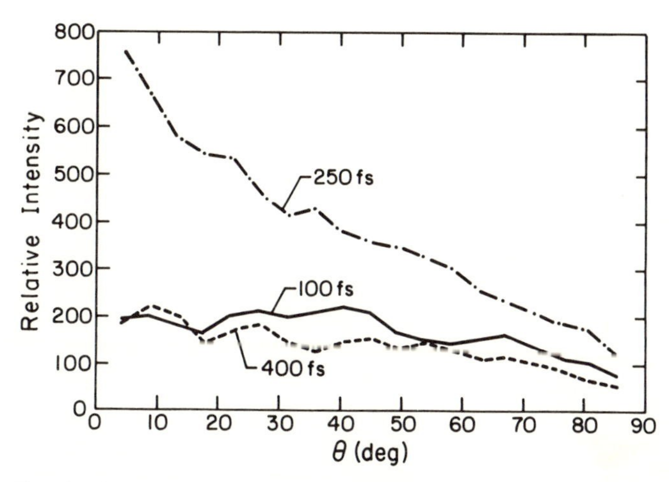

Figure 9. The development of the angular distributions in the first surface layer. The majority of particles eject between 100 and 250 fs (2 fs = 10^{-15} s). (From Ref. 49.)

quantities.[31] However, Husinsky[50] has pointed out that the values of U fitted to experimental energy distributions are sometimes larger than the ΔH_s value of the material.

One possibility is that the energy cost to remove an atom from a solid is actually greater than the heat of sublimation.[51–53] The roots of this idea lie in two virtually unnoticed papers by Jackson.[54,55] He simulated the ejection of a surface atom by giving it an initial kinetic energy. The final kinetic energy (for a pairwise additive potential approximation) corresponds to an energy loss greater by 30%–40% than the bulk ΔH_s. One would think that this implies that by ripping the entire solid apart it would take more energy than it had in the first place.

This dilemma is best described by examining a diatomic molecule that has a bond strength of D_e. Under the rules used for solids, the binding energy of each atom is $D_e/2$. If, however, one atom is clamped fixed and one asks how much kinetic energy must be supplied to the other atom so that the bond can be ruptured, then the answer is D_e—twice the "binding energy". Note that it now costs nothing to remove the second atom. An analogous situation occurs in the solid. In one extreme in the ion bombardment procsss, all atoms but one are fixed. The energy cost for this atom to escape the solid is greater than ΔH_s, in line with recent experimental results. It is not clear either theoretically or experimentally precisely what is the value of the energy cost to remove an atom from a surface. It is also not clear which factors influence the magnitude of this energy cost. Should it be the same for amorphous Rh, Rh{111}, Rh{110}, and Rh{100}, for example?

Another problem with the picture as painted by the transport theory has recently been pointed out by Shapiro.[56] From molecular dynamics simulations he found that a significant portion of ejected atoms actually lost more kinetic energy in their escaping trajectory than was required to overcome the surface binding energy. This conclusion is in conflict with the assumptions made in the transport theory.[52,57–59] This energy loss phenomenon has recently been observed by Reimann *et al.*[60] in a series of EARN experiments on clean Rh{111} and oxygen covered Rh{111}. The energy distribution of the neutral Rh atoms ejected from the oxygen covered substrate was found to peak at a *lower* energy than the distribution from the clean surface. Computer simulations (Section 3.2) on the same systems confirm this trend. Although the relative binding energies of the Rh atoms in the real system are not known, it is known from the simulations that the Rh{111} surface atoms with an oxygen overlayer have a *larger* binding energy than those of the clean surface. This trend is in contradiction to the transport theories which predict that the peak in the energy distribution should be proportional to the binding energy. Obviously there are other factors such as collisions that influence the peak position.

In addition to the problems with describing the details of the energy and angular distributions, the transport theories are not equipped to describe the ejection of clusters. For example, the linear cascade approximation is inconsistent with the ejection of clusters such as Ni_2, which, as described below, forms via a recombination mechanism involving two interlinking cascades. There is no way to incorporate into the transport models explicit molecules (e.g., benzene) and examine the relative ejection yields of the fragment, C_2H_2, and the molecular, C_6H_6, species. Finally, the transport theories do not include single-crystal structures, an important aspect of the studies which is of interest to many surface scientists.

Our intent here is not to denigrate the transport theory, as it has been a reasonably good qualitative guide for experimentalists over the years. However, we wish to point out that if one wants a detailed qualitative and quantitative picture of the particle bombardment process one must look beyond transport theory. The approach that we feel has had considerable success is described in the next section.

3.2. Molecular Dynamics Calculations

Many of the difficulties associated with transport theory can be overcome by utilizing a molecular dynamics calculation on a large ensemble of atoms to compute actual nuclear positions as they change in time subsequent to the primary ion event. Classical dynamics calculations have, of course, been very successful in explaining trajectories in atom–diatom scattering,[61] properties of liquids,[62] gas–surface scattering,[63] and even the solvation of large molecules like dipeptides.[64] For describing the desorption process this approach has the distinct advantage of utilizing many fewer approximations than required for the statistical theories. On the other hand, no simple equation falls out of the calculations, although important concepts may emerge from the resulting numbers. The calculations often prove very useful in testing the validity of possible analytical theories. In general, these calculations require considerable computer time. With the recent surge in computer efficiency and decrease in costs, it would appear that this is becoming less of a difficulty.

The computation of the classical trajectories using the molecular dynamics procedure rests on Hamilton's equations of motion. For a particle i of mass m_i, the equations of motion are

$$m_i d\mathbf{v}_i/dt = \mathbf{F}_i = -\nabla_i V(r_1, r_2, \ldots, r_N) \tag{7}$$

and

$$d\mathbf{r}_i/dt = \mathbf{v}_i \tag{8}$$

where $\mathbf{r}_i$, $\mathbf{v}_i$, and $\mathbf{F}_i$ are the position, velocity, and force of the ith particle. The interaction potential V depends on the positions of all the N particles. It is the quality of V that determines the reliability of the results of the molecular dynamics simulations. Since V is such an integral part of the calculation an entire section, 3.3, is dedicated to a discussion of the various types of available interaction potentials. For the moment, we will assume that there is a suitable interaction potential and examine the results of the classical dynamics calculations.

The presence of N particles results in $6N$ coupled first-order differential equations. There is a plethora of numerical schemes for solving these equations.[65] The general idea is to take sufficiently small time increments Δt such that the force is approximately constant. The positions and velocities of all the particles are then generated through time. For the ion bombardment process the timestep is typically 0.1 fs ($1\text{ fs} = 1 \times 10^{-15}\text{ s}$) at the beginning of the bombardment event when the primary particle is moving rapidly and increases to about 3–10 fs at the end of the event. Most of the ejection events occur in less than 500 fs, which means that there are 50–1000 time steps.

To integrate Hamilton's equations of motion, the initial positions and velocities of all the particles must be specified. In most simulations to date the substrate atoms are placed at their equilibrium positions with zero velocity. These values could be chosen from a Maxwell–Boltzmann distribution at the appropriate temperature but there is no overwhelming evidence that ejection events are strongly temperature dependent. One can thus choose to mimic a Rh{111} crystal,[66,67] an amorphous In sample,[49] or an overlayer of benzene on Ni{100}.[68] That leaves the primary particle's initial conditions to be chosen. In general the experimental conditions are such that the initial kinetic energy, E_{in}, polar angle, θ_{in}, and azimuthal angle, φ_{in}, are prescribed. These values then determine the initial velocity of the primary particle as

$$v = (2E_{in}/m)^{1/2} = (v_x^2 + v_y^2 + v_z^2)^{1/2} \tag{9a}$$

where

$$v_x = v \sin \theta_{in} \cos \varphi_{in} \tag{9b}$$

$$v_y = v \sin \theta_{in} \sin \varphi_{in} \tag{9c}$$

$$v_z = -v \cos \theta_{in} \tag{9d}$$

and m is the mass of the particle. To ensure that the incident particle does not interact with the crystallite at $t = 0$, the z component of the position should be infinite. Most interaction potentials have a finite range (2–6 Å), however, so that infinity is not that large. The lateral positions (x, y) must be chosen to mimic the experimental conditions. Since in the laboratory reference it is not possible to choose an impact coordinate, the data are actually averages over many primary particle impacts on the surface. In the calculations one must also perform numerous simulations in which different sets of initial (x, y) aiming points are chosen. This leads to "statistical noise" in the calculated observables. The yield is calculated from the average number of particles that eject per incident particle. It is usually possible to obtain a statistically reliable value with 50–100 different impact positions. For energy and angle resolved distributions, several thousand impacts may be required to obtain a respectable signal-to-noise ratio.

The remaining initial condition to choose is the number of atoms, N, in the sample. Ideally, N should be as large as possible. Depending on the nature of the interaction potential, however, the required computer time scales somewhere between N^1 and N^3. Thus, a practical number of atoms must be chosen by performing test calculations on samples of various sizes and making a determination of what is sufficiently large to accurately reproduce the observables. The crystal size will depend on exposed crystal face, overlayer unit cell size, and primary particle kinetic energy and angle. Typically for calculations below 1 keV of energy we have used crystals of 1000–2000 atoms. Normally, a flat and wide-shaped crystal rather than a cubic one yields more accurate results with the same number of atoms. Typical execution times on an IBM 3090 for one primary particle impact are 30–900 s for a 1450-atom Si{110} target and 100–1000 s for a 700-atom Rh{111} target. There is a large variation in time since various aiming points on the surface may result in rather simple trajectories or extremely complex ones.

A unique feature of the sputtering simulations, in contrast to other simulations of solids and liquids, is the choice of boundary conditions on the sides and bottom of the crystallite. There can easily be one or two very energetic (20–500 eV) particles that reach the edge of the crystallite. These particles in a real solid penetrate further into the bulk, damaging the sample along their path. By enlarging the crystallite in the simulations it can be verified that these energetic particles do not substantially contribute to the ejection of atoms and molecules into the gas phase. Thus, these atoms are truncated from the simulation once they leave the side or bottom edge. Periodic boundary condition should *not* be employed as it is nonphysical to have the energy enter the other side of the crystal. Likewise the generalized Langevin[69,70] prescription or a rigid layer would cause reflection of the energy back into the crystal, again a nonphysical phenomenon.

Finally, the atomic motion subsequent to the primary particle impact is computed for enough time such that no particles have sufficient total energy to escape the solid. For a metal such as Rh, we have found that after the most energetic particle left in the solid (not counting those ejected or those that left via the sides of the crystal) has less than 0.2 eV of energy (kinetic plus potential energy) that integration for longer times does not produce any more ejected particles.

After the integration has been terminated, a number of quantities can be calculated from the final positions and velocities of the atoms. These range from experimentally observable quantities such as the energy distribution to unobservable features such as ejection mechanisms. Below we give examples of the various output data that can be obtained from computer simulations of the sputtering process.

3.2.1. *Yields*

The average number of particles ejected per incident ion is one of the easiest properties to determine numerically yet one of the hardest to calculate reliably. The calculated value depends on the crystal face of the sample, the primary particle–substrate interaction potential, the substrate–substrate interaction potential, etc. In addition, reliable experimental values are almost impossible to obtain under the same conditions as the calculations are performed. Virtually all the simulations are performed in the low-dose mode, i.e., each primary particle impacts a fresh sample. Most experimental measurements of yields are carried out under high-dose conditions where there is a sufficient dose of primary particles such that portions of the sample are rebombarded. This means that the crystal structure has been altered during the course of the measurement and that primary particles have been implanted. Shown in Fig. 10 are yields calculated by Stansfield *et al.*[71] for the sputtering of Si{100} as a function of Ar^+ kinetic energy. They have used different Ar–Si and Si–Si repulsive interaction potentials along with the attractive many-body potential for silicon as given by Stillinger and Weber.[72] They have compared their yields to experimental values obtained using a weight-loss procedure.[73] The calculated yields depend strongly on the type of potential function. Moreover, the yields for unreconstructed

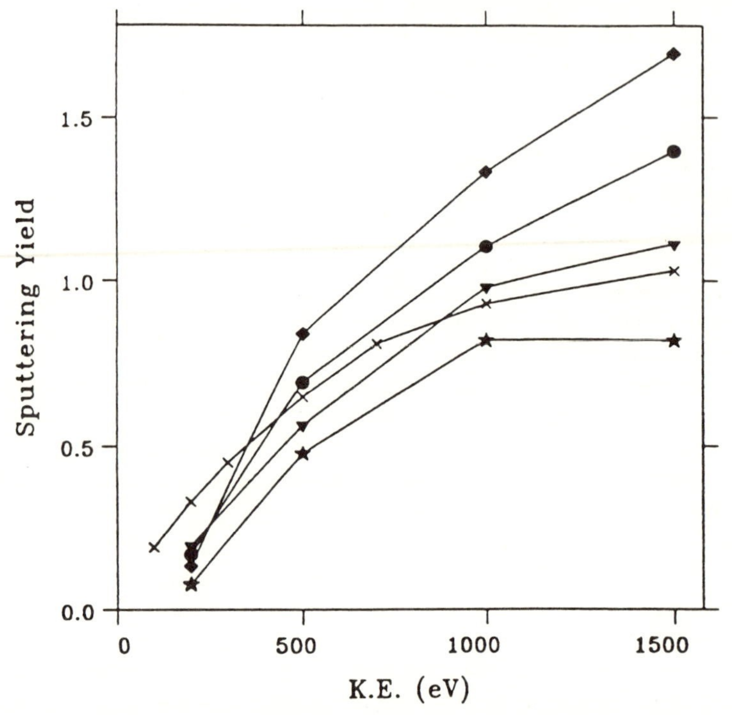

Figure 10. Experimental and calculated Ar^+ +Si{100} sputtering yields. ×, Experiment[73] (steady-state yield); •, the SDCI and SCF potentials (unreconstructed surface); ◆, the Smith modified Wedepohl potentials (unreconstructed surface); ▼, the Biersack-Ziegler universal potentials (unreconstructed surface); ★, the SDCI and SCF potentials (reconstructed surface).

Si{100} and the dimer reconstructed face vary even with the same potential. As stated in Section 1 an average quantity like yield does not help to discriminate among the various assumptions inherent in any given model.

3.2.2. Energy and Angular Distributions

The final velocities yield sufficient information to calculate the kinetic energy (E) of each ejected particle as

$$E = \tfrac{1}{2}mv^2 \tag{10}$$

where

$$v^2 = (v_x^2 + v_y^2 + v_z^2) \tag{11}$$

with angles

$$\theta = \arccos(v_z/v) \tag{12}$$

and

$$\varphi = \arctan(v_y/v_x) \tag{13}$$

To obtain either an energy or angle distribution, the particles from a multitude of different primary particle hits on the surface are collected. The energy and angle distributions are mutually interdependent. Shown in Fig. 11 are the polar angle distributions from Rh{111} from both the EARN experiment described in Section 2 and from calculations using a pairwise additive potential and an embedded atom method (EAM) potential (Section 3.3). The azimuthal angles for the Rh{111} crystal face are defined in Fig. 12. The peak intensity along the azimuthal direction $\varphi = -30°$ is greater than along $\varphi = +30°$ or 0° for all secondary-particle ranges. As the Rh atom energy increases, the $\varphi = +30°$ intensity increases relative to the $\varphi = 0°$ intensity. Moreover, both theory and experiment find that the intensity at $\theta = 0°$ increases relative to the peak intensity at $\theta = 25$–$40°$ as the energy increases . Of note is that the polar distributions are quite complex and as stated in Section 1 are an aid in discriminating among various theories and models. The relative agreement between the data from two calculations and the experiment will be discussed in Section 3.3. The reasons behind this structure will be unraveled in Section 5.

In a similar fashion it is possible to calculate the secondary particle kinetic energy distributions. As shown in Fig. 13, these distributions for Rh{111} agree well with those determined from EARN experiments.[67] Both the angle-integrated distributions and the distribution at $\theta = 40°$

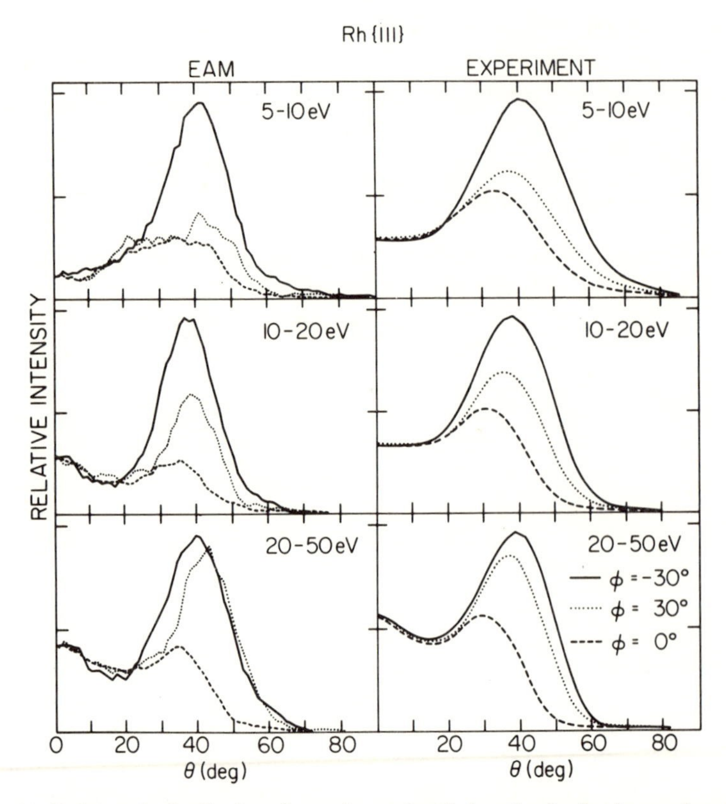

Figure 11. Polar angle distributions for various azimuthal angles for fixed secondary kinetic energy of the Rh atoms. In each frame the data are normalized to the $\phi = -30°$ peak intensity. The calculated data using the EAM potential are reported with a full width at half maximum (FWHM) resolution of 15° in the polar angle. A constant solid angle is used in the histogramming procedure. The experimental resolution is also approximately 15°. The surface normal corresponds to $\theta = 0°$.

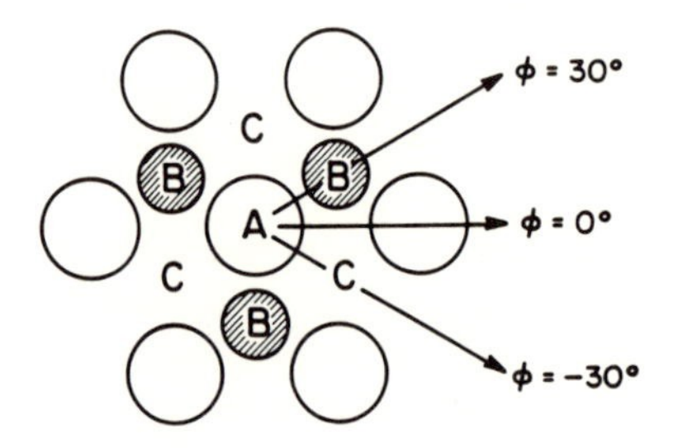

Figure 12. The azimuthal directions are defined above. A polar angle of 0° is normal to the surface and 90° is parallel. Open circles designate first layer atoms and shaded circles second-layer atoms. The letters A, B, and C designate possible adsorption sites for oxygen atoms.

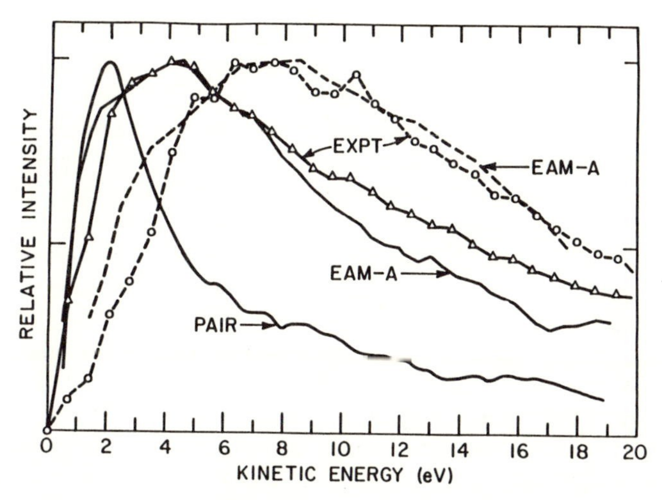

Figure 13. Experimental and calculated kinetic energy distributions. In all cases the curves are peak normalized. The two experimental curves are the angle-integrated distribution and one at $\phi = -30°$ and $\theta = 40 \pm 3°$. The EAM-A curves are the angle-integrated distribution and one at $\phi = -30°$ and $\theta = 38 \pm 7.5°$. Only the angle-integrated distribution is shown for the pair-potential calculation. The angle-integrated distributions are shown as solid lines and the ones at $\theta \approx 40°$ as dashed lines. (From Ref. 67.)

and $\varphi = -30°$ have been determined. It is quite clear that in a fashion similar to the polycrystalline surfaces (shown in Fig. 7), the energy distribution depends on the take-off angle of the secondary particles. As we shall see below the molecular dynamics calculations using the EAM potential do an excellent job of reproducing the experimental distributions.

3.2.3. *Clusters*

One of the intriguing aspects of the particle bombardment process is the appearance in the mass spectra of molecules or clusters. These species can be of the type Ni_n, where $n \geq 2$ from a Ni substrate or C_6H_6 which is observed from a bombarded film of benzene on Ni. Do these clusters actually exist on the surface? Initially one may think that the bombardment process is too catastrophic to preserve the fragile information encoded in chemical bonds. On the other hand, if everything is dissociated into atoms, how does the C_6H_6 molecule get back to its original composition? If the clusters do arise from contiguous surface atoms, can their presence provide key information regarding the local atomic structure of surfaces of alloys and supported metal catalysts?

A powerful aspect of classical mechanics is that the microscopic motions of the atoms arise *from* the calculations. That is, one does not have to *assume* that one particular collision sequence is more important than another. This allows formulation of the mechanisms of cluster formation directly from the simulations. To check for the appearance of the clusters, the total internal energy is computed. This energy consists of the relative kinetic energy plus potential energy of an aggregate of m atoms above the surface. If the total internal energy is greater than zero, then the aggregate is not bound. If it is less than zero, then the cluster is at least temporarily stable. If m is greater than 2, then the cluster may decompose on the way to the detector as has been observed in alkali–halide clusters.[(74)] The center-of-mass velocity of the cluster can be used to predict energy and angular distributions.

Computer simulations have clearly demonstrated that molecules such as CO and C_6H_6 can survive the ion bombardment process without fragmentation. A graphic representation of a computer simulation of molecular ejection is illustrated in Fig. 14. There are two principal reasons for this result. First, the roughly equivalent effective sizes of metal substrate atoms and adsorbed hydrocarbon molecules facilitates the ejection of molecules without significant fragmentation. For example, the hard-sphere diameter for a single Rh atom in a crystal is 2.7 Å and the van der Waals diameter for a four-atom CH_3 group is 4 Å.[(75)] This means that a moving Rh atom encounters this group as a single species and can eject it without dissociating the CH bonds. The second reason for the intact ejection of these molecules is that they contain many vibrational degrees of freedom that can absorb kinetic energy that would otherwise lead to fragmentation. This contribution becomes more important for intact ejection as the size of the molecule increases.

The fact that molecules can survive the ion impact is encouraging for correlating the composition of the desorbed species with surface composition. Unfortunately, sputtered atoms and molecules have also been observed in the simulations to recombine into bound molecules in the near-surface region. This mechanism can lead to the detection of stable molecules that were not present on the surface, and can complicate the interpretation of results. For example, the observation of NiCO and Ni_2CO clusters from the sputtering of a CO covered Ni surface could be interpreted as indicating that chemisorbed CO molecules are bound both to single Ni atoms (atop binding sites) as well as pairs of Ni atoms (bridge-binding sites).[(76)] This could be an erroneous interpretation because recombination processes can also lead to the formation of multiply bonded CO molecules.[(77)] The recombination mechanism, however, does not completely destroy the ability to evaluate the significance

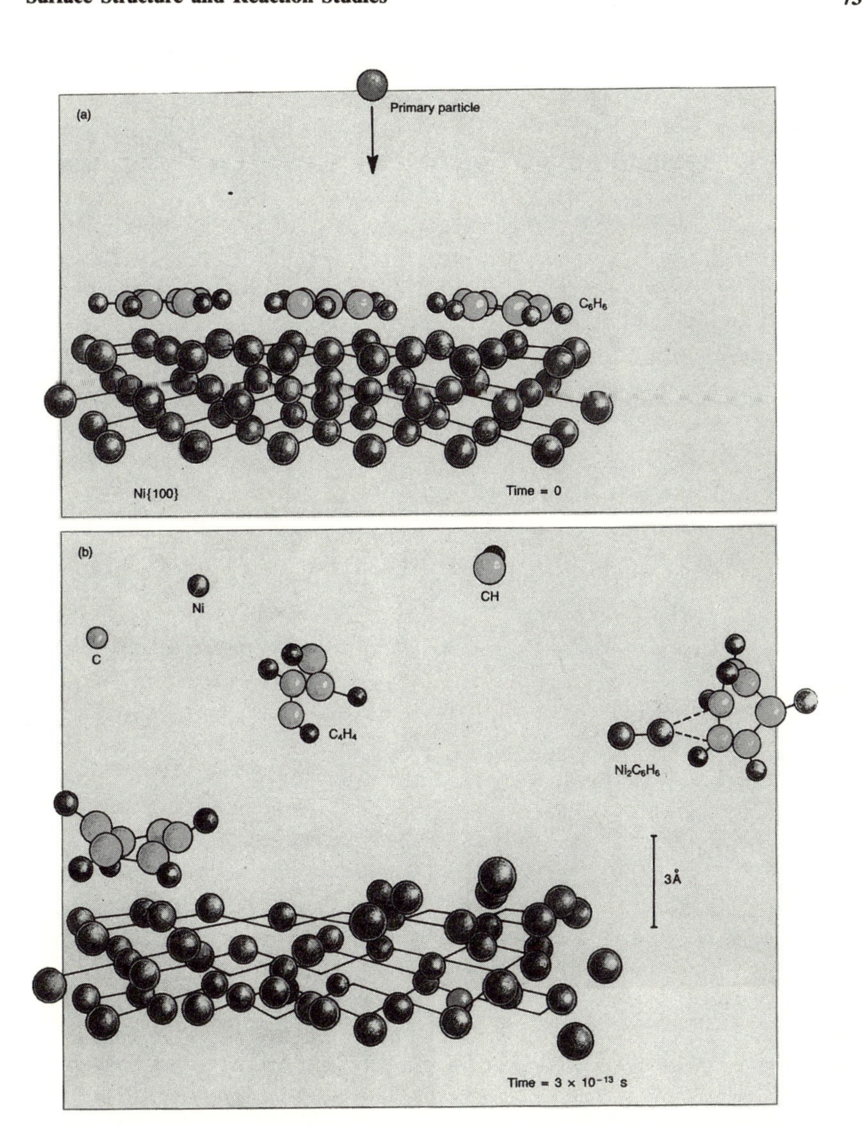

Figure 14. Ion bombardment event on a benzene covered Ni{100} surface. (a) Before the primary particle has stuck the surface. (b) After 3×10^{-13} s. The benzene is initially bonded parallel to the surface. Frame (b) shows several possible outcomes of the collisional event. The center benzene molecule was hit sufficiently hard that it fragmented with parts of the molecule ejecting and one of the H atoms implanting into the solid. In this particular hit on the surface, the primary particle also implanted. One benzene molecule (on the left) remained bound to the surface although it did get distorted. Another benzene molecule ejected, in this case in close proximity to two Ni atoms, thus there is the possibility of a $Ni_2C_6H_6$ formation.

of the composition of desorbed particles because it only occurs between species that are originally in close proximity on the surface. The general picture from simulations is that the identity of desorbed material can be used to assign surface species, but that care must be exercised in making such assignments.

An example of how two surface atoms that are not contiguous might recombine in the near-surface region to form a dimer or diatomic molecule is shown in Fig. 15. In this case[78] both experiment and calculation found that Ni_2 dimers are ejected from a Ni{001} surface preferentially along the $\varphi = 0°$ azimuth or ⟨100⟩ direction. An analysis

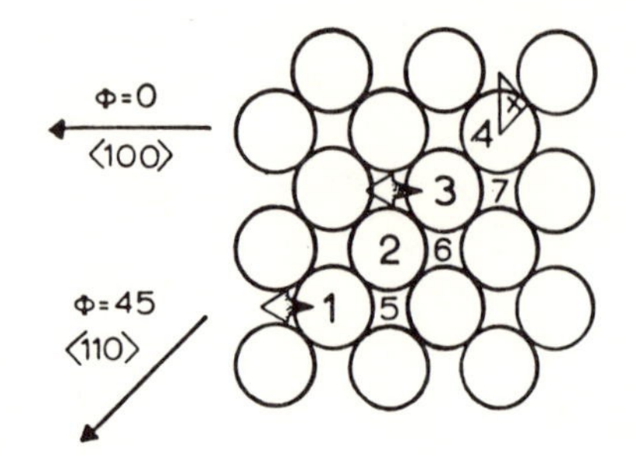

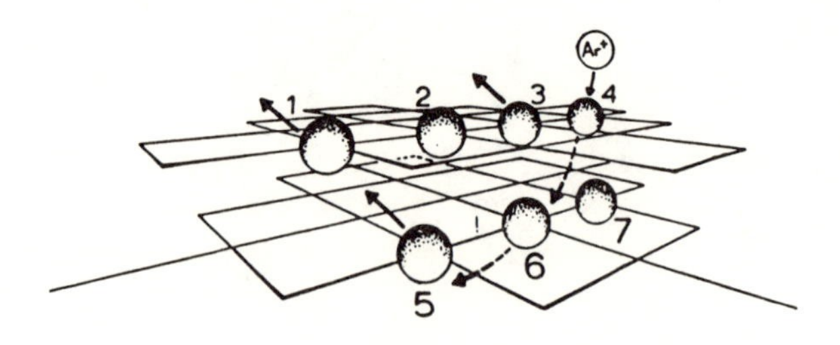

Figure 15. Mechanism of formation of the Ni_2 dimer, which preferentially ejects in the ⟨100⟩ directions, contributing the majority of intensity to the peak in the angular distribution. (a) Ni{001} showing the surface arrangements of atoms. The numbers are labels while the × denotes the Ar^+ ion impact point for the mechanism shown in (b). Atoms 1 and 3 eject as indicated by the arrows forming a dimer, which is preferentially moving in a ⟨100⟩ direction. (b) Three-dimensional representation of a Ni_2 dimer formation process. The thin grid lines are drawn between the nearest-neighbor Ni atoms in a given layer. For graphical clarity, only the atoms directly involved in the mechanism are shown.

of the dimers that eject in the simulation show that one mechanism of formation is dominated in the peak of the angular distribution of the dimers. This mechanism is shown schematically in Fig. 15b. The Ar^+ ion strikes the target surface atom, No. 4, which moves under No. 3, ejecting it. Atom No. 3 is channeled through the fourfold hole in the surface and thus escapes in the ⟨100⟩ direction. Concurrently, atom No. 4 is moving toward the second layer pushing atoms No. 6 and No. 7 down. In the same manner in which atom No. 4 ejected atom No. 3, No. 6 pushes No. 5 up toward the first layer. Atom No. 1 is subsequently ejected by atom No. 5. Again, No. 1 is channeled through the fourfold hole in the ⟨100⟩ direction. These two atoms, No. 1 and No. 3, are moving parallel to each other, are in close proximity, and are thus susceptible to dimer formation. In this case, the relative kinetic energy is less than 1 eV, the center-of-mass kinetic energy is 20 eV and the potential energy −1 eV. When ejected as monomers, these two atoms (No. 1 and No. 3) exhibit the same azimuthal anisotropy; however, these are a relatively minor part (~10%) of the corresponding energy selected monomer peak.[78]

There is an important ramification of the prediction that the dimers that give rise to the maxima in the angular distribution are formed primarily from constituent atoms whose original relative location on the surface is known. If this result were extrapolated to alloy surfaces such as CuNi, the relative placement of the alloy components on the surface would be determined. Recent calculations indicate that other originating sites of ejected dimers can be preferentially enhanced by varying the angle of incidence of the primary Ar^+ ion.[79] It is hoped that information on the surface structure of alloys can be obtained from an analysis of the angular distributions of the multimers. Preliminary experiments and calculations of other clusters which form via a recombination mechanism also display angular anisotropies. These include such species as Ni_3, NiO, and NiCO. The ejection direction tends to be toward the surface normal and thus is difficult to measure with the existing experimental setup.

The interpretation of the meaning of the cluster composition is obviously a complex issue. The molecular dynamics calculations provide insight into the basic formation mechanisms. These mechanisms have forced investigators to devise special experiments to test for the presence of recombination. Examples of these tests will be presented in Section 5. In the future, it will be valuable to see if more quantitative predictions will be possible. Perhaps by measuring the vibration and/or rotational excitation of neutral molecules as a function of kinetic energy and take-off angle, it will be possible to formulate more specific models. In the meantime these calculations provide at least a critical guide to the experimentalist.

3.2.4. Damage to the Substrate

Concurrent with the ejection of particles into the vacuum there is considerable action in the solid which leads to damage to the remaining substrate. This damage ultimately limits the resolution of depth-profiling analyses and is a limiting factor in the applications of ion-beam processing of microelectronic materials. These processes have typically been modeled by BCA codes such as TRIM[80] since the crystal sizes

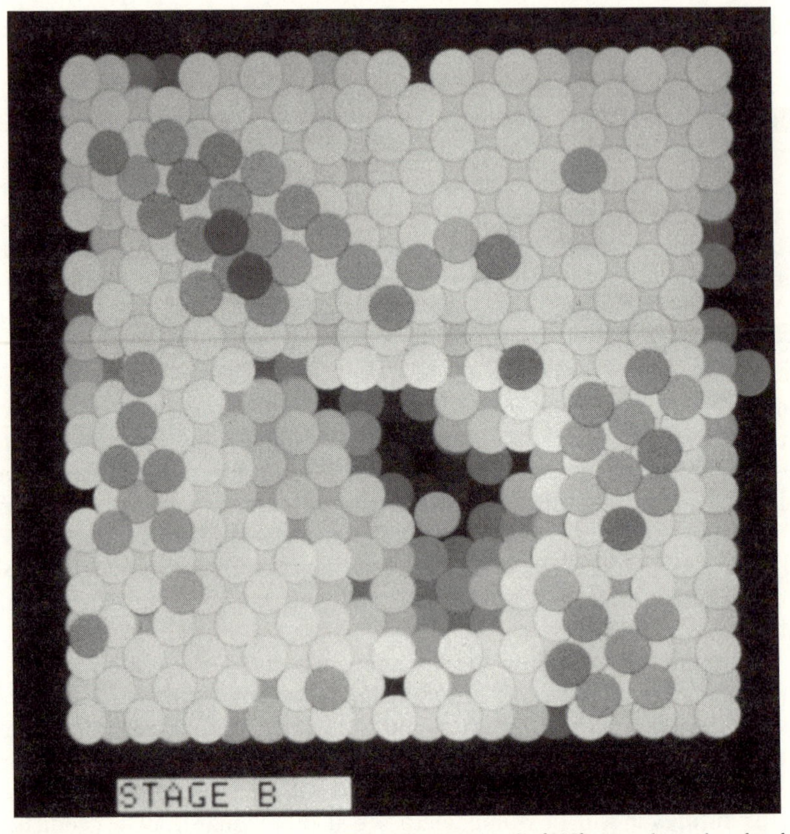

Figure 16. Computer simulations of radiation damage. A Cu{100} crystal was bombarded by a single Ar^+ ion with 500 eV of kinetic energy. The dynamics of the energy relaxation are then followed for several picoseconds as described in the original article. The size of this crystal is approximately 35 $Å^2$. This picture was provided by D. E. Harrison, Jr.

needed to describe the damage region are too large for molecular dynamics simulations. Webb and Harrison,[81] however, did model the formation of pits that form during bombardment of a Cu{100} surface. The surface rearrangement which occurs after a single bombardment event is shown in Fig. 16. The surface is about 35 Å wide. The open circles represent surface atoms that have not changed their lattice sites. The atoms in the layers below the surface are shown in increasing shades of gray. Note that a few atoms have been deposited on top of the surface during the collision cascade. Recently, it has been possible to observe these pits directly on PbS with a scanning tunneling microscope (STM).[82] The STM image is shown in Fig. 17. The agreement between the qualitative features of the surface damage is remarkable. Note that the dimensions of the surfaces are the same for both examples. In both cases a pit is formed with a buildup of atoms around the edges of the pit. To

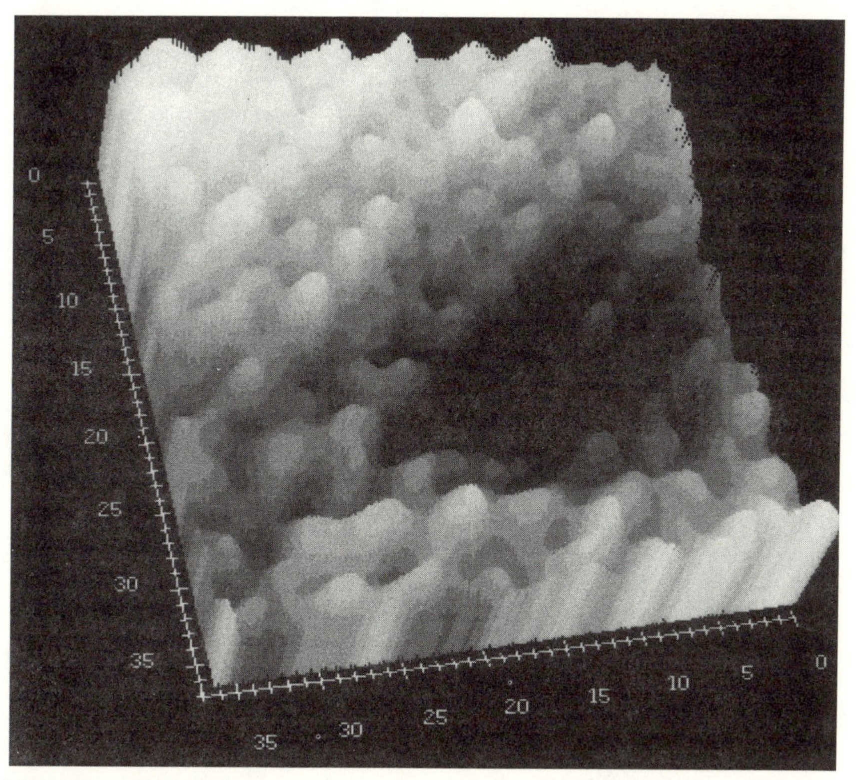

Figure 17. A three-dimensional STM image of freshly cleaved PbS{001} bombarded by 8-KeV Kr^+ ions.

date, there have not been many systematic studies of lattice damage with molecular dynamics owing to the extensive computer time required. As this technology is rapidly improving, we expect this area to be quite active in the future.

3.3. Interaction potentials

The reliability of the results from any molecular dynamics simulation depends on the quality and appropriateness of the interaction potential that is assumed in the calculation. For example if one wanted to examine the dissociative chemisorption of O_2 molecules on Ni{001}, then the potential must include the proper interactions that allow the molecules to separate into atoms on the surface and must not assume that the molecule is a point mass of 32 amu. Moreover, the repulsive part of the potential must be suitable for representing keV interactions. For example, several many-body potentials have recently been developed to examine the equilibrium properties of silicon.[72,83–86] All except one[84] of the potentials have very weak repulsive walls and are inappropriate for simulations of keV particle bombardment. It is well known that the desorption yields are in some sense inversely proportional to the binding energy of atoms to the solid. Thus, if a potential has a drastically incorrect binding energy, the predicted yields and energy distributions will also be incorrect.

In the most rigorous sense, we require a complete many-body description of the interaction potential of the model microcrystallite. The dynamical calculations must then be performed within this framework by including the interaction of each particle with all other particles. The integration of the classical equations of motion are inexorably intertwined with the form of the interaction potential function. During the early years, many attempts were made to simplify this interrelationship using the BCA. Interestingly, the BCA itself imposes certain restrictions on the potential function. We next review how the sophistication of molecular dynamics calculations has increased in recent years, beginning with the simplest models and finishing with the most complex ones. In each case we will concentrate on the potentials appropriate for the description of the substrate interactions. These interactions involve a wide kinetic energy range and are most influential in leading to particle ejection.

3.3.1. Repulsive Pair Potentials and the BCA

The simplest interaction among a group of atoms is one that is pairwise additive. In this case, the total potential energy, E, can be written as

$$E = \sum_{i} \sum_{j>i} V(r_{ij}) \tag{14}$$

where V is a pair-potential, r_{ij} is the scalar distance between atoms i and j, and the sums are over all the particles in the simulation. The assumption of pairwise additivity of the potentials physically means that the interaction between two atoms is independent of the presence of a third or fourth atom. This approximation is chemically not correct. For example, if pair potentials were used to describe ozone, O_3, then the molecule would be an equilateral triangle with a binding energy three times that of O_2. Both the qualitative and quantitative descriptions are wrong. Why then, has a multitude of simulations of the keV particle bombardment process used pair-potentials quite successfully?

For atomic collisions of energies greater than typical bond energies (a few eV's), strong repulsive forces dominate the dynamics. In other words, the motion of the atoms and molecules are governed by the presence of other species in a given geometrical arrangement. This structural dependence is quite strongly seen in the angular distributions of the ejected particles discussed elsewhere in this chapter. The angular distributions from different fcc materials are strikingly similar, again confirming the idea that the crystal structure dominates the collision process. Chemical effects such as the bonding of O_3 are not overly important, as least for a first-order description of the sputtering process. Thus for many years, pair potentials have been used in sputtering simulations.[13,55,80,87–93]

The pair potentials generally come in two forms—one is purely repulsive and the other includes an attractive portion at nuclear separations corresponding to bond lengths. Popular analytic forms for the repulsive portion at small internuclear distances include the Molière and Born–Mayer potentials. Both of these are described in detail elsewhere.[93] Some researchers[13] also include an attractive portion, often a Morse potential, which is splined to the repulsive potential in order to incorporate the binding interactions in the solid.

The use of a repulsive pair potential allows one to make an approximation in the integration of the classical equations that greatly reduces the required computer time. The BCA assumes that the atoms interact only with the nearest-neighbor atom. In addition, the atoms are followed in turn from the most energetic atom downward. The fact that only nearest-neighbor atoms interact means that tables of deflection angles for a given interaction potential can be stored and used from a look-up table in the computer program. This eliminates the need for actually integrating the equations of motion and the computer time is greatly reduced. There are really two approximations in the implementations of the BCA, one that involves collision dynamics and one that involves the interaction potential. We will discuss them in turn.

The BCA assumes that only two atoms interact at a time during the collision cascade. This seems quite reasonable when the atoms have hundreds to thousands of eV of energy. For example, the BCA has been used quite successfully to model the reflection of the primary particle where there are just one or two collisions at a high energy.[94] Even for this case, however, it has been shown that the BCA is not always appropriate.[95] For sputtering, the peaks in the energy distributions (Figs. 7 and 13) occur in the range 5–10 eV. At these energies the atoms are hitting more than one neighbor at a time. Thus, for the majority of the collisions important for the ejection process, the BCA is not appropriate. The other inherent assumption in the implementation of the BCA is that the cascades are "linear." That is, the trajectory of the most energetic atom is followed until the atom "stops" by some criteria. Then the trajectory of the next most energetic atom is followed, etc. It is assumed that these motions are independent. Processes such as the ejection of dimer molecules as given in Fig. 15 above cannot be treated directly.

Although we seriously question whether the BCA is appropriate for modeling the sputtering process, especially in this era when computer resources are bountiful, computer codes such as TRIM[80] and MARLOWE[93] have been successful in modeling some aspects of the sputtering process. In addition, TRIM is routinely used to model the depth of implantation of ions into substrates. In this case the crystal sizes needed for a proper description of the ion implantation are too large for a molecular dynamics simulation.

Tied to the BCA is the use of purely repulsive pair potentials. It must be pointed out, however, the repulsive pair potentials can and have been used with molecular dynamics simulations that do not utilize the BCA.[55,87,88,90] The lack of any attractive interaction means that assumptions must be made about the surface binding energy or there will be no peak in the energy distribution of the ejected atoms. Likewise, energetic thresholds for displacements of the atoms in the solid must be assumed or the atoms would never stop moving through the solid. The incorporation of molecules such as CO or C_6H_6 on the surface is not possible if one only uses repulsive pair potentials. Some of these difficulties may be removed by including attractive interactions in the pair potentials, the next level of calculational complexity.

3.3.2. Attractive Pair Potentials

In the molecular dynamics simulation, it is relatively straightforward to move from pair potentials that are purely repulsive to those that include an attractive interaction. The logic in the computer program is the same. The only additional factor is that more computer time is required. Generally the interaction potentials are cut off at some distance beyond which it is assumed that the potential and force are negligible. If only the repulsive interactions are included, then this cutoff distance is typically smaller than the nearest neighbor distance in the solid. If attractive interactions are included then this distance is increased to between the second and third neighbor atom distances. Thus interactions between more pairs of atoms are included and the computer time required is increased.

The advantage of using attractive pair potentials is that displacement energies, surface binding energies, and bond energies in molecules are automatically included in the simulation. The comparisons between the experimental and the calculated data include such detailed features as the angular distributions of Rh atoms ejected from Rh{111} as we shall see in Section 5. Many of the examples discussed in this chapter are from simulations using these potentials.

As mentioned above with the O_3 example, it is not possible to fit pair potentials to both the energy of a diatomic molecule and the energy of the corresponding bulk phase. For example, pair potentials that have been fit to the bulk heat of sublimation of Rh (5.75 eV)[96] have a well depth of 0.82 eV. The diatomic molecule, Rh_2, however, has a bond strength of 2.92 eV.[97] Conversely, if one uses the pair potential for Rh_2 to describe the bulk metal, the predicted heat of sublimation is a factor of 3.5 too large. Conceptually this same problem arises when one wants to describe the energetics of atoms in the surface layer. The pair potential formulation is not equipped to describe the interactions of atoms in a variety of bonding configurations.

Although we know that pair potentials contain inherent deficiencies, it is logical to inquire whether the experimental data are sufficiently good to find the flaws. The first evidence that pair potentials were not sufficient to describe the ejection process arose from the EARN experiments on Rh{111}. Concurrent with these experimental observations was a flurry of activity into the development of many-body potentials to describe bulk phases. Below we describe various approaches for many-body interactions, some of the first sputtering calculations using many-body potentials and our projections for the future.

3.3.3. Many-Body Potentials—Metals and the Embedded Atom Method

The glaring deficiency of using attractive pair potentials in the simulations of the keV particle bombardment process is shown in Fig. 13. Here the angle integrated energy distributions of Rh atoms ejected from Rh{111} from the EARN experiments and the simulations using attractive pair potentials are shown. The normalization of the curves is explained below. The most obvious problem is that the peak in the calculated curve is 2–3 eV below that of the experimental curve. No reasonable adjustment of the parameters in the pair potentials could be made to bring the peaks into alignment.

Fortuitously, about the same time that these experiments were being carried out, a functional form for describing the many-body interactions among metal atoms was presented by Daw and Baskes.[98] As with the pair potentials this embedded atom method (EAM) potential is empirical and the parameters must somehow be determined. However, the functional form is inherently many-body in nature. The first step in determining the potential is to define a local electron density at each atomic site in the solid. A simple sum of atomic electron densities has been shown to be adequate in many cases, and often a sum of free atom densities is used.[90,99] The second step is to determine an embedding function that defines the energy of an atom for a given electron density. Finally, the attractive contribution to the binding energy produced by this embedding function is balanced by pairwise additive repulsive interactions. The expression for the total binding energy is given by

$$E = \sum_i F_i(\rho_i) + \tfrac{1}{2}\sum_i \sum_j \Phi(r_{ij}) \tag{15}$$

where ρ_i is the electron density at each atomic site, $F_i(\rho_i)$ is the embedding function, and $\Phi(r_{ij})$ is the pair term arising mainly from the core–core repulsions. The function ρ_i is given by the expression

$$\rho_i = \sum_j \rho(r_{ij}) \tag{16}$$

where $\rho(r_{ij})$ is the contribution of electron density to site i from atom j and is a function of the distance r_{ij}. This dependence of the electron density on the interatomic distances facilitates the determination of the forces that are simply derivatives of the energy with respect to distance.

If free atom densities are used in the sum from Eq. (16) above, the embedding function is left to determine the properties of the condensed phase. An accurate determination of this function is therefore important for modeling realistic systems. The approach commonly used is to fit this function to a large number of properties. For example, experimentally determined values of the lattice constant, sublimation energy, elastic constants, and vacancy formation energies are often combined with theoretically determined relations such as the universal equation of state[(100)] to provide an extensive data base.[(99)] This formalism has proven to provide both a realistic and easily evaluated potential which is suitable for describing a large range of properties of various pure metals and alloys.[(67,101–111)] To date, the EAM potentials have been found to be successful with close-packed metals although some progress has been made towards modeling covalent bonding by the introduction of angle-dependent electron densities.[(112)]

A preliminary fit of the embedding function and the core repulsive term was made to the properties of Rh metal in order to determine whether the EAM description of the interaction predicts the EARN data of Rh atoms ejected from Rh{111} better than the pair-potentials.[(67)] The most dramatic change in the predicted distributions is observed in the angle-integrated energy distributions. As shown in Fig. 13, the experimental and calculated distributions using the EAM interaction are in excellent agreement, while the calculated distribution using pair potentials is quite different from the experimental curve. The peaks in the polar angle distributions as calculated using the EAM potential are also found to increase by about 10° from those predicted by the pair potentials (Fig. 11). The agreement between the EAM and the experimental energy distributions is most convincing and the improvement in the accuracy of the polar distribution is encouraging.

Is the closer agreement fortuitous or is there a sound basis for it? It has been known that the pair-potential description is inadequate in the surface region, but the detailed data that exposed the nature of the deficiencies were not available. There are several fundamental differences between the EAM and the pair potentials. First, the surface binding energy of the EAM potential is larger (5.1 eV) than that of the pair potential (4.1 eV). Of note is that both potentials were fit to the bulk heat of atomization of Rh (5.76 eV). The peak position in the energy distribution is proportional to the binding energy,[(21)] and thus it is logical that the peak in the EAM energy distribution occurs at a higher value than for the pair potential. In addition to the larger binding energy at the equilibrium site, the EAM potential is relatively flat in the attractive portion of the entire surface region as shown in the contour plots of Fig. 18. There is more than a 4-eV attraction for the ejecting atom even above a neighboring atom, while the pair potential exerts only ~1 eV overall attraction. Thus, particles that eject at more grazing angles will ex-

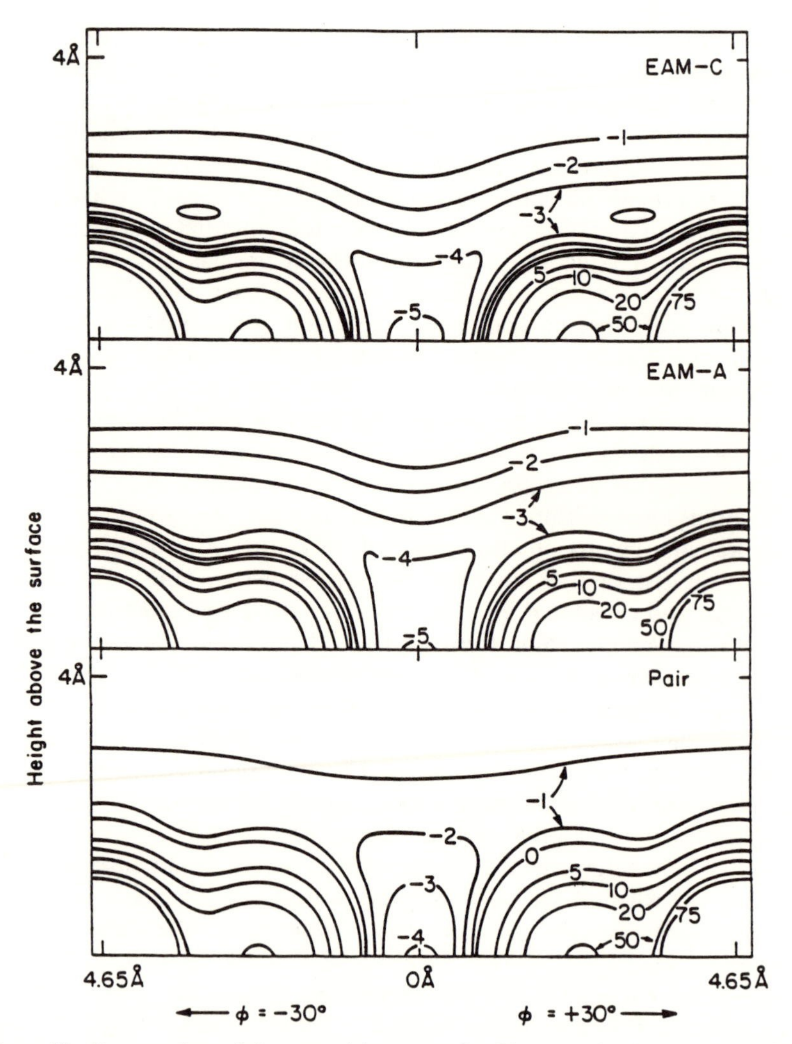

Figure 18. Contour plots of the potential energy of a Rh atom ejecting from a Rh{111} surface for the pair potential, EAM-A and EAM-C. The ordinate is the height of the atom above the surface (Å). The abscissa is the position of the atom (Å) along the surface in the $\phi = \pm 30°$ azimuths.

perience a larger attraction to the surface in the EAM potential than in the pair potential. This increase will shift the peak in the energy distribution to larger energies, reduce the ejection yield, pull the particles away from the surface normal, and move the peak in the polar distribution as seen in Fig. 13. It is the low-energy particles therefore that are not well described by the pair potentials. The experimental and calculated energy distributions using the EAM shown in Fig. 13 are peak normalized. The area under the pair-potential curve is about 2 times that of the EAM curve. This ratio is obtained from the ratio of the calculated yields. This normalization quite clearly shows that it is the low-energy particles that have been inhibited from ejecting when the EAM potential is used in the simulations.

Recently the previously developed Rh{111} EAM potential has been employed to model the ejection process from Rh{331}, a stepped surface that consists of {111} terraces three atoms wide with a one-atom step height.[113] In this surface there are atoms that are both more and less coordinated than on the {111} surface. As we shall see in Section 5, the agreement between the experimental and calculated angular distributions is excellent. This same EAM potential was used for the Rh interactions in the O/Rh{111} study and is also discussed in Section 5.

In a similar study, Lo *et al.* have compared the characteristics of atoms sputtered from copper surfaces in simulations that used both pair potentials and EAM potentials.[114] Significant differences were found for many properties of interest, including the peak in the energy distributions. Although adjustment of the potentials to fit experiment was not attempted, this study concluded that many-body potentials are required to realistically model much of the sputtering process.

The EAM approach appears to provide a formalism within which realistic potentials describing atomic dynamics can be developed. It should also provide a method for realistically incorporating adsorbates into dynamics simulations. Both of these applications can be considered significant advances, and will help molecular dynamics simulations to continue to contribute to the understanding of keV particle bombardment processes.

3.3.4. Many-Body Potentials—Silicon and Covalent Solids

Even more difficulties arise when attempting to describe covalently bonded systems such as silicon or even proteins. The crux of the problem is that there is a strong angular dependence in the bonding since the number of neighbors is smaller than in close-packed systems. For improperly constructed potentials, the system will tend to collapse into a close-packed array of atoms during the molecular dynamics simulation. There has been a flurry of activity recently centered mainly on describing solid silicon. Silicon serves as a good model system since it is homonuclear.

A convenient starting point for developing covalent interactions is to write the energy as a many-body expansion of the form

$$E_{\text{tot}} = V_2 + V_3 + V_4 + \cdots \tag{17}$$

where the first term represents a sum over pairs of atoms, the second term represents a sum over triads of atoms, etc. A well-known example of this type of expansion, even though restricted to atomic displacements near equilibrium, is the valence-force field. While this expression is exact if all terms are included, computational restrictions demand that it be truncated. In most applications, it is truncated at three-body interactions. This is partly for computational convenience and partly because the three-body term can be written in the form of a bond-bend—a concept that is physically appealing. While this approach is well developed for few-body, gas-phase reactions,[115–116] it has only recently been extended to condensed phases.

Silicon has been the test case for potentials of this type, and so this discussion will be restricted to this element. The most widely used silicon potential was developed by Stillinger and Weber.[72] The interactions used are composed of a sum of two-body and three-body terms, with the three-body interactions serving to destabilize the sum of the pair terms when bond angles are not tetrahedral. The parameters for this potential were determined by reproducing the binding energy, lattice stability, and density of solid silicon and also by reproducing the melting point and the structure of liquid silicon. Although this potential was originally developed to model liquid–solid properties, subsequently studies have demonstrated that it also provides a good description of the Si{001} surface.[117,118] The wide applicability of this potential can be considered a testament to the care (and computer time) invested by Stillinger and Weber in its development.

A simpler potential of the form of Eq. (17) has been used by Pearson *et al.* to model Si and SiC surfaces.[83] The two-body term is of the familiar Lennard–Jones form while the three-body interaction is modeled by an Axilrod–Teller potential.[119] This potential form is restricted to weakly bound systems, although it apparently can be extended to model covalent interactions.

Brenner and Garrison introduced a potential that was derived by rewriting a valence force expression so that proper dissociation behavior is attained.[84] Because the equations were extended from a set of terms that provided an excellent fit to the vibrational properties of silicon, this potential is well suited for studying processes that depend on dynamic properties of crystalline silicon. For example, Agrawal *et al.* have studied energy transfer from adsorbed hydrogen atoms into the surface using this potential.[120]

While these potentials have been successful in modeling dynamic processes on silicon surfaces, the many-body expansion as applied in this case suffers from several drawbacks. Because each of the three potentials above has been fit to properties of the crystalline silicon solid, they implicitly assume that the bonding is tetrahedral in nature. Atoms on the surface of silicon are known to exhibit nontetrahedral hybridizations, and so the results for surfaces are at best uncertain. Also, none of these potentials reproduce accurately the properties of the Si_2 diatomic molecule. This again inhibits a complete description of surface reactions.

A related potential form, primarily developed to reproduce structural energetics of silicon, was introduced by Tersoff[85,121] and later by Dodson[122] and was based on ideas discussed by Abell.[123] The binding energy in the Abell–Tersoff expression is written as a sum of repulsive and attractive two-body interactions, with the attractive contribution modified by a many-body term. The stability of the diamond lattice is achieved by modifying the attractive pair terms according to local coordination, so that the atomic binding energy is at a minimum when each atom has four nearest neighbors. The parameters in this potential were determined by fitting to the properties of the Si_2 molecule, and by reproducing the binding energies and lattice constants of several crystal structures of silicon. The Tersoff style potentials are expected to yield a good overall expression for silicon because they correctly describe the isolated dimer, and because they are fit to nontetrahedral structures. This means that they should provide an adequate description of silicon surfaces, although thorough testing is still being carried out.[121]

Additional silicon potentials have also been introduced, but they appear to be cumbersome for studying large numbers of atoms.[86] Also, several of the silicon potentials mentioned above have been modified to represent germanium, carbon, and the heteronuclear compounds SiC and SiGe.[124–125]

To date two simulations on the keV particle bombardment of silicon using many-body potentials have been performed. Stansfield *et al.*[71] have modeled Ar bombardment of Si{100} and the dimer reconstructed Si{100}(2 × 1) surfaces using the Stillinger and Weber potential.[72] They found that, for bombardment at energies between 100 and 1500 eV, Si atoms ejected from 1–4 layers in the solid. This is in contrast to bombardment of the low index faces of metals where the ejection is primarily from the first atomic layer. Smith *et al.*[126] modeled the Ar bombardment at 1000 eV of Si{110}, Si{100}, and Si{100}(2 × 1) using the Tersoff-II potential and showed that the angular distributions are very sensitive to the structure of the original surface. Both studies showed that the important collision mechanisms in the solid were different than for close-packed systems because of the openness of the crystal. This is especially apparent in the energy distributions as shown in Fig. 19.[126] The sublimation energy of Si is 4.63 eV, yet the peak in the energy distribution occurs at about 7.5 eV. The openness of the crystal prevents randomization of the energy and thus the distributions are not described by the transport theory expressions.[21] On a technical note, the Smith study shows that a moving atom approximation can be used reliably to reduce the required computer time for the simulation.[126]

The silicon and carbon–germanium potentials are certainly a significant step forward in terms of having realistic many-body potentials for describing bulk phases. However, the reactions of molecules ranging from H_2 and C_2H_4 to DNA on the surface are also of interest. Many-body potentials for these systems are just now beginning to be developed and have not as yet been used in simulations of the energetic particle bombardment event.

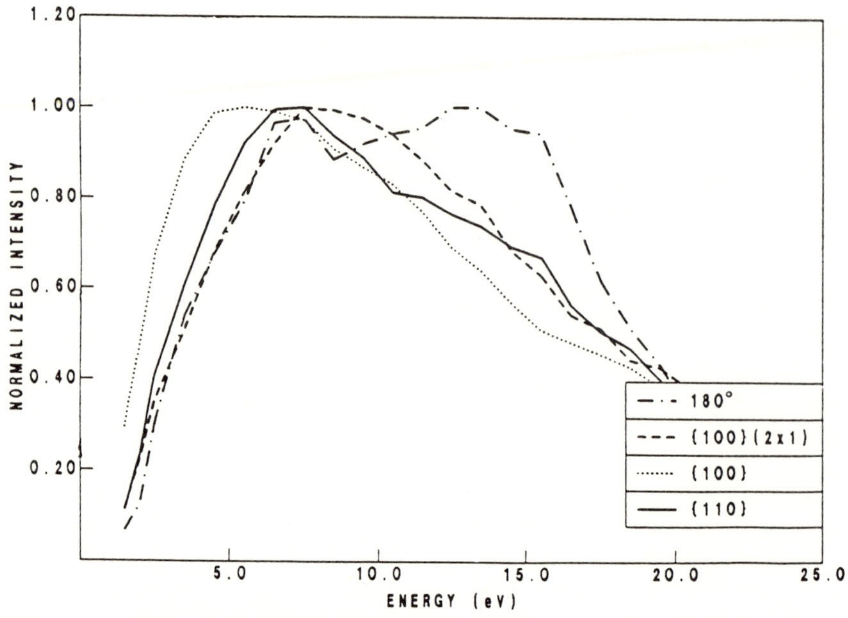

Figure 19. Angle-integrated distributions as a function of crystal face. The fourth curve is for Si_h atoms that eject from Si{110} at $\theta = 40° \pm 10°$ and $\varphi = 180°$, i.e., the most intense feature of the angular distribution. In all cases the Ar had 1000 eV of kinetic energy.

3.3.5. Many-Body Potentials—Reactions on Surfaces

The main many-body interaction that has been used for reactions of gases (mainly diatomic molecules) with surfaces has been a modification of the London–Eyring–Polanyi–Sato (LEPS) method used successfully for gas-phase potential surfaces.[127] The spirit of this type of approach is to consider the entire surface as one body of a few-body reaction. This method has been used for examining the reaction of H_2 on W{100},[128] H_2 on various Cu faces,[129–131] the dissociative chemisorption of N_2 on W{011},[132] the anisotropic chemisorption of O_2 on Ag{110},[133] and the abstraction of a C atom adsorbed on the Pt{111} surface by a gas-phase O atom.[134] There are two problems with using the LEPS potential for simulations of keV particle bombardment. First, the LEPS approach requires diagonalization of a matrix at each integration step. The size of this matrix increases dramatically with the number of particles. For large systems the computer time will be quite large. Secondly, in the reactions studied to date, the surface has just been a heat bath and surface atoms have not participated directly in the reaction. In the sputtering process, the entire surface cannot be thought of as one body of a many-body system. The LEPS matrix would be astronomically large. Although this formulation of a potential for gas reactions with solid surfaces is not quite optimal for sputtering simulations, in the very near future some formulations may well be.

3.3.6. Many-Body Potentials—Molecular Solids

Recently Brenner has developed a many-body potential for modeling the chemical reactions in shock waves.[135] The protypical reaction is

$$2NO(g) \rightarrow N_2(g) + O_2(g) \tag{18}$$

Brenner's potential is based on the Tersoff–Abell approach for many-body interactions described above. The key feature is that one potential surface describes the bonding interaction in NO, the van der Waals interaction among NO molecules, the bonding interactions in N_2 and O_2, and the van der Waals interactions among the product molecules. This is real reaction chemistry. This type of potential could well describe the events that take place when water ice is bombarded in the high-dose regime and eventually O_2 is ejected. Obviously during the course of the experiment, reactions are taking place that cause the formation of O_2 in the solid. As with LEPS potentials, this type of potential may not prove to be the best for use in sputtering calculations. However, there is no doubt that potentials will be developed to examine complex reaction systems useful in simulations of the particle bombardment process.

3.3.7. *The Future*

We are very optimistic that in the near future very realistic potentials will be available for examining chemical reactions in bulk phases and at surfaces. Already the EAM potentials have been used to predict which of the face-centered-cubic {110} surfaces will reconstruct and which will not.[100,136,137] We have performed molecular dynamics calculations of the microscopic reaction mechanisms that are important for the opening of the dimers on the Si{100}(2 × 1) surface during molecular beam epitaxy.[138,139,140] As mentioned above, Brenner has examined chemical reactions in shock waves. The ability to model the reaction of ethylene on Rh{111} to form ethylidyne is not that far away. There is considerable activity in modeling the van der Waals interactions among protein and DNA molecules. These groups are also interested in incorporating chemical reactions into their approach. A merging of their potentials with others for the bulk substrate will make the modeling of the sputtering of a protein from a surface possible. Of note is that as the potentials improve, it will also be possible to examine events such as damage quite realistically. The nature and distribution of the defects could be reliably predicted.

There is one broad class of potentials that we have omitted from this chapter. In all the cases discussed above, we have assumed that the reaction occurs on one potential surface, that is, electronic effects have been omitted. Potentials to describe both the neutral and charge transfer states are still a long way from being developed for a large number of atoms. In addition, if there are electronic effects, then some quantum mechanics is needed in the dynamical description. Performing calculations to describe the charge transfer among many particles is still far off.

4. Electronic Effects

There is ample evidence to believe that the dynamics of neutral particle motion subsequent to the ion impact event is reasonably well understood for quite a few different types of systems. Electronic excitations that may accompany the classical motion, however, add an enormous degree of complexity to the theory. It is extremely important to be able to predict how important these excitations may be. In SIMS, for example, it has been known for many years that the number of ions detected is more dependent on the electronic properties of the matrix than on the concentration of the species on the surface. This very practical annoyance, however, presents a challenging fundamental problem to investigate and understand. What are the electronic events that occur in the solid due to the keV particle bombardment and how do these influence the properties of the ejected particles?

In this section we focus on those theories and experiments that help to extract the underlying mechanism of the excitation/ionization/neutralization process. Although an understanding of these mechanisms will ultimately be crucial to performing better quantitative measurements with SIMS, it is not certain that such a goal is yet possible, given the potential complexities of the electronic events. It is not clear, first of all, that one theory is applicable to all types of electronic processes. Should a theory appropriate for metals where there is significant electron delocalization be also appropriate for an oxide where there are localized electron densities? Can one always assume that the electronic states of the substrate connected to the ejected species is the same as the original quiescent surface? The figures in Section 3 would lead one to doubt this assumption.

A final consideration, and a nontrivial one at that, is to directly compare experimental with theoretical results. In the laboratory, it would be desirable to measure the intensity of each electronic state of the desorbing particle as a function of velocity (v) and angles of ejection (θ, φ). For simplicity, assume that the ion and neutral atom distributions are under consideration. The measured ion intensity is $I^+(v, \theta, \varphi)$ and the measured neutral distribution is $I^0(v, \theta, \varphi)$. The temptation is to define the ionization (or excitation) probability as

$$R^+(v, \theta, \varphi) = I^+(v, \theta, \varphi)/[I^0(v, \theta, \varphi) + I^+(v, \theta, \varphi)] \qquad (19)$$

Here v is the measured velocity of each particle, θ is the polar angle as measured from the surface normal, and φ is the azimuthal angle.

The problem with this definition is that ions and neutral atoms depart the surface region with trajectories that are influenced by different potential energy surfaces.[141] For a particle near the solid surface, two schematic potential curves are shown in Fig. 20. At the time of collision, assume the particle is at the bottom of the lowest surface-potential well($-D$) and has a kinetic energy E' and velocity v'. There is a curve crossing or decision point at a distance z^* from the surface. At this point, the particle has energy $E^+(=E' - D)$ and corresponding velocity v^*. If the particle ejects along the neutral surface (lower one) the measured velocity will also be nearly v^*. However, if the particle is ionized, there is still a barrier of energy ΔE that must be overcome before the particle finally reaches the detector. An image charge in the solid is one possible source of this energy barrier.

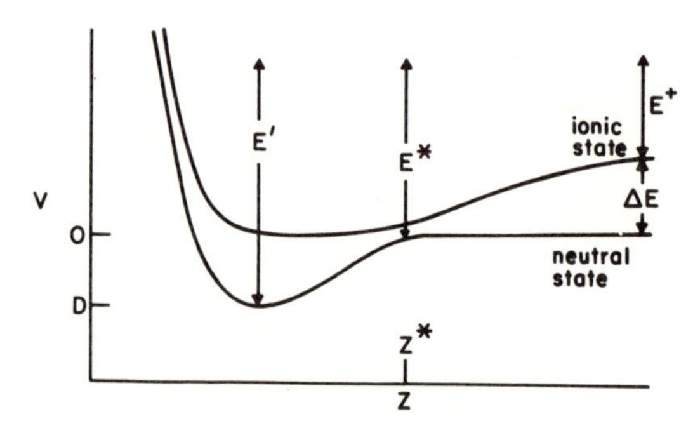

Figure 20. Schematic potential energy curves for a particle at a distance z from the surface.

The easiest way to analyze the experimental data in order to extract the ionization probability is to determine the ratio of the ion to neutral yield as a function of the measured velocity of each species. However, to compare the neutral and ionic particles that had analogous histories of motion in the solid, the neutral species that eject with v^* must be compared with ions that eject with velocity v^+ where $(v^+)^2 \propto E^+ = [(v^*)^2 - 2\Delta E/m]$. The appropriate ratio for ejection normal to the surface thus becomes

$$R^+(v^*) = I^+(v^+)/[I^0(v^*) + I^+(v^+)] \tag{20}$$

or

$$R^+(E^*) = I^+(E^* - \Delta E)/[I^0(E^*) + I^+(E^* - \Delta E)] \tag{21}$$

Furthermore these expressions might be expressed in terms of the velocity associated with the energy E'.

Given the above complications, we proceed with some highlights of experiments and theories that help to elucidate the fundamental processes of electronic motion. The two most fundamentally sound theories are the resonant tunneling model and the bond-breaking model. A number of reviews[142–150] on both of these have appeared over the years so we will only give an overview. In fact, these two models are quite similar; however, they are generally applied to different types of systems. The tunneling model is generally used to describe charge exchange in systems where there are delocalized electrons (e.g., metals) whereas the bond-breaking model is more applicable to systems where there are localized electron densities (e.g. oxides). In addition to the above models, which are primarily for describing ionization events, we also describe a model that explains the velocity distributions of neutral atoms in excited electronic states. Our particular focus in this review is to emphasize the current need to connect the ionization theories which involve macroscopic parameters such as work function to the dynamics of nuclear motion in the solid.

4.1. Tunneling Model

The tunneling model has been one of the most widely applied explanations for the work function and velocity dependences of the ionization probability of species ejected from metal surfaces. A number of workers[141–154] have given excellent descriptions and reviews of the topic so a brief summary will be given here along with experimental examples that illustrate different features of the predictions of each model.

The main concepts behind the tunneling model are shown schematically in Fig. 21. The free-electron-like metal surface is assumed to be smooth with no inhomogeneities and has a Fermi energy, E_F, which is equal to the work function ϕ. The sputtered atom (assumed for this discussion to be one that will form a cation) has an energy level E_a. When the atom is at infinity, this energy is the ionization energy. At closer distances to the surface, the energy level shifts upward due to the image potential. Finally the electron in the atomic energy level has a finite lifetime, so the level is broadened with a half-width of $\Delta(z)$. The

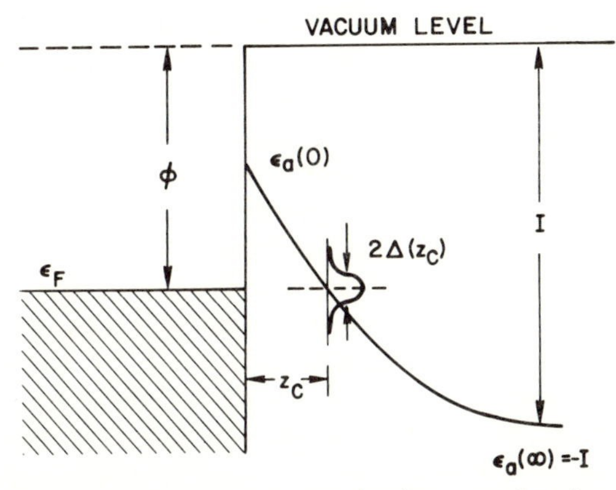

Figure 21. Schematic energy diagram of an atom leaving a metal surface. The Fermi level E_F lies below the vacuum level by the work function value ϕ. Initially the atomic level E_a is broad and may lie above E_F before the atom is sputtered off. The variation of the image potential causes E_a to lower with separation until $E_a = E_F$ at the crossing point. Electrons in the metal can tunnel out to fill the atomic level once $E_a < E_F$ beyond the crossing point. (From Ref. 147.)

essence of the tunneling model then is that electrons can tunnel between the atomic level and levels in the solid of the same energy.

Lang[151] has made the assumption that the particle moves in a straight line trajectory

$$z = v_{\perp} t \tag{22}$$

where $v_{\perp}$ is the velocity component perpendicular to the surface and t is the time. This assumption, coupled with the idea that the critical-level width is determined at the point when the atomic energy level equals the Fermi energy, yields the following expression for the ionization probability:

$$R^{+} \propto \exp[-2\Delta(z^{*})/\hbar\gamma v_{\perp}] \tag{23}$$

where γ is the decay length of the width of the atomic level. Since the crossing point, z^{*}, depends on the Fermi energy or the work function, the above equation reduces to

$$R^{+} \propto \exp[-c(I - \phi)/v_{\perp}] \tag{24}$$

or

$$R^{-} \propto \exp[-c(\phi - A)/v_{\perp}] \tag{25}$$

where A is the electron affinity of the sputtered anion. The energy difference ΔE of Fig. 20 is $(I - \phi)$ or $(\phi - A)$.

This model predicts that the ionization probability depends on the difference in the ionization energy and the work function and on the perpendicular velocity of the sputtering atom. Some of the more detailed experiments that test the ideas in this model have been performed by Yu and co-workers. In one experiment Cs was adsorbed onto a substrate at a fractional monolayer coverage. The work function of the surface was then systematically varied by coadsorbing small amounts of Li. The intensity of Cs^{+} ions ejected as a function of the work function change was then carefully measured.[155] As shown schematically in Fig. 22, if the work function is larger than the ionization energy, then the probability of an ion being ejected is unity. The electron remains in the metal which has a lower energy state. For smaller work functions, the atomic level crosses the Fermi level and thus the ionization probability depends exponentially on the difference between the ionization energy and the work function. Shown in Fig. 23 is the measured Cs^{+} ion intensity as a function of work function along with the predictions of Eq. (23).

In another set of experiments Yu measured the O^{-} ion intensity for different angles of ejection.[156] This approach allowed a clever means of changing $v_{\perp}$ without changing v. As shown in Fig. 24 the ion intensity

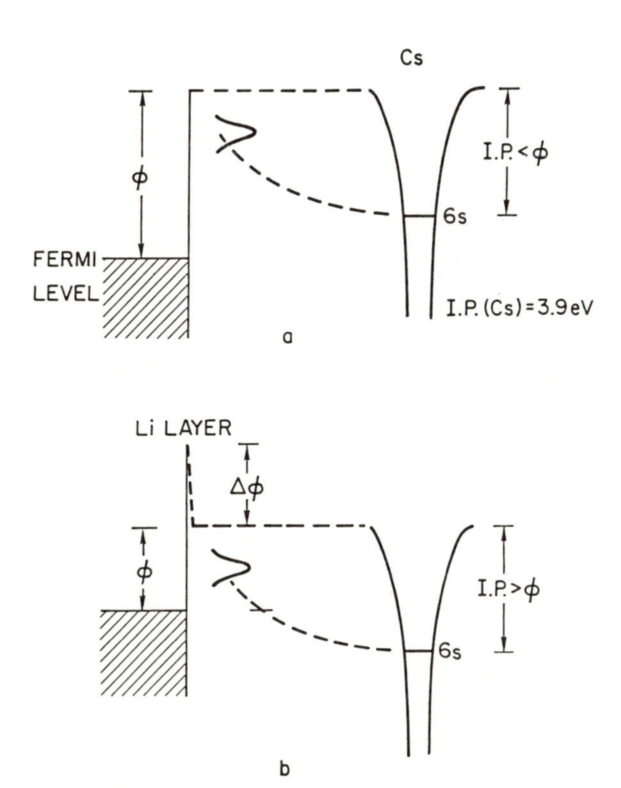

Figure 22. Schematic energy diagram for the sputtering of (a) $\phi > I$, (b) $\phi < I$ after the work function is lowered by the deposition of Li. (From Ref. 147.)

depends exponentially on the inverse of the perpendicular velocity except for small velocities. This seems to indicate that the tunneling model is appropriate for predicting the ionization probability of ions that eject from metal surfaces.[156–159] There are a number of other studies, both experimental and theoretical, however, which predict that R^+ should be proportional to $v_\perp^n$ where n is between 0 and 3.[160–162] It is not clear how well in many cases the ionization probability has actually been measured. Often only the energy distribution of the ions has been measured at one exiting angle. The neutral distribution is assumed to follow that predicted by Thompson.[21] As discussed above the assumptions of isotropy inherent in this model and the ad hoc choice of the heat of sublimation as the energy cost to remove an atom are not fully established.[47,48,51,52,53] In addition, the choice of velocities at which to compare the ion and neutral distributions must be carefully evaluated.[141]

Although the tunneling model appears to do quite well in

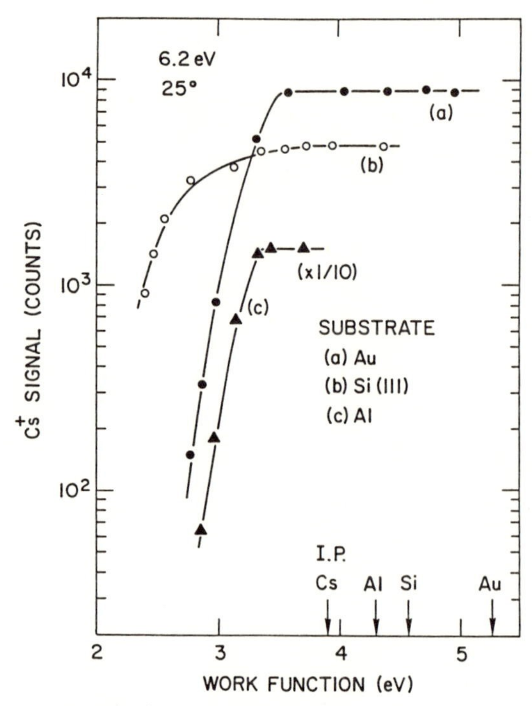

Figure 23. A 1-nA, 500-eV Ne^+ beam was used to sputter in the static mode. The Cs^+ ions emitted have a $v_\perp$ of 2.7×10^5 cm/s. (From Ref. 155.)

accounting for the experimental observations, the pictures presented by the dynamics simulations leave one concerned about several of the fundamental assumptions of these approaches. One feature that the dynamics simulations point out is that the velocity of the ejecting atom is not constant as is assumed in Eq. (23). In model cluster calculations of an ionization model, it is possible to systematically vary the binding energy of the atom to the surface and also to monitor the initial velocity and the measured velocity.[163] As shown in Fig. 25, the ionization probability scales with the initial velocity (v' of Fig. 20) rather than the measured velocity (v^*). This effect was used by Yu and Lang to explain the low velocity deviation of the O^- intensity from the tunneling model prediction.[155]

The other assumption of concern is that there is a homogeneous electron density in the metal. In the pictures shown in the previous section, the environment around the departing atom can be severely distorted. There are some ion impacts, however, where there is significantly less damage around the ejection site. Harrison[30] and

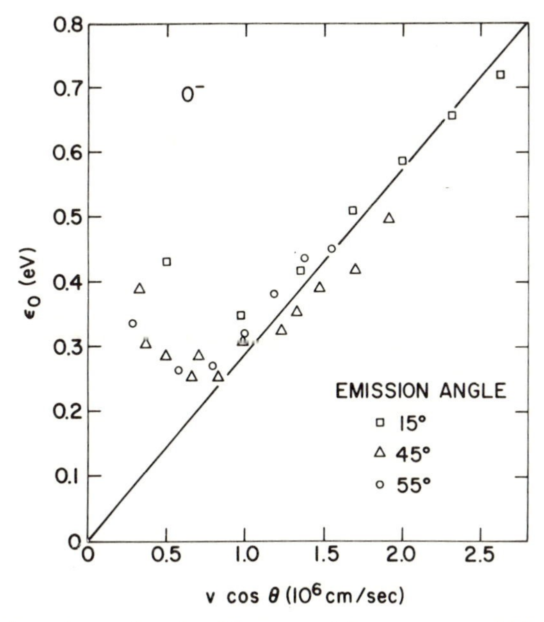

Figure 24. The dependence of ϵ_0 for O^- sputtered from oxygenated Nb surfaces on the normal component $v_\perp$ of the emission velocity. Here $\epsilon_0 = v_\perp/c$ of Eq. (24). (From Ref. 156.)

Williams[143] have suggested that *megaevents,* that is, impacts where there is significant damage in the surface region at the time of particle ejection, may require a different model of ionization. In this regime, the electron density may be severely distorted. In a similar vein, Nourtier[164] has assumed that for metal ion ejection from the same substrate, specific collision sequences may preferentially give rise to ion ejection.

4.2. Bond-Breaking Model

To account for the presence of an inhomogeneous electron density, Slodzian[145] and Williams[143] have proposed a bond-breaking model which is based on the Landau–Zener model.[165] The process is depicted in Fig. 20 and is similar in spirit to the tunneling model except that at the crossing point (z^*) there are discrete levels rather than a continuum of levels in the substrate.[148] As developed by Yu[166] the ionization probability for the bond breaking model is given by

$$R^+ \propto \exp(-2\pi H_{12}^2/v\,|a|) \tag{26}$$

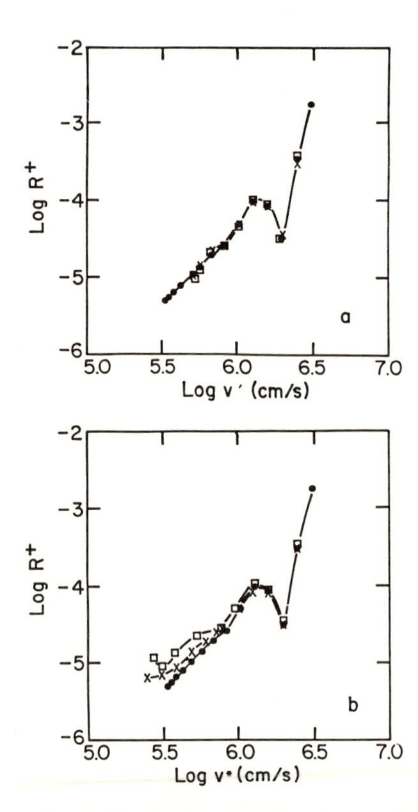

Figure 25. Log R^+ vs. $\log v$ for different binding energies, D: •, $D = 0.0$ eV; ×, D = 0.8 eV; □, D = 1.6 eV; (a) $\theta = 0°$, $v = v'$; (b) $\theta = 0°$ $v = v^*$. (From Ref. 163.)

where H_{12} is the overlap integral between the diabatic ion and covalent curves, v is total velocity, and $|a|$ is the difference in the slopes of the two diabatic curves, all evaluated at the crossing point. The velocity at the crossing point can be written as

$$v = [2(E^* + I - A)/m]^{1/2} \tag{27}$$

where $I - A$ is the energy difference between the two potential curves (Fig. 20) at infinity. For positive ion emission, I is the ionization energy of the departing cation and A is the electron affinity of the vacancy left behind. Of extreme importance is that this electron affinity is a *local* property and not the *work function* of the substrate. It is, in fact, the observation that the ion yield does not depend on the work function that allows one to assume that the bond breaking model is valid.(148)

The fact that the ionization probability in the bond-breaking model is local implies that the ion yield is proportional to coverage as long as the environment of the departing ion is constant.(167) Shown in Fig. 26 are the Si^+ and O^- SIMS intensities from a silicon sample that has been

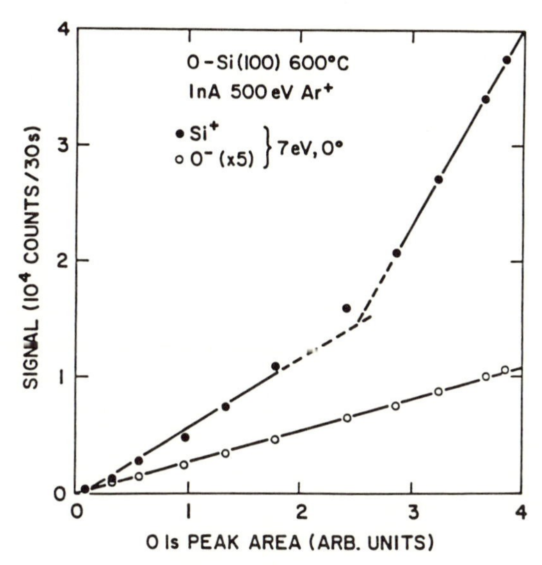

Figure 26. Sputtered Si^+ and O^- yields as a function of oxygen coverage (≤2 monolayers). (From Ref. 167.)

exposed to varying amounts of oxygen. The oxygen concentration on the surface is assumed to be proportional to the O 1*s* signal as measured by x-ray photoelectron spectroscopy (XPS). The signals are clearly linear in oxygen surface concentration. The O^- signal has only one slope. indicating that it maintains a constant environment. The Si^+ signal, however, changes slope. At the same oxygen coverage where the slope changes, the Si 2*p* signal also shifts peak position, indicating that the Si atom environments have changed.

The basic ideas behind the tunneling and bond-breaking models are quite similar. As an ion departs the surface, it reaches a decision or crossing point. In the case of the tunneling model, this crossing point occurs when the energy of the atomic level crosses the fermi level. In the case of the bond-breaking model it is the distance at which the ionic and neutral adiabatic curves cross. Both of these models have attained widespread acceptance and a multitiude of experimental data can be explained by them. However, there are constants in each that cannot be predicted theoretically and are thus used as fitting parameters to the data. It would ultimately be desirable to be able to predict the ion yields *a priori* from a first-principles theoretical basis. Factors such as the effect of individual collision sequences or different local environments (a terrace site versus a kink site) on the ionization probability are not known. In addition, these effects are quite difficult to examine experimentally. Of note is that development of theories and experiments to predict the ion yield from a single element substrate (e.g., Ni^+ from Ni) are only in very premature stages of development.

4.3. Deexcitation Model for Sputtered Excited Neutral Atoms

One of the consequences of both the tunneling and bond-breaking models is that energy distributions of the ions peak at a larger value than the neutral distributions. A similar observation has also been made for the excited fine structure states of Ba ^1D and ^3D,[168] Ca $^3P_2$,[169] and Fe $^5F_5$[170] relative to the ground state distributions. On the other hand, the excited state distributions of Zr[171] and Fe[172] were found to be similar to the ground state distributions. Recently the energy distributions of In atoms in both the ground ($^2P_{1/2}$) and excited ($^2P_{3/2}$) states have been measured.[173] In this case the excited state distribution peaks at a *lower* energy than the ground state distribution.

A model to predict the amount of relaxation of an excited fine-structure state has recently been proposed.[173] In one limit are fine-structure states such as the ^5D_J, $3d^44s^2$ manifold of Fe and the ^3F_J, $4d^25s^2$ manifold of Zr. The fine-structure states are determined by d-electron coupling and not by s-electron coupling. In these cases the s orbitals are larger than the d orbitals, and effectively shield the d electrons from interaction with the metal conduction band as the atom departs from the surface. Consequently, relaxation processes have long lifetimes and the velocity distribution is independent of atomic state.

In the intermediate regime are the ^5F, $3d^74s^1$ state of Fe, the $^3P_2$, $4s^14p^1$ state of Ca, and the $^{1,3}D$, $6s^15d^1$ states of Ba. Here the velocity distributions of the excited states are broader than the ground-state distribution. Of note is that the Ba distribution depends on the electronic environment of the original matrix,[168] i.e., Ba metal, BaO, or BaF_2. The $^{1,3}D$ distributions from Ba metal and BaF_2 are broader than the one for the ^1S ground state, indicating that there is interaction of the $6s$ or $5d$ electron with the substrate. For BaO the distributions of the D states are similar to the ground state. The calcium process is very similar to the Ba system except that the $^{1,3}D$, $4s^13d^1$ states of Ca are not metastable in vacuum and will radiatively decay. In this intermediate case, there is partial shielding of, for example, the $3d^7$ electrons in Fe(^5F_J) by the $4s^1$ electron.

The first example of the third regime is In(^2P_J). In this case, the fine-structure states are determined by L-S coupling of the outermost electron. The interaction of the $5p$ electron with the conduction band is strong and the atoms in the $^2P_{3/2}$ state are effectively deexcited to the $^2P_{1/2}$ state.

There are implications of this proposed model. First, for the well-shielded cases, the populations of atoms in each state cannot be influenced by the interaction of the departing atom with the surface, but rather by some event that occurs in the solid. It is possible that the probability of initial excitation of the *d* electron structure is determined by the degree of the electron–gas disturbance induced by the incoming bombarding particle.[174] The experimental results of the populations in the high-shielding cases show a dependence on the kinetic energy of the incoming ion.[171,172] Since there is no deexcitation as the atom departs from the surface, as determined from the velocity-distribution shapes, then the measured excited-state populations reflect the excitation process. For the moderate-shielding and exposed cases, the final-state populations reflect both the initial excitation process and the deexcitation as the atoms depart from the surface region.

There are several aspects of these proposed ideas that can be tested and examined further. Obviously, there are other elements and associated fine-structure states that will fit into the three categories. These systems can be investigated. The *a priori* determination of the interaction strength should be examined. It would be desirable to find other examples like In($^2P_{3/2}$) where the excited state is completely deexcited.

Our understanding of the ionization process for metal substrates has increased substantially in the past decade. The need is for more detailed measurements of both the ion and neutral distributions and for theoretical models that combine both the nuclear and electronic motion.

In the next section, we move forward with this knowledge to what we can learn about the structure and chemical reactivity of surfaces using these ion-beam methods.

5. Surface Characterization with Ion Bombardment

At this point, it is hopefully clear that much is known about the details of the ion–solid interaction event, of the significance of energy and angular distributions of desorbed particles from single crystals, of the mechanisms of cluster formation and of the excitation processes that lead to ionization. With this important theoretical base, it is now particularly appropriate to examine the potential application of ion beams to a variety of surface characterization experiments. There are a number of aspects to these applications that merit special consideration. First, the theoretical modeling has suggested that the angular distributions reflect the near-surface structure. Secondly, these same calculations indicate that a large majority of the particles eject from the first layer to the target. And finally, the models suggest that the composition of the ejected clusters should embody in some fashion the chemical nature of molecular species that may exist on solid surfaces. Experimentally, as noted in Section 2, there are a number of options for realizing these applications. The detection of desorbed neutral particles is possible via several sensitive detection schemes. The detection of desorbed ions is also straightforward, both with high mass resolution and with high sensitivity. In this section, then, a sampling of the applications of importance to surface science will be reviewed. We shall restrict our examples, however, in two important ways.

Only those experiments performed using low doses of primary particles will be emphasized. Moreover, we shall focus mainly on reasonably well-defined experimental systems—those performed in UHV and/or with single-crystal substrates. These experimental aspects are critical for comparisons to theory and for elucidation of elementary surface chemical processes.

5.1. Surface Structure Studies

5.1.1. Trajectories of Substrate Species

It is of fundamental interest to accurately measure the trajectories of an atom in its ground electronic state desorbing from an ion bombarded single-crystal substrate and to compare these trajectories with those calculated using the molecular dynamics simulations. These measurements have become possible in the last few years using the EARN technique and are extremely important for validating the reliability of any theory. The Rh{111} surface bombarded at normal incidence by 5 keV Ar^+ ions has served as a convenient model system for these studies.[45] The structure of this surface, as in Fig. 12, has been well characterized using LEED and is known to exhibit little surface relaxation or reconstruction that might complicate the analysis. Note that for the {111} face of this face-centered-cubic lattice, the surface atoms exhibit sixfold symmetry. The three second-layer atoms, however, create bulk threefold symmetry with characteristic azimuthal directions along $\langle 111 \rangle$ ($\phi = 0°$ in our notation), $\langle 211 \rangle$, $\phi = -30°$, and $\langle 112 \rangle$, $\phi = +30°$. The experimental EARN distributions for this crystal face along $\phi = \pm 30°$ are shown in Fig. 27. The measurements were recorded by bombardment at normal incidence with a series of 200-ns pulses of 5 keV Ar^+ ions

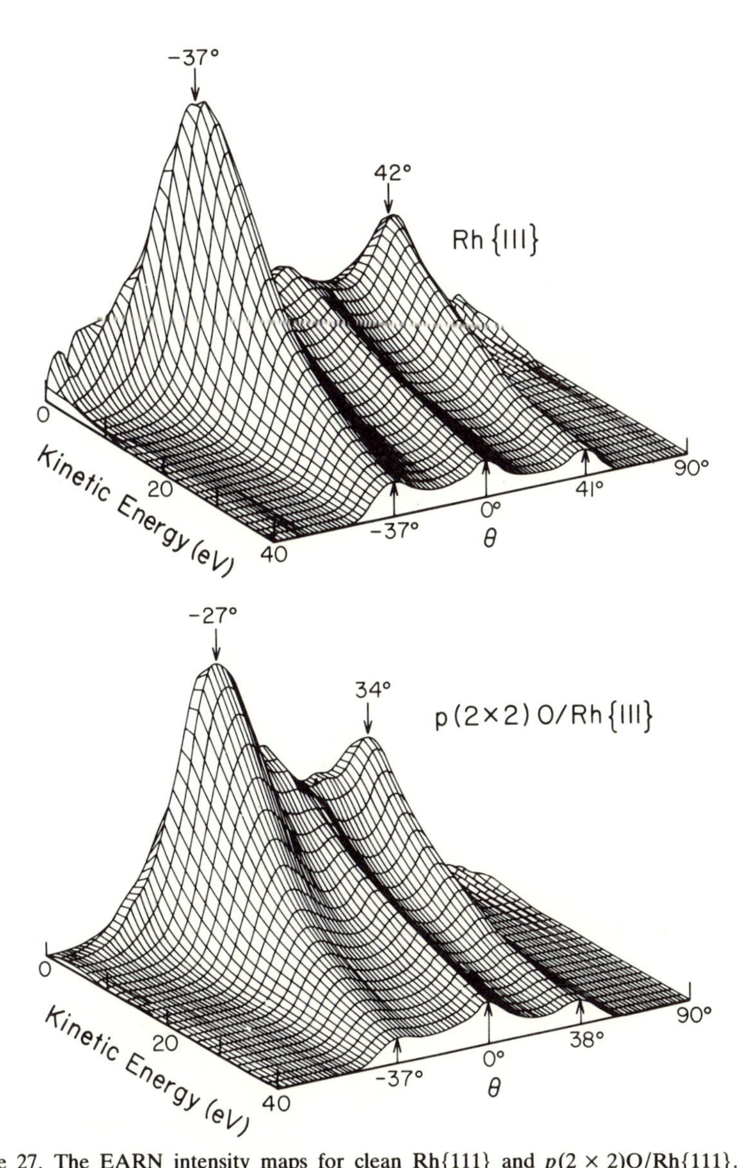

Figure 27. The EARN intensity maps for clean Rh{111} and $p(2 \times 2)$O/Rh{111}. The plots are normalized to the highest intensity peak in both cases. The positive values of θ are recorded along $\phi = 30°$ and the negative values of θ are recorded along $\phi = -30°$. (From Ref. 45.)

($\sim 2 \times 10^6$ ions/pulse). The desorbed Rh atoms were then ionized above the surface using the laser-based MPRI method described in Section 2. It is important to note that these data were recorded with a total dose of less than 10^{11} Ar^+ ions/cm^2 so that the influence of surface damage is minimized.

It is possible to gain a great deal of mechanistic information from these trajectories by comparison to molecular dynamics calculations performed using either pair potentials or many-body (EAM) potentials. These sequences begin with the alignment of atomic motions inside the solid. As these motions cause ejection of first-layer atoms, further focusing is caused by channeling or blocking by other first-layer atoms. For example, the highest intensity is observed along the open crystallographic directions ($\phi = \pm 30°$ in our case) and the minimum intensity is observed along the close-packed crystallographic direction ($\phi = 0°$). If only surface processes were important, the peaks at $\theta = -37°$, $\phi = -30°$ and $\theta = 42°$, $\phi = 30°$ should be equal in intensity and not unequal as shown in Fig. 27. The additional intensity at $\theta = -37°$, $\phi = -30°$ arises mainly from the ejection of atom A by atom B (Fig. 12) with first-layer focusing by other surface atoms. The peak at $\theta = 42°$ and $\phi = 30°$ is lower in intensity by a factor of 0.5 at low kinetic energy (KE) since no such mechanism is available along this azimuth. The peak at $\theta = 0°$ arises mainly from ejection of the second-layer atom B which is focused upward by three surface atoms. As noted in Section 3, these curves may be accurately calculated using the EAM–molecular dynamics approach, a powerful testament to the validity of the channeling and blocking concepts.

A simple test of the consistency between experiment and theory is to examine other crystal faces of Rh and to look for similar channeling and blocking mechanisms. Although there has not yet been a great deal of data of this type, a recent study on Rh{331}, an atomically stepped surface, strongly supports these ideas.[113] The Rh{331} structure is shown in Fig. 28. The corresponding EARN data and EAM calculations are shown in Fig. 29. The most interesting features of the angular distribution plot are that the ejection is strongly peaked at $\theta = 15°$ along the $\phi = +90°$ azimuth, and that the desorption along other crystallographic directions is considerably reduced. Note that these features are well reproduced by the EAM calculations. The parameters in the potential function were unchanged from those used in the Rh{111} calculation. This agreement is quite important to see since the excess electron density present at the step edge might have altered the trajectory of the departing Rh atoms. These calculations conclusively show that the peak at $\theta = 15°$ arises from the same channeling mechanism operative

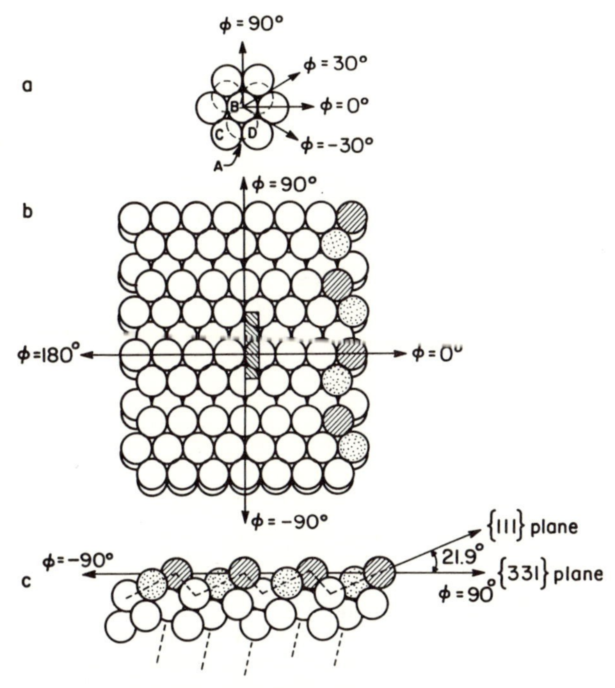

Figure 28. (a) Rh{111} surface indicating azimuthal directions used in the text. The dashed atoms are in the second layer of the crystal. (b) Rh{331} surface indicating the definition of azimuths used in the text. The $\phi = 0°$ azimuth is the same as that in (a). The impact zone used in the classical dynamical simulation is shown at the center of the crystallite as a shaded box. (c) Rh{331} surface viewed from the $\phi = 0°$ direction. The shaded atoms correspond to the shaded atoms in (b). The dotted near-vertical lines indicate a major channeling direction. (From Ref. 113.)

for {111}. This polar angle corresponds to the $\theta = 37°$ peak observed from {111} except that the crystal orientation is simply tilted by 21.9° about $\phi = 0°$ as seen in Fig. 28. The reduced intensity along the other directions arises either from the presence of open channels in the crystal (near $\theta = 0°$) or from blocking due to the atomic step along $\phi = -90°$. Overall, the structure in the angular distributions is quite striking for both clean Rh{111} and Rh{331}, especially when compared to the simple $\cos^2 \theta$ dependence exhibited by the polycrystalline Rh surface.

Since crystal structure via channeling and blocking strongly influences the EARN distributions, it follows that adsorbate atoms or molecules on single-crystal substrates should systematically alter the trajectories of the desorbing underlayer species. This concept has been tested in detail for the $p(2 \times 2)$ O ordered overlayer on Rh{111} with

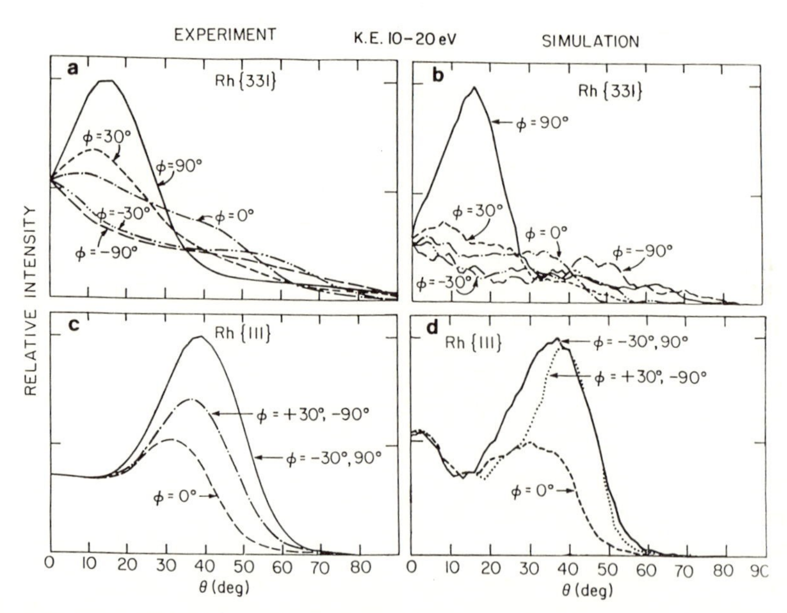

Figure 29. Polar angle Rh atom distributions of keV ion-induced desorption from Rh single crystals. The 10–20-eV kinetic energy range is shown. (a) Rh{331}, experimental data; (c) Rh{111}, experimental data; (b,d) calculated curves employing the EAM potential. (From Ref. 113.)

the EARN experiment.[67] For an atomic adsorbate of this sort, there are a number of possible high-symmetry binding sites including two different three-fold hollow sites (often referred to as the **B** or fcc site and the **C** or hcp site) and an on-top site (or **A** site) (Fig. 12). The simplest idea is that adsorption of oxygen atoms at a **B** site should preferentially block Rh atoms desorbing along the $\phi = +30°$ azimuth, adsorption at a **C** site should alter the path of Rh atoms leaving along $\phi = -30°$, while **A** site adsorption may have very little effect, except perhaps on the particles emitted at $\theta = 0°$. Preliminary polar angle measurements[45] as seen in Fig. 27 clearly show that the $\phi = -30°$ direction is preferentially reduced relative to either $\theta = 0$ or $\phi = +30°$, strongly suggesting **C** site adsorption.

These results are quite interesting since they clearly show how sensitive the desorption angular distributions are to small overlayer coverages. In this case, in addition to the observed perturbations in the EARN distributions, the total ground state neutral Rh atom yield was observed to decrease by a factor of about 2. These observations further illustrate that trajectory measurements that are to be compared with theory must be performed on well-cleaned and chracterized surfaces *and* under low-dose conditions.

It would be gratifying to obtain the same excellent agreement between the $p(2 \times 2)$ O/Rh{111} EARN measurements and the EAM–molecular dynamics calculations as was found for the clean surface. The development of the EAM is not yet directly applicable, however, to chemisorbed overlayers such as oxygen. As an approximation, it has been possible to take a fairly rigorous look at overlayer effects by using EAM forces to describe the Rh–Rh interactions, but using pairwise additive potentials to describe the Rh–O surface interactions. These simulations clearly confirm that the simple blocking ideas are correct, and yield EARN distributions for the Rh atoms that are in quite good agreement with experiment.[67]

Usually, as pointed out in Section 3.1, kinetic energy distributions of ejecting atoms are not terribly relevant to studies involving surface structure. There is a situation associated with chemisorbed species, however, that merits closer attention. The KE distributions for clean Rh{111} and oxygen covered Rh{111} are shown in Fig. 30.[67] Note that the peak in the KE spectra shifts by 2–5 eV to *lower* KE. The computer simulations suggest that this lowering arises from the fact that the underlying Rh atoms lose some energy as they escape through the overlayer oxygen atoms. The effect might be misinterpreted to mean that oxygen decreases the Rh surface binding energy since analytical models predict that the peak is proportional to 1/2 of this value. The effect is even more clearly noticeable when benzene is adsorbed on Rh{111} as seen in Fig. 31.[175] Notice that the benzene overlayer is dramatically reducing the intensity of the highest-energy Rh atoms and is shifting the peak energy to lower values. These types of experiments may eventually prove valuable in following the flow of energy between the substrate and surface molecules as they are coaxed to desorb intact from the surface.

5.1.2. Trajectories of Overlayer Species

If channeling and blocking are important desorption mechanisms for substrate atoms, even more dramatic effects should be visible for the overlayer species themselves. To date, EARN experiments have not yet yielded significant information aimed toward this problem. This lack of progress arises, in part, as a result of the difficulty of ionizing typical adsorbate species such as oxygen, hydrogen, sulfur, or carbon layers. It has been possible, however, to obtain this information on the desorbing secondary ions. Although it is dangerous to directly compare these data with computer simulations, every experiment to date suggests that at

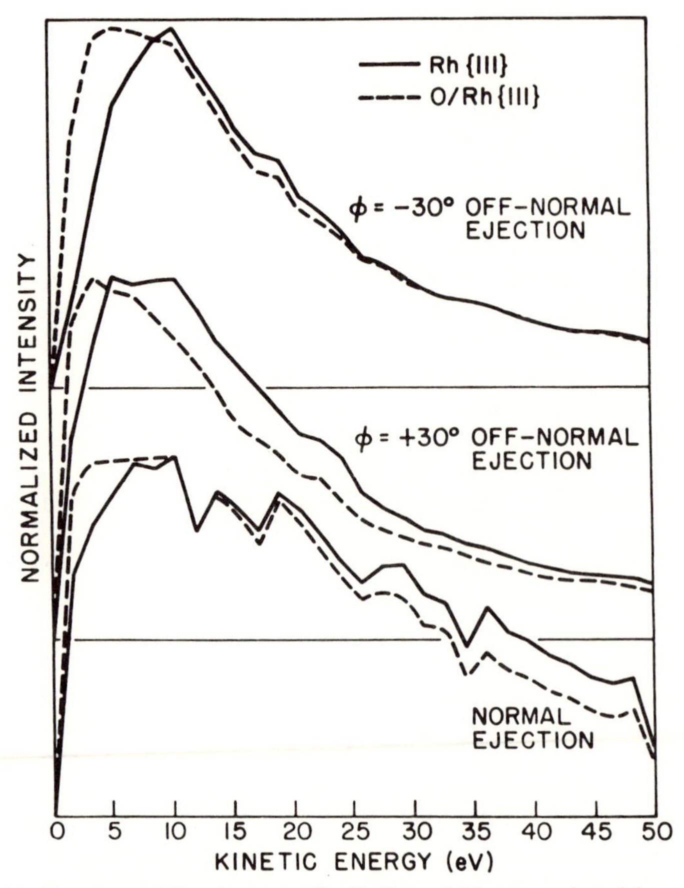

Figure 30. Experimental kinetic energy distributions of Rh atoms ejected from clean and oxygen-covered Rh{111}, taken at ejection angle $\theta = 45°$. $p(2 \times 2)$ LEED pattern was observed for the oxygen overlayer. The data are normalized to the same peak signal intensity. The polar angle resolution is ±3°. (From Ref. 67.)

least in a qualitative sense, channeling and blocking are also applicable to understand secondary ion distributions.

The first studies of this sort were reported for oxygen adsorbed on Cu{100}.[176] For this example, a quadrupole mass spectrometer was positioned to detect either Cu^+ or O^- ions ejected at a polar angle of $\theta = 45°$. The Cu crystal was exposed to 1200 L (Langmuirs) of oxygen at 300 K to yield a $c(2 \times 2)$ oxygen overlayer. The measured intensity for both Cu^+ and O^- ions as the crystal was rotated through 360° is shown in Fig. 32. Note that the Cu^+ ion yield maximizes along $\phi = 0°$, 90°, 180°

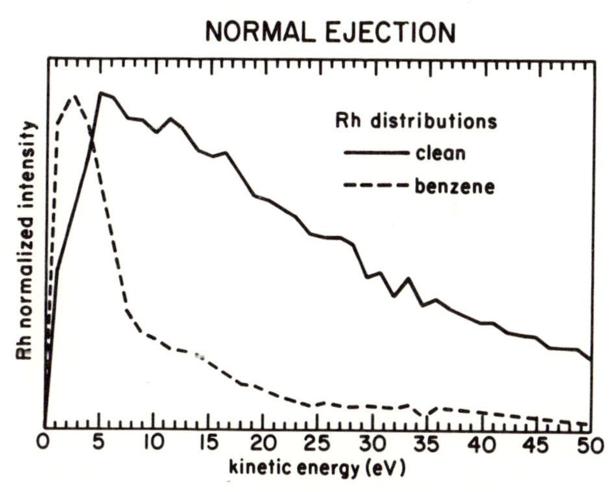

Figure 31. Experimental angle-integrated kinetic energy distribution for Rh atoms ejected from Rh{111} for a clean surface and a surface covered with approximately 1 monolayer of benzene. The sample was bombarded by 3-keV Ar^+ ions at normal incidence. (From Ref. 175.)

and 270°, corresponding to the open crystallographic directions. The $\phi = 0°$ direction is equivalent to $\langle 001 \rangle$. As has been shown for the EARN experiments for Rh{111}, the surface channels the desorbing Cu atoms specifically in these directions. The Cu^+ ion yield is a minimum along $\phi = 45°$, 135°, 225°, and 315° owing to blocking along the close-packed row. The $\phi = 45°$ direction is equivalent to $\langle 011 \rangle$.

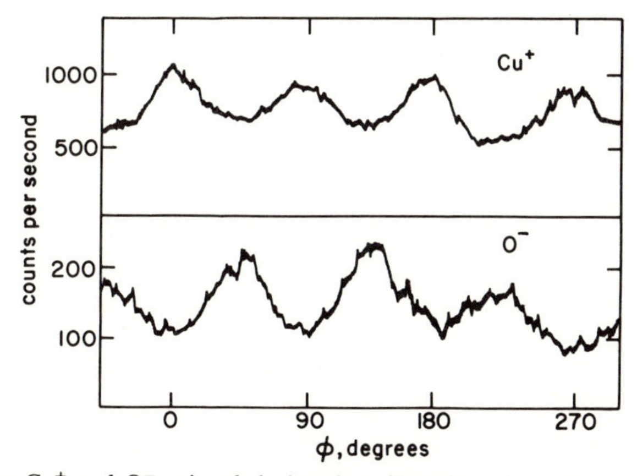

Figure 32. The Cu^+ and O^- azimuthal plots for $c(2 \times 2)$—O/Cu{100}. The sample was bombarded with 1500 eV Ar^+ ions at normal incidence. (from Ref. 176.)

Using this simple concept, it is possible to understand the azimuthal behavior of the O^- ion yield. For adsorption in the four-fold hollow site, oxygen is channeled in a direction rotated by 45° from the Cu substrate. It experiences blocking, however, in the $\phi = 0°$ direction. For adsorption in the A-top site, the oxygen azimuthal direction would be expected to match that of the substrate since the oxygen angular distributions would be controlled mainly by the desorbing substrate species. As is seen from Fig. 32, the results strongly support the assignment of the adsorbed oxygen atom to the four-fold hollow site. There are a number of complications associated with this simple interpretation. First, the magnitude of the azimuthal anisotropies is dependent upon the kinetic energy of the desorbing ion. For the very low-energy particles, there has been sufficient damage to the crystal structure near the impact point of the primary ion that the channeling mechanisms are no longer operative. On the other hand, at higher kinetic energies, say greater than 10 eV, the desorbing ion leaves the surface early in the collision cascade while there is still considerable order in the crystal. The channeling mechanisms are much stronger and the angular anisotropies are larger.

A second complication involves the determination of the height of the adsorbate atom above the surface plane. Calculations have been performed using pair potentials where this bond distance has been varied over several angstroms in order to find the best fit with experiment.[176] These studies have also shown that the polar angle distribution is sensitive to the effective size of the adsorbed atom. Thus, it is important to know more about the scattering potential parameters if this distance is to be determined accurately.

Early attempts at quantitatively calculating the O–Cu bond length were apparently unsuccessful, perhaps owing to the complexity of this system. Angle-resolved photoemission experiments indicated that although the bonding was indeed occurring in the four-fold hole site, at least a portion of the oxygen atoms were found directly in the plane of the Cu{100} surface.[177] Nevertheless, the channeling and blocking ideas were established as important for desorbing ions as well as neutral species. Apparently, effects such as azimuthally dependent ionization probabilities and bending forces such as the image force are not strong enough to mask these dominant factors, particularly if only the high-kinetic-energy ions are detected.

A second example of the importance of channeling and blocking has more recently been developed for the case of Cl adsorbed on various crystal faces of Ag. The azimuthal distributions for Cl^- ejected from Ag{110} covered with various amounts of Cl atoms are shown in Fig. 33.[178] As for oxygen on Cu, the ions were detected at $\theta = 45°$, and only

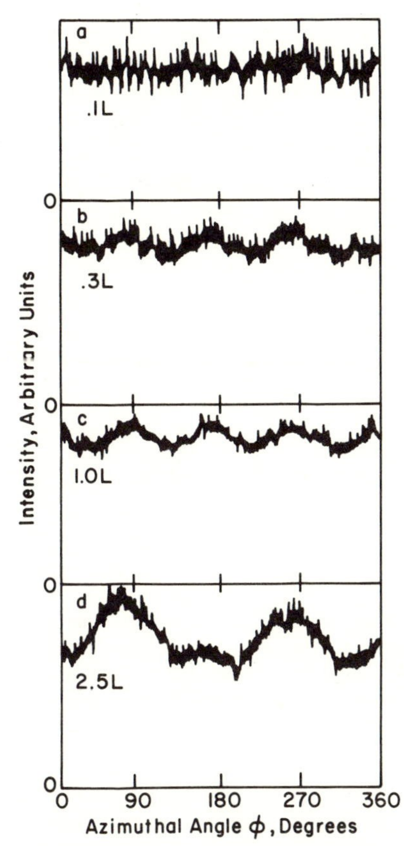

Figure 33. Azimuthal angle distributions of high-energy (18–22 eV) secondary Cl^- ions as a function of Cl_2 exposure. In our notation, $\phi = 0°$ and $\phi = 90°$ correspond to the $\langle 001 \rangle$ and $\langle 110 \rangle$ directions, respectively. The Ar^+ ion is incident at $\theta = 0°$ and the particles are collected at $\theta = 45°$ (From Ref. 178.)

those ions with high kinetic energy were recorded. For this case, it was found for values of $\theta < 45°$ that Cl–Cl scattering in the overlayer itself is the dominating mechanism for Cl^- desorption. As seen from the plots, rotation of the Cl-covered Ag{110} crystal about 360° yields systematic signal variations. The magnitude of the anisotropy increases from near zero after only 0.1 L exposure, but increases to about 12% of the total signal after 1.0 L exposure. In each case, the position of the peaks occurs along the $\langle 001 \rangle$ ($\phi = 0°$) and along the $\langle 110 \rangle$ ($\phi = 90°$) azimuthal directions. Channeling and blocking concepts lead to a straightforward assignment of the adsorption site of the Cl surface atom. At the lowest coverage point, it is proposed that the Cl atom is quite large in size and therefore resides above the trough of the {110} surface as seen in Fig. 34. At this position, it would not experience significant channeling by the substrate atoms. At higher coverages, however, it is proposed that the adsorbed Cl atom becomes significantly smaller owing

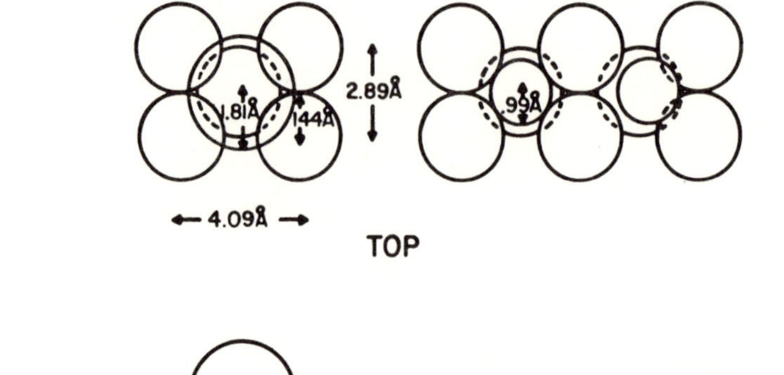

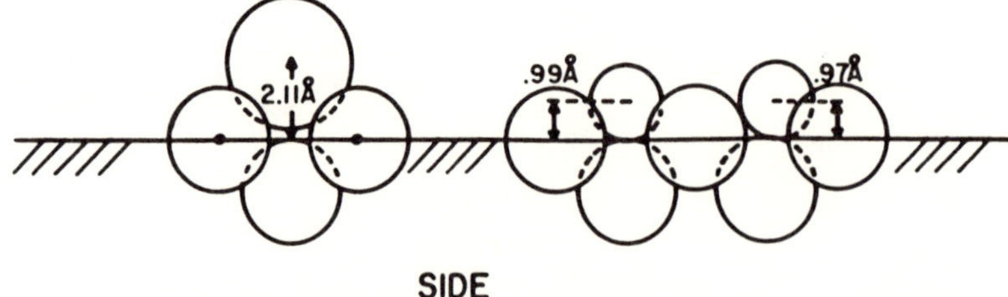

Figure 34. Possible bonding geometries of differently charged Cl. (a) Fully charged Cl^- adatoms and (b) atomic Cl. Note that the surface plane is defined by the nuclear centers of the topmost Ag atoms. In addition, note that there is a possible asymmetric bonding site in (b) which allows the Cl atom to approach to within 0.97 Å of the surface plane. (From Ref. 178.)

to the presence of dipole–dipole interactions between neighboring adsorbate atoms. As it shrinks, it literally falls into the valley of the {110} substrate and can experience channeling either along the valley direction ($\phi = 90°$ or 270°) or perpendicular to the valley ($\phi = 0°$ or 180°). In this latter case, the Cl^- ion is channeled by two Ag atoms comprising the valley wall. Thus, the adsorption site is defined as a four-fold bridge site with the Cl atom binding directly over a Ag atom in the second layer. Note that at the highest Cl coverage corresponding to an exposure of 2.5 L, the desorbing ion can more easily escape along the direction of the valley, yielding a two-fold symmetric azimuthal angle scan. As we shall see in the next section, these simple observations are entirely consistent with ion scattering studies of this system.

It will be quite interesting to examine the influence of the polar angle on the simple nature of these azimuthal scans. Very little research has yet been completed in the area, but some fascinating data for Cl on Ag{100} are available. This system is quite complicated and many different models have been proposed for the structure of the overlayer. The representation of the simple overlayer model (SOM)[179–181] and the mixed

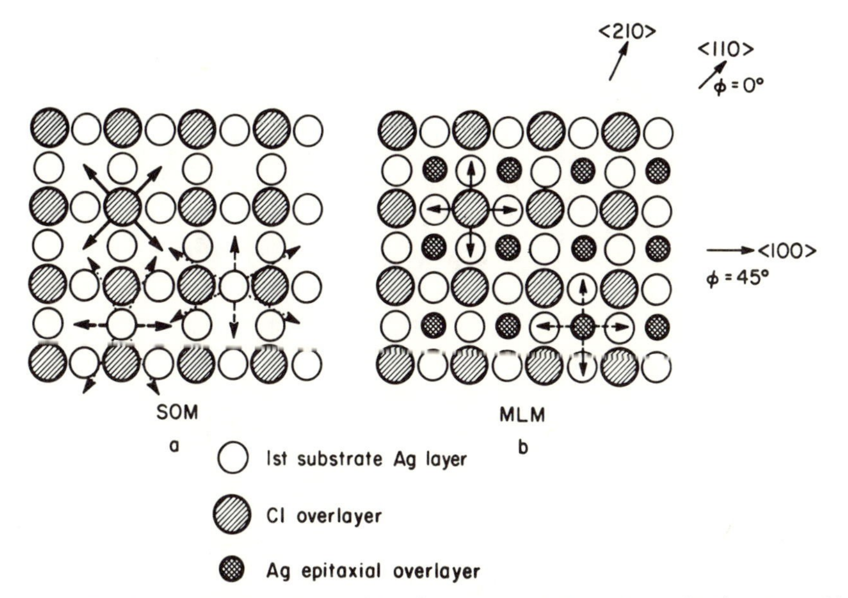

Figure 35. Schematic picture of two $c(2 \times 2)$ structures of Cl on the Ag{100} surface. (a) Simple overlayer model (SOM). (b) Mixed layer model (MLM).

layer model (MLM)[182] are shown in Fig. 35. The LEED data indicate that after 5.0 L exposure to Cl a $c(2 \times 2)$ structure is observed.[179] The angle-resolved SIMS experiments shown in Fig. 36 support the idea that the Cl adsorbates bond in the C_4 symmetry sites.[183] Minimum ion intensities are observed along ⟨110⟩ azimuths, which correspond to the direction of close-packed rows of atoms in the surface plane. Particles ejected along these directions are obstructed at small values of θ. For $\theta = 45°$ and a Cl exposure of 1.5 L, the maximum ion intensities arise along ⟨100⟩ directions. These peaks result from a strong channeling effect (dashed arrow of Fig. 35) exerted on the ejected Ag^+ ions by the nearest-neighbor Ag atoms and/or by the adsorbed Cl atoms. At large values of θ, however, the departing Ag^+ ion is scattered by the Cl adsorbate, resulting in the complex azimuthal pattern seen in Fig. 36 and explained by the diagram in Fig. 35a.

Channeling and blocking are not only observed when atomic adsorbates are found to a surface, but also are seen in the presence of molecular adsorbates. Polar angle distributions of benzene and pyridine ions ejected from ion-bombarded Ag{111} exhibit features that indicate that a vertical channeling process is occurring.[184,185] For a Ag surface covered with a monolayer of benzene or exposed to 0.15 liter of

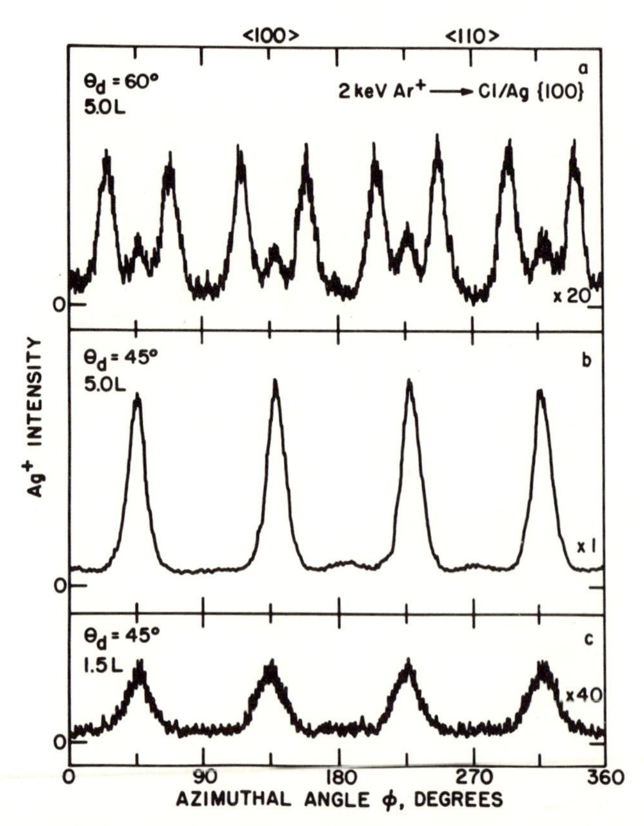

Figure 36. Azimuthal angular distributions of high-energy (20 ± 3 eV) Ag^+ secondary ions from a Ag{100} surface after exposure to (a and b) 5.0 L of chlorine, and to (c) 1.5 L of chlorine. The Ar^+ ion of 2 keV is incident normal to the surface and the particles are collected at (a) $\theta_d = 60°$ and (b and c) $\theta_d = 45°$, respectively. The angles are defined in Fig. 38. (From Ref. 183.)

pyridine, the molecules are believed to be lying parallel to the surface in a π-bonding configuration. The polar angle distributions of $(M - H)^+$ ions ($C_6H_5^+$ for benzene) and $(M + H)^+$ ions for pyridine are broad and exhibit a peak at $\theta = 20°$ as shown in Fig. 37. At high exposures of pyridine, however, the polar angle distribution is sharper and peaks at $\theta = 10°$. This result suggests that at higher coverages, the pyridine molecules are forced into a vertical σ-bonding configuration through the N atom lone pair. This bonding configuration allows more molecules to interact with the surface. This σ-bonded interaction is, of course, not possible with benzene and the polar angle distribution is invariant with

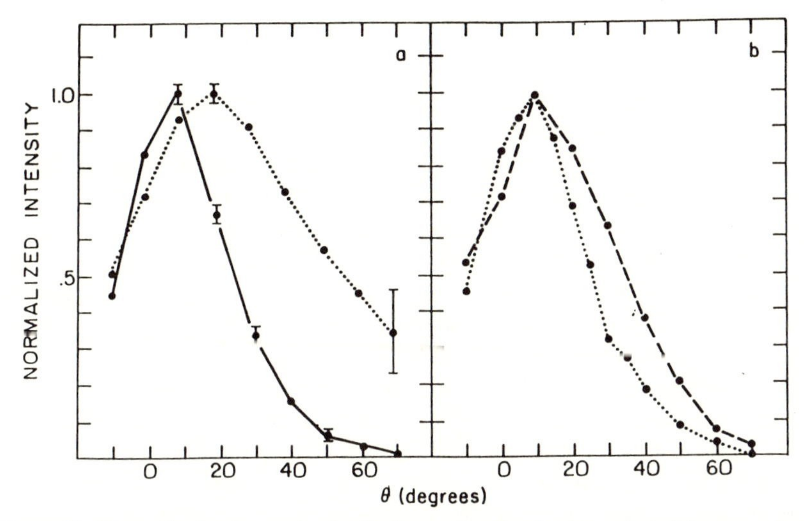

Figure 37. Normalized polar angle distributions of molecular ions ejected from overlayers of organic molecules adsorbed on Ag{111} at 153 K. The polar angle is defined with respect to the surface normal. (a) —, 4.5 L pyridine $(M + H)^+$; · · ·, 2.5 L benzene $(M - H)^+$. (b) 4.5 L pyridine $(M + H)^+$; – – –, 6–10 eV secondary ions; · · ·, 3–7 eV secondary ions. (From Ref. 185.)

coverage. The vertical array of pyridine molecules acts as a channel for directing the desorbing pyridine molecules in a direction close to the normal. It is interesting that a wide range of other surface spectroscopies including electron energy loss[186] and EXAFS methods[187] confirm that these bonding changes are, in fact, occurring in pyridine. In addition, classical dynamics calculations for this molecular desorption system reveal that such channeling mechanisms are quite commonly found.[183]

In summary then, angular distributions contain much detailed information about single-crystal surface structures which are, in fact, quite difficult to extract using other techniques. The symmetry of azimuthal scans can yield straightforward site determinations using desorbing substrate atoms or ions or using desorbing overlayer ions. As discussed in earlier sections, it has been possible in some cases to obtain quantitative agreement between calculated and measured angular distributions. In many other instances, however, the qualitative features of the angular distributions add a powerful approach to examining the bonding configuration of surface species. The next step, of course, is to see if these ion-bombardment methods could be used to obtain accurate bond-length measurements.

5.1.3. Shadow-Cone Enhanced Desorption

It has recently been discovered that the desorption yield of all particles is enhanced when the shadow cone created by a surface atom intersects a nearby atom.[188] The shapes of these shadow cones are quite well characterized since they have been utilized in Rutherford backscattering[189] and impact-collision ion scattering[190] experiments for many years. For ion-bombardment experiments, we take advantage of the fact that the flux of incident particles along the edge of the shadow cone is large. With knowledge of the angle of incidence of the primary beam, it is then feasible to determine surface bonding configurations using a strategy similar to that developed for ion scattering.

The basic experimental scheme and angle definitions are shown in Fig. 38. In principle the detector should be configured to collect all desorbing ions, regardless of their kinetic energy or angle. Most experiments to date, however, have been performed with a fixed angle β and with a detector set to collect only the high-kinetic-energy particles. With this configuration, the shadow-cone enhanced desorption concepts may be more carefully evaluated.

Representative results for Ag{110} are shown in Fig. 39.[191] There are two major features in these distributions which correspond to the

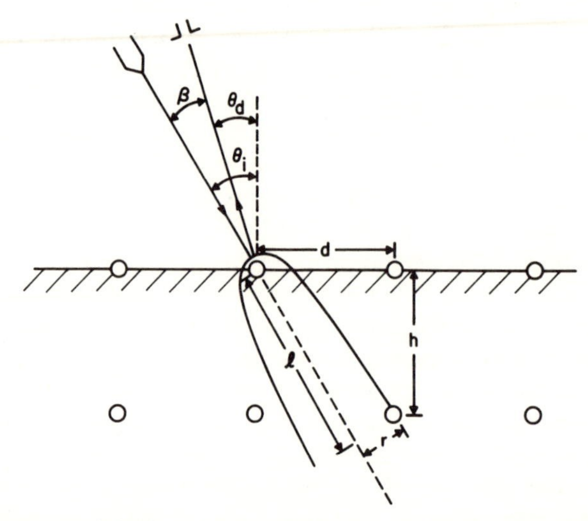

Figure 38. Parameter definitions for the shadow-cone-enhanced desorption experiment. The ion beam is incident at θ_i, the desorbed particles are detected at θ_d and $\beta = \theta_i - \theta_d$. The shadow cone is described by a radius r at a distance l behind the target atom. The d and h values describe the surface bond lengths.

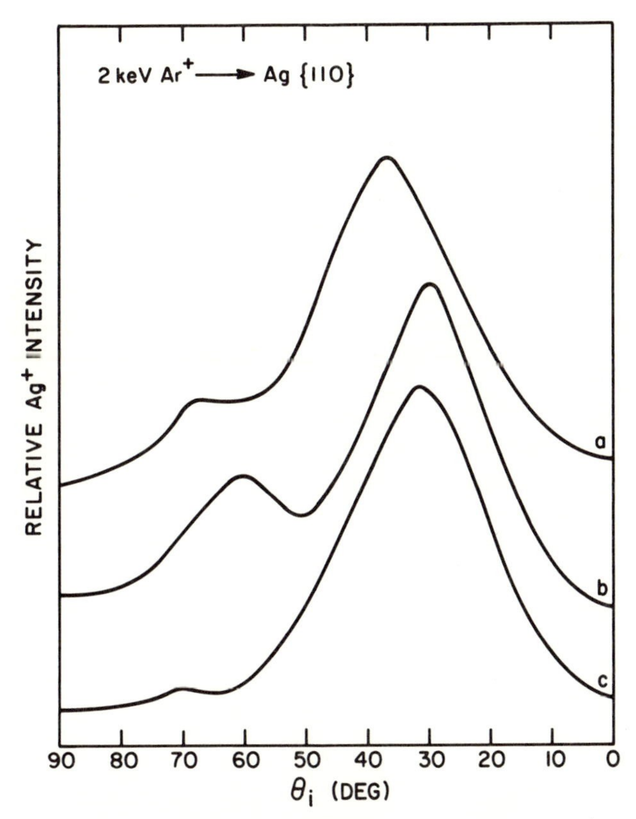

Figure 39. The shadow-cone enhanced SIMS spectra of the desorbed Ag^+-ion yield as a function of the angle of incidence. The particles are collected along the (a) ⟨100⟩, (b) ⟨110⟩, and (c) ⟨211⟩ azimuth during 2-keV Ar^+-ion bombardment at a current of ~5 nA. Only particles desorbed in plane with $\beta = 25°$ are collected. (From Ref. 191.)

intersection of the shadow-cone with a second-layer atom ($\theta \sim 35°$) and with a first-layer atom ($\theta \sim 70°$). The fact that these peak maxima actually correspond to the intersection of a shadow-cone edge with a substrate atom has been tested by using a full three-dimensional computer simulation like that discussed in Section 3. The position of this peak has been shown to be unaffected by the image interaction or other forces related to the secondary ionization process since the peak position is found to be independent of β for θ_d values between 0° and 45°. From the position of the peaks, it is possible to calculate **d** using a Molière potential and to compare this calculated value to the known interatomic spacing for Ag. The results compare to within ±0.5°, yielding an uncertainty of about ±0.06 Å in the bond length. Similar procedures for determining **h** suggest that the spacing between the first and second layer is relaxed by (7.8 ± 2.5)% and the spacing between the first and third layer is relaxed by (4.1 ± 2.1)% relative to the bulk spacing. These values agree within the same error limits with those bond lengths found by Rutherford backscattering measurements.[192]

Adsorbate atoms will also create shadow cones that can intersect nearby substrate atoms, opening the possibility of measuring the bond length of chemisorbed species. An excellent case to test this idea is the chemisorption of Cl on Ag{110} discussed in the previous section. The channeling and blocking experiments suggest that the binding site is invariant over a wide coverage range, allowing the Ag–Cl bond length to be examined under a variety of experimental conditions. The results are shown in Fig. 40.[188,193] Of particular note is that at extremely low Cl coverages, the observed bond length is extended over that observed in the high coverage limit by nearly 0.4 Å. This change in length, accompanied by a change in the shape of the Auger electron emission spectra, has been explained as being due to a shift from highly ionic bonding at

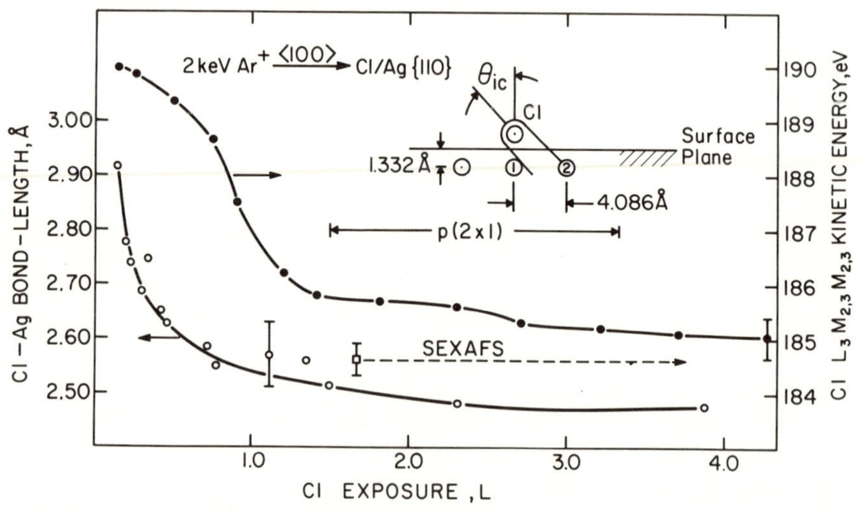

Figure 40. Ag–Cl bond-length change (○) and Auger Cl L_3 $M_{2,3}$ $M_{2,3}$ kinetic energy (•) as a function of Cl_2 exposure at 300 K. The reported value refers to the distance between the center of silver atom 1 and the Cl atom. The $p(2 \times 1)$ LEED pattern was observed in the exposure region as shown. The bond lengths from the SEXAFS experiments are associated with the LEED pattern; the same value was also obtained at a coverage beyond 4 L and associated with the $c(4 \times 2)$ LEED pattern. The shadow-cone induced desorption mechanism used to calculate the bond length is shown in the inset. The interplanar spacing refers to the relaxed clean-surface Ag{110} value. The shadow-cone shape was calculated using a Thomas–Fermi–Molière potential with a scaling factor of 0.86. (From Ref. 193.)

low coverage to more covalent bonding at high coverage. If the Ag–Cl interaction is highly ionic, the bond should have a significant dipole moment. Presumably dipole–dipole interactions between nearby Cl^- atoms force charge back into the Ag substrate. The absolute value of the bond length at high exposures agrees quite well with surface-EXAFS experiments obtained from the $p(2 \times 1)$ ordered overlayer as seen in Fig. 40.[194] The results are also consistent with the channeling and blocking experiments discussed previously, which suggest that the Cl adsorbate could more closely approach the surface at higher coverages.

These types of surface structure determinations may be extended to more complex systems provided that several different shadow-cone intersections can be found without being influenced by distortion effects. Materials such as GaAs and Si representing interesting test solids since their lattice spacings are generally large relative to the shadow-cone radius. One example has been reported using GaAs{110}.[195] In this case, the As surface atoms are forced up from their bulk-terminated lattice positions. Since lateral displacements are also possible, any given peak in the desorption yield would correspond to a large number of possible lattice positions. This degeneracy may be resolved by tracing out the locus of points associated with each desorption mechanism and then searching for intersections of the resulting lines using several different desorption mechanisms. For GaAs{110}, for example, nearly 40 different features could be discerned from the various polar angle distributions. The location of the surface Ga and As atoms relative to the second layer positions could then be uniquely determined as seen from Fig. 41. The positions are in quite good agreement with extensive LEED calculations performed over the last 15 years or so.

The fact that these simple angular distributions provide such microscopic information about surface structure is really remarkable from several points of view. Certainly, the primary beam is creating lots of damage to the surface, and yet it seems possible to obtain accurate surface bond lengths. The experimental configuration is exceedingly straightforward, requiring only a simple ion source, quadrupole mass filter and polar angle rotation capabilities for the sample holder. The potential sensitivity of this approach to low concentrations of overlayers is indeed unprecedented. The three examples presented in this review are very promising ones, but much research remains in this area. It will be interesting to examine how the distributions change with shadow-cone radius and whether it will be possible to enhance certain desorption mechanisms by changing particle mass and energy. In any case, it is clear that angular distributions tell us a great deal about the ion-impact event in particular and provide detailed surface structural information in

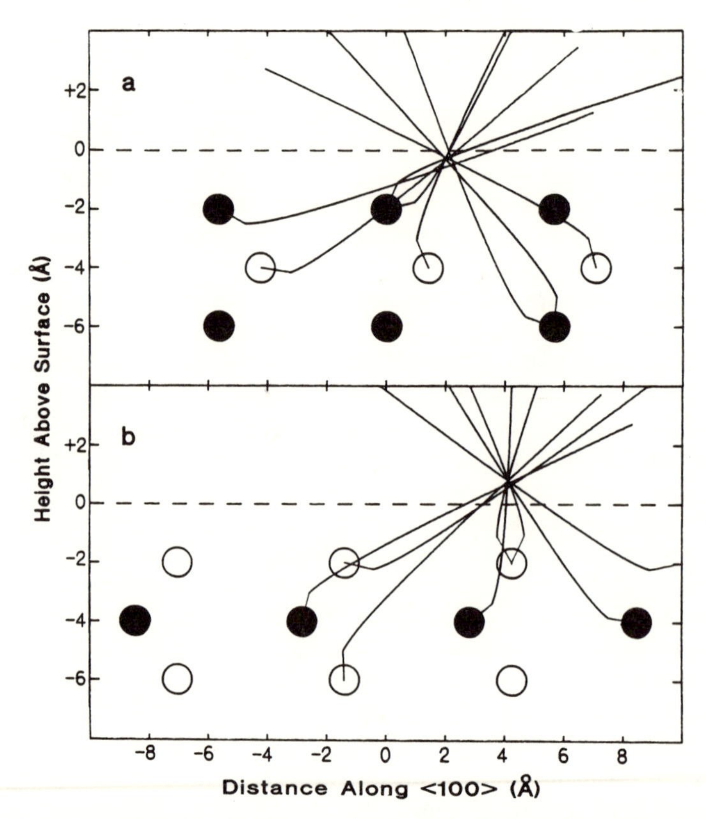

Figure 41. Cross section of GaAs{110} with Ga ○ and As ● atoms delineated. The top row of atoms has been removed. Panel (a) is the Ga terminated plane of atoms while Panel (b) shows the As terminated plane. Each curve represents the calculated set of surface atom positions that can result in an experimentally observed interaction. The region of the crossing of these curves represents the experimentally determined Ga(As) surface atom position. (From Ref. 195.)

general. We have concentrated so far on the desorption of atomic species. It is next of interest to see if these ion-beam techniques are useful in surface chemical applications by detection of desorbed molecular components of the top layer of the sample.

5.2. Molecular Composition Studies

There has been intense interest in measuring the mass of molecular cluster ions ejected from ion bombarded materials. These experiments can be important for the compositional analysis of solids and surfaces if these clusters can be shown to reflect in some fashion the original make-up of the sample. It is clear from the study of a vast array of species that it is possible to desorb molecules intact with high efficiency. This observation has led to the birth of the whole field of high-molecular-weight mass spectrometry. The use of this technique for protein sequencing is now widespread. It is ironic that most of the molecules that desorb intact would decompose during thermally induced desorption.

As we have discussed in Section 3.2, the theory of molecular desorption is rather complex, suggesting that a detailed interpretation of SIMS spectra may be difficult. The molecular dynamics simulations tell us, for example, that recombination of surface atoms to form new molecules is highly probable when the solid itself does not possess a specific molecular identity as for clean metals or metals covered with a chemisorbed oxide overlayer. When the intramolecular forces are particularly strong, however, the molecule can soak up energy from the collision cascade and desorb intact. In this case, in fact, the simulations suggest that the larger the molecule, the less fragmentation will be observed. Finally, it appears that desorbing molecules can recombine with other desorbing ions to create new cationized molecules. This scheme has been used to increase the number of molecular ions relative to the number of neutral species.

Now to the point of this discussion: It is clear that the desorbed molecules in some way reflect the surface composition. In many experimental situations, this result can provide powerful surface chemical information. In other situations, however, recombination in various forms may lead to faulty data interpretation. At this point, it is not possible to quantitatively predict which mechanism of cluster formation will dominate unless all of the bond energies for the system are well known. It is imperative, then, to devise additional experimental approaches to maximize the information content contained within the mass spectra of these desorbed molecules.

5.2.1. Intact Molecular Ejection

Perhaps the least complicated system from a chemical viewpoint is the chemisorption of hydrogen on solid surfaces. In general, however, hydrogen is quite difficult to detect experimentally and is known to bind to surfaces in many atomic or molecular states. For example, three distinct hydride phases (mono-, di-, and trihydride) have been found on the important Si(111)-(7 × 7) surface, and the elucidation of their properties is important in Si etching technology. Ion beam techniques are particularly powerful for this system. Several SIMS spectra recorded at different deuterium adsorption temperatures are shown in Fig. 42. Note

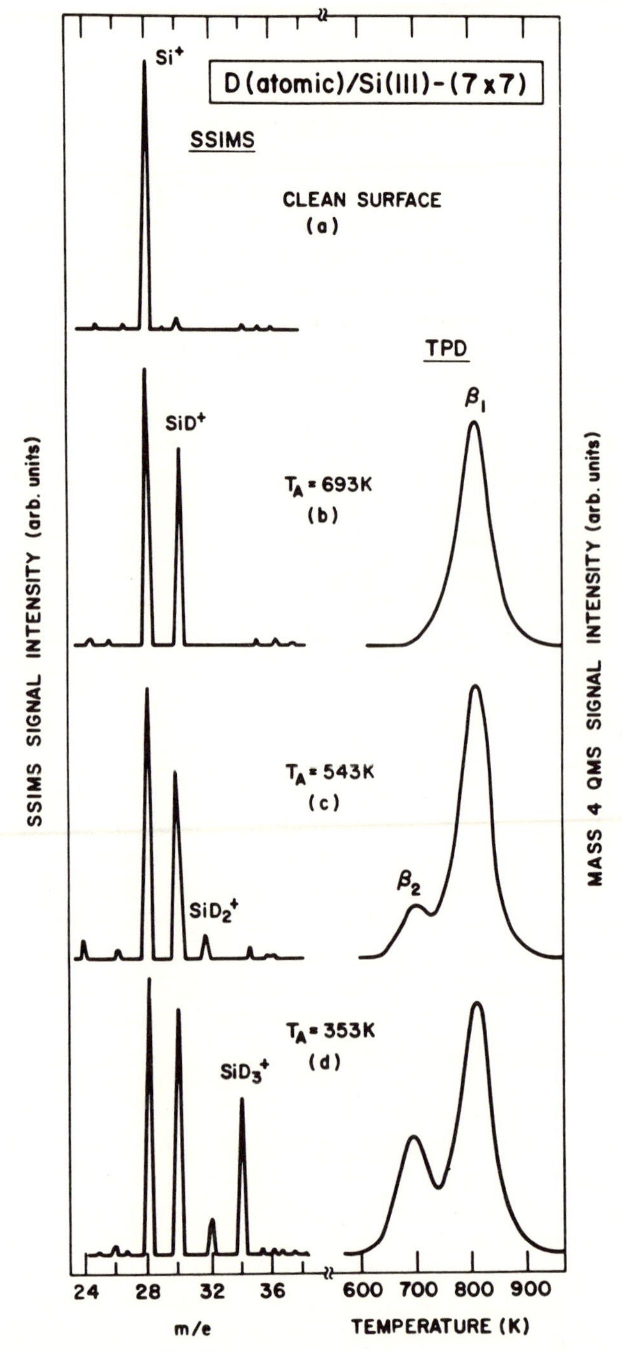

Figure 42. (a) SIMS (left side) spectrum for a clean Si(111) – (7 × 7) surface. TPD (right side) and SIMS (left side) spectra for a deuterium-saturated Si(111) – (7 × 7) surface. The adsorption temperatures are (b) 693 K, (c) 543 K, and (d) 353 K.

that at $T_A = 693\,K$ that the SiD^+ signal correlates well with the single β_1 thermal desorption peak.[196] At lower adsorption temperatures, SiD_2^+ and SiD_3^+ ions are observed which also correlate with thermal desorption features. From these data and from the known total D coverage, it has been possible to assign each of these cluster ions to the respective hydride phase and to quantify the coverage of each phase as a function of temperature. In this instance, then, apparently recombination and ion beam induced fragmentation of the cluster ion are not significant and the hydrides are being ejected from the surface intact.

If it is possible to desorb hydrides intact, it becomes of interest to examine a series of surface molecules consisting of just a few C—H units such as CH_x and $CH_2{=}CH_2$ species. A difficult case involves the commercially relevant methanation reaction via the Fischer–Tropsch process as

$$CO + 3H_2 \xrightarrow[\text{Ni}]{450\,\text{K}} CH_4 + H_2O \qquad (28)$$

In the production of CH_4, the C—O and H—H bonds must clearly be broken, but what sort of reaction intermediates are possible in the process? In a series of experiments, CO and H_2 were allowed to react over Ni{111} at peassures near 100 Torr until CH_4 could be detected by gas chromatography.[197] The crystal was then rapidly transferred into a SIMS system for analysis. Room temperature SIMS spectra are shown in Fig. 43. Note that each CH_x fragment ion is observed directly and attached to a substrate Ni atom. It was hypothesized that these reaction intermediates were indeed present on the surface since the intensity of each cluster ion decreased at a different rate as the substrate temperature was raised. If the surface concentrations of the CH_x intermediates are similar even to within an order of magnitude, a mechanism involving sequential hydrogenation of surface carbon seems most reasonable.

As the size of the adsorbate molecule increases, the SIMS spectra do get considerably more complicated. The spectrum of a condensed layer of CH_3OH on Pd{111} at 110 K is shown in Fig. 44.[198] This spectrum is characterized by a strong CH_3^+ signal, a CH_2OH^+ ion peak and a series of $(CH_3OH)_nH^+$ and $Pd(CH_3OH)_nH^+$ cluster ions. Protonation of the molecular ion is commonly observed with organic SIMS.

The presence of CH_3^+ (and the lack of CH_2^+, CH^+ or C^+ ions) is quite interesting in that it is the only CH_x fragment ion observed. Similar behavior has been reported for the adsorption of $CH_2{=}CH_2$ on Pt{111} by White and co-workers.[199] In this cae the $CH_2{=}CH_2$ converts at room temperature to ethylidyne ($\equiv$C—CH_3), where a carbon atom is bound directly to the metal surface and the CH_3 functionality is pointing

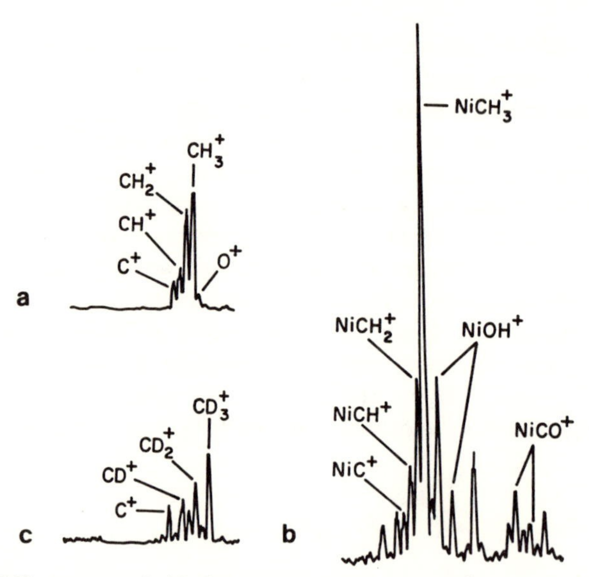

Figure 43. SIMS spectra of (a) low mass region, (b) Ni^+ region after $CO + H_2$ methanation, and (c) low mass region after $CO + D_2$ methanation (from Ref. 197).

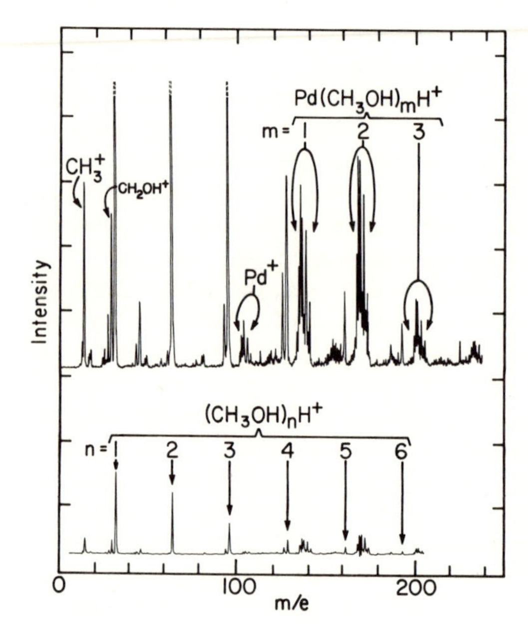

Figure 44. Positive ion SIMS spectra for Pd{111} exposed to 6 L CH_3OH at 110 K. The spectra were recorded using a 3-keV Ar^+ primary ion beam with a current density of $0.4\ nA\ cm^{-2}$. The top spectrum is shown with a 10× magnification. (From Ref. 198.)

upward. These workers found that CH_2^+ was the most intense peak at 200 K, but that the CH_3^+ ion dominated at 300 K, consistent with $\equiv C-CH_3$ formation from $CH_2=CH_2$.

If the CH_3^+ SIMS ion is a useful tag in hydrocarbon decomposition, it would be of interest to follow the decomposition of molecularly adsorbed CH_3OH as the temperature is raised. For this experiment, the SIMS ion signal as a function of temperature, combined with XPS measurements, yield a definitive picture of the decomposition pathway. When the temperature reaches 120 K, all cluster ion intensities begin to decrease as CH_3OH desorbs from the surface.[200] By 175 K, however, the CH_3^+ signal increases. The XPS experiments suggest that at this temperature, the CH_3OH decomposes to CH_3O_{ads} plus a hydrocarbon fragment of unknown composition. Since no CH_3^+ ion signal is seen from pure layers of CH_3O_{ads}, it was concluded that the rise in the CH_3^+ ion signal is due to the formation of a surface methyl intermediate, CH_{3ads}. The initial decomposition pathway, then, is

$$2CH_3OH \rightarrow CH_{3ads} + CH_3O_{ads} + H_2O_{ads} \tag{29}$$

Similar results have been obtained on Pt{111} surfaces.[201] These results are particularly important since they illustrate how multicomponent surfaces may be followed with SIMS and that methyl fragments can be desorbed intact. This species may be quite important catalytically, and not many techniques are sensitive to its presence.

Similar important diagnostic surface chemical information has been found for ketene (CH_2CO) adsorbed onto Ru{111}.[202,203] In this case, at 105 K and at low exposures, the adsorption is largely dissociative. The dominant SIMS ion is found at $m/e = 14$ (CH_2^+) and the ions associated with molecular ketene ($m/e = 42$ and 43) are the smallest. At higher exposures where molecular adsorption dominates, the ion yields of $m/e = 15$ and 43 increase abruptly while those of $m/e = 14$ and 42 increase only slowly. With its sensitivity to hydrogen, these types of SIMS measurements promise to yield important new mechanistic data regarding hydrocarbon surface chemistry, particularly C_1 chemistry. This research area appears to be a rich one for the near term, since theory and experiment are in an advanced state of development.

As the adsorbed molecules get heavier, the surface chemistry becomes more complex, but still directly amenable to SIMS studies. The simplest system is CO adsorption on metal surfaces. This molecule is an especially important model compound since computer simulations have been performed for CO desorption, CO generally enhances positive ion yields by several orders of magnitude, and it has a strong intramolecular bond making recombination unlikely. The computer simulations suggest that only a small fraction of the molecules are dissociated by the incident beam and that nearly all of the CO is ejected intact. This aspect of the problem has been quite conclusively demonstrated by Lauderback and Delgass.[204] Using isotopically labeled equal amounts of ^{13}CO and $C^{18}O$, no isotope mixing was found for the observed $Ru{-}CO^+$ cluster ions.

With this information, it is possible to employ SIMS cluster ions to determine whether CO adsorbs molecularly or dissociatively. Dissociative adsorption to carbon and oxygen atoms is, of course, an important preliminary step for promotion of the reactivity of CO on surfaces to build more complicated molecules. An interesting example involves the interaction of CO with the same Rh{111} and Rh{331} crystal planes discussed previously.[205] Does the presence of the atomic step enhance the dissociation probability of the CO molecule? This question has been debated for a number of years in the catalysis literature. The SIMS experiments, along with concomitant XPS measurements, clearly show that under low-pressure conditions, CO remains intact on both surfaces at room temperature. The $RhCO^+$ and Rh_2CO^+ ions are the only significant molecular cluster ions. They presumably form via a recombination mechanism involving Rh and CO species.

The NO molecule has a bond strength that is only ~75% of that of CO. Since it is still a tightly bound diatomic molecule, the classical dynamics models would certainly predict that NO should eject intact if it existed in its molecular state on the surface. The SIMS spectra of NO on Rh{111} at 300 K are shown in Fig. 45.[205] Note that although the $RhNO^+$ ion and Rh_2NO^+ ion are quite intense species, there are significant signals arising from Rh_2N^+, Rh_2O^+, and, to a lesser degree, RhN^+. For this case, the XPS data revealed that at least 90% of the NO is adsorbed molecularly. These fragments must therefore be at least partially created during the ion-induced desorption event. The enhanced fragmentation of NO relative to CO is in line with its slightly weaker bond strength. During thermal desorption, however, the Rh_2NO^+ signal drops in precisely the same fashion as the N 1*s* XPS peak associated with molecular NO, while the RhN^+ ion signal rises to a maximum value at 400 K before falling, mimicking the N 1*s* XPS peak intensity of dissociated NO. Thus, it appears that even though some dissociation during desorption is observed, the molecular specificity of the ejected cluster ions to the chemical state of the surface molecule is preserved. Similar experiments for NO adsorbed onto Rh{331} clearly show that approximately 50% of the molecules dissociate upon adsorption, while the remainder dissociate at higher temperatures before desorption.[206] Thus, the presence of the atomic step on Rh{331} exerts a sufficiently strong effect to influence bond cleavage for NO, but not for CO.

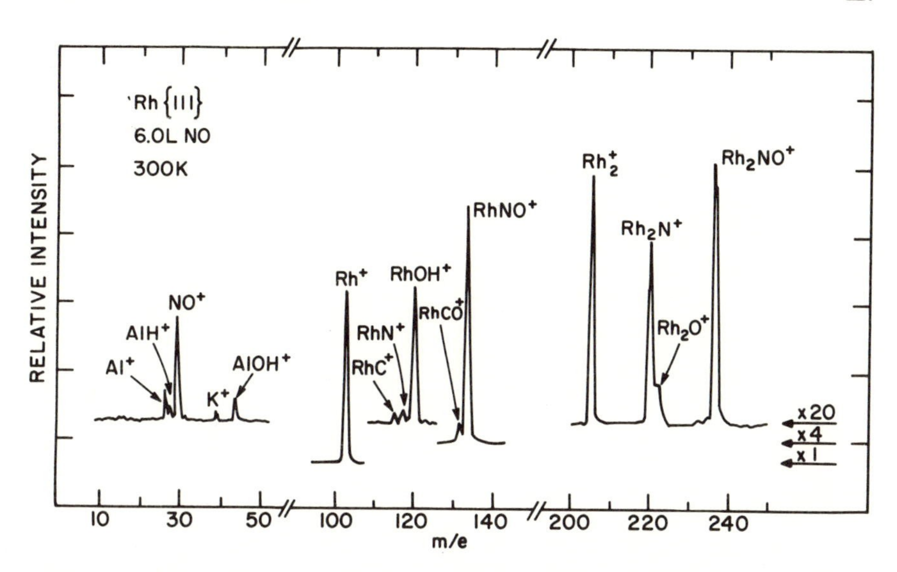

Figure 45. SIMS result of the adsorption of 6.0 L NO on the clean Rh{111} surface at 300 K. Data were taken using a 3-keV Ar^+ ion beam of ≈50 pA focused into a 2 × 2-mm spot. The incident angle $\theta_i = 25°$ and the polar collection angle $\theta_d = 20°$ are measured from the surface normal. Molecular bonded NO_{ads} is monitored by the NO^+ (*m/e* 30) $RhNO^+$ (*m/e* 133), and Rh_2NO^+ (*m/e* 236) secondary ions. Dissociated products of NO_{ads} are detected by the Rh_2N^+ (*m/e* 220) and Rh_2O^+ (*m/e* 222) ion yields. Equal amounts of N_{ads} and O_{ads} determined by XPS yield a Rh_2N^+/Rh_2O^+ ion ratio of ≈4. Carbon monoxide and H_2O surface contaminants adsorbed from the residual gas are seen as the Rh_2CO^+ and $RhOH^+$ ions. (From Ref. 205.)

From this discussion, it is apparent that the combination of XPS and SIMS is a particularly powerful one. There are always questions regarding quantitation with SIMS, while XPS is especially well suited to evaluating surface concentrations. The major difficulty with XPS, however, is that the BE's are often not highly specific to molecular structure, whereas the cluster ions in SIMS yield detailed compositional information. A good example of this combination is the study of the reduction of NO on a carbon pretreated Rh{331} surface.[207] The XPS measurements for this system yielded significant changes in the shape of the N 1*s* peak assigned to dissociated NO when the surface was heated from room temperature to 500 K. However, the nature of this new N-containing species could not be ascertained from the data. The SIMS spectra, shown in Fig. 46, clearly reveal that the dissociated N atoms are reacting with adsorbed C atoms to produce a surface CN^- species. The ability to characterize the surface chemical behavior of these species in the

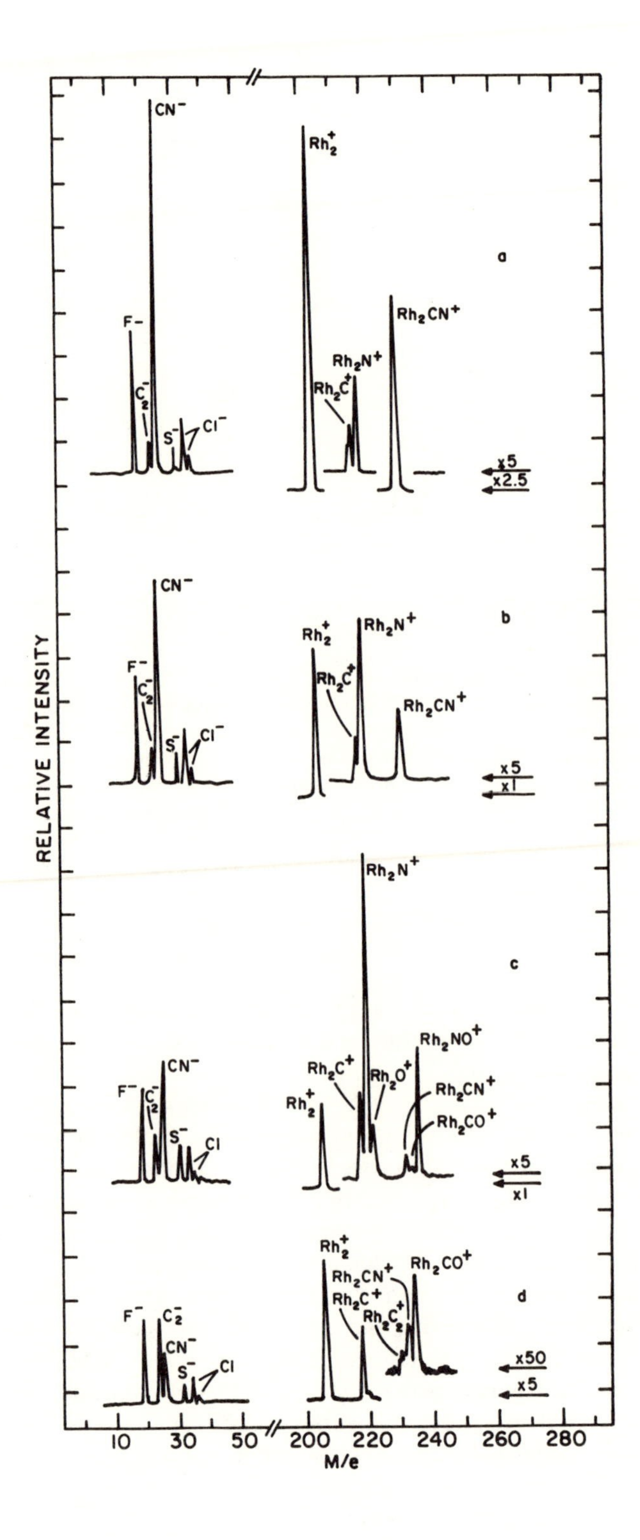

RELATIVE INTENSITY
M/e
a
b
c
d
CN⁻
F⁻
C₂⁻
S⁻
Cl⁻
Cl
Rh₂⁺
Rh₂C⁺
Rh₂N⁺
Rh₂CN⁺
Rh₂O⁺
Rh₂NO⁺
Rh₂CO⁺
Rh₂C₂⁺
x5
x2.5
x1
x50
10
30
50
200
220
240
260
280

presence of other carbonaceous deposits is, of course, important in many catalytic reactions.

Thus, there are many convincing examples where intact ejection of surface molecular ions can be observed from metal and semiconductor surfaces. There are many hundreds more beyond the scope of this review where even large molecules are seen to desorb intact from a variety of complex matrices ranging from thick films to glycerol. The examples from surface science experiments, however, provide not only useful models for understanding the ion–solid interaction but also add a unique tool to the arsenal of surface analysis techniques which are specific to rather complicated surface molecules.

5.2.2. Molecular Recombination During Ejection

There have been fewer systematic efforts to quantitatively characterize the recombination process predicted by the classical dynamics simulations, perhaps due to the complexities introduced by the presence of the ions themselves. As an example of this problem, consider the case of chemisorbed oxygen on single-crystal transition-metal surfaces. The computer simulations show that O_2 formation is possible by $O + O$ recombination. As for the case of Ni_2 discussed in Section 3.2, the O atoms that comprise the oxygen molecule do not need to be contiguous. If one of the oxygen atoms is ionized as O^+ or O^-, however, then the observed molecule would be O_2^+ or O_2^-. The probability of recombination would then be influenced by the potential surface of the charged molecule. Using these surfaces, the calculated number of dimers differs by less than a factor of 2 for the three cases O_2^+, O_2, and O_2^-. The most interesting aspect of these results, however, is that the thermodynamically most stable species, O_2^+, produces the lowest number of dimers. The reason for this fact is that the range of the potential is the smallest for this species. Apparently, the number of possible two-body interactions that a particular atom might experience is more important than the ultimate stability of a dimer in its equilibrium configuration. If it is possible to generalize this idea further, we would expect oppositely charged species such as Na^+ and Cl^- to have a high probability of forming an NaCl

Figure 46. Static SIMS spectra of the carbon covered Rh{331} surface before and after exposure to NO at 300 K. Heating to 500 K (a) and 400 K (b) produces cyanide as evidenced by the increase of the CN^- and Rh_2CN^+ ion yield. (From Ref. 207.) Exposing the carbon-covered surface to 10 L NO at 300 K results in the spectra shown in (c). Exposure to 10 L C_2H_4 at 300 K and heating to 770 K produces the spectra shown in (d). The carbon residue is evident by the C_2^-, Rh_2C^+, and $Rh_2C_2^+$ ions.

molecule owing to the infinite range associated with purely electrostatic interactions.

A test of the formation of cluster ions by an atom–ion collision mechanism was attempted using the Ni{001} surface during exposure to oxygen.[208] In this case, the ions Ni^+, $NiO^\pm$, $O^\pm$, and $O_2^\pm$ could be monitored as a function of the oxygen dose. The results as shown in Fig. 47 are striking, as the shapes of the ion yield versus coverage curves are quite different depending upon the ion involved. Note, however, that the

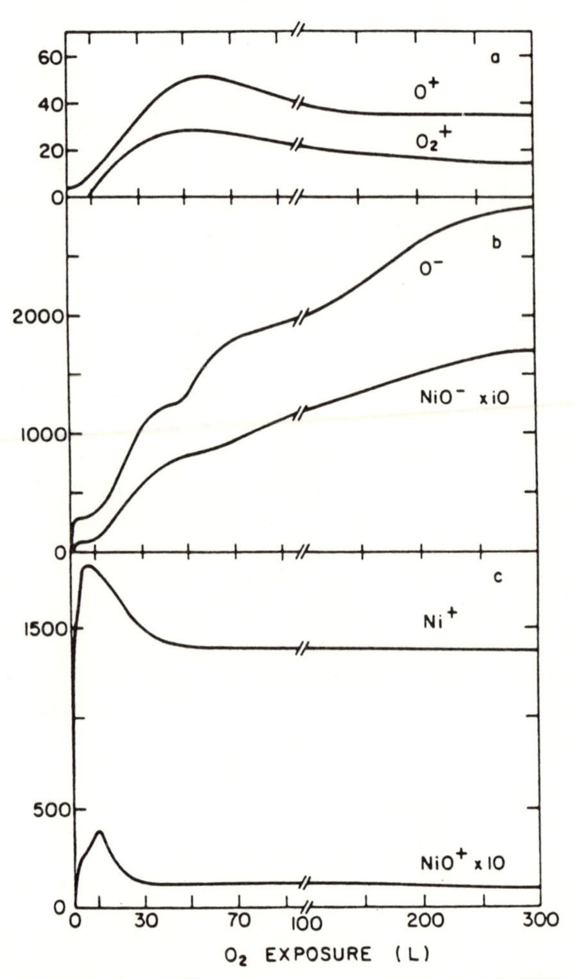

Figure 47. Experimental ion yields versus oxygen exposure for Ni{001} bombarded by 2-keV Ar^+ ions. The ordinate gives the number of ions/s detected. (From Ref. 208.)

NiO^+ yield has a shape similar to the Ni^+ yield whereas the NiO^- and O^- and the O_2^+ and O^+ curves also seem to track each other. If one invokes a cluster formation mechanism as discussed above, then the cluster ions could form as a result of collisions between neutral and ionized atoms. For example, the NiO^+ ion should form by interactions between Ni^+ ions and O atoms. The reaction of Ni atoms and O^+ ions would not be important owing to the low levels of O^+ ions. Furthermore, if the number of ions above the surface is much less than the number of neutrals (which is the case for most materials except perhaps alkali halides), the intensity of the resulting cluster ion will be controlled by the amount of the corresponding atomic ion. Thus for Ni, the clusters would form by the following reaction:

$$Ni^+ + O \xrightarrow[\text{surface}]{} NiO^+ \tag{30}$$

$$Ni + O^- \xrightarrow[\text{surface}]{} NiO^- \tag{31}$$

and

$$O^+ + O \xrightarrow[\text{surface}]{} O_2^+ \tag{32}$$

The mechanism of cluster formation for ions would therefore be very similar to that of the neutrals.

There are other complications that have yet to be checked on single crystals and will only be briefly mentioned here. When Ti is oxidized, a variety of $Ti_xO_y^{\pm}$ cluster ions are observed by SIMS.[209] When the work function of the surface is altered by Cs deposition, the positive cluster ion intensities change in a manner consistent with recombination. The negative ions deviate considerably from the predicted behavior. Yu has proposed, however, that since the electron affinity of negatively charged ions is usually low, unless the cluster constituents are in ground vibrational states electron detachment is very favorable.[210] To produce ground state molecules he feels it is necessary to invoke the condition that they form from nearest neighbors on the surface. Thus for the Ti–oxygen system, Yu proposes that the positive ions form by recombination while the negative ions are produced directly by lattice fragmentation.

To make matters even more complicated, there can be molecular ions formed by electron ejection from neutral molecules. Snowden and co-workers[211] have suggested, for nitrogen implanted into silicon targets, that during bombardment

$$N + N \xrightarrow[\text{surface}]{} N_2 \rightarrow N_2^+ + e^-$$

This type of mechanism may indeed be quite common since unbound N_2 molecules can be stabilized as N_2^+ by electron ejection. Finally, it has been proposed that large cluster ions may be created through "solvation" of ejecting atomic ions and through a sort of Joule–Thompson condensation mechanism. All of the above ideas merit consideration, although there is yet only scant experimental evidence to back them up.

5.2.3. Prospects for Detection of Desorbed Neutral Molecules

There is clearly an intimate relationship between ionization and cluster formation, although the details are elusive. Because of the success of the classical dynamics models it would be valuable to be able to directly monitor the neutral molecules in a manner analogous to that with static SIMS. There have been many attempts at performing such measurements, but most postionization detection schemes are simply not sufficiently sensitive to perform static-mode measurements.

The one exception to this situation involves the use of either resonant or nonresonant laser ionization of the desorbed molecules. So far the MPRI approach has only been applied to desorbed atoms. Recently, however, Becker[(212)] has studied the adsorption and thermal decomposition of CH_3OH on Ni{110} using 193-nm laser light for nonresonant ionization. The initial data from these experiments are quite encouraging. Using ion doses of $<5 \times 10^{12}/cm^2$ they found several molecular signals in their spectra that were characteristic of surface chemical species. After adsorption of CH_3OH at 120 K, for example, a peak at m/e 32 could be monitored that was clearly associated with the molecular adsorbate. As seen in Fig. 48, as the temperature reaches about 240 K, all of the m/e 32 has disappeared. The slower fall of m/e 13 (—CH) is believed to be associated with the formation of CH_3O_{ads} before this intermediate decomposes to CO. The CO is seen at m/e 28. Although ion-beam-induced fragmentation as well as laser-induced fragmentation are probably both influencing the mass spectra, these experiments yield hopeful indications that static measurements on desorbed neutral molecules will soon be possible.

6. Conclusions and Prospects

In this chapter, we have attempted to present an atomic and molecular view of the interaction of keV particles with solids. The

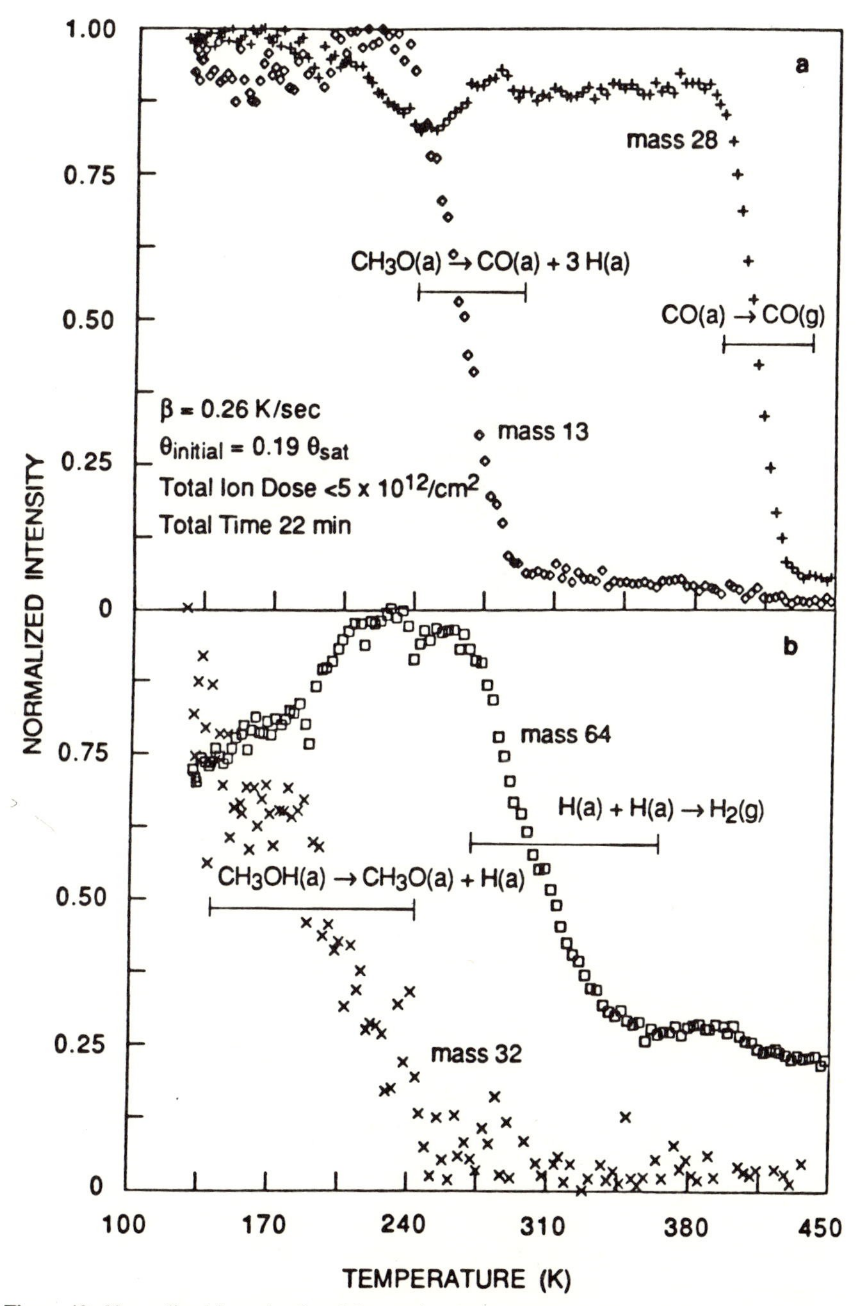

Figure 48. Normalized ion stimulated desorption intensities for masses 13 and 28 (a) and 32 and 64 (b) during a linear temperature ramp of 0.26 K/s following methanol adsorption, $\theta_{\text{initial}} = 0.19\ \theta_{\text{sat}}$, on Ni(110) at 120 K. Each point consists of an accumulation of 400 spectra obtained at a repetition rate of 40 Hz. The temperature resolution is therefore ~3 K. (From Ref. 212.)

process is clearly a complicated one. In the early days, we were forced to rely on analytical models, with so many approximations that the equations had very little predictive value. At first, the complexity of the atomic motion as seen even on primitive computers using molecular dynamics concepts was discouraging. With the dramatic improvement in computational speed, however, these early approaches have been extended to consume the resources of the most powerful supercomputers. With many-body potentials such as the EAM, accurate computer simulations of this complex process are at hand. Ironically, the massive computer print-outs have led to reasonably simple concepts regarding the important momentum dissipation mechanisms. The advances in the theory of electronic excitation presented here provide an impressive solution to understanding the mixing of the classical motion with other inelastic processes. It is satisfying, indeed, to see such dramatic progress in a field once considered beyond help.

Experimental advances have been equally impressive. The last 15 years or so have seen the ushering in of a new era of surface science. Since the ion–solid interaction event is largely a surface process at keV energies, the advance has had a direct beneficial effect on many fundamental studies. We can now trace the motion of atoms in the solid from their initial positions to their final resting locations. Measurement of the trajectories of state-selected desorbing atoms or molecules, if not yet a *fait accompli,* is at least on the drawing board. The atomic resolution images of the lattice damage provide important new experimental data which must be quantitatively understood.

If there have been great strides in recent years, there is an even more exciting future for this field. Many of the emerging new applications are, of course, outlined extensively in later chapters. The increased fundamental understanding of ion–solid interactions, however, promises new advances in reactive ion etching where the energy in chemical bonds act together with the momentum from the primary ion. The development of TOF detectors and angstrom size ion-beam sources is really in its infancy. Both of these tools should continue to inspire the construction of more sensitive and higher-resolution microscopes. All of these developments may spur another round of technological advances using, for example, ion implantation. We have been working in this general area since 1976. It has been an exciting and challenging 14 years. It is difficult to stray to other research fields.

Acknowledgments

Our venture into the detailed understanding of the ion–solid interactions began in 1977 with our initial collaboration with the late Don

E. Harrison, Jr. It is the interaction with Don that led to the further development of the classical dynamics model, and it is the belief in this model that led to the design of many of the experiments described in this chapter. To him we owe our deepest thanks. Throughout the years there have been numerous students and postdoctoral associates that have contributed to the ideas and to the execution of many of the ideas. These collaborators include Che-Chen Chang, Dae-won Moon, Lisa DeLouise, Mark Kaminsky, Robert Levis, Rik Blumenthal, John Olson, Don Brenner, Jung-hui Lin, Curt Reimann, Paul Kobrin, Alan Schick, David Deaven, and Brian Craig. In addition, there are numerous scientists from other institutions that have stimulated and refined our ideas about the particle bombardment process. These investigators include Ming Yu, Peter Williams, Drew Evans, Tom Tombrello, Mark Shapiro, Roger Smith, Roger Kelly, and Stan Williams. Our secretaries, Sabrina Glasgow and Chris Moyer, have graciously spent innumerable hours helping us with this and other projects. Finally we gratefully acknowledge the financial support of the Office of Naval Research, the National Science Foundation, the Air Force Office of Scientific Research, the IBM Program for the Support of the Materials and Processing Sciences, the Shell Development Corporation, the Camille and Henry Dreyfus Foundation, the Sloan Foundation, and the Research Corporation. Three people at these agencies, Henry Blount (NSF), Larry Cooper (ONR), and David Nelson (ONR), have been particularly supportive. Penn State University has supplied a generous grant of computer time for these studies.

References

1. J. F. Zeigler, ed., *Ion Implantation, Science and Technology,* 2nd Ed., Academic Press, Boston (1988).
2. H. F. Winters and J. W. Coburn, *J. Vac. Sci. Technol.* **B3,** 1376 (1985).
3. D. E. Hunton, *Sci. Am.* **261,** 92 (1989).
4. E. M. Sieveka and R. E. Johnson, *Icarus* **51,** 528 (1982).
5. J. J. Thompson, *Phil. Mag.* **20,** 252 (1910).
6. Y. Homma, S. Kurosawa, Y. Yoshioka, M. Shibata, K. Nomura, and Y. Nakamura, *Anal. Chem.* **57,** 2928 (1985).
7. J. M. Chabala, R. Levi-Setti, and Y. L. Wang, *Appl. Surf. Sci.* **32,** 10 (1988).
8. M. J. Bernius and G. H. Morrison, *Rev. Sci. Instrum.* **58,** 1789 (1987).
9. A. Benninghoven, *Phys. Status Solidi* **34,** k169 (1969).
10. M. Barber, R. S. Bordoli, R. D. Sedgewick, and A. N. Tyler, *J. C. S. Chem. Commun.* 325 (1981).
11. I. V. Bletsos, D. M. Hercules, D. Greifendorf and A. Benninghoven, *Anal. Chem.* **57,** 2384 (1985).

12. P. Steffens, E. Niehuis, T. Friese, D. Greifendorf, and A. Benninghoven, *J. Vac. Sci. Technol.* **A3,** 1322 (1985).
13. D. E. Harrison, Jr., P. W. Kelly, B. J. Garrison, and N. Winograd, *Surf. Sci.* **76,** 311 (1978).
14. W. R. Grove, *Trans. R. Soc. (London)* **142,** 87 (1852).
15. H. Oechsner and W. Gerhard, *Z. Phy.* **B22,** 41 (1975).
16. D. Lipinsky, R. Jede, O. Ganshow, and A. Benninghoven, *J. Vac. Sci. Technol.* **A3,** 2007 (1985).
17. N. Winograd, J. P. Baxter, and F. M. Kimock, *Chem. Phys. Lett.* **88,** 581 (1982).
18. *Our assumption is that fast atom bombardment or FAB is generically equivalent to ion bombardment. The consequences of the bombardment are most critically dependent on the kinetic energy of the primary particle. Note that a neutral atom differs from its positive ion by the mass of the electron.*
19. G. Carter and J. S. Colligan, *Ion Bombardment of Solids,* American Elsevier, New York (1968).
20. R. E. Honig, *J. Appl. Phys.* **29,** 549 (1958).
21. M. W. Thompson, *Phil. Mag.* **18,** 377 (1968).
22. G. K. Wehner, *Phys. Rev.* **102,** 690 (1956).
23. G. Carter, M. J. Nobes, I. V. Katardjiev, J. L. Whitton, and G. Kiriakidis, *Nucl. Instrum. Methods* **B18,** 529 (1987).
24. A. Benninghoven, D. Jaspers, and W. Sichterman, *Appl. Phys.* **11,** 35 (1976).
25. R. A. Gibbs and N. Winograd, *Rev. Sci. Instrum.* **52,** 1148 (1981).
26. D. Hammer, E. Benes, P. Blum, and W. Husinsky, *Rev. Sci. Instrum.* **47,** 1178 (1976).
27. R. B. Wright, M. J. Pellin, D. M. Gruen, and C. E. Young, *Nucl. Instrum. Methods* **170,** 295 (1980).
28. C. H. Becker and K. T. Gillen, *Appl. Phys. Lett.* **45,** 1063 (1984).
29. P. H. Kobrin, G. A. Schick, J. P. Baxter, and N. Winograd, *Rev. Sci. Instrum.* **51,** 1354 (1986).
30. D. E. Harrison, Jr., *Crit. Rev. Solid State Material Sci* **14,** Suppl. 1, 1–78 (1988).
31. P. Sigmund, *Phys. Rev.* **184,** 383 (1969).
32. P. Sigmund, *Phys. Rev.* **187,** 768 (1969) (Erratum).
33. P. Sigmund, *Rev. Roum. Phys.* **17,** 1079 (1972).
34. P. D. Townsend, *Rad. Effects* **37,** 235 (1978).
35. P. Sigmund, in *Inelastic Ion–Surface Collisions* (N. H. Tolk, J. C. Tully, W. Heiland, and C. W. White, eds.), Academic, New York, p. 121 (1977).
36. I. S. T. Tsong and D. J. Barber, J. Mater. Sci. **8,** 123 (1973).
37. J. Lindhard, V. Nielsen, and M. Scharff, *Mat. Fys. Medd., Dan Vid. Selsk.* **36**(10), (1968).
38. K. S. Kim, W. E. Baitinger, and N. Winograd, *Surf. Sci.* **55,** 285 (1976).
39. D. Almen and G. Bruce, *Nucl. Instrum Methods* **11,** 257 (1961).
40. M. F. Dumke, T. A. Tombrello, R. A. Weller, R. M. Housley, and E. H. Cirlin, *Surf. Sci.* **124,** 403 (1983).
41. H. H. Andersen, B. Stenum, T. Sorensen, and H. J. Whitlow, *Nucl. Instrum. Methods* **B6,** 459 (1985).
42. S. Kundu, D. Ghose, D. Basu, and S. B. Karmohapatro, *Nucl. Instrum. Methods* **B12,** 352 (1985).
43. J. P. Baxter, G. A. Schick, J. Singh, P. H. Kobrin, and N. Winograd, *J. Vac. Sci. Technol.* **4,** 1218 (1986).
44. G. A. Schick, J. P. Baxter, J. Singh, P. H. Kobrin, and N. Winograd, in: *Secondary*

Ion Mass Spectrometry—SIMS V, Springer Series in Chemical Physics Vol. **44,** p. 90, Springer-Verlag, New York (1986).
45. N. Winograd, P. H. Kobrin, G. A. Schick, J. Singh, J. P. Baxter, and B. J. Garrison, *Surf. Sci. Lett.* **176,** L817 (1986).
46. J. P. Baxter, J.. Singh, G. A. Schick, P. H. Kobrin, and N. Winograd, *Nucl. Instrum. Methods* **B17,** 300 (1986).
47. B. J. Garrison, *Nucl. Instrum. Methods* **B17,** 305 (1986).
48. B. J. Garrison, *Nucl. Instrum. Methods B* **40/41,** 313 (1989).
49. D. Y. Lo, M. H. Shapiro, T. A. Tombrello, B. J. Garrison, and N. Winograd, *Proc. Matl. Res. Soc. Mtg.* **74,** 449 (1987).
50. W. Husinsky, *J. Vac. Sci. Technol.* **B3,** 1546 (1985).
51. B. J. Garrison, N. Winograd, D. Lo, T. A. Tombrello. M. H. Shapiro, and D. E. Harrison, Jr., *Surf. Sci. Lett.* **180,** L129 (1987).
52. R. Kelly, *Nucl. Instrum. Methods* **B18,** 388 (1987).
53. G. Falcone, R. Kelly, and A. Oliva, *Nucl. Instrum. Methods* **B18,** 399 (1987).
54. D. P. Jackson, *Radiat. Effects* **18,** 185 (1973).
55. D. P. Jackson, *Can. J. Phys.* **53,** 1513 (1975).
56. M. H. Shapiro, *Nucl. Instrum. Methods B,* **42,** 290 (1989).
57. G. Falcone, *Surf. Sci.* **187,** 1212 (1987).
58. G. Falcone and F. Gullo, *Phys. Lett. A* **125,** 432 (1987).
59. G. Falcone, *Phys. Lett. A* **129,** 188 (1988).
60. C. T. Reimann, M. El-Maazawi, K. Walzl, B. J. Garrison, and N. Winograd, *J. Chem. Phys.* **90,** 2027 (1989).
61. See, for example, D. G. Truhlar and J. T. Muckerman, in *Atom Molecule Collision Theory* (R. B. Berstein, ed), Plenum Press, New York (1979).
62. See, for example, P. Lykos, ed., *ACS Symp. Ser.,* No. 86 (1978).
63. D. W. Brenner and B. J. Garrison, *Adv. Chem. Phys.,* **281,** (1989).
64. D. J. Rossky and M. Karplus, *J. Am. Chem. Soc.* **101,** 1913 (1979).
65. R. Smith and D. E. Harrison, Jr., *Computers in Physics,* Sep/Oct, 68 (1989).
66. B. J. Garrison, C. T. Reimann, N. Winograd, and D. E. Harrison, Jr., *Phys. Rev. B* **36,** 3516 (1987).
67. B. J. Garrison, N. Winograd, D. M. Deaven, C. T. Reimann, D. Y. Lo, T. A. Tombrello, and M. H. Shapiro, *Phys. Rev. B* **37,** 7197 (1988).
68. B. J. Garrison, *J. Am. Chem. Soc.* **102,** 6553 (1980); **104,** 6211 (1982).
69. S. A. Adelman, *Adv. Phys.* **44,** 143 (1980).
70. J. C. Tully, *J. Chem. Phys.* **73,** 1975 (1980).
71. R. A. Stansfield, K. Broomfield, and D. C. Clary, *Phys. Rev. B* **39,** 7680 (1989).
72. F. H. Stillinger and T. A. Weber, *Phys. Rev. B* **31,** 5262 (1985)
73. P. C. Zalm, *J. Appl. Phys.* **54,** 2660 (1983).
74. W. Ens, R. Beavis, and K. G. Standing, *Phys. Rev. Lett.* **50**, 27 (1983).
75. J. E. Huheey, *Inorganic Chemistry,* Harper and Row, New York (1978).
76. R. S. Bardoli, J. C. Vickerman, and J. Wolstenholme, *Surf. Sci.* **85,** 244 (1979).
77. N. Winograd, B. J. Garrison, and D. E. Harrison, Jr., *J. Chem. Phys.* **73,** 3473 (1980).
78. S. P. Holland, B. J. Garrison, and N. Winograd, *Phys. Rev. Lett.* **44,** 756 (1980).
79. K. E. Foley and B. J. Garrison, *J. Chem. Phys.* **72,** 1018 (1980).
80. M. T. Robinson and I. M. Torrens, *Phys. Rev. B* **9,** 5008 (1974).
81. R. P. Webb and D. E. Harrison, Jr., *Phys. Rev. Lett.* **50,** 1478 (1983).
82. I. H. Wilson, N. J. Zheng, U. Knipping, and I. S. T. Tsong, *Appl. Phys. Lett.* **53,** 2039 (1988).

83. E. Pearson, T. Takai, T. Halicioglu, and W. A. Tiller, *J. Crystal Growth* **70,** 33 (1984).
84. D. W. Brenner and B. J. Garrison, *Phys. Rev. B* **34,** 1304 (1986).
85. J. Tersoff, *Phys. Rev. Lett.* **56,** 632 (1986).
86. R. Biswas and D. R. Hamann, *Phys. Rev. Lett.* **55,** 2001 (1985).
87. J. B. Gibson, A. N. Goland, M. Milgrim, and G. H. Vineyard, *Phys. Rev.* **120,** 1229 (1960).
88. D. E. Harrison, Jr., W. L. Moore Jr., and H. T. Holcombe, *Radiat. Effects* **17,** 167 (1973).
89. M. H. Shapiro, P. K. Haff, T. A. Tombrello, D. E. Harrison, Jr., and R. P. Webb, *Radiat. Effects* **89,** 234 (1985).
90. V. E. Yurasova, A. A. Sysoev, G. A. Samsonov, V. M. Bukhanov, L. N. Nevzovova, and L. B. Shelyakin, *Radiat. Effects* **20,** 89 (1973).
91. J. P Biersack and L. G. Haggmark, *Nucl. Instrum. Methods* **174,** 257 (1980).
92. J. P. Biersack and W. Eckstein, *Appl. Phys. A* **34,** 73 (1984).
93. I. M. Torrens, *Interatomic Potentials,* Academic, New York (1972).
94. E. P. Th. M. Suurmeijer and A. L. Boers, *Surf. Sci.* **43,** 309 (1973).
95. C. C. Chang, N. Winograd, and B. J. Garrison, *Surf. Sci.* **202,** 309 (1988).
96. L. Brewer and G. M. Rosenblatt, *Adv. High Temp. Chem.* **2,** 1 (1969).
97. K. P. Huber and G. Hertzberg, *Constants of Diatomic Molecules,* Van Nostrand Reinhold Company (New York, 1979).
98. M. S. Daw and M. I. Baskes, *Phys. Rev. B* **29,** 6443 (1984).
99. S. M. Foiles, M. I. Baskes, and M. S. Daw, *Phys. Rev. B* **33,** 7983 (1986).
100. M. S. Daw, *Surf. Sci.* **166,** L161 (1986).
101. F. Ercolessi, E. Tosatti, and M. Parrinello, *Phys. Rev. Lett.* **57,** 719 (1986).
102. F. Ercolessi, M. Parrinello, and E. Tosatti, *Surf. Sci.* **177,** 314 (1986).
103. M. Garofalo, E. Tosatti, and F. Ercolessi, *Surf. Sci.* **188,** 321 (1987).
104. D. Tomanek and K. H. Bennemann, *Surf. Sci.* **163,** 503 (1985).
105. B. W. Dodson, *Phys. Rev. B* **35,** 880 (1987).
106. S. P. Chen, A. F. Voter, and D. J. Srolovitz, *Phys. Rev. Lett.* **57,** 1308 (1986).
107. S. M. Foiles, *Phys. Rev. B* **32,** 7685 (1986).
108. T. E. Felter, S. M. Foiles, M. S. Daw, and R. H. Stullen, *Surf. Sci.* **171,** L379 (1986).
109. D. M. Deaven, Honors Thesis, The Pennsylvania State University, (1988).
110. J. M. Eridon and S. Rao, *Mat. Res. Soc. Symp. Proc.* **141,** 285 (1989).
111. *Atomic Scale Calculations in Mateirals Science,* Mat. Res. Soc. Symp. Proc., Vol. 141 (1989).
112. M. I. Baskes, *Phys. Rev. Lett.* **59,** 2666 (1987).
113. C. T. Reimann, M. El-Maazawi, K. Walzl, B. J. Garrison, and N. Winograd, *J. Chem. Phys.* **90,** 2027 (1989).
114. D. Y. Lo, T. A. Tombrello, M. H. Shaprio, B. J. Garrison, N. Winograd, and D. E. Harrison, Jr., *J. Vac. Sci. Technol.* **A6(3),** 708 (1988).
115. N. Sathyamurthy, *Computer Phys. Rep.* **3,** 1 (1985) and references therein.
116. J. N. Murrell, S. Carter, S. C. Farantos, P. Huxley, and A. J. C. Varandas, *Molecular Potential Energy Functions,* Wiley, New York (1984) and references therein.
117. K. E. Khor and S. Das Sarma, *Phys. Rev. B* **36,** 7733 (1987).
118. F. F. Abraham and I. P. Batra, *Surf. Sci.* **163,** L752 (1985).
119. B. M. Axilrod and E. Teller, *J. Chem. Phys.* **11,** 299 (1943).
120. P. M. Agrawal, L. M. Raff, and D. L. Thompson, *Surf. Sci.* **188,** 402 (1987).
121. J. Tersoff, *Phys. Rev. B* **37,** 6991 (1988).
122. B. W. Dodson, *Phys. Rev. B* **35,** 2795 (1987).

123. G. C. Abell, *Phys. Rev. B* **31,** 6184 (1985).
124. K. Ding and H. C. Anderson, *Phys. Rev. B* **34,** 6987 (1986).
125. M. H. Grabow and G. H. Gilmer, in *Initial States of Epitaxial Growth,* (J. M. Gibson, R. Hull, and D. A. Smith, eds), Materials Research Society, Pittsburgh (1987), p. 15.
126. R. Smith, D. E. Harrison, Jr., and B. J. Garrison, *Phys. Rev. B,* **40,** 93 (1989).
127. J. T. Muckerman, *J. Chem. Phys.* **57,** 3388 (1972).
128. J. H. McCreery and G. Wolken, *J. Chem. Phys.* **63,** 2340 (1975).
129. A. Gelb and M. Cardillo, *Surf. Sci.* **59,** 128 (1976).
130. A. Gelb and M. Cardillo, *Surf. Sci.* **64,** 197 (1977).
131. A. Gelb and M. Cardillo, *Surf. Sci.* **75,** 199 (1977).
132. A. Kara and A. E. DePristo, *J. Chem. Phys.* **88,** 2033 (1988).
133. J.-H. Lin and B. J. Garrison, *J. Chem. Phys.* **80,** 2904 (1984).
134. J. C. Tully, *J. Chem. Phys.* **73,** 6333 (1980).
135. D. Brenner, C. T. White, M. L. Elert, and F. E. Walker, *Int. J. Quantum Chem.,* **23,** 333 (1989).
136. S. M. Foiles, *Surf. Sci.* **191,** L779 (1987).
137. B. W. Dodson, *Phys. Rev. Lett.* **60,** 2288 (1988).
138. D. W. Brenner and B. J. Garrison, *Surf. Sci.* **198,** 151 (1988).
139. B. J. Garrison, M. T. Miller, and D. W. Brenner, *Chem. Phys. Lett.* **146,** 553 (1988).
140. D. Srivastava, B. J. Garrison, and D. W. Brenner, *Phys. Rev. Lett.,* **63,** 302 (1989).
141. B. J. Garrison, *Surf. Sci.* **167,** L225 (1986).
142. K. Wittmaack in, *Inelastic Ion–Surface Collisions* (N. H. Tolk, J. C. Tully, W. Heiland, and C. W. White, eds.) Academic, New York (1977), p. 153.
143. P. Williams, *Surf. Sci.* **90,** 588 (1979).
144. G. Blaise and A. Nourtier, *Surf. Sci.* **90,** 495 (1979).
145. G. Soldzian, *Phys. Scr.* **T6,** 54 (1983).
146. P. Williams, *Appl. Surf. Sci.* **13,** 241 (1982).
147. M. L. Yu and N. D. Lang, *Nucl. Instrum. Methods Phys. Res.* **B14,** 403 (1986).
148. M. L. Yu, *Nucl. Instrum. Methods Phys. Res.* **B18,** 542 (1987).
149. Z. Sroubek, *SIMS VI* (A. Benninghoven, A. M. Huber, and H. W. Werner, eds.), J. Wiley, New York (1987), p. 17.
150. M. L. Yu, *SIMS VI* (A. Benninghoven, A. M. Huber, and H. W. Werner, eds.) J. Wiley, New York (1987), p. 4.
151. N. D. Lang, *Phys. Rev. B* **27,** 2019 (1983).
152. A. Blandin, A. Nourtier, and D. W. Hone, *J. Phys.* **37,** 396 (1976).
153. J. K. Norskov and B. I. Lundqvist, *Phys. Rev. B* **19,** 5661 (1979).
154. R. Brako and D. M. Newns, *Surf. Sci.* **108,** 253 (1981).
155. M. L. Yu and N. D. Lang, *Phys. Rev. Lett.* **50,** 127 (1983).
156. M. L. Yu, *Phys. Rev. Lett.* **47,** 1325 (1981).
157. G. A. v.d. Schootbrugge, A. G. J. de Wit, and J. M. Fluit, *Nucl. Instrum. Methods* **132,** 321 (1976).
158. M. Bernheim, and F. LeBourse, *Nucl. Instrum. Methods Phys. Res.* **B27,** 94 (1987).
159. M. J. Vasile, *Phys. Rev. B* **29,** 3785 (1984).
160. R. G. Hart and C. B. Cooper, *Surf. Sci.* **94,** 105 (1980).
161. A. R. Krauss and D. M. Gruen, *Surf. Sci.* **92,** 14 (1980).
162. R. A. Gibbs, S. P. Holland, K. E. Foley, B. J. Garrison, and N. Winograd, *J. Chem. Phys.* **76,** 684 (1982).
163. J.-H. Lin and B. J. Garrison, *J. Vac. Sci. Technol A* **1**(2), 1205 (1983).
164. A. Nourtier, J. P. Jardin, and J. Quazza, *Phys. Rev. B* **37,** 10628 (1988).
165. L. Landau, *Z. Phys. Sov.* **2,** 46 (1932); C. Zener, *Proc. R. Soc.* **A137,** 696 (1932).

166. K. Mann and M. L. Yu, *SIMS V,* Springer Series in Chemical Physics (A. Benninghoven, R. J. Colton, D. S. Simons, and H. W. Werner, eds.), Springer-Verlag, New York (1986), p. 26.
167. M. L. Yu and K. Mann, *Phys. Rev. Lett.* **57,** 1476 (1986).
168. M. L. Yu, D. Grischkowsky, and A. C. Balant, *Phys. Rev. Lett.* **48,** 427 (1982).
169. W. Husinsky, G. Betz, and I. Girgis, *Phys. Rev. Lett.* **50,** 1689 (1983).
170. B. Schweer and H. L. Bay, *Appl. Phys. A* **29,** 53 (1982).
171. M. J. Pellin, R. B. Wright, and D. M. Gruen, *J. Chem. Phys.* **74,** 6448 (1981).
172. C. E. Young, W. F. Callaway, M. J. Pellin, D. M. Gruen, *J. Vac. Sci. Technol. A* **2,** 693 (1984).
173. B. I. Craig, J. P. Baxter, J. Singh, G. A. Schick, P. H. Kobrin, B. J. Garrison, and N. Winograd, *Phys. Rev. Lett.* **57,** 1351 (1986).
174. B. I. Craig and B. J. Garrison, unpublished.
175. C. Reimann and N. Winograd, unpublished results.
176. S. P. Holland, B. J. Garrison, and N. Winograd, *Phys. Rev. Lett.* **43,** 220 (1979).
177. S. Kono, C. S. Fadley, N. F. T. Hall, and Z. Hussain, *Phys. Rev. Lett.* **41,** 117 (1978).
178. D. W. Moon, R. J. Bleiler, and N. Winograd, *J. Chem. Phys.* **85,** 1097 (1986).
179. E. Zanazzi, F. Jona, D. W. Jepsen, and P. M. Marcus, *Phys. Rev. B* **14,** 432 (1976).
180. M. J. Cardillo, G. E. Becker, D. R. Hamann, J. A. Serri, L. Whitman, and L. F. Mattheiss, *Phys. Rev. B* **28,** 494 (1983).
181. P. H. Citrin, D. R. Hamann, L. F. Mattheiss, and J. E. Rowe, *Phys. Rev. Lett.* **49,** 1712 (1982).
182. S. P. Weeks and J. E. Rowe, *Solid State Commun.* **27,** 885 (1978).
183. C. C. Chang and N. Winograd, *Surf. Sci.,* **230,** 27 (1990).
184. D. W. Moon, R. J. Bleiler, E. J. Karwacki, and N. Winograd, *J. Am. Chem. Soc.* **105,** 2916 (1983).
185. D. W. Moon, N. Winograd, and B. J. Garrison, *Chem. Phys. Lett.* **114,** 237 (1985).
186. J. E. Demuth, K. Christmann, and P. N. Sanda, *Chem. Phys. Lett.* **76,** 201 (1980).
187. R. Dudde, E. E. Koch, N. Ueno, and R. Engelhardt, *Surf. Sci.* **178,** 646 (1986).
188. C. C. Chang, G. P. Malafsky,and N. Winograd, *J. Vac. Sci. Technol A* **5,** 981 (1987).
189. J. F. Van der Veen, *Surf. Sci. Rep.* **5,** 199 (1985).
190. J. A. Yarmoff and R. S. Williams, *Surf. Sci.* **127,** 461 (1983).
191. C.-C. Chang and N. Winograd, *Phys. Rev. B* **39,** 3467 (1989).
192. Y. Kuk and L. C. Feldman, *Phys. Rev. B* **30,** 5811 (1984).
193. N. Winograd and C.-C. Chang, *Phys. Rev. Lett.* **61,** 2568 (1989).
194. D. J. Holmes, N. Panagiotides, C. J. Barnes, R. Dus, D. Norman, G. M. Lamble, F. Della Valle, and D. A.King, *J. Vac. Sci. Technol. A* **5**(4), 703 (1987).
195. Rik Blumenthal, S. K. Donner, J. L. Herman, Rajender Trehan, K. P. Caffey, B. D. Weaver, Ehud Furman, and Nicholas Winograd, *J. Vac. Sci. Technol B* **6,** 1444 (1988).
196. C. M. Greenlief, S. M. Gates, and P. A. Holbert, *Chem. Phys. Lett.* **159,** 202 (1989).
197. M. P. Kaminsky, N. Winograd, G. L. Geoffroy, and M. A. Vannice, *J. Am. Chem. Soc.* **108,** 1315 (1986).
198. R. J. Levis, J. Zhicheng and N. Winograd, *J. Am. Chem. Soc.,* **111,** 4605 (1989).
199. J. R. Creighton and J. M. White, *Surf. Sci.* **129,** 327 (1983).
200. R. J. Levis, Z. C. Jiang, N. Winograd, S. Akhter, and J. M. White, *Catalysis Lett.* **1,** 385 (1988).
201. S. Akhter and J. M. White, *Surf. Sci.* **167,** 101 (1986).
202. M. A. Henderson, P. L. Radloff, J. M. White, and C. A. Mims, *J. Phys. Chem.* **88,** 4111 (1988).

203. M. A. Henderson, P. L. Radloff, C. M. Greenlief, J. M. White, and C. A. Mims, *J. Phys. Chem.*, **88,** 4120 (1988).
204. L. L. Lauderback and W. N. Delgass, *Phys. Rev. B* **26,** 5258 (1982).
205. L. A. DeLouise, E. White, and N. Winograd, *Surf. Sci.* **147,** 252 (1984).
206. L. A. DeLouise and N. Winograd, *Surf. Sci.* **159,** 199 (1985).
207. L. A. DeLouise and N. Winograd, *Surf. Sci.* **154,** 79 (1985).
208. N. Winograd, B. J. Garrison, T. Fleisch, W. N. Delgass, and D. E. Harrison, Jr., *J. Vac. Sci. Technol.* **16,** 629 (1979).
209. M. L. Yu, *Phys. Rev. B,* **26,** 4731 (1982).
210. M. L. Yu, *Appl. Surf. Sci.,* **11/12,** 196 (1982).
211. K. J. Snowden, W. Heiland, and E. Taglauer, *Phys. Rev. Lett.* **46,** 284 (1981).
212. J. J. Vajo, J. H. Campbell, and C. H. Becker, *J. Am. Vac. Soc. A,* **7,** 1949 (1989).

3

Particle-Induced Desorption Ionization Techniques for Organic Mass Spectrometry

Kenneth L. Busch

"When your favorite tool is a hammer, every problem looks like a nail."

1. Introduction

The use of new ionization methods in organic mass spectrometry has undergone a rapid expansion in recent years.[1,2,543,544] The need for a reliable and sensitive analytical method for nonvolatile and thermally labile compounds, including those of high molecular weight such as biomolecules, has encouraged such development. The prospect of large-scale manufacture of synthetic biomolecules, with its concomitant need for analytical methods, ensures the continued refinement of such methods. The new-found methods through which ions can be formed has also revitalized such areas as chromatography/mass spectrometry, and fundamental thermodynamic and kinetic studies of gas phase ions. Justifiably, however, considerable attention has been focused on the analysis of large biomolecules by *plasma desorption* (PD) with time-of-flight mass spectrometry,[3,4,544] with the mass range extending as high as 60,000 daltons. Within the past year, advances in sample preparation and matrix selection have allowed laser desorption to reach masses as high as

Kenneth L. Busch School of Chemistry and Biochemistry, Georgia Institute of Technology, Atlanta, Georgia 30332.

Ion Spectroscopies for Surface Analysis (Methods of Surface Characterization series, Volume 2), edited by Alvin W. Czanderna and David M. Hercules. Plenum Press, New York, 1991.

120,000 daltons[545–547] and electrospray ionization has broken the 100,000 mark in the creation of ions directly from solution samples. The high-mass challenge has also been taken up by the manufacturers of sector instruments, who have optimized and redefined magnet technology; with *fast atom bombardment* (FAB) ionization, analyses of biomolecules of up to about 17,000 daltons have been carried out.[5,6,548] Cluster ions of inorganic salts such as the ubiquitous cesium iodide have been observed to near 100,000 atomic mass units. The quest for higher masses has obscured some of the more fundamental advances in technology that promise a more substantial long-term expansion of mass spectrometry, and fortunately, significant effort has been directed toward extending the useful range of experiments as well. For instance, peptide sequencing via FAB mass spectrometry has proven sufficiently viable after only a few years to compete with the more classical analyses.[7–10,550,551] Products formed as a result of enzymic reactions occurring within the instrument have been monitored in real time,[11,12,552,553] or at discrete intervals from reaction mixtures then separated by liquid chromatography.[554] Meanwhile, the new methods are not limited to biochemical problems; exciting applications to the analysis of organometallic and coordination compounds have also been developed.[13,14,555,556] It is the goal of this chapter to place these techniques and their applications in historical perspective, to point out the similarities and the differences in the processes involved, and to outline their analytical capabilities, both realized and potential, in organic mass spectrometry. The development of all of these techniques has been such that the pace of applications has outrun the refinement of the mechanistic understanding. That this is true is apparent both from a historical perspective as well as from a survey of current uses. Such dichotomy will also undoubtedly characterize the future as well. Thus the contents of this chapter must be considered in the context of our incomplete understanding of many of the processes involved. There are inconsistencies in the literature, and the many "unified" models are not necessarily in mutual harmony. The excitement generated when a method unexpectedly (and against all theoretical odds) provides a required result is a hallmark of the development of the methods of particle-induced desorption mass spectrometry.

Desorption ionization will be used as a general term to describe techniques that, as a group, involve the creation of sample ions directly from a condensed phase (solid or liquid) by the impact of an energetic particle. The contrast with electron or chemical ionization can be drawn from the fact that those methods require that the sample be evaporated before ionization. Ionization methods that involve the formation of a sample aerosol (at atmospheric pressure or under some vacuum), and transport of ion-containing droplets into the vacuum of the mass spectrometer, will be called *nebulization ionization methods,* and include thermospray, ionspray, electrospray, electrohydrodynamic ionization, and the various liquid ionization methods. A third group includes the field/plasma/discharge methods, including the classical methods of field desorption, spark source, and glow discharge mass spectrometry. This chapter reflects the preponderance of current usage by concentrating on the techniques of particle-induced desorption ionization for the analysis of organic and biological samples.

In a volume devoted to techniques used for the characterization of surfaces, the relationship of a chapter on organic mass spectrometry must be carefully considered. Although the instrumentation and the experiments may be quite similar, fundamental differences in approach exist. What exactly is an *organic* surface? To the catalytic chemist, an organic surface may represent a metallic catalyst surface, the performance of which is degraded by some fraction of a monolayer of organic poison. To a chemist using SIMS in inorganic analysis, an organic surface may be the matrix in which the analyte of interest (alkali or metal ions) is found, as in studies of the spatial distribution of such inorganic species in retinal tissue. To the organic mass spectrometrist, the surface represents the interface between the bulk of the organic sample and the vacuum environment in which the mass spectrometric measurement must take place. The physics of the sampling process (that is, sputtering) generates a surface sensitivity that engenders concerns about surfactancy, and differential diffusion through the liquid reservoir that contains the sample. The high vacuum of the surface scientist is not the vacuum of the organic mass spectrometrist, who may be, for example, feeding a constant supply of glycerol and other volatile solvents into the vacuum system. The sample preparation of the scientist analyzing peptide solutions by FAB differs from the scientist who painstakingly mounts a selected crystal face for analysis.

Mutual appreciation of the researches carried out by these disparate groups is the key to continued application of "surface spectroscopies" to new analytical problems. This overview will attempt to support that hypothesis by tracing the roots and relationships of the particle-induced desorption ionization methods.

1.1. Ionization Overview

1.1.1. Desorption Ionization

The techniques of particle-induced desorption ionization can be subdivided into secondary ion mass spectrometry (SIMS), fast atom bombardment (FAB), laser desorption (LD), plasma desorption (PD), and other novel methods such as the dust particle-induced method described by Krueger.[15,16] In what follows, each method will be afforded a brief instrumental description.

SIMS and FAB are the most closely related of the two particle-induced desorption technique. When reduced to the mechanistic fundamentals, they are, in fact, indistinguishable. Organic SIMS was introduced in a systematic way by Benninghoven in a prescient series of papers in the mid 1970s;[17–19] an isolated report had appeared earlier.[20] In organic or molecular SIMS, a primary ion beam, usually composed of argon ions at an energy of from 500 to 5000 eV, is directed onto the surface of an organic material. Keeping the flux of primary ions below the static SIMS[19] limit, defined as 10^{-9} A/cm^2, limits the extent of damage due to the irradiation. Within the several minutes required to obtain the spectrum, evidence of this irradiation damage does not appear in the spectrum. Typical organic secondary ion fluxes are 500–10^4 ion counts per second. Benninghoven,[20–22] Cooks,[23–27] Colton[28–30] and Standing[31–32] have continued the development of organic SIMS.

Fast atom bombardment (FAB) had been available in various forms for many years, but was brought into focus by Barber in 1983.[33] The technique uses a beam of energetic atoms rather than ions to bombard an organic material and create secondary ions. This is of subsidiary importance to the use of a liquid matrix to overcome the static SIMS limit, and to provide a stable and persistent source of organic secondary ions. Barber's first paper omitted mention of the requisite use of a liquid glycerol matrix in which the organic sample is dissolved. As originally described, the method was identical to that described earlier by Devienne,[34–35] Tantsyrev[36,37] and several other workers. For organic and biochemical analysis, the use of glycerol or other liquid matrixes[557,558] in FAB has become a standard procedure. For samples that are themselves liquids, the additional matrix is not necessary. Matrixless FAB has also been used in the analysis of neat surfaces of inorganic composition,[38,39] duplicating experimental methods previously carried out under the auspices of another acronym.[40,41] Such an approach has elicited comment and response.[42,43] One notes that some organic samples can be analyzed without a liquid matrix, and outside of the normal static SIMS limits,[422] and that many recent developments in flow-FAB[559–561] substantially reduce the amount of matrix solvent present at the sputtered site.

In organic analysis, the introduction of the liquid glycerol matrix for FAB avoided the static SIMS requirement of a reduced primary particle flux.[44–46] In FAB, the damage is actually greater in magnitude because of the generally higher particle flux. The liquid matrix provides a reservoir through which undamaged sample molecules can diffuse to the surface from the bulk, and into which damaged organic material from the surface can be submerged. The result is an increase in the absolute intensity of organic secondary ions, and in their persistence in the mass spectrum. The glycerol may also act to "soften" the sputtering process; through a process of desolvation, the glycerol molecules themselves may remove excess energy from sputtered cluster ions that contain both glycerol and sample molecules.[47] That such processes can occur has been shown explicitly.[562] The "self-cleaning" characteristics of the glycerol surface refer to the ability of the glycerol surface to be eroded by the primary particle beam at a rate that itself ensures the continued presentation of a fresh supply of sample for sputtering.[563,564] Despite a lack of understanding of the subtleties of the chemical and physical mechanisms through which the secondary ions are formed, the FAB technique gained rapid acceptance in biological laboratories eager to determine the molecular weight of nonvolatile or thermally fragile biomolecules. More recently, careful studies of mechanism are being pursued, and these should provide significant gains in capabilities. A liquid matrix for the preparation of samples has been used in ion beam bombardment (SIMS) with the same advantages accrued, and this variant of the procedure has become known as "liquid SIMS".[48] Barber[49] has objected to the use of this term, perhaps as Devienne[35] may prefer an acronym other than FAB.

Plasma desorption (PD) uses as primary beam particles the heavy and energetic (MeV) fission fragments from the decay of ^{252}Cf as described originally by Macfarlane.[50,51] The mass and energy of the primary particle responsible for each sputter event is unknown, although the set of fission fragments from californium is well characterized. The small californium source is mounted behind the sample in a time-of-flight (TOF) mass spectrometer. A fission fragment passes through a thin foil coated on the far side with the organic analyte, and the fragment emitted from the californium source in the opposite direction starts the timing acquisition for the mass spectrometer. Noteworthy advances in the analysis of high-molecular-weight samples have been made with the PD-TOF instrument, taking full advantage of the sensitivity and mass range of the time-of-flight mass spectrometer, and underscoring the secondary nature of mass resolution (see Section 4.3). Recently, beams of defined mass and energy from linear accelerators have been used as the source of primary ions[52] and the mechanisms of sputtering in this energy range are being carefully investigated (see Section 2).

Laser desorption mass spectrometry uses as its primary particle beam the intense photon flux from a laser. A distinction is drawn here between laser desorption instruments used for general organic mass spectrometry, and those developed for research studies in photoionization,[565,566] although the occurrence of gas phase photoionization and photodissociation processes in laser desorption instruments is not discounted. Both continuous wave and pulsed lasers can be used; the spectra obtained with the former closely resemble those obtainable with thermal desorption. A diverse assortment of custom-built instruments have been used for laser desorption, with assorted combinations of lasers and mass analyzers. In both of the commercially available systems,[53,54] a sighting laser is used to pinpoint a spot of 1–10 μm diameter on the surface of a thin organic sample. The sighting laser is swung out of the way and a higher-power laser (often a frequency-quadrupled Nd-YAG laser) is then used to bombard the organic surface with a short 10-ns pulse of light. The sputtered secondary ions are extracted into a time-of-flight mass spectrometer and mass-analyzed. Hillenkamp[55] has described many of the applications of the commercial device. Hercules has further described organic applications, and has suggested some general mechanisms of ion formation.[56,57] More recently, fairly complete models of desorption and ionization (supported in each case by specific spectral examples) have been described.[569–571]

The use of dust particles for the sputtering of small organic molecules has been described.[15,16] In this technique, a direct descendant of space research, ferrous "dust" particles of masses of 10^9–10^{15} daltons and high keV energies are used as the primary beam. The sputtered secondary ions are analyzed by a time-of-flight mass spectrometer. The analytical advantage of such a system over FAB or SIMS is not established for general analysis, but the origin of the experiment in space erosion studies is clear. The development of such a method underscores the fact that the rapid transfer of energy to an organic surface can be the central process underlying methods in many instrumental guises. Commonplace and commercial methods once seemed far-fetched, and future developments will certainly occur in unpredictable directions. A case in point is the development of spontaneous desorption mass spectrometry.[572–575] Here, results were obtained under instrumental conditions in which no ion signal at all could reasonably be expected. Recent work[574–577] focused on the mechanisms, and after the first results were obtained, provided evidence for the occurrence of a low voltage gradient field-desorption process combined in some degree with sputtering by back-scattered ions of keV energy.

1.1.2. *Nebulization Ionization*

A thorough review of the methods of nebulization ionization techniques has been provided by Vestal.[58] Inclusion of an overview in this chapter is justified because the mechanisms of ion ejection from small droplets formed in an aerosol nebulization, or formed by energetic particle impact, must be similar. In the final analysis, the processes that determine the nature of the ultimate population of ions generated by these disparate routes may be quite similar. Alternatively, significant differences may suggest a divergence in mechanism.

Electrohydrodynamic ionization mass spectrometry (EHMS) was developed by Evans.[59,60] The source was described in 1972; applications to liquid metals[60] were followed by the first investigations of organic compounds in 1974.[61] Evans and Cook have studied a number of compound classes with EHMS.[62–65] In EHMS, ions are generated directly from an electrolyte-doped liquid (most often glycerol) by the application of a high electric field and simultaneous evaporation of the solvent. Dominating the spectra are clusters of glycerol molecules, combinations of glycerol and supporting electrolyte, and sample, glycerol, and electrolyte. Many of these same solvated species are generated in fast atom bombardment. Although EHMS employs a field gradient to aid in the formation and extraction of ions from the glycerol solution, the interactions between sample and glycerol are of practical interest. The development of EHMS has been relatively slow, although the work of Cook continues,[578,579] and others have now initiated research in this area.[580,581] Additional impetus for development may be provided by recent experiments which have shown that EHMS can sample ions directly from aqueous solutions[66] or by the development of alternate matrices with more amenable properties.

Electrospray is a method that has been applied to the analysis of nonvolatile and thermally fragile molecules. The technique was originally developed by Dole[67] for the creation of ions from polymeric macromolecules. A dilute solution of the macromolecule was sprayed through a fine needle held at some high potential into a bath of nitrogen near atmospheric pressure. Ions formed in the electrospray processes were passed through a lens system and several stages of differential pumping into a mass-analyzing device. The first mass measurements of the ions produced by electrospray were carried out with retarding potential differences, since the ions produced were well beyond the mass range of the analyzers available at that time, or indeed, most present-day mass analyzers. The successes of the method, and the limitations of the apparatus as originally conceived, have been reviewed by Dole.[68]

Recently, the electrospray technique has received renewed attention from Fenn[69,70] and Bruins[71] for the most part as the interface between a liquid chromatograph and the mass spectrometer (LC/MS), for which purpose its simplicity and sensitivity are advantageous. It is a variation of the theme of ion expulsion from charged aerosol droplets, a phenomenon that underlies many of the direct LC/MS interfaces, and relates to the process of desorption and desolvation thought to occur in FAB and SIMS sputtering of liquid surfaces. Electrospray and ionspray (essentially pneumatically assisted electrospray) have had extraordinary success in the past two years in the determination of molecular weights of high molecular weight biomolecules.[582–585] The method has the capability for production of multiply charged ions (especially demonstrated for peptides). The distribution of basic residues in peptides is such that a proton can be accepted about every 10 residues in a polypeptide chain, and these charges do not generally interact with each other. Accordingly, the molecular ions (there are many, corresponding to the addition of different numbers of protons) are generally distributed across a mass range of 1100–1200 daltons, well within the capabilities of quadrupole mass analyzers.

The thermospray ionization process[72–75] involves the nebulization of a sample solution through an electrically heated fine capillary directly into the vacuum chamber of the mass spectrometer. Charge is distributed statistically, with an overall neutral balance, but with some droplets carrying an excess positive charge and others an excess negative charge. Solvent evaporation from the droplets as they expand into the vacuum produces progressively higher field gradients at the droplet surface. Ultimately, the gradient can become sufficiently high to force the expulsion of ions from the remaining solvent. Thermospray is in widespread use as an interface for LC/MS.[76–78,586–588]

Several researchers, notably Thomson and Iribarne[79,80] and Tsuchiya[81–83] have reported on several other ionization methods loosely grouped under the term "liquid ionization," although the formation of an aerosol of fine droplets is not a prerequisite for these methods. These works form the basis for much of the electrospray and ionspray work discussed above. Many of the ions found in "liquid ionization" mass spectra of liquid samples parallel those found in the nebulization and desorption ionization techniques. Further development of all of the liquid ionization methods will be a near-future focus. Similarities rather than differences will likely form the theme of the fundamental work; the breadth of applications, especially in biotechnology, is extremely promising.

1.2. Historical Perspective

The previous section introduced the various methods of particle-induced desorption ionization. The development of each technique proceeded more or less independently, and it is only recently that common features have been recognized and appreciated. In this section, a short history of each technique will be presented in order to lend perspective to the hectic pace of present development. Since most applications to organic mass spectrometry have occurred only within the last 15 years, this "history" is relatively short.

The use of secondary ion mass spectrometry (SIMS) for organic analyses was the first of the particle-induced desorption ionization methods to be discussed. Application of SIMS to *inorganic* problems enjoys a long anteriority, and it was common knowledge that organic samples were volatile (at least at the 10^{-6}-Torr level), and that these samples compromised the integrity of the vacuum system, and should therefore be avoided. Further, methods (based on energy discrimination) to suppress the observation of signals from molecular ions (contaminants) were often used in inorganic SIMS. However, in 1958, Honig[84] used an energetic ion beam to probe the surface of ethylated germanium. The observation of $GeC_2H_5^+$ clusters was interpreted as indicating the formation of Ge–C bonds at the surface. Optimistically, the next sample reported in the paper was coal. Small hydrocarbon clusters with up to four carbon atoms were sputtered from the surface, and observed in both the positive and negative ion spectra, but ions corresponding to more complex organic structures were not observed. After this initial exploration, the use of SIMS in the analysis of organic molecules languished, at least in the literature, for the next 15 years. The renaissance of organic SIMS applications can be attributed to the efforts of Benninghoven, who in the early 1970s began a systematic study of the behavior of organic compounds on metal supports. Benninghoven also introduced the concept of static SIMS in 1973.[19] In such experiments, the primary ion flux is held to a low level so that damage to an organic monolayer would not contribute to a spectrum obtained over the course of a few minutes. Over the next few years, the positive and negative secondary ion mass spectra of biologically important compounds such as the amino acids, peptides, pharmaceuticals, and vitamins were obtained.[18,21,22,85–89] Early mass spectra were obtained with a quadrupole mass spectrometer with a mass range of up to 300 daltons, and secondary ion fluxes of 10^3–10^4 cps were maintained for at least 10^3 s, indicating the persistence of the secondary ion mass spectra under these conditions. Since background signal was low, the interpretation of these spectra in terms of the structures of the compounds present at the surface was possible even with the relatively low secondary ion currents obtained.

Cooks demonstrated the generality of the process of cationization in SIMS in 1977.[23,24] Cationization is defined as the complexation of the neutral sample molecule M with a cation other than a proton, such as sodium, potassium, or a metal, to form ions such as $(M + Na)^+$, $(M + K)^+$, or $(M + met)^+$. Cooks also used a relatively simple quadrupole mass spectrometer. By coparison of the data obtained in SIMS to that obtained with daughter ion MS/MS spectra of the parent molecular ion, it was postulated(90) that most of the major fragment ions observed in SIMS spectra are the result of gas-phase rather than surface reactions. Similar conclusions are reached again in recent work with laser desorption.(570,571) The preferential sputtering of preformed ions from the surface of the organic solid was noted in the initial work with SIMS of quaternary ammonium compounds in 1979;(91) by 1980, a review article(26) listed three mechanisms by which secondary ions are formed in the sputtering process, and ranked them according to their relative efficiencies. The preferential creation of secondary ions from preformed ions was explicitly described at this time; this concept led to the use of derivatization reactions to create such ions from neutral sample molecules.(92) Honig has assembled an excellent retrospective on the development of SIMS for the American Society for Mass Spectrometry.(93)

The development of LD for the analysis of organic compounds took place during the same period. This technique was first reported by Vastola and Pirone in 1968.(94) Using a ruby laser, these workers generated ions from a variety of organic and organometallic samples. Their work continued for several years, gradually becoming more inorganic in character.(95,96) Application to organic analysis was comprehensively reintroduced by Kistemaker in 1978.(97) Using a magnetic sector mass spectrometer and a pulsed (150 ns, 0.1 J/pulse, energy density of about 1 MW/cm^2) CO_2 laser, spectra of oligosaccharides, glycosides, nucleotides, and other compounds were reported. Cationized ions [$(C + M)^+$ where C is the cation and M the organic molecule] provided molecular weight information, as in SIMS. The fragmentations were reasonably correlated with the known structure of the compound. Experiments indicated that the processes by which the ions were formed were *not* wavelength dependent, and characterized the process as a fast nonequilibrium heating of the condensed phase; this concept has received renewed attention.(98)

By the end of the 1970s, several other research groups were involved in studies of LD. The first commercial LD instrument, the LAMMA 500, was described.(99,100) Hillenkamp and Heinen have been most closely associated with the development of this instrument, now in its second generation form as the LAMMA 1000,(100) Rollgen began investigations using a continuous wave laser for the desorption of organic compounds.(102) Antonov used a UV laser at several different wavelengths to irradiate organic materials on metal surfaces; this work was to provide the first evidence of wavelength-dependent *desorption* of organic compounds.[103] This concept has resurfaced recently with the design of support matrices that absorb energy from the primary laser beam, and facilitate the desorption of adsorbed organic and biochemical films.[589,590] In conjunction with time-of-flight mass spectrometers, truly extraordinary mass ranges and sensitivities have been documented.

Other particle-induced desorption ionization techniques were simultaneously advancing. As mentioned, the technique of plasma desorption (PD) first surfaced in 1974 when Torgerson, Skowronski, and Macfarlane analyzed small amino acids deposited as a thin film on a metallized support.[50] Arginine and cystine were the first samples investigated, but almost immediately, PD was applied to the most challenging biochemical problems, including samples of molecular weights higher than those otherwise attempted (say, up to several thousand daltons).[64–66] Although the time-of-flight mass analysis typically used in the PD instrument is of low resolution, the ability to average the spectra obtained over a period of several hours proved advantageous. For several years, Macfarlane's PD instrument was unique, but PD instruments were soon in place at Darmstadt (Furstenau, Wien, and Krueger),[107,108] Orsay (LeBeyec),[109] and at Marburg (Jungclas).[110] By 1980, several other research groups were involved, and mass- and energy-analyzed beams from linear accelerators were used as the primary beam in order to more precisely understand the process of sputtering and desorption ionization.[52] Torgerson has reviewed the "early days" of PD as part of the Texas Symposium on Particle Induced Desorption.[111] A commercial instrument for californium-based plasma desorption mass spectrometry is available,[112] and plasma desorption is now in place in nearly 100 laboratories worldwide.

As mentioned, the term "fast atom bombardment" was coined in 1981 to describe an ionization method in which atoms rather than ions were used as bombarding particles. In present usage, FAB usually connotes the use of a liquid matrix for sample dissolution, rather than a beam of atoms as opposed to ions. The introduction of the liquid matrix, and the new capabilities for the analysis of organic and bioorganic molecules, belongs unquestionably to Barber. However, the vagueness of the introductory papers[33,113] led to unnecessary confusion. The misleading emphasis on neutral beam bombardment stood in contrast to seemingly similar experiments reported as early as 1962,[114] and continued in neutral beam bombardment of inorganic surfaces.[34,35] Such neutral beams were also used in the sputtering analysis of insulators to avoid charge accumulation on the surface.[40,41] Since the secondary ions sputtered from a surface are oblivious to the charge state of the incident primary particle, the spectral distinction between SIMS and FAB was and is tenuous. The methods of sample preparation can be identical. The details of the device which brings a beam of energetic ions or atoms into a mass spectrometer source held at high potential are straightforward. The time for distinction between SIMS and FAB, at least for organic samples, is long past. In fact, "particle-induced desorption" is sometimes used as an all-encompassing term to deemphasize differences in methods used to sputter organic ions from a liquid solution. In this review, discussions of the relationship between SIMS and FAB emphasize the synergistic nature of the advances in these methods for the production of ions from previously intractable biomolecules.

The 1970s also included the development and refinement of other techniques for the analysis of thermally fragile samples. Although the practice of field desorption (FD) mass spectrometry has been eclipsed in recent years because of the rapid commercialization of FAB, for many years, FD was the sole tool available. Since its introduction by Beckey in 1969,[115] it has developed into a powerful tool for both inorganic and organic analysis.[116–118] The preparation of the requisite emitters and loading of the sample is crucial to the success of the FD experiment, and proficiency is obtained only with a sufficient investment of time. The technique acquired a reputation for difficulty that is not entirely deserved, although the transient nature of the ion currents produced from most samples is a limitation in many analytical applications. Nevertheless, the experience gained in a decade of FD experiments was invaluable in assessing the merits of the new particle-induced desorption ionization techniques, and provided some of the theoretical concepts necessary to the understanding of their mechanisms. One notes that field desorption mass spectrometry is alive and well in 1990,[59] and in many cases (through tradition or capability) is still the method of choice in many situations.

1.3. Instrumentation

Instrumentation developed for particle-induced desorption ionization constitutes an eclectic collection; almost every conceivable combination of source design and mass analyzer has been used at one time or another. Devices that can be described with some generality are the FAB source and the commercial LAMMA instrument used for laser desorption. For review of the basic principles of mass analysis using sectors, quadrupoles, ion cyclotron resonance, or time-of-flight mass spectrometers, the reader is directed to more specialized literature.[119–124] The details of the overall instrument and the experimental conditons used during analysis determine the quality of the mass spectra obtained, almost always making interlaboratory comparisons of data difficult. Specific experimental parameters and their effect upon the mass spectrum measured will be discussed in Section 2.

1.3.1. Source Designs

1.3.1.1. Secondary Ion Mass Spectrometry. Many early SIMS sources were based on the Duoplasmatron ion source originally developed by von Ardenne.[125] These early sources have been gradually replaced with electron ionization sources for noble gas ions[126] or thermal ionization alkali ion emitters,[127,128] although plasma sources are still constructed for use in SIMS instruments.[129] Alkali ion emitters generate a beam of Cs^+ by thermal emission from a heated surface of a salt such as cesium alumina silicate, or other alkali ions from the appropriate salt.[130–132] Microamp beams can be obtained from the larger sources. Restrictions on the size of the ion source encountered in retrofitting mass spectrometers with new SIMS sources has prompted redevelopment of the cesium ion source in a smaller version.[133]

Sources that bombard organic materials with mercury ions or atoms[134] (following the experiments of Wehner in the analysis of inorganic surfaces),[153] and with large organic ions have been described.[136] Intense beams of metal ions generated from liquid metal guns[137–141] have been used in organic SIMS.[142–145] Ionization source requirements for molecular SIMS are not stringent. In contrast to the finely focused ion guns developed for ion microprobes[146–148] or ion beam lithography, defocused beams with spot sizes of up to several millimeters in diameter can be used for the analysis of discrete organic samples. The primary ion current should be variable over several orders of magnitude. Microamperes of current are used for depth profiling in inorganic SIMS, and in dynamic SIMS with appropriate sample preparation. Lower fluxes are used in static SIMS experiments both to extend the period of sampling and to minimize the degree of irradiation damage. The mass of the primary ion is easily varied by choosing among the various inert gases (argon to krypton to xenon, increasing in mass and cost). Rollgen's work[136] emphasizes that organic ions themselves can be used as bombarding species with interesting chemical effects. High-intensity beams of any organic species generated by electron or chemical ionization[149] can be used as the primary particle beam, although there has been little work in this area. A tandem SIMS system has been described in which a primary surface is bombarded by argon ions to create secondary ions (organic or inorganic). These can be collimated and mass analyzed to be used as the bombarding beam for the secondary surface onto which the actual sample is placed.[150] Alternatively, this setup can be used as a method of surface preparation by the "soft landing" of reagent beams[151] in an organic analogy to the well-characterized methods of ion implantation.[152–154] Nebulization ionization methods (thermospray or electrospray) may eventually be used to generate energetic macromolecular ions for use as primary ion beams; the experiment using macromolecular iron cluster ions has already been completed.[15,16] Recent results of the use of polyatomic primary ion species have been reported.[592–594]

Commercially, noble gas ion sources at several levels of performance are available, ranging from moderate focus, relatively low cost guns, to the fine focus guns (less than 10-μm beam diameter) complete with rastering systems for surface imaging. These guns must be differentially pumped, as the pressure within their source is generally 10^{-5} Torr and upwards. An advantage of the cesium and the liquid metal ion guns is the very low gas load added to the system. Cesium guns with low focus capabilities are commercially available. Typical specifications for a cesium ion gun are a beam energy of 100 eV to 10 keV, with an ion beam current up to 20 μA. An angular divergence of two degrees is usual, which translates to a spot size of 0.68 cm for a target 10 cm distant. Restricting apertures can be used to reduce the spread of the ion beam, but they also reduce the primary ion flux. These ion guns are available mounted on a standard 2.75-in. vacuum flange, and are also available with lithium, sodium, and potassium charges. Breck[155] and Heinz[156] have discussed the emission of ions from heated zeolite compounds. In recent commercial instrumentation, high-voltage alkali metal ion guns (up to 40 keV primary ion energy in some cases) are fitted in place of the noble gas guns offered just a few years ago.[595,596] These high-velocity, high-flux ion sources typically produce a higher absolute ion signal for higher-mass biomolecular samples, in concordance with the higher mass ranges of modern instruments (see Section 2.4.2.1). Although these sources may provide better sensitivity at higher masses, substantially greater damage to lower-mass molecules is evident if the primary particle energy density is not attenuated.

Liquid metal ion guns have been in development for many years, but have only become commercially available since 1982. The operation of these ion sources has been described in some detail.[137–141,157–159,597–601] The most common metal used is gallium, owing to its low melting point, although the list of available sources also includes gold, tin, lead, aluminum, bismuth, zinc, copper, and cesium. Two basic designs handle first the relatively low melting metals (below 200°C) and then the remaining metals with melting points of up to 1200°C. The source metal is held in a heated reservoir around the shaft of a fine needle. As it melts, it coats the needle with a thin liquid film. A high potential is applied between the needle and an extractor electrode. The meniscus of the metal is drawn into a cone shape, and individual metal atoms are evaporated into the gas phase, and then ionized through a process of field ionization. The liquid film is replenished from the liquid reservoir so that a continuous emission of metal ions can be created. Typically, energies of the ions are from 1 to 25 keV, depending on the manufacturer and the metal. Ion currents of up to 100 nA are typically quoted. Since the field emission of ions constitutes a point source, the focus of the liquid metal ion gun can be made very small; a spot diameter of as low as 50 nm can be created. Typically, a gallium gun will focus 0.3 nA of ion current into a spot of 200 nm diameter when operated at 25 keV. This translates into an ion current density of about 1 A/cm^2. Such a high ion flux can be used for very fast surface imaging, or for very fast depth profiling,[160,161] but is generally several orders of magnitude too high for organic analysis. For molecular SIMS, the spot is defocused, and the ion current is limited to a smaller value. In the study of Barofsky, a typical value was 4×10^{-7} A/cm^2.[143] The beam emanating from the field emission source is not usually mass analyzed, and consists of mixtures of metal isotopes in varying charge states. Relatively large droplets are also emitted from such sources.[162–165] Barofsky has described a liquid metal ion source that uses a Wien filter to select the mass and energy of the bombarding particle.[166] Intense beams of finely focused primary ion beams have been recently used in the analysis of damage-resistant organic compounds.[601–603]

The advantages of the liquid metal ion gun as a primary ion source are (1) it can be finely focused for imaging analysis, (2) it has a large depth of focus and a long working distance, (3) it adds no gas load to the system and can be mounted in almost any configuration, and (4) the secondary ion yields from organics may be enhanced[142] over those expected from noble gas ion bombardment. Disadvantages include the relatively high initial cost of these sources, the use of high voltage, which can break down and disrupt the system electronics, and the instability of the metal ion emission in the presence of organic contaminants, including the solvents usually used for sample preparation in FAB and liquid SIMS. Finally, laboratory experience with liquid metal ion guns indicates a typical source life less than expected from experience in "cleaner" vacuum chambers, which may be due to a contamination of the surface of the molten metal with an organic film.

With any ion gun system, consideration must be given to the system with which it must operate. For instance, in sector mass spectrometers, the source is normally held at a potential far from ground; for positive ions, this can vary from 3000 to 10,000 V. If a 5000-V beam of incident primary ions is required, then the ion gun must be floated at a potential 5000 V above the source potential, i.e., at 8000–15,000 V. This can represent a considerable problem in instrument operation and operator safety. Consider reversing the source potentials and analyzing for negative ions from the same source. If the source is then held at −8000 V, then a 5000-V ion source must be insulated to a potential difference of 13,000 V, and the ions of the primary beam will bombard the surface with 13,000 V. This argument holds whenever ions are used as the primary beam. In addition, there are defocusing effects on the beam profile with acceleration or deceleration. A practical advantage of the neutral beam source is the fact that the primary beam can be reproducibly directed into a high-voltage source regardless of the source potential, although neither the beam profile nor its flux can be as easily measured. For sources operating at low potentials, such as those used with quadrupole or ion cyclotron resonance instruments, the potential at which the ion gun must operate is not a significant problem. Additionally, for primary ion guns that operate at very high voltages (40 keV), even the maximum accelerating potential of the source (10–15 kV) is easily surmounted.

A number of miniature ion sources have been developed that facilitate the addition of particle-induced desorption sources to mass spectrometers. The flanging problems associated with the addition of an external source are avoided if the ion gun is mounted directly to the ion source within the vacuum chamber, close to the sample. Many older source designs are sufficiently voluminous to allow for such an addition. For primary ion sources, appropriate insulation must still be provided. Rudat has described a miniature ion/atom source based on a gas discharge source.[(167)] Hass has described an in-chamber focused primary ion source of much the same design.[(144)] An advantage of the latter is that the irradiation of the sample occurs only in an area from which the sputtered secondary ions are likely to be accepted into the mass spectrometer, and bombardment damage to the sample is therefore minimized. A miniature cesium ion source mounted in various configurations has been described by Aberth.[(168)]

Novel ion sources developed for specialized applications may be appropriated for use in particle-induced desorption ionization mass spectrometry. Klaus[(169)] has recently described a primary particle gun for secondary ion mass spectrometry which produces either an ion or a neutral beam. Most ion sources do produce neutrals, and most neutral sources ions; the characteristic of note is that the particle density and homogeneity of the neutral beam is the same as that of the ionic beam, with a neutral beam diameter down to 0.4 mm FWHM. Su[170] has described an ion source that can be operated in either pulsed or continuous modes so that it can be used with a time-of-flight mass spectrometer or with the more conventional sector instruments. Orient[171] has described a "reversal" ion source that produces relatively intense negative ion beams, and this source can also be operated in a pulsed or a continuous mode.

Following the need for intense ion beams for surface modification and ion beam lithography, several high-flux ions sources have recently been described. Field emission cathodes[172] can be used to create very intense but fintely focused beams of electrons for an electron ionization source. Brown[173] has described a high-current ion source in which a metal vapor vacuum arc is used to create the plasma from which the ion beam is extracted. Currents of up to 1 A at 25 keV acceleration voltage can be generated for metals as diverse as lithium and uranium. Multiply charged metal ions of charge state as high as +6 can be extracted from such sources. Ultimately, such sources may be used to investigate the energy and charge state dependence of organc secondary ion sputtering in the keV regime, in analogy to the experiments that have been performed for plasma desorption mass spectrometry (*vide infra*). Other high-intensity ion beam sources based on radiofrequency[174] and microwave frequency[175] induction plasma sources have been described in the literature.

Time-of-flight mass analysis requires that the ions are produced in a pulse from the ion source. Benninghoven and co-workers have described the development of a pulsed ion source of moderately high resolution for TOF mass analysis.[604,605] Liquid metal ion guns can also be pulsed in a time sequence short enough to allow for TOF analysis.[597]

1.3.1.2. Fast Atom Bombardment Mass Spectrometry. The designs of FAB sources are as diverse as the ion beam sources of SIMS from which most are derived. Noble gas ion guns can be fitted with a charge exchange cell[34] to create a neutral beam from an initially charged beam. Although charge exchange can be a resonant process, the conversion of ions to atoms is not complete, and the output from the charge exchange cell is a mixture of ions and neutrals. Construction of FAB sources was the subject of a workshop in 1981[176] sponsored by the American Society for Mass Spectrometry. Commercial FAB sources have been modified to increase the efficiency of the charge exchange process, and to increase the percentage of neutral components in the output of the gun. This contrasts with some early neutral bombardment sources which simply deflected the ionic component out from the ion/neutral particle mix formed by a standard ion source.[177] A popular neutral particle source has been the saddle field ion source[178,179] (Fig. 1). A good description of the use of this source in mass spectrometry has been given by Franks.[180] In this source, electrons oscillate between two cathodes under the influence of a dc field. Ions formed from noble gases in the plasma discharge are neutralized both by resonant charge exchange and by an additional process of electron–positive ion recombination. When this basic design is configured as an ion gun, the beam output was contaminated with 30% neutrals.[181,182] With appropriate modifications, the ionic content is claimed to be reduced to less than 1%. Ligon has characterized the ion flux produced by such a saddle field source, and concluded that both ions and neutrals produced have a broad energy distribution, which may range up to 30 kV, in part because of the initial formation of multiply charged ions in the plasma [Ref. 183; see also W. V. Ligon and S. B. Dorn, *Int. J. Mass Spectrom. Ion Proc.* **72,** 317 (1986) for corrected distributions]. These will not be deflected by the source potentials of the sector instruments (which are normally below 10,000 V). The inability of ions of 5000 V energy from an ion gun to enter a source held at 8000 V is otherwise a convenient means of cleansing the mixed ion and neutral beam of its ionic component. Saddle field ion sources are characterized by a high heat production and undesirable sputtering of the cathodes (normally made of aluminum). Such sources must be disassembled and cleaned, and some internal components replaced, every few months, depending on the usage. In addition, the erosion of internal components may result in contamination of the sample by metal ions. Morris[184] has used a discharge gun of modified design that also produces a mixture of

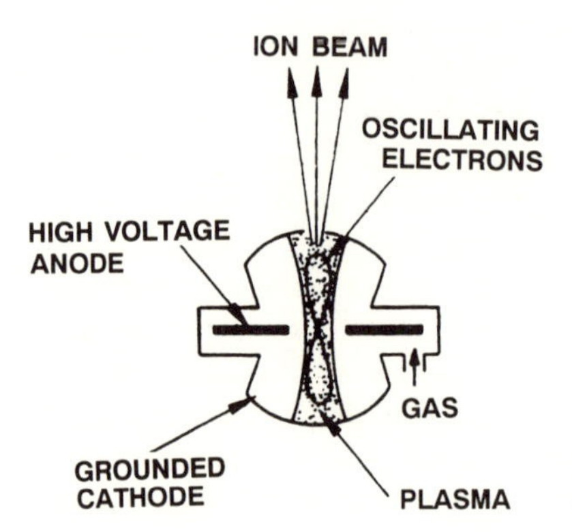

Figure 1. Saddle field ion source used for FAB mass spectrometry (adapted from Ref. 178).

ions and neutrals. The heat output of the source is reduced, and metal ion contamination of the sample is avoided. Since this source is operated at 10 keV, both ions and neutrals enter the source as bombarding species. For the organic samples analyzed, no differences between the spectra obtained with this source and a true neutral beam were noted.

A third type of "neutral beam" source is based on a plasma discharge creation of ions and efficient charge transfer in a coaxial flow of fast ions and neutral gas atoms (Fig. 2). The residual ionic component is removed by a series of electrostatic deflection plates. The source construction[182] and its applications[186–188] have been described. This source has been miniaturized to fit within the direct probe of many mass spectrometers so that source modification becomes unnecessary.[189] A gaseous discharge system within the source block for a quadrupole mass spectrometer is described by the commercial acronym DISIMS.[190,191] Again a mixture of ions and neutrals are used to bombard the sample to create secondary ions from biological samples held in a glycerol matrix.

A general disadvantage of the FAB source is the inability to manipulate the beam of atoms. Focusing cannot be accomplished, and the translational energy of the beam, once formed, cannot be changed. Significantly, the flux of the beam at the sample is only estimated and not measured, while a simple picoammeter established the flux for an ion beam. Finally, because of the requirement for a charge exchange process in most sources, there is not as wide a choice of primary particles as in SIMS, and all work reported to date has been limited to atoms of the noble gases. Current applications literature does not generally include a description of the source in terms of the relative ion/neutral content. The sensitivity of the FAB source is a function of the spread of the neutral beam, the angular spread of the sputtered secondary ions, and the angular and energy acceptances of the mass analyzer. The sensitivity of the first generation of FAB sources retrofitted to mass spectrometers can in general be improved by decreasing the gun–sample distance, and more carefully matching the ion emission and acceptance angles.

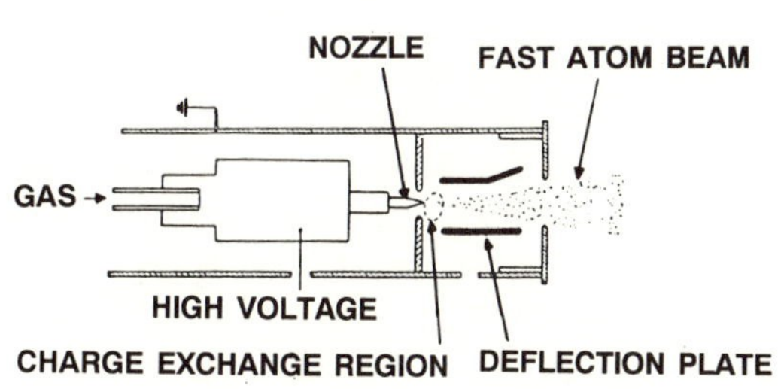

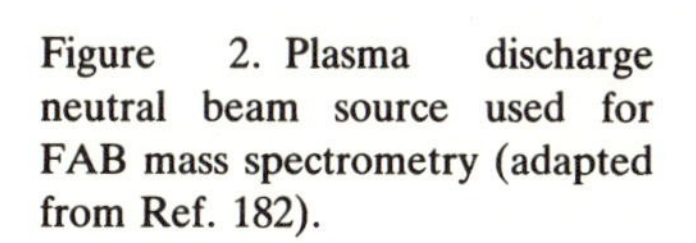
Figure 2. Plasma discharge neutral beam source used for FAB mass spectrometry (adapted from Ref. 182).

Most manufacturers modify existing sources in order to accommodate the FAB gun and the sample platform. Clearly, the sample platform for the liquid glycerol reservoir in FAB should be level. Most designs, however, force the sample stage to be held at an angle so that the geometrical relationship between the primary beam and the surface changes over the course of the experiment. In extreme cases, this can drastically reduce sensitivity because secondary ions can no longer be efficiently extracted from the source. Glycerol or other liquid matrices sputtered by the primary beam can quickly contaminate the source of the mass spectrometer and optical components further down the flight tube. This problem is most severe in high-energy instruments because such contamination can cause breakdown within the source. The contamination from the glycerol can extend as far back as the source slit, the cleanliness of which is crucial in obtaining higher resolution. Mass spectrometers of extended geometries (built for the purpose of extending the mass range of analysis) increase the distance between the FAB platform and the source slit and minimize this contamination.

Consistent with modern needs for fine focus neutral particle beams, such devices have been developed by Eccles *et al.*[606,607] and by Appelhans.[608,609] The latter device is based on the autoneutralization of SF_6^- along an extended flight path. The SF_6 neutral gun has been fitted to microprobe SIMS instruments as well as to an FTMS instrument.[610]

1.3.1.3. Plasma Desorption Mass Spectrometry. As developed by Macfarlane,[51] PD employs a ^{252}Cf radioactive source. This radioisotope is available in source strengths of up to several microcuries, encapsulated in a small metal disk of 1.25 cm diameter, which can be easily fitted into a source design. Although the californium is diffusion bonded to the platinum metal support by heat treatment at 1000°C, and then covered by a thin layer of gold, handling precautions must still be exercised. Energy deposition in the gold overlayer increases its porosity, and the integrity of the disk is not guaranteed. The ^{252}Cf nucleus decays with a half-life of 2.6 years; 97% of the emission consists of alpha particles and the remainder are massive energetic ions. Approximately 40 different fission fragment ion pairs are produced; each ion of the pair is emitted in the direction opposite to its partner. A fresh source (about 2 μCi) emits about 5000 ion pairs per second. Neither the mass, charge, energy, nor direction of each individual primary ion can be specified, and thus the identity of the primary ion which initiates a sputter event is unknown. A typical pair of fission fragments is $^{106}Tc^{22+}$ and $^{142}Ba^{18+}$, with energies of 104 and 79 MeV, respectively.

Californium is an inexpensive and reproducible source of energetic heavy particles for PD. The secondary ion fluxes from such bombardment are very low owing to the low primary ion flux, but use of a time-of-flight (TOF) mass analysis reduces the severity of this problem. TOF analysis provides a complete mass spectrum for each individual sputter event. The spectra can be integrated over a long period of time because of the low rate of sample consumption and the stability of the radioactive source. If the spectral integration time is relatively short, the sample can be recovered after analysis, a characteristic PD shares with static SIMS and some FAB work. Typically, the fission fragments irradiate the reverse side of the sample and support (Fig. 3). The primary ion traverses completely through the thin foil and causes desorption of secondary ions from the obverse face. Use of high-energy primary ion beams prepared and characterized in a linear accelerator allows the effects of primary ion velocity, mass charge, and angle of incidence on secondary ion yields to be determined, and the mechanisms of the secondary ion ejection can therefore be studied (Section 2). Although higher primary ion fluxes can be extracted from the accelerator than from the typical californium source, the beam is attenuated to a primary ion flux of about the same level to match the throughput of the time-of-flight mass analysis. The beam from an accelerator can be brought to a sharp focus by means of apertures and lenses, and imaging experiments with primary particles in this energy regime have been reported, albeit for inorganic ions.[192]

1.3.1.4. Laser Desorption Mass Spectrometry. The nature of the laser source used in organic mass spectrometric analysis has been quite varied. An excellent overview of applications in organic mass spectrometry is given by Hillenkamp;[193] a comprehensive description of the use of lasers in mass spectrometry has been prepared by Conzemius.[194] Both continuous wave and pulsed lasers of various wavelengths have been used

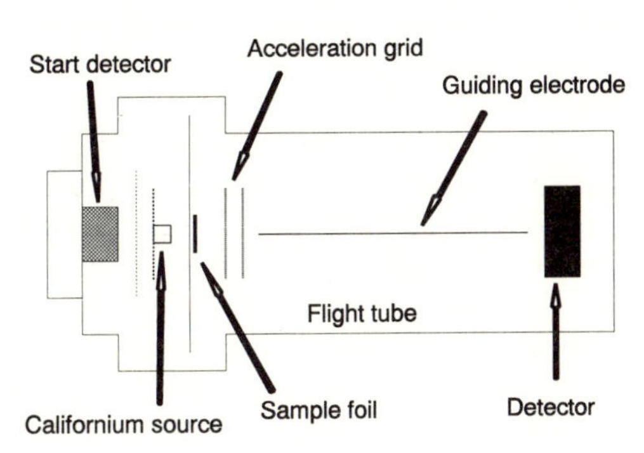

Figure 3. Schematic of a plasma desorption time-of-flight mass spectrometer.

as irradiation sources. Continuous irradiation with the 10.6-μm beam of a CO_2 laser has been used[195] for the desorption of sucrose (via cationization with sodium), citric acid, and quaternary ammonium salts,[196] and for saccharides, amino acids, nucleotides, and nucleosides.[197] Pulsed CO_2 lasers at the same 10.6-μm wavelength have also been used[198–200] for the analysis of organic compounds. The 1.06-μm wavelength of the Nd:YAG laser has been frequently used in LD experiments.[201,202] This excitation source, also used in the frequency tripled or quadrupled mode, is used in the commercial LAMMA instrument and LIMA instruments.[203,204] Beams from pulsed ruby lasers have been used in LD studies of organic compounds; 347-nm,[205,206] 483-nm,[207] and 694-nm[94,208,209] wavelengths have been used. Although there is some evidence of a wavelength dependence[103,210] in general, all of the laser sources used have provided satisfactory mass spectra of organic compounds, and a wavelength of choice has not emerged. Commercial availability of lasers has been reflected in the preponderance of work using the 1.06-μm wavelength. Pulsed lasers consistent with TOF mass analysis are used in most of the commercial instruments, even those based on scanning mass analyzers. A combination of a fast pulse repetition with a slow scan rate provides the illusion of a continuous wave source, at least as long as the sample survives (see Section 1.3.1.6). In recent work, much more attention has been given to selection of the wavelength, for the laser is also selected based on the matrices used to support the samples.[611,612]

The intensity of the laser irradiation can be easily varied over several orders of magnitude, and several workers have defined two regimes of secondary ion formation.[211,212] In general, lower irradiation leads to quasimolecular ions and a minimum of fragmentation. Higher irradiation fluxes lead to additional fragmentation, which can be used to deduce structural information about the molecule. The same general effects are observable in SIMS and FAB, and the effect of the primary beam flux will be discussed more fully in the next section. Successful analyses of organic samples have involved the use of energy densities ranging from 0.1 to 500 J/cm^2.

A distinction can be drawn between those laser sources with an imaging capability and those that produce a more diffuse irradiation of the sample. The Nd:YAG lasers and ruby lasers used in the LAMMA instrument can be focused to a spot as small as 3 μm in diameter. Imaging for inorganic components using such an instrument has been reported,[213–215] and more recently, a few applications to biological materials.[216–218] In an interesting application, the ion microprobe capabilities of the laser source was used to differentiate two related strains of bacteria.[219] More recent applications detail microprobe-based work.[613,614] This work deals mostly with compounds or substrates that resist degradation induced by the high primary beam energy density. In low-focus laser desorption sources, the spot size ranges up to about 1 mm in diameter to lower the energy density in the sample, and differences in sample morphology on a smaller scale cannot be determined.

1.3.1.5. Other Sources. Mention has already been made of the dust-particle-induced desorption method of Krueger.[15,16] The mass spectra obtained to date are similar to those created by the impact of more humble particles. Droplets such as those generated in electrospray can also be used as bombarding particles. If these droplets are composed of solvent molecules, the addition of these molecules to the surface at the point of impact may affect the spectrum obtained. This is in contrast to the SIMS and FAB experiment, where the primary particle is itself seldom observed in the secondary ion mass spectrum, with the exception of easily ionized cesium. If sufficient energy from any source is added suddenly to the sample, fracturing may occur. This in itself leads to the emission of ions from the surface[220,221] by a process known as fracto-emission.

1.3.1.6. Pulsed versus Continuous Sources. Pulsed sources for particle-induced desorption include PD, most LD, and a few SIMS instruments.[222,604,605] Most SIMS and FAB sources provide continuous irradiation of the sample. The operation of the mass analyzers must be appropriately matched to the operation of the source, and this requirement has been most clearly emphasized in PD and LD experiments. The LAMMA, as mentioned, uses pulsed irradiation in combination with a TOF mass analysis. Thus, as with PD, each pulse yields a complete mass spectrum. Continuous wave lasers can be easily used with either sector or quadrupole instruments, as they produce secondary ions continuously as the analyzer is scanned across the mass range. If the laser is pulsed fast enough, and the sector or quadrupole scanned slowly, this combination can also be used successfully. Persistent ion emission has been noted from many organic surfaces bombarded with pulsed laser irradiation,[223–225] especially at higher source pressures,[226,227] and this property eases the requirements for synchronization. Cotter[228–230,615,616] has developed an instrument that takes advantage of the extended neutral molecule emission from a bombarded surface. The experiment involves an adjustable time delay between the laser pulse and a pulse to an ionizing filament, allowing the time evolution of the desorption to be followed.

1.3.2 Mass Analyzers

The development of particle-induced desorption ionization has not only extended the accessible mass range in biomolecular analysis, but through the novelty of the ions formed, has consequently increased the need for careful mass spectrometric measurements and exacting spectral interpretation. It is appropriate to review briefly the basic properties of the mass analyzers that form the core of the instrument. The four types of mass analyzers described here are the quadrupole mass filter, magnetic and electric sectors, time-of-flight mass analysis, and ion cyclotron resonance, also known as Fourier transform mass spectrometry. These various types of mass analyzers form the great majority of commercial and custom-built instruments for particle-induced desorption ionization. Ion traps, developed in the past few years, have been joined to external FAB and SIMS ion sources, and now to electrospray ion sources, but a commercial instrument would still seem to be a few years in the future.

1.3.2.1. Quadrupole Mass Analyzers. The quadrupole mass filter analyzes the mass of ions by subjecting them to the action of electric fields (rf and dc) projected onto four colinear rod structures.[231–233] Ions follow an oscillating trajectory through the rod assembly, the spatial parameters of which depend on the m/z ratio. At a given ratio of the rf to dc fields, only ions of a given mass pass through the quadrupole; the trajectories of other ions intersect the rods or supporting structure. As commonly used, the quadrupole is a unit-mass-resolution analyzer. The low cost, small size, large ion entrance aperture, and wide tolerance of ion energy acceptance of quadrupole mass filters all contribute to their popularity. Quadrupole mass analyzers have been extensively used in SIMS instruments for inorganic analyses. Here their relatively limited mass range and inability to provide exact mass measurements are not impediments. Particle-induced-desorption ionization sources for organic analysis are now fitted to commercial quadrupole and multiple quadrupole instruments with a mass range of 1000 daltons, and with new high-performance rods, 4000 daltons, still with unit mass resolution through the range to about mass 3000. Analysis of still higher mass ions with unit mass resolution is not available with present quadrupole instruments, but ions of much higher mass can be crudely analyzed if the resolution of the mass filter is degraded to low values.[234] Quadrupole instruments operate with low source voltages of 10–50 eV, and the velocity of high-mass ions is thus relatively slow. These ions undergo fewer oscillations under the influence of the quadrupole fields, and limit the ability of the quadrupole to perform the mass analysis. Focusing of such ions into and out of the quadrupoles is also proportionately more difficult. These factors conspire to reduce the apparent transmission of high-mass ions through the quadrupole, which is also due to the fact that the slower-moving high-mass ions are less efficiently detected in a standard electron multiplier. The higher velocity of high-mass ions through a sector instrument aids in their detection in those systems.[235] Postacceleration detectors (with impact potentials of up to 30 keV) are fitted to most commercial instruments to increase the sputtered ion yield for higher-mass molecules. While the higher acceleration detectors are used with sector instruments, multipliers with acceleration dynodes of 5–10 kV are used on quadrupole mass spectrometers.

Quadrupole mass analyzers do enjoy the advantage of passing positive or negative ions simultaneously (their paths are identical except for the direction of rotation). With appropriate switching of source voltages, and the use of a dual detector system, ions of both polarities can be monitored with near simultaneity.

The recent use of electrospray ionization for the creation of multiply charged ions from high-mass biomolecules in the liquid phase has placed the quadrupole mass filter in a more competitive position for high-mass analysis. The multiply charged ions formed with high abundances from peptides fall in a distribution well within the mass range of the quadrupole mass filter. For singly charged ions of high mass, and for exact mass measurements, other mass analyzers must be chosen.

1.3.2.2. Sector Instruments. Magnetic sectors separate ions by momentum, and electric sectors by kinetic energy analysis; in conjunction with the value of the accelerating potential in the source, a mass-to-charge analysis is obtained. The mass range of magnetic sectors depends on the radius of their curvatures, the magnetic field strengths that can be generated, and the time ions spend under the influence of that field, related to the source acceleration voltage. For magnetic sectors, the equation that describes the interaction of these parameters is $m/z = B^2r^2/2V$, where B is the strength of the magnetic field, r is the radius of the sector, and V is the value of the accelerating voltage; m is the mass of the ion and z its charge. The development of particle-induced desorption ionization techniques, and FAB and SIMS in particular, has pressured manufacturers into providing mass ranges that were unheard of even a few years ago. Usually, increased range could be obtained by operating at a lower accelerating voltage V. However, the transmission of sector instruments falls off rapidly with decreased accelerating voltage for a variety of reasons. To maintain sensitivity, the accelerating voltage should be kept high. New technology in laminated magnet design has pushed the field strengths available to new highs. For many years, electromagnets could be expected to develop around 0.8 Tesla (1 tesla = 10,000 gauss). The top of the line magnets now available provide field strengths of up to 2.45 T. Since the dependence on B is squared, the mass range climbs rapidly. To further increase the mass range, the radius of the magnetic sector has also been increased. In some of the newer instruments, this value has more than doubled, increasing the mass range by a factor of more than 4. For FAB, mass ranges at an accelerating voltage of 10,000 V can now be specified to 12,000 daltons, rising to an incredible 40,000 daltons at 3000 V accelerating voltage. Until the last few years, only those with PD/TOF instruments could approach these ranges, which fall into the lower ranges of the mass range accessible through gel permeation chromatography. Implications of operation in the higher mass ranges of organic molecules are discussed fully in Section 5.3.

Magnetic and electric sectors can be combined in a double focusing mass spectrometer to provide exact mass measurements of ions. Only certain empirical formulas of atomic composition will sum to this exact mass, and this capability is especially valuable in the identification of unusual ions formed in particle-induced desorption ionization. Again, the value of the measurement of exact mass diminishes with increasing ionic mass. Further, exact mass measurements in overlapping isotopic envelopes of multiply charged ions (such as generated in electrospray ionization) is a complex experiment.

1.3.2.3. Time-of-Flight Mass Analyzers. The time of flight (TOF) mass analyzer is a system in which a packet of ions in the source is accelerated by a high potential into a field free flight tube. Lighter ions move with higher velocities, and thus arrive at the detector sooner than do the slower-moving heavier ions. The equation for TOF mass analysis in a simple linear system is $m/z = 2Vt^2/l^2$, where V is the accelerating voltage, t the flight time, and l the length of the ion path from source to detector. Typical flight times range from 10 ns up to 800 μs. Mass analysis is not a continuous process as in quadrupole or magnetic sector instruments, but takes place only as the ions are accelerated from the source. Processes that occur after this acceleration (such as dissociation or neutralization) do not alter the flight time of the ion, and can be observed with special techniques. Otherwise, the product ions appear in the same time channel as the precursor ion.

The TOF analyzer is used almost exclusively in PD, and has been used in pulsed SIMS and pulsed LD, including application in the commercial LAMMA and LIMA instruments. There are reports[236] of a TOF mass analyzer used in conjunction with a neutral primary beam. To a great extent, the success *and* the limitations of PD have been linked with the use of TOF mass analysis. The TOF provides a complete mass spectrum of secondary ions for each primary ion event that trips the timing start. Usually, the spectra are integrated over the time necessary to accumulate several hundred ion counts in the time channel. In combination with the relatively long cycle time (the time it takes for the highest mass ion of interest to traverse the analyzer), the overall result is a relatively long data acquisition. For instance, the PD spectra of palytoxin (a powerful marine toxin) and derivatives with masses of 2700–2800 daltons were acquired in times from 2.7 to 31 hours.[237] Later work with improved instrumentation has reduced this period; 1.5 h was sufficient to obtain a spectrum that contained the bovine insulin molecular ion.[3] With higher energy densities from a laser desorption source, acquisition times for even high-mass biomolecules (over 20,000 daltons) is reduced to a few minutes at most.

The TOF analyzer has several other special characteristics. The mass resolution is inherently low, on the order of several hundred. The positive molecular ion of insulin, for instance, is reported at a mass of 5730 ± 10 daltons.[3] In the mass range of up to several thousand daltons, it suffers by comparison with sector instruments which provide at least unit mass resolution. At higher masses, this distinction loses significance. Ion intensity for a molecular ion at a mass of several thousand daltons is distributed over an envelope as a simple consequence of the various isotopic contributions, primarily by ^{13}C, but also by ^{34}S, ^{15}N, and ^{18}O. The TOF spectrum contains a peak the centroid of which corresponds to the isotopically averaged mass of the molecule under study. It cannot generally be ascertained whether the "envelope" of the isotope peaks corresponds to that expected from a proposed empirical formula, as is possible with a sector instrument. Finally, the TOF analyzer is by nature unlimited in mass range (requiring, however, a pause of some milliseconds between primary ion impacts). As the mass range moves higher, however, the time window for detection must be lengthened. After acceleration, up to several hundred microseconds can pass before detection. This is often sufficient time for a number of reactions to occur which result in changes in mass and charge of the original ion. Recent TOF work by Chait and Standing in SIMS,[238] and Chait and Field in plasma desorption[239] show that the majority of ions of mass 2000–3000 daltons have undergone a change in mass or charge by the time they reach the detector.

Not all TOF mass analyzers used with PID are linear in configuration. Benninghoven has described a SIMS instrument in which time-of-flight analysis is combined with an electric sector analysis.[240,241] The design follows that of Poschenrieder.[242,243] In these combination instruments, the analysis of ions by mass-to-charge ratio is continuous through the electric sector. The processes that contribute to the broadening of ion peaks observed in the linear TOF instruments occur, but the products are not passed through the sector. The resolution is increased to up to about 5000. This increase is offset by a decrease in the absolute intensity of the signal recorded, since only those ions sufficiently stable to traverse the entire instrument will be recorded. The second generation of LD-TOF instruments includes what is termed a Mamyrin reflector, which decreases the effect on resolution of the initial kinetic energy spread of the ions formed in a pulsed source.[244,245] In simple terms, ions with slightly lower kinetic energy follow an appropriately shorter flight path through the instrument, while those that move faster are provided with a longer path. The device is operated so that the slightly slower and the slightly faster ions arrive at the detector at the same time. Such a device can also be used to investigate the degree to which various large molecular ions decompose along the flight path, as described recently by Nowak and Hercules.[246]

1.3.2.4. Ion Cyclotron Resonance Instruments. The development of Fourier Transform ion cyclotron resonance (FT-ICR) mass spectrometry (also termed FTMS) took place during the same period as did the developments in ionization methodology outlined in the historical survey. FTMS provides a mass analysis in a manner fundamentally different from that of the other mass analyzers. A small cell (typically 2.5 cm on a side) is placed between the poles of an electromagnet. Ions formed within the cell, and absorbing energy from an external rf field, move in circular orbits because of the magnetic field. The frequency of the orbit, termed the cyclotron frequency, is given by $w_c = zB/m$, where m is the mass of the ion, z is the charge, and B is the strength of the magnetic field. The frequency of the orbit can be easily related to the mass of the ion (Fig. 4). Orbiting ions within the small cell induce an alternating signal on receiver plates. Individual frequencies, and thus masses, are extracted from the composite signal derived from that plate via the application of the Fourier Transform.[247,248] This mathematical process transforms the signal in the time domain into the frequency domain from which the mass information can be deduced. Note that each transform provides the entire mass spectrum. Again, in practice, many transforms are averaged together to provide the spectrum.

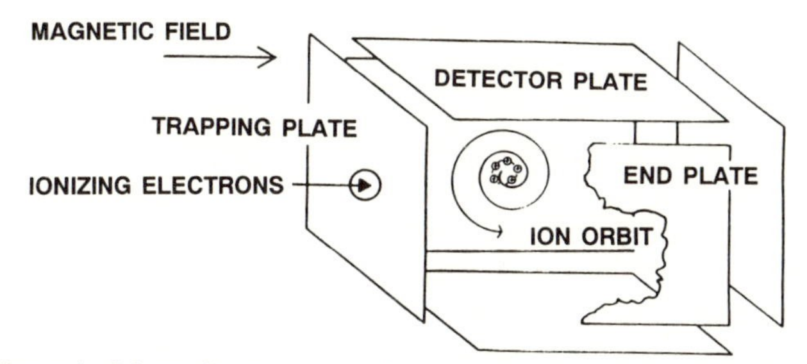

Figure 4. Schematic of an ion cyclotron resonance mass spectrometer.

An advantage of FTMS is that the magnet need not be scanned, but need only remain stable in field strength. Superconducting magnets are now available that provide high uniform field strengths across the cell, and the mass range of the FTMS instrument is expected to rise to near 20,000 daltons; the demonstrated mass range has been steadily increasing.[249,250] The basic obstacle to very high mass analysis would seem to be the electronics necessary to generate signals in the appropriate frequency regime.

By its nature, the FTMS instrument must operate at low pressures. This rules out the use of glycerol as a solvent, as is commonly practiced in FAB, within the primary analyzer cell. However, several other ionization methods operate successfully with FTMS. Lasers have been used to sputter metal ions into the gas phase for studies of gas phase organometallic chemistry, and also to ionize specific organic compounds supported on a surface.[251,252] SIMS using an ion beam as the primary particle has been reported recently by Russell.[253] The system must be set up so that the paths of the Cs^+ primary ions are parallel to the magnetic lines of force, but this is not a problem. Neutral beam bombardment has not been used because neutral beams are created by charge exchange processes within a relatively high-pressure cell. The gas load of such a cell is inconsistent with the pressure requirements of the FTMS. However, a finely focused SF_6 neutral atom gun (see Section 1.3.1.2) is now being used on an FTMS instrument.[610] Plasma desorption with a simple ^{252}Cf source has also been reported in conjunction with FTMS mass analysis.[254] This source provides no gas load to the cell, and can easily be mounted in a direct probe arrangement that also allows rapid sample introduction, in a manner similar to the miniature FAB gun in a probe described earlier. The ability of the FTMS to integrate signals over a long period of time makes this combination particularly attractive. In addition, the possibility for storage of high-mass ions within the cell for an extended period of time is attractive in compensating for low ion yields from the source itself.[255] Many researchers feel that the mechanisms of PD offer the best chance for the creation of very high-mass ions, and thus the PD FTMS combination may very well be the instrument of choice for high-mass analysis, especially in conjunction with accurate mass measurements.

A dual-cell configuration has been described for a commercial FTMS instrument which establishes a differential pressure between the analysis cell and the region of ion formation.[256,257] By so doing, many of the ionization techniques that operate in the higher-pressure regimes (that is, anything above 10^{-8} Torr) can now be used in conjunction with the FTMS mass analysis. For example, ions created by sputtering of a glycerol solution are conducted into the analyzing cell while the relatively high pressure of the glycerol itself is handled by a high-conductance pump in the ion source. Although there can be discrimination problems in the transfer of ions from the source cell to the analyzer cell through the conductance-limited ion aperture, such a dual-cell arrangement allows great freedom in the use of a number of ionization methods on the high-pressure side of the divided cell. Chemical ionization, fast atom bombardment, and even electrospray ionization have been used with the dual-cell design.

1.3.3. Detection of Ions

Most mass spectrometers use a standard electron multiplier to detect mass-analyzed ions after passage through the instrument. Positive or negative ions are accelerated into a surface held at the appropriate high voltage. Impact releases several electrons that then initiate a cascade amplified through the multiplier, yielding a measurable current which is taken as signal and fed directly to a recorder or data system. As a group, the methods of particle-induced desorption ionization can be used to produce higher-mass ions than were previously possible. The efficient detection of high-mass ions, however, requires that specific attention be given to modification to the detection system. Ions of large mass-to-charge ratios do not reach a velocity sufficiently high to initiate an electron cascade at the first dynode of a conventional electron multiplier. This realization led Friedman[234,235] to construct an elaborate post-mass-analysis detection system in order to accelerate very high-mass ions to threshold velocity and beyond; voltages of up to 400 kV could be used. The threshold velocity for secondary electron emission was determined to be between 10^6 and 10^7 cm/s for positive ions of m/z up to 6×10^4 impacting on a copper dynode. Practically, commercial mass spectrometers solve the detection problem with more modest potentials. It was mentioned earlier that mass ranges of about 12,000 daltons at a full accelerating voltage of 10,000 V are now available, and that at 3000 V accelerating voltages, ions of m/z 40,000 could be analyzed. Postacceleration detection systems employing voltages up to 30,000 V are now in use to ensure the detection of high-mass ions in sector instruments. A simple calculation (kinetic energy = $\frac{1}{2}mv^2$) reveals that the impact velocity of an ion of 40,000 daltons accelerated at 3000 V (source acceleration potential) will be 4×10^5 cm/s. Even with substantial postacceleration at the detector, such ions appear to be on the border of the detection threshold. The situation is further complicated by the fact that the efficiency of ion detection varies not only with the kinetic energy of the impacting ion, but also with its chemical structure.[258] The recent incorporation of conversion dynode electron multipliers[259] and postacceleration detection systems[260] into commercial instruments has allowed organic ions up to mass 10,000 daltons or so to be detected, but the efficiency is low and the extension to still higher masses not assured. Recent work is still exploring the fundamental aspects of operation of electron multipliers and channelplates.[261] Time-of-flight mass analyzers, which are presently producing ions of m/z 25,000 from organic compounds, must also deal with the efficient detection of high-mass secondary ions. A recent design[262] uses a microchannel plate detector in conjunction with a simple electron converter. A factor of 10 increase in sensitivity was observed using the electron converter with incident ions (m/z 1255) at 10 keV; the multiplication factor falls off at higher incident ion energies.

In assessing the efficiency of detection for electron multipliers, a distinction should be drawn between a signal derived from the impact of a high-mass organic ion, and that measured for the impact of a high-mass cluster ion such as those of the ubiquitous cesium iodide. The detection of ions of the latter type is facilitated by the fact that the cesium-ion-containing cluster ions readily dissociate on impact with the first dynode of the electron multiplier, leading to a large number of secondary ions that can be further accelerated to create the electron cascade within the electron multiplier. Further cesium atoms are implanted into the dynode surfaces with each impact event. Cesium and its salts exhibit a high secondary electron yield. Greater concentrations of cesium lead to a higher electron yield with each subsequent impact event. In fact, cesium-doped surfaces are now available in a line of commercial electron multipliers promising higher sensitivity of detection at higher masses. On the other hand, large organic ions produce fewer charge carriers in the dissociation that follows initial impact with the surfaces of the electron multiplier. Further, the carbon content of the ions can lead to fouling of the surfaces. Sputter cleaning keeps the first few dynodes of the system clean, but moves the organic contamination to the back dynodes of the electron multiplier, where the change in the surface exhibits a cumulative effect in the decrease of the gain of the system. A simple calculation shows that each day of normal operation places a few nanograms of organic material onto the surfaces of the electron multiplier. For low-mass organic ions, most of the organic material is removed as volatile gases. For large organic ions, such transformation processes are much slower, and detection efficiency inevitably suffers. The recent introduction of commercial electron multipliers in which replacement of the *last* dynode restores original gain specifications is evidence for this contamination problem. Some commercial manufacturers have introduced photomultiplier tube detectors joined to a system that includes a material that scintillates under ion impact. The relative efficiencies of these devices for the detection of high-mass ions is yet to be determined.

2. Spectral Effects of Primary Beam Parameters

2.1. Obverse or Reverse Irradiation

From the preceding sections, it is clear that FAB and SIMS are relatively low energy techniques, dealing with particle energies in the 5000–40,000-V range, while PD deals with particles with kinetic energies in the MeV range. The depth of penetration of the low-energy particles into a solid can be estimated as perhaps as a few hundred angstroms at most, although the actual effect of energy deposition may extend beyond this, especially at longer times and in liquids. In contrast, the high-energy particles in PD can penetrate solids to depths of up to several tens of microns. In practical terms, this distinction means that the former analyses *must* bombard the sample from the same direction into which secondary ions are extracted, while in the latter the primary particle can pass completely through the sample support, and that secondary ions can be extracted from either sample face. Variation in secondary ion intensity with irradiation of the obverse (front) or reverse sides of organic samples has not been systematically investigated. Practically, it has not been feasible to fashion a sample thin enough to allow such studies for SIMS or FAB.

A few suggestive studies have been completed. If the energy of the primary ion beam in a SIMS experiment is raised to 25 keV, sputtering of organic ions from quaternary ammonium organic films can be observed from the obverse upon irradiation of the reverse.[263] In standard nomenclature,[264] the angle of incidence in this arrangement is 180°. In a "liquid SIMS" experiment, the highest angle of incidence usually reported is about 85°[265] in an "on line" primary beam configuration, and for neutral beam bombardment, a 70° angle of incidence. Several advantages might accrue for reverse irradiation with either a primary ion or a neutral beam. Momentum transfer within the sample could be directed to create secondary ions that could be efficiently extracted into the mass analyzer, as compared to the wide angular divergence of ions emitted from a surface sputtered from the obverse. Complete traversal of the sample might also produce ions that would not survive formation via a more complex spike or cascade mechanism. A thin film of the liquid matrix could be drawn into a thin film over a support, and the primary beam irradiating the sample from the reverse side would travel completely through this thin film. After passage of the primary particle, the film would heal itself rapidly and reform in time for the next primary particle passage.

In PD, the energy of the primary beam (MeV) is sufficient to completely traverse the sample and thin support (total thickness of 2–5 μm), and a sample thickness of 50–100 ng/cm^2. In PD, both obverse[266,267] and reverse irradiation has been used to sputter ions of organic compounds, although the majority of instruments use the latter mode. Apparently, both of these geometries provide satisfactory spectra. Although a great deal of energy is deposited within the solid (10 keV of energy is lost in passing through a sample of 1 μg/cm^2 thickness), the energy density at the far side of a sample foil as the primary ion exits the foil is not significantly different from that at the front side.

In LD work, custom built instruments typically irradiate the obverse of the sample. The original LAMMA 500 instrument was based on a geometry in which the reverse of a thin sample foil was irradiated ("transmission" mode). The laser pulse was so intense that a small (20 μm diameter) hole was created in the sample foil.[268] Ions that are formed along the track of the laser beam and in the discharge that was created were extracted into the TOF mass analyzer. The newer LAMMA 1000 was designed for the analysis of thicker samples which could not be punctured by the laser beam, but is otherwise identical in execution. (Obverse irradiation is termed the "reflection" mode.) The properties of the laser-induced discharge are essentially independent of the direction of the irradiation. Thus the spectra of organic samples can be expected to be the same from the LAMMA 500 or the LAMMA 1000. An instrumental complication is the fact that the ions produced by obverse irradiation of thick samples usually have a broader kinetic energy than those produced by perforation, and a more sophisticated time-focusing TOF analyzer must be employed to maintain the resolution of the secondary ion mass analysis. Such an effect has not been noted in the PD experiments described above. The independence of the LD spectrum from the geometry of irradiation suggests the intermediacy of common secondary processes within the discharge plasma, as explicitly noted in recent papers that interpret a wide variety of spectra to generate a model of laser desorption and ionization.[569,570]

2.2. Angle of Incidence of Primary Beam

The angle of incidence is defined as the angle between the path of the primary beam and a line normal to the surface of the point of impact.[264] For obverse irradiation, then, the angle of incidence lies between 0° and 90°, and between 90° and 180° for reverse irradiation. Many earlier reports in the literature used the opposite definition of this angle, but most now conform to this convention. The true effect of the angle of incidence on the yield of secondary ions from organic compounds is ascertained only when the detailed morphology of the surface is controlled. This is seldom the case, even when solid rather than liquid samples are investigated. Many samples are prepared by burnishing the solid onto a support; on the microscopic scale, the angle of incidence is indeterminate. In an attempt to prepare more uniform sample layers, samples can be electrosprayed[269,270] onto the support at concentrations chosen to give thin coverage of samples on the supper. However, even with careful sample preparation, agglomeration of polar sample molecules occurs spontaneously[271] and may occur even more readily under sample irradiation. When samples are dissolved in a liquid matrix, the situation becomes even more complicated. Glycerol on a tilted sample platform will distort under the force of gravity. In addition, a constant evaporation of the liquid matrix occurs over the course of the experiment. Both of these factors will alter the effective angle of incidence for each primary ion impact. On the microscopic scale, electrohydrodynamic and sputtering phenomena will change the angle of incidence almost at random.

The extraction efficiency of secondary ions into the mass analyzer also contributes to the experimentally determined "optimum" angle of incidence. The cone of angular acceptance for sputtered secondary ions varies from instrument to instrument. Determination of the optimum angle of incidence, as reported in many publications, cannot be expected to reflect directly the mechanism of the sputtering process. Practically, the angles of incidence in molecular SIMS instruments ranges from 20° to 70°. An angle of incidence of 45° seems to be a reasonable value arrived at by commercial manufacturers. The instrument of Winograd[272,273] was constructed to determine the angular anisotropy of secondary ion emission.[274,275] The results obtained with such investigations are described in another chapter in this book. Aberth has completed a preliminary study of the angular dependence of the spectra obtained with Cs^+ ion irradiation of an organic sample in a glycerol matrix (the "liquid SIMS experiment).[265] An angle of incidence of 10° was compared to an angle of 15°. However, in addition, the extraction optics of the mass spectrometer were changed. In positive ion analysis of vitamin B_{12}, for geometry A, greater fragment ion abundances are observed relative to the base peak, but the background signals from the glycerol matrix at these masses were also more abundant. Background ions at low masses (less than 700 daltons) are more intense in geometry B. The main features of the spectrum are the same in either case.

As mentioned, the effective angle of incidence depends upon the ability to control the morphology of the surface. Benninghoven has developed methods for the precise construction of organic surfaces.[276,277] The leveling effect of the surface morphology might be avoided in such a situation, but a study of angle of incidence of the primary beam has not been carried out. Of particular interest are SIMS studies which show that the yields of organic compounds could be increased using highly porous silver supports.[278] These results are analogous to those that report increased yields via mechanical abrasion of the surface.[279] Many methods are available for the preparation of thin films of organic compounds,[280,281] but are usually optimized for one particular type of organic sample. For a general sample preparation method, wide applicability is required. A most promising approach is the preparation of thin films or organic cations supported on an ion exchange resins of known structure.[282] The support of thin organic films on a liquid metal support of known shape may also allow the surface morphology to be controlled.[283]

In FAB, early reports emphasized the optimum angle of incidence of 70° for secondary ion formation from organic compounds.[44] However, using a different instrument, other workers have found the optimum angle of incidence to be 60°.[284] In this latter work, the range of angles from 30° to 90° was investigated; the difference in absolute ion intensities was within a factor of 5 across this entire range.

In PD using the ^{252}Cf source, no precise correlation of secondary ion yield with angle of incidence can be drawn because of the inherent spread in angular emission of fission fragments. Typically, the geometrical configuration involves either a colinear arrangement of ^{252}Cf source and sample[285–288] or with small angles of incidence ranging from about 20° to

45°.[289,290] When an accelerator source is used for PD, the secondary ion yield as a function of the angle of incidence of the primary ion beam can be followed quite closely.[291,292] A range of incident angles from 10° to 78° was investigated with an angular resolution of 0.5°.[291] Secondary ions from glycylglycine $(M + H)^+$, ergosterol M^+, and insulin $(M + H)^+$ were used as models. The primary ions $^{16}O^+$, $^{32}S^+$, $^{63}Cu^+$, and $^{127}I^+$ were used at energies ranging from 2 to 42 MeV. In general, the secondary ion yield increases with an increasing angle of incidence. More detailed analysis showed that for some sample–primary ion beam combinations, the distribution could be approximated by a $\cos^{-1}$ relationship, while other combinations produced data more accurately described by a $\cos^{-2}$ curve. The relevance of the data to various theoretical models of MeV particle-induced sputtering and ionization is discussed by these authors. Using 9-MeV O^{n+} ions incident (obversely) upon a homogeneous layer of valine, Nieschler[293] also concluded that the secondary ion yield increases rapidly with the angle of incidence, and that it deviates strongly from a $\cos^{-1}$ distribution at large angles.

Changes in the yield of organic secondary ions with changes in the angle of incidence of a laser beam has not been systematically investigated. The geometry of the LAMMA instrument is set to an angle of incidence of 30° (LAMMA 1000) or 180° (LAMMA 500). Custom-built LD instruments use angles of incidence of 0°[97,224] or 45°.[207] Heresch[223] found that secondary ion yield was maximized when the angle of incidence reached 45°. The effect observed is again a convolution of the effects of ion extraction and the angle of incidence of the primary ion beam. More detailed experiments need to be carried out with an instrument designed to minimize changes in secondary ion abundance due solely to ion extraction. Again, one notes that the angle of incidence is a fairly crude *instrumental* parameter. The use of a sighting laser in commercial laser microprobe instruments allows the operator to target individual particles of the sample for analysis. These particles vary in size and shape, and of course, the angle of incidence of a 2-μm diameter laser beam on a 1-μm diameter particle encompasses all possible angles of incidence.

The empirical nature of the secondary ion yield as a function of the angle of incidence is apparent from this discussion. In instruments that bring the sample into the source on a direct probe assembly, the operator determines the optimum angle of incidence by rotating the sample in the vacuum lock until a maximum in secondary ion current is obtained. Reproducible positioning of the probe within the source of the mass spectrometer is difficult, although some of the newer direct insertion probes are fitted with a calibrated rotation barrel.

2.3. Charge State Dependence

When FAB was first introduced, much discussion centered on the difference between charged and uncharged particle bombardment of surfaces. This argument had its origins in work with inorganic SIMS, which had shown that positive ion bombardment can lead to charging of insulating surfaces, while neutral particle bombardment did not. Practical solutions to this problem had been devised, and are well known in both inorganic SIMS and electron microscopy. Charging of insulating surfaces can be reduced to some extent by bombarding with a negative ion beam.[299] Additionally, the surface can be irradiated with a diffuse beam of electrons from an electron "flood gun" in order to neutralize any positive charge built up on the surface.[295–296] Surface pretreatment can convert an insulating surface into a surface that will resist charge buildup; one of the strategies adopted was the incorporation of a thin metallic layer on the surface,[297] or a fine metal wire grid.[298] Antistatic organic compounds have also been sprayed onto surfaces that otherwise might accumulate an electrical charge.[299]

Bombardment with either a charged ion or a neutral atom of keV energy makes no difference in the creation of secondary ion mass spectra from organic compounds. As discussed earlier, a neutral beam can be directed into a mass spectrometer source regardless of the operating potential, and this is a considerable practical advantage. The advantage of the ion beam is the focusing that can be obtained and the direct measurement of the flux of primary particles bombarding the surface. However, in terms of the deposition of energy into the condensed phase, the initial charge of the particle is irrelevant. Aberth[133] has compared the secondary ion mass spectra of bioorganic samples obtained by bombarding with xenon neutral particles or cesium ions. No change in the spectra are observed. The advantage of ion beam focusing is reflected in a threefold enhancement of the intensity of the secondary ions. This study is noteworthy in that both primary beam sources are mounted within the source; most of the instrumental factors that might otherwise affect the comparison are canceled. A comparative study between SIMS and FAB using organic polymers[300] came to a similar conclusion with the caveat that sample charging must be avoided, as these surfaces can act as insulators. Liquid polymers have been studied by both SIMS (Cs^+) and FAB (Xe°), and the spectra obtained are identical.[301]

The indistinguishability of charged versus neutral particle bombardment in secondary ion mass spectra has also been discussed by Magee.[302] There may be subtle differences that have not yet been explored. Diffusion of ions through a solid matrix may be influenced by the charge state of the primary particles. The studies of McCaughan,[303] among many others, show that ion bombardment can cause impurity ion migration in insulating films. Briefly, the argument is as follows. As a primary ion approaches the surface of an insulating material and is neutralized, it induces a polarization of the material. This effect is modified in strength by the dielectric constant of the insulator. Ionic species initially strongly bound within the insulating material are freed during the electron transfer required for primary ion neutralization. Ion migration depends on the mobility of the species, which depends upon their initial charge state and their chemical environment. Primary ion beams that cannot undergo neutralization at the insulator surface do not cause migration, and neutral particle bombardment also reduced the amount of ion migration. Organic ions can be expected to have a lower mobility through a solid matrix than alkali metal ions. Their diffusion under the influence of a charged beam may become significant when highly organized surfaces (such as monolayers) are investigated. In such situations, preferential channels of ion movement (along an ion channel, for example) may be available. When liquid matrices are used (as long as they are not insulating), charged-particle-induced migration is insignificant compared to normal diffusion processes.

Although the charge state of the primary particle is of no consequence in the mechanisms of sputtering of most organic compounds, there may be consequences of the fact that the distribution of energetic photons from ion sources and from atom sources differ.[618] The fact that an atom source *must* be in a line-of-sight configuration with the sample platform can lead to photoinduced reactions of the sample,[619,620] particularly since there are many reactive species in the solution produced by the impact of the primary particle beam itself.

The dependence of the secondary ion yield on the charge state of the primary particle has been carefully investigated in PD. Using well-defined ion beams extracted from accelerators, the charge-state dependence of organic secondary ion desorption has been studied by the Uppsala and the Erlangen plasma desorption groups. The first work of Hakansson in 1981[304] investigated the desorption of ergosterol, glycylglycine, and cesium iodide with ^{16}O ions in charge states varying from +2 to +8, at a constant 20 MeV energy. The fragmentation patterns in the spectra of ergosterol and glycylglycine did not change with different charge states of the incident primary ion. A pronounced charge state dependence of yields was found and observed to be similar for the M^+ of ergosterol, the $(M + H)^+$ of glycylglycine, and Cs^+ from cesium iodide. Yields at higher charge states (>+5) were observed to vary with q^4 (q is the charge of the incident ion), in accord with several descriptive models that describe sputtering yields as a function of electronic stopping power. At lower charge states, yields higher than those predicted by the same model were observed. Duck[305] used ^{16}O at 20 MeV and 25 MeV to investigate the charge state dependence of secondary ion yield from valine, using the various ions $(M + H)^+$, $(M + Na)^+$, and $(M - COOH)^+$. Charge states between +4 and +8 were investigated; the data show an essentially linear dependence of secondary ion yield on charge state.

In recent work,[306] the secondary ion yield of phenylalanine was investigated when bombarded with ^{16}N in states from +2 to +7, and ^{238}U in states from +26 to +49. In varying the charge states, a three orders of magnitude increase in secondary ion yield could be obtained, but with some changes in the fragmentation pattern as well. A unique q^n dependence was not observed; the exponent required to fit the data varies from 0.8 to 4.3, and depends both on the primary ion chosen and the nature of the secondary ion monitored. The secondary ion yields of valine were investigated using ^{32}S, ^{16}O, and ^{12}O MeV ions of various charge states.[306] The yield was found to vary in a simple way with the primary ion charge, but this dependence was found to be different for positive and negative secondary ions. Previous work had been limited to the investigation of positive secondary ions.

In these PD experiments, obverse irradiation must be used to preserve the chosen charge state of the primary ion beam. When the beam is passed through a thin foil before striking the organic target, a steady charge state distribution is obtained;[307] yields of organic secondary ions are then found to be invariant. No definitive charge state dependence in the MeV range of particle bombardment has been established in these studies. In MeV as in keV particle bombardment, the charge of the primary ion changes as it penetrates the surface. The rate of charge equilibration is a parameter that complicates the interpretation of the data. Models of the MeV sputtering process combine the charge of the primary ion with its mass and velocity into a multiparameter function[308–311] that is then used to describe the creation of secondary ions as a function of primary ion parameters.

2.4. Energy Dependence

2.4.1. Wavelength Dependence in Laser Desorption

The wavelength of lasers used in LD studies of organic molecules ranges from 250 nm to 10.6 μm. Systematic variations in organic secondary ion production with wavelength have not been satisfactorily established, although a relationship with the properties of underlying support substrates is now clear. In laser desorption mass spectrometry as used in organic mass spectrometry, there are two processes occurring that are difficult to separate. The first is the sputtering of organic ions by a true laser desorption process, which is defined in Hillenkamp[193] as a collective, nonequilibrium process in the condensed phase. The second process is the absorption of laser energy by the substrate on which the organic material is placed. If enough energy is absorbed, a thermal desorption of the organic ions can occur. The former process should not exhibit a wavelength dependence, while the latter may exhibit the expected dependence based on the absorbance properties of the support material. Again, these effects are convoluted with the ion analysis parameters of the instrument, specifically the energy range of ions accepted and the time window of their acceptance, as well as with the spectral properties of the sample support itself.

Antonov[103] first reported on the desorption of adenine molecular ions at laser wavelengths of 249, 308, and 337 nm, and reported a decrease in molecular secondary ion yield with increasing wavelength. The maximum yield of sputtered ions was found at a wavelength for which the adenine molecule had a strong absorption. In irradiating crystals of anthracene, a 10^2 enhancement in molecular secondary ion yield was obtained using 249 nm rather than 308 nm wavelength laser irradiation at equal intensities.[103] Later results[312,313] suggest that selective desorption of samples may be possible via incorporation of wavelength-specific chromophores into the organic molecule. Investigations with tunable lasers are expected to yield the crucial results. A persistent experimental difficulty is the separation of the wavelength dependence of the condensed phase energization (be it support or organic sample) from the dependence on wavelength of the gas-phase photoionization process. Selective ionization of molecules in the gas phase with laser radiation is a well-known experiment,[314–317] and has recently been elaborated with the photoionization of neutrals sputtered from a surface by an incident particle beam.[318–320]

Hillenkamp[321,322] expanded the concept of wavelength-dependent desorption of the sample to include the spectral properties of the sample support itself, and this has been a very active area of recent research.[611,612] Tryptophane was originally used as an energy-absorbing matrix for 266 nm light, and higher mass nonvolatile compounds such as stachyose, valinomycin, and cyclodextrin were successfully desorbed from this support. The LD mass spectra contained only the protonated or sodium-cationized molecular ions. Nicotininc acid was used as an alternative solid energy-absorbing matrix, while *o*-nitrophenyloctylether was used when a liquid solvent was required. It is thought that the matrix is the primary receptor of the energy imparted by the laser burst, and transfer then occurs to the organic sample overlayer in a more gentle process that reduces the degree of fragmentation observed. Several new matrices specifically developed for the matrix-assisted laser desorption of proteins have been described.[611,612] Nicotinic acid matrices previously used in LD methods produced photochemically induced adduct peaks in the mass spectra that could confuse the assignment of the molecular weight of a truly unknown sample. Cinnammic acid derivatives (ferulic, caffeic, and sinapinic acids) are presently in use to absorb laser energy at a wavelength of 355 nm, and to facilitate the desorption of large (up to 65,000 daltons in molecular weight) proteins supported on the surface.

2.4.2. Particle Mass and Velocity Dependences

2.4.2.1. Mass Dependence in SIMS and FAB. In SIMS, the dependence of atomic secondary ion yields on the mass and velocity of the incident ion has been the subject of extensive research.[323–328] Investigations into organic secondary ion yields have not been carried out with the same rigor. This section summarizes the few reports that deal with the sputtering of organic secondary ions as a function of the mass of the bombarding particle. Since the total kinetic energy remains the same, the velocity of the more massive particle will be reduced proportionately.

Argon is often used in SIMS and FAB studies because of its chemical inertness and the facility of its neutralization by charge exchange. However, both SIMS[329] and FAB[330] studies have demonstrated an increase in organic secondary ion yield with use of xenon (a mass of 131 daltons) rather than argon (40 daltons). The observed molecular species are 2–4 times more intense for xenon particle bombardment, but the mass spectra are nearly identical. The increased cost of xenon over argon is not a factor because of the low rate of consumption by primary gun (in the absence of leaks).

Rollgen[134] has used a mixture of mercury ions and atoms to bombard organic samples. Mercury was used because of its high mass (200.5 daltons) and its high vapor pressure at room temperatures. Samples of saccharides, amino acids, and a peptide showed an enhancement factor of about 10 for quasimolecular ions $(M + H)^+$ or $(M + Na)^+$ when mercury rather than argon was used as the bombarding particle. No changes in the pattern of fragmentation were observed. SIMS spectra have also been obtained by bombardment of an organic in glycerol with the molecular ions of organic compounds.[136] A mixture of ions formed from a siloxane diffusion pump oil in the mass range of about 200–600 were accelerated to 7 keV and used as primary ions. A secondary ion yield enhancement factor of 10 was observed for stachyose, but was not consistent for all of the oligosaccharides. Primary ions from perfluorotributylamine (mostly below m/z 200) were also used to bombard organic samples, again with varying results.[136] Interestingly, here the fragmentation processes in the SIMS spectrum vary with the mass and nature of the primary ion. Siloxane ion bombardment increased the abundance of fragment ions resulting from loss of water in the spectrum of at least two saccharides; other fragment ions became less abundant. Although the number of reported experiments are few, it can be seen that the secondary ion yield and the mass of the primary ion are not simply related.

The most systematic study of this primary beam parameter has been completed by Standing, using the TOF SIMS instrument at the University of Manitoba. This work describes the secondary ion yields for the $(M + H)^+$ ion of alanine as a function of primary ion mass at a number of energies.[331] At a fixed kinetic energy of 6 keV, the relative yield was 1(Li^+), 12.5(Na^+), 37.5(K^+), and 500(Cs^+). These factors remained essentially constant across the energy range 1–10 keV for the primary ion beam. Some changes in the fragmentation pattern were noted, with cesium ion bombardment producing the least fragmentation.

Barofsky and co-workers have investigated the use of liquid metal ion guns as primary beam sources in molecular secondary ion mass spectrometry.[142.143] Using the $(M + Na)^+$ of stachyose as a model, the relative yields for Ar^0, Ga^+, In^+, Xe^0, and Au^+, were reported to be 1, 8–12, 35–40, 3, and 45–55, respectively. The masses of these primary particles are 40, 70, 115, 13, and 197 amu, respectively. There appears to be no simple correlation between the mass of the primary particle and the yield of secondary ions. The authors suggest that bombardment with heavier particles is more conducive to spike formation within the condensed phase, and that more efficient sputtering occurs within this regime.

It is likely that future developments will include the bombardment of organic samples with primary particles of very high mass. Inorganic halides form cluster ions which can be selected and used as a primary ion beam. Clusters of organic molecules, such as water or glycerol clusters, might also be used. Finally, rare gas molecular clusters can be formed by sputtering frozen surfaces of these gases.[332,,333] A molecular beam/cluster ion source[334,335] might also be used to generate a primary particle beam.

In summary, an increase in organic secondary ion yield is noted as the mass of the bombarding ion is increased, although the total kinetic energy remains constant. Standing's results[38] also show that for an ion of given mass, the secondary ion yield increases as a function of the velocity for small organic ions, a trend confirming that which had been established in the study of inorganic surfaces. For organic samples, the difficulty in establishing the mass and velocity dependence of the secondary ion yield rests ultimately on the ability to reproducibly create organic surfaces of the same morphological character.

Hunt and co-workers[592–594] have followed earlier work of Thomas *et al.*[621] in studying the secondary ion yields of organic molecules with the use of polyatomic ions as primary particles. Primary ions in the energy range of 400 keV to 4.5 MeV were used. Valine, among several other organic samples, was used to prepare a model surface for these studies. A collective effect on the secondary ion yield was noted that results in greatly increased secondary ion yields for all of the polyatomic incident ions across this energy range. The greater-than-linear enhancement is thought to be related to an enhanced electronic stopping power for the polyatomic incident particles.[622]

2.4.2.2. Velocity Dependence in PD. The yield of organic secondary ions on the velocity of the primary particle has been investigated in PD. Two early studies have yielded somewhat different results. Albers[336] has studied the yield of valine $(M + H)^+$ and $(M - COOH)^+$ ions as a function of the velocity of incident $^{16}O^+$ ions. The yield reaches a maximum around a velocity of 1 cm/ns, and perhaps a second maximum at a lower velocity. A different dependence was noted in the studies of Sundqvist,[337] who used ergosterol and glycylglycine as samples, and primary ions of $^{16}O^+$, $^{32}S^+$, $^{63}Cu^+$, and $^{127}I^+$.

A recent study[338] investigates organic secondary ion yield in PD as a function of primary ion energy, and uncovers different dependences for positive secondary ions desorbed from polar and ionic compounds, as well as for positive and negative secondary ions desorbed from the same polar sample. Such differences were found when the initial studies with valine were expanded to a range of organic molecules, including phenylalanine, the dyes crystal violet and malachite green, the antibiotic chloramphenicol, and the antiarrhythmic drug verapamil. Samples were electrosprayed onto a thin aluminized Mylar foil, and bombarded with $^{16}O^+$ ions in the energy range from 10 to 40 MeV. For polar molecules, the organic secondary ion yield falls off at higher energies, but the positive ion decay can be described as a n^2 dependence, whereas the falloff ion negative secondary ion intensity follows an n^4 dependence. For preformed ions, positive ions shown an n^4 decay. The authors interpret these data to reflect surface–sample interactions of various forms and strengths, including Coulomb effects for ionic compounds and various polar interactions for the other molecules investigated.

The yields of organic ions under high- and low-energy bombardment have been compared.[339] Several organic molecules were used to compare yields under bombardment by either 54-MeV $^{63}Cu^{9+}$ or 3-keV $^{133}Cs^{+}$. The faster heavy ions produce a greater secondary ion yield for all of the organic systems studied. The ratio of the high-energy/low-energy yield increases with the mass of the *sample* molecule, ranging from 1.5 for a secondary ion of Cs^{+} to about 200 for the secondary ion $M^{+\cdot}$ from bleomycin. Note that the difference in velocity of the primary ions in this experiment is quite large, ranging from 7×10^{-2} cm/ns for Cs^{+} to 3.9 cm/ns for Cu^{9+}

2.5. Primary Particle Flux and Dose

In low-energy particle bombardment, static conditions are chosen so that the probability of sampling a surface site previously disturbed by primary particle impact is low.[19] This calculation presumes that the surface is itself static and thus the limit does not apply to samples in which diffusion can be significant (liquid SIMS or FAB). The transition from static to dynamic SIMS for organic samples has been demonstrated to result in increased fragmentation.[340] Typically, the absolute yield of a given secondary ion increases with an increasing primary ion flux. This can be considered as a simple summation of individual sputter events within the static limit. At higher fluxes, increased sputtering is partially offset by increased damage to the sample, leading to less of an increase than otherwise expected. The exact nature of the dependence depends on the secondary ion chosen and the nature of the sample preparation.

In FAB or SIMS that involves samples in solution, sample molecules damaged by irradiation are free to diffuse throughout the solution, or across its surface. Thus reactive fragments formed at the surface can diffuse into the bulk of the solution, and circulate as stable species or until reaction takes place. As expected, many of these reactive species are free radicals, and can be tracked with electron spin resonance techniques.[341,342] This situation is in direct contrast to that of a reactive fragment in a solid matrix in which diffusion is more limited. The reactive fragment may be trapped at the initial point of formation, and thus it may not be able to react with other components in the solution. The reactive nature of radical species in an irradiated glycerol matrix was initially studied by Field,[343] who noted that each primary particle impact generates 100 molecules of new product. More recently, Keough[623] has provided further results on the distribution of compounds formed in glycerol on irradiation by a primary particle beam. The persistence of background signal even at high masses in FAB spectra provides unambiguous evidence of the propensity and speed of these reactions. That radiation-induced chemistry plays an important role in establishing the chemical environment within the liquid matrix is an emerging aspect of the mechanistic study of SIMS and FAB. Diffusion across the surface of an irradiated glycerol solution, driven by differences in surface tension[624] generated as the composition of the surface changes, is an effect at least as large in magnitude as bulk diffusion processes. The involvement of electrons formed at the surface of the solution by the primary particle bombardment is implicated in reduction processes of organic compounds, and has been explicitly studied by Collins[625] for doubly charged organic compounds. One notes that the extent of these reactions is also a direct function of primary particle dose.

Several examples of the effects of extended irradiation on the spectra of organic compounds have been provided by Kambara.[344] The SIMS spectra of bradykinin in glycerol were obtained at primary ion (Xe^+, 5 keV) currents of 5×10^{-8} A and 3×10^{-7} A. The integrated total secondary ion current was greater in the latter case, as was the amount of fragmentation. Once exposed to the higher primary ion flux, the spectrum afterwards retained the greater level of fragmentation. Irreversible changes could also be induced in the spectrum of chymostatin; glycerol adducts not formed under low fluxes are formed under high flux conditions, and their appearance persists even when the primary ion current is later reduced. On the other hand, a low primary ion current does not always result in *reduced* fragmentation. The SIMS spectrum of a complex biological compound (a cord factor) contained an $(M + H)^+$ ion at m/z 760 that was enhanced relative to the fragment ions at a higher primary ion current (7×10^{-8} A versus 1×10^{-8} A).

Other workers have noted time (dose-) dependent phenomena in particle-induced desorption ionization mass spectra of organic molecles. Cationization of dimethyl phthalate by silver ions occurs more readily with prolonged ion bombardment.[345] More complicated reactions that involve recombinations of molecular fragments with intact molecules of porphyrins have also been demonstrated.[346] In FAB mass spectra of corrins, time dependence of the relative abundance of both molecular and fragment ions has been observed.[347] Lehman has quantified Schiff base formation from samples dissolved in glycerol as a function of irradiation time.[348] The potential for complex chemical interactions in the sample is increased with increasing irradiation time, or with increasing chemical complexity of the matrix itself. One notes that rapid changes in the nature of the glycerol solution of the sample is evident from first principles; Barofsky has reported that at primary ion current densities of 10^{-5} A/cm^2 for a sample dissolved in glycerol, stable organic secondary ion emission is established only after 10–30 s of irradiation.[349]

It is tempting to speculate that the time dependence of the spectra is inherent to the rapid diffusion and trapping of organic molecules within the liquid matrix, and that the use of solid sample supports and sublimation matrices may offer an alternative means of preparing organic samples for analysis.[350,351] High ion and atom fluxes will also result in the development of new features of surface topography on solid organic surfaces,[352] and the critical dose for the formation of new surface features may be several orders of magnitude lower for organic surfaces than for metals.[353,354]

In plasma desorption, the primary ion flux is held low (less than about 2000 impacts per second, corresponding to a current of 3×10^{-16} A for singly charged ions) to accommodate the throughput of the TOF mass analyzer. At these low levels of irradiation, there is a very small amount of damage to the organic material. Le Beyec[355] reports that the sputtering yields for ions from mica increase shraply after ion implantation with deuterium. The paper suggests that two different types of radiation-induced defects in the mica contribute to the enhanced yield. Such a result has not been verified for organic samples.

In laser desorption mass spectrometry, increased fragmentation with increasing laser power has been noted, as previously discussed. This effect represents the transition between the laser-induced desorption mode and the plasma mode. Hercules[57] has suggested that at least five processes must be considered in the discussion of volatilization and ionization by the laser pulse: direct ionization by the laser beam, vaporization and ionization of the region immediately surrounding the beam impact, surface ionization, ion–molecule reactions in the gas phase, and emission of neutral particles. Variation in the power of the laser beam will vary the relative contributions of all of these processes. The morphology of the surface, and the chemical environment of the sample, will also affect the relative contributions. Dose relationships in the analysis of organic compounds have not in general been reported.

A differentiation between sample damage and otherwise predictable chemical reactions in an energized matrix is difficult in any of the particle-induced desorption ionization methods (for a discussion, see Section 4.2.2.1). In a simple analysis, it is apparent that the energy carried within the primary beam is sufficient to break many bonds in the same thermally fragile molecules that are successfully analyzed by these techniques. Models that describe the dissipation and dilution of the primary beam energy in the condensed phase must be developed to explain the survival of these species. As energy distributes through the condensed phase, it also transfers surface material into the gas phase (aided by pressure and potential gradients), but no longer can cause extensive sample damage. It is suspected that the area from which the desorption and ionization of organic molecules can take place is far greater than the actual cross-sectional area of particle impact. Damage in a high energy density region invariably occurs, but sample ions that carry the useful information in the mass spectrum originate from the less-excited adjacent surfaces.

3. Properties of Secondary Ions

In this section, the properties of the secondary ions emitted from surfaces under primary particle bombardment are discussed. Only for simple inorganic systems have such parameters been adequately explored. As discussed in the previous section, variation in the surface morphology, or changes induced in it during bombardment, can be expected to obscure all but the most dominant properties of the secondary ion beam. When reproducible methods of organic surface preparation are used, it is possible to deduce these dependences for secondary ions from organic compounds.

3.1. Energy Distribution

It has been established in inorganic SIMS (and the same argument holds for neutral particle bombardment of simple surfaces) that the greater the structural complexity of the sputtered ion, the lower will be the initial kinetic energy. For example, the Si_4^+ ion has an energy distribution that peaks at a lower value than does the Si^+ ion.[356,357] In many organic SIMS instruments, this difference in the energy distribution at one time formed the basis of a method for discriminating against the passage of organic "contaminant" ions into the mass analyzer.[358] Most quadrupole instruments are fitted with an energy prefilter of some sort, often a Bessel Box,[359] that allows only ions within a certain range of energy values to pass. The purpose of this prefilter is to prevent higher kinetic energy ions from passing into a quadrupole mass analyzer, where they would be poorly resolved because of the low number of oscillations. These high-energy ions would thus provide a background level at all masses in the spectrum. At the same time, however, it is possible to set the bandpass of this filter so that only ions within a specified kinetic energy bandpass pass into the mass analyzer.

There have been several studies that establish the kinetic energy distribution of organic secondary ions sputtered from surfaces in SIMS experiments. Benninghoven has carried out experiments using retarding field measurements for organic secondary ions formed via bombardment of leucine on a silver support.[360] Most sputtered organic ions have kinetic energies of less than 3 eV. The kinetic energy distributions for higher-mass ions is expected to be biased towards lower values. High-mass organic secondary ions will also possess a lower velocity; an ion of 100 daltons and 2 eV has a velocity of 1.4×10^3 m/s, and an ion of m/z 10,000 of the same kinetic energy a velocity of 1.4×10^2 m/s. The extraction voltage (100 V or so) makes it possible to extract higher-mass ions from the surface with equal facility.

Kelner has recently studied the energy distribution of secondary organic ions in both SIMS and FAB.[361,362] A Bessel Box energy filter was used to measure the energy of ions sputtered from a variety of surfaces. Experimental parameters that were varied included the energy of the primary ion beam, the nature of the secondary ions, and a bias potential at which the target itself was held. Using the ion at m/z 133 from polyethylene glycol, it was found that the energy of the secondary ions maximized at a primary ion energy of 4 keV, and reached a most probable kinetic energy value of about 3 eV. For a constant primary ion energy of 8 keV, the energy of larger glycerol clusters was found to maximize at a lower value than smaller glycerol clusters. It was found that some molecular ions of organic compounds had a lower overall energy distribution than did cluster ions of the same mass, but others exhibited a shift in the opposite direction. Figure 5 compares the energy distribution of a sputtered organic ion with that of a glycerol cluster ion

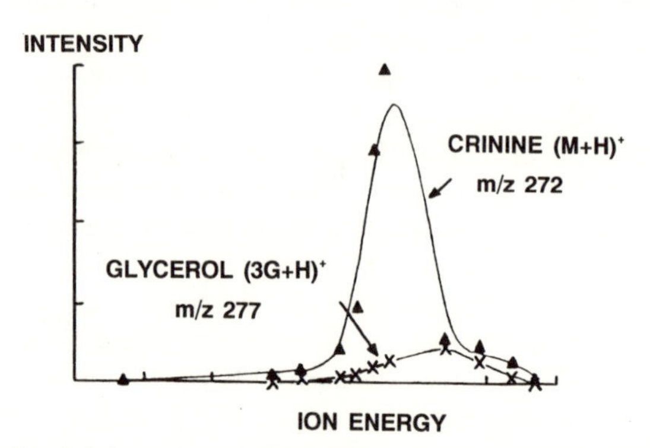

Figure 5. Comparison of the energy distributions for the protonated molecular ion of an organic compound and a glycerol cluster ion of similar mass using SIMS ionization (adapted from Ref. 361).

of similar mass. Finally, it was observed that increasing the potential on the target increased the efficiency with which organic secondary ions could be extracted into the mass analyzer.

Kistemaker[363] has also studied the kinetic energy distributions of organic ions sputtered from solids and liquids using both SIMS and FAB ionization. The instrument used in these studies was composed of a magnetic sector mass analyzer preceded by a hemispherical electrostatic condensor for energy analysis of the sputtered ions. A position-sensitive channelplate detector provided a simultaneous detection of mass-selected ions across the full energy distribution. The specified energy resolution of the instrument is 0.24 eV full width half maximum. Charging of the sample was noted with both ion and neutral bombardment, but could be minimized with liquid samples by spiking the sample with an electrolyte such as sodium chloride. The absolute kinetic energy values of sputtered ions could not be accurately determined due to the sample charging. Full-width-at-half-maximum energy distribution values of 0.3–0.7 eV for organic ions sputtered from a glycerol matrix are reported, with the values increasing slightly with the mass of the ion in the range of 150–400 daltons. Ions sputtered from solid samples are indicated to have a much higher value of absolute kinetic energy, and a larger range. The authors discuss the significance of these results, but conclude that a mechanistic model of the sputter process cannot be unequivocally discerned from the data.

In PD, especially when the irradiation is from the reverse side, the kinetic energy distribution of organic secondary ions from the surface can be expected to be wider than in the SIMS or FAB experiment, as the momentum of the primary ion can be expected to contribute directly to the kinetic energy of the secondary ions. In most instruments, the ions sputtered into the gas phase are immediately accelerated to a high energy (say 10 keV) to pass into the flight tube. New time-of-flight instruments are designed to increase the resolution of the mass analysis[364] and confirm that the energy spread, along with the time and spatial profile of ion formation (see next section), all contribute to the low overall mass resolution obtained with most TOF instruments. The energy spread of the secondary ions formed in a PD experiment has been estimated as a few eV.[365]

In LD, the energy distribution will undoubtedly depend on the direction of illumination, and, as in PD, may be submerged in acceleration voltages used to extract the ions into a TOF or a sector mass analyzer. van der Peyl *et al.*[366,367] have described an instrument developed to measure kinetic energy distributions of ions produced by laser desorption, and reports results for sodium ions. The data show that the energy distribution of laser-desorbed sodium ions is the same (within experimental error) as the distribution of sodium ions produced via a thermal desorption mechanism.

3.2. Angular Distribution

The angular distribution of secondary ions emitted from a surface has been the subject of numerous studies, particularly in the SIMS analysis of metallic surfaces. Sputtering models suggest that secondary ions are emitted from the sample surface with a cosine distribution centered about an axis normal to the surface.[368] The validity of this model for the sputtering of either atomic or small molecular species from surfaces has been established.[368–370] An assumption in establishing the angular distribution of the sputtered ions is that the structure of the surface itself is known and controlled, that is, that an axis normal to the surface at which the sputtering is occurring can be established. Given that restriction, mechanistic information about the structure and orientation of surface adsorbates can be discerned, as shown by Winograd.[274]

For organic analysis, as previously discussed, the detailed nature of the surface from which the organic secondary ions originate is not known. More importantly, even simple organic molecules can be attached to a surface in a number of discrete ways. Experiments have shown, for instance, that pyridine can attach to a surface in either an "end-on" or a "face-on" position, depending on the surface coverage and the nature of the surface.[371] How might more complicated molecules attach to a surface? In the absence of covalent bonds, the interaction is thought to arise in induced dipole interactions and in hydrogen bonding. These bonds are both readily formed and readily broken on a very short time scale; the organic molecule may thus continually vary its orientation with respect to the surface. When the sample is a solution, the question of angular distribution of the secondary ions is moot.

3.3. Time Distribution

In laser desorption mass spectrometry, much evidence has been reported for a time-dependent emission of organic secondary ions from surfaces. Cotter has investigated this phenomenon most thoroughly,[372] and the same phenomenon has been noted by Krueger[373] and Cooks.[374] In 1981, Cotter[375] reported that the $(M + Na)^+$ ion of glycocholic acid could be observed for a full 0.5 ms after a 40-ns irradiation of the probe onto which the sample was coated. The key to sample ion retention within the source of the mass spectrometer was the introduction of a gas into the source at a normal chemical ionization pressure of 0.5 Torr. Note, however, that the ionization filament was off in these experiments. This persistence made it easy to couple the pulsed laser source with a scanning sector mass spectrometer. Cooks[(374)] observed that ion currents for analysis of sucrose persisted for 300 μs at a source pressure of 0.010 Torr, rising to 1.5 ms at 0.5 Torr of source gas. These effects are a combination of the persistent ion emission from the surface itself and the efficient storage of secondary ions within the source, as has been shown previously in chemical ionization mass spectrometry. The phenomenon is of great practical use, as it has been developed by Cotter into the technique of time-delay focusing with the time-of-flight mass spectrometer. After the initial laser pulse, the ions that are directly sputtered are of very high energy. After a period of some microseconds, however, the ions emitted (now probably as a result of thermal processes) are of much lower energy, and are less likely to fragment. If the ions in this time regime can be selectively extracted from the source, then the resultant spectrum can be of significantly better quality.

Hercules has described a time profile of the events that occur after the impact of a laser beam on a surface,[(376)] which is reproduced in Fig. 6. In a series of experiments that investigated the time-resolved desorption of organic ions into a chamber held at 10^{-7} Torr, persistent ion emission for organic compounds of up to several hundred microseconds was observed.[(377)] The explanation for this observation may be based in a model of thermal desorption of ions and molecules from surfaces. Cotter[(378)] and Rollgen(379) have shown that many quaternary salts, normally considered to be nonvolatile or thermally fragile, can be

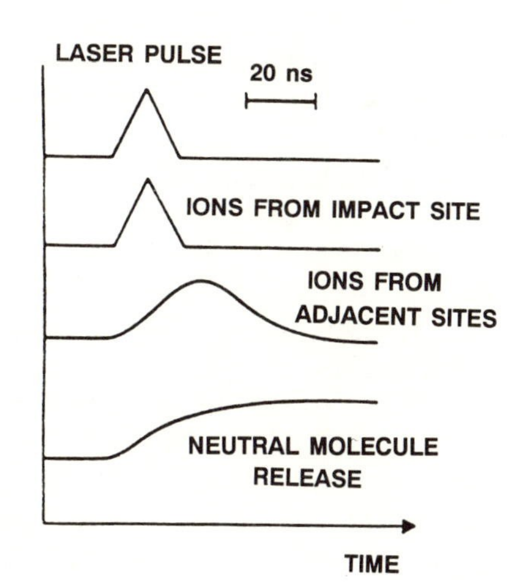

Figure 6. Time profile of events following impact of a laser beam pulse on a surface (adapted from Ref. 376).

thermally desorbed. The elegant experiments of Kistemaker have shown that even sucrose can be desorbed as a neutral sample molecule, and that formation of the $(M + Na)^+$ ion occurs as the result of a gas phase ion/molecule reaction.[380,381] The efficiency of the laser source for producing sample and substrate heating may account for the preponderance of thermal effects noted in LD mass spectrometry. Bombardment with ion or atom beams does not generally deliver enough power to the surface to induce a rapid rise in temperature, although there is some evidence for extensive thermal damage induced by sputtering of inorganic crystalline surfaces.[354] The expanding use of highly focused primary ion beams from liquid metal ion guns may allow thermal effects in SIMS to be similarly explored.

3.4. Charge Distribution

It is generally acknowledged that the charge state of the organic secondary ions found in the particle-induced desorption ionization mass spectrum reflect the charge state of the organic species in the condensed phase. Cations or anions present in a liquid glycerol matrix tend to appear as abundant ions in the mass spectra. This premise has also been demonstrated for laser desorption, but less clearly for plasma desorption, although the same even-electron ions predominate in PD spectra as they do in all of the other mass spectra. This predilection for the sputtering and desorption of "preformed ions" forms the basis of derivatization reactions and sample preparation methods that are discussed in Section 4.2. In 1980, Cooks[26] set forth three ionization mechanisms in SIMS that seem to hold true for the other particle-induced desorption ionization methods as well. The three mechanisms by which secondary ions could be formed were, in decreasing order of efficiency, direct desorption of preformed cations and anions, cationization and anionization reactions (in which cations and ions complex with neutral sample molecules to form a stable complex), and finally electron ionization, in which the low-energy electrons formed at the surface during the bombardment can ionize molecules of low ionization potential or high electron affinity. In the first mechanism, there is no need for ionization but only a transfer of ions between phases, while in the latter mechanism, both ionization and phase transfer must occur.

The relevance of the direct desorption model to the sputtering of multiply charged organic ions is unclear. In inorganic SIMS, however, it has long been known that cations such as Ba^{2+}, initially present in the solid phase as the doubly charged ion, are sputtered into the gas phase as singly charged ions, a reduction having taken place in the sputtering process. The desorption of multiply charged organic species could be studied to unravel some of the mechanism of the desorption process. In a study of the SIMS spectra of the doubly charged quaternary ammonium salts,[382] it was found that the observation of the doubly charged cation in the spectrum could be correlated with the distance between the charged localized on the nitrogen atoms in the dication. Increased distance between the charge sites resulted in a decrease in the repulsion energy due to simple Coulombic repulsion, and a higher probability that the ion would survive the sputtering process. Typically, the doubly charged ion fragments by loss of small *charged* species such as H^+ or CH_3^+, forming the more stable singly charged species.

These same compounds have been investigated by several other particle-induced desorption ionization methods, and by electrohydrodynamic ionization as well. Barofsky[383] has investigated the FAB spectra of a series of organic diammonium salts dissolved in a glycerol solution. The equivalent of a 20–40 μA beam of neutral argon atoms was used to bombard the sample. In contrast to the SIMS results of Cooks, most samples did not produce multiply charged ions in the FAB spectra. The intercharge distances were at least as great as those involved in the SIMS study. However, the primary particle flux was three orders of magnitude higher in the FAB experiment, and this may explain the increased fragmentation. Cotter[384] has also investigated the FAB spectra of doubly charged organic salts. Although cluster ions of the type $(C^{2+}A^-)^-$ are observed, the intact C^{2+} was not observed. The fragmentation processes observed were losses of small charged particles to create the more stable singly charged fragment ion. In an LD study of the quaternary ammonium dications, doubly charged species were not observed in the spectra.[385] A relatively high energy flux was used in these experiments. Laser desorption with TOF mass analysis preferentially extracts those ions that are formed in the first 100 ns after the laser pulse; the multiply charged ions may desorb during the extended period of thermal desorption following the laser pulse.

As the masses of the organic sample molecules increase, the probability that a doubly charged positive ion will be found in the mass spectrum also increases. Many peptides with molecular weights above m/z 1000 will provide a detectable doubly charged ion in the positive ion FAB or SIMS mass spectrum.[386,387] In PD, still more oxidized forms of the sample molecule have been identified, including M^{3+}, M^{4+}, and M^{5+}.[388] Nebulization ionization techniques such as thermospray and electrospray seem to produce much higher abundances of the multiply charged negative ions,[389] as well as multiply charged positive ions, with attendant advantages in the analysis of samples of high molecular weight.

It has been suggested that the sputtering/desorption of multiply charged ions may be the result of a field effect. The argument posits that doubly charged ions should be more difficult to remove from the condensed phase than are singly charged ions for two reasons. The first is the ionic force that binds the cation to anions within the condensed phase, which is doubled. The second reason is the image charge generated in the surface during the process of phase transfer. The magnitude of this image force is given by $E_i = -q^2/4r^2$. This force is four times larger for a doubly charged ion as compared to its singly charged analog. The high electric fields used in field desorption mass spectrometry lower the activation energy of the desorption so that the image force is insignificant, and doubly charged organic ions are often noted in the mass spectra produced by that ionization method. In methods that do not use an external field, the image force effectively decreases the probability that the multiply charged ions will be sputtered from the condensed into the gas phase. This factor rather than the tendency for small multiply charged species to fragment in the gas phase explains the inability to observe doubly charged ions. Indeed, multiply charged ions *are* often observed in field desorption mass spectra and in the nebulization ionization methods that involve high fields localized on small droplets of solvent. The observation of multiply charged ions in PD may be in accord with this mechanism, since there is an electric field gradient generated by MeV particle bombardment that may be sufficient to induce the same effects.

4. Sample Preparation

4.1. Neat Samples

Since FAB almost invariably involves the use of a liquid matrix, this section will discuss only the means for preparing neat samples for SIMS, LD, and PD. In SIMS, solid samples can be simply prepared by burnishing the sample onto a roughened support planchette, often consisting of a metal such as silver, gold, or copper, but a carbon foil is also satisfactory.[390] With silver planchettes, a small piece of the foil is roughened with a fine grit sandpaper. After rinsing with solvent, a small solvent of the sample powder is placed directly on the scored surface, and rolled onto the surface with a glass pipet. Excess solid sample is then removed by tapping the support against a hard surface. Sufficient sample adheres to the roughened surface for SIMS analysis. If the sample is in solution, an appropriate amount can be added directly onto a similarly prepared planchette with a syringe or micropipet. The advantage of the latter method of sample preparation is that a more uniform film can be created on the surface of the planchette. For very thin films, the planchette can be dipped in a solution of the sample. In each of these methods, the roughness of the surface aids in the retention of the sample, and seems to increase the secondary ion yields as well. Thick layers of organic compounds prepared as solid samples may charge under continued ion bombardment. A practical solution to this problem is to etch a series of fine lines in the surface to expose the underlying metal. Since the diameter of the primary beam in these experiments is generally at least a millimeter, the exact dimensions of the etched lines is not critical. In general, the amount of sample introduced into the vacuum chamber is far in excess of that necessary for the analysis. Since SIMS is a surface sensitive method, only a few nanograms of material may be consumed during analysis. The remaining sample can be removed from the mass spectrometer, and stored on the planchette for reanalysis at a later time, or can be scraped from the metal backing and used in another experiment. Electrospray can also be used to prepare uniform thin films of sample material.

The nature of the metal backing does not seem to be crucial in this method of sample preparation, although it may affect the degree of fragmentation for organic molecules deposited onto it. In this respect, increased fragmentation can be induced by changing the nature of the metal support. Thin silver foils seem to be the support of choice, as the secondary ion yields of organics from a silver support is high, and the identification of the silver-cationized ion $(M + Ag)^+$ is aided by the distinctive isotopic doublet. However, copper, palladium, platinum, lead, zinc, nickel, and a variety of other metals have been used, leading to a diversity of metal-cationized ions of the organic compound. Simultaneous exposure of both the metal and the organic sample to the primary ion beam results in the sputtering of both organic and inorganic secondary ions. The metal (met) ions serve as cationizing reagents, forming ions of the type $(M + met)^+$, and also as convenient mass markers within the spectrum. When other cationized species are desired, the metal foil is changed, or the sample is mixed with the appropriate metal salt. MS/MS experiments of several cationized species for small organic molecules are distinct,[626] indicating different sites of cationization within the organic structure. Further, the simultaneous presence of both metal atoms and a coordinating ligand has been shown to increase the secondary ion yields in the analysis of neat samples.[627]

Benninghoven has described methods for the preparation of organic overlayers of amino acids by a molecular beam technique.[276,277] Basically, organic layers are formed in vacuum on a metal substrate by exposing the metal surface to a molecular beam that emerges from a heated container containing the sample of interest. The crucible that holds the sample to be deposited is similar in design to a Knudsen cell, so that an equilibrium between vapor and the solid should be attained. A series of apertures and cooled diaphragms are used to ensure that the vapor species emanating from the heated cell is in the form of a well-defined molecular beam. The substrate is placed at a distance of 60–270 mm away from the cell exit aperture, and is hooked directly to a microbalance that monitors the amount of material deposited (see Fig. 7). The initial studies focused on three concerns: (a) thermal decomposition of organic compounds within the heated crucible; (b) the amount of material that could be deposited on the substrate in a reasonable time; (c) the evenness of the coverage of the substrate. Amino acids have reasonable vapor pressures, and were used as model compounds. Benninghoven concluded that for these samples, mono- and multilayer coverages of the amino acids on copper and nickel metal foils could be created. Such well-defined surfaces are used in experiments of secondary ion yield as a function of primary beam parameters.

Laxhuber[391] has described a method through which molecular layers of fatty acids were assembled and transferred onto gold supports for SIMS analysis. Layers of fatty acids are well defined and provide characteristic crystalline layers of known thickness that are transferable onto a solid support. In addition, they serve as model organic systems for more complicated biological membrane materials. A 95% coverage is claimed for the production of a single monolayer, and by repeating the dipping process, the number of overlying monolayers could be increased. Using these well-defined surfaces, the time dependence of the secondary ion signals from both the organic and the inorganic components of the surface were studied. Interestingly, the secondary ion yields of most organic ions first increased, and then decreased with increasing primary particle dose, suggesting a combination of mechanisms for their forma-

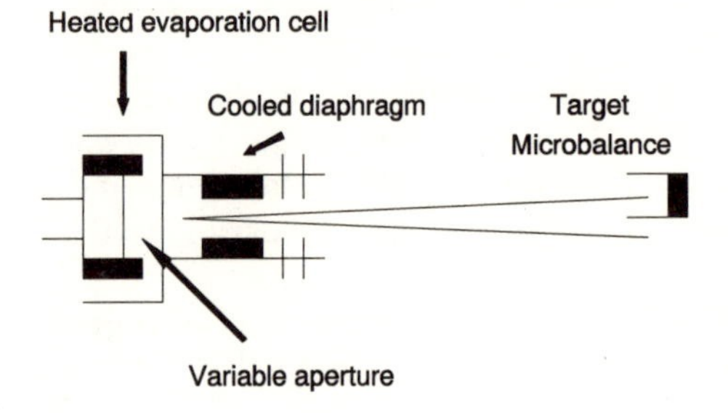

Figure 7. Microbalance technique for the creation of thin layers of organic compounds (adapted from Ref. 276).

tion. The authors conclude that the ion escape depth in the organic matrix for a solid sample is less than 25 Å, in accord with data from inorganic SIMS studies.

Samples can also be pressed into pellets[392–394] and analyzed by SIMS. Seidel[392] has claimed that a greater sample lifetime is thereby obtained, presumably because of the higher sample loading and the ability of a high-flux primary ion beam to depth profile through the sample to progressively uncover more sample. Mention should also be made of the SIMS analysis of frozen organic samples, for which the surface structures should be simple and reproducible. Early work in the SIMS analysis of frozen organic gases was carried out by Rabalais[395,396] and by Michl.[397,398] Barber[399] has studied the SIMS spectra of several frozen organic surfaces. Frozen methanol was studied first by De Pauw[400] and later by Field.[401] Field described a change in the particle-induced desorption ionization mass spectrum of methanol that reproducibly and reversibly occurs as the temperature of the frozen methanol surface is varied between 170 K and 100 K. The suggestion was made that the chemical damage to the methanol from primary beam irradiation was prevalent at the higher temperature, but is "frozen out" at the lower sample temperature. Additional work on the SIMS and FAB analysis of frozen organic vapors can be expected in the next few years. Distinction is drawn between this work dealing with frozen matrices, and that which uses a cooled direct probe to increase the lifetime of a volatile solvent in FAB or to study reactions in the cooled matrix.[628,629]

In LD, the LAMMA 500 requires reverse irradiation of the organic sample, and some special methods of organic sample preparation have been developed. Hercules[402,403] grinds solid samples into fine powders, dissolves them in a volatile solvent, and then evaporates the sample onto Formvar-filmed copper grids, such as used in electron microscopy. The laser beam penetrates the thin film coated with the sample, but the Formvar itself does not contribute significantly to the LD spectrum. For polymer analysis, thin films of the material were used directly.[404] In the LAMMA 1000, with obverse irradiation, sample preparation follows the methods developed for SIMS. Other reported LD configurations are also obverse by design, and use similar sample preparation methods.

Prior to the experiments of Benninghoven, the most careful methods of sample preparation for neat samples were those followed in PD. Although a few examples could be found where a crude dipping technique was used, electrospray became the early method of choice for the preparation of thin films of organic molecules. This method was first described by Brunnix and Rudstam in 1961[405] for the preparation of thin films of inorganic compounds. These researchers studied the parameters affecting the performance of the electrospray, including the design of the delivery electrode, and the influence of solvent, voltage, and the compound deposited. In PD, electrospray was first reported by Krueger.[269] McNeal[270] has described the electrospray apparatus in some detail. A solvent containing the sample of interest is held in a hypodermic delivery electrode fitted with a stainless steel wire used to regulate the flow of solvent. A voltage of between 5000 and 7000 V is applied to this needle with respect to the substrate, which consists of a thin nickel or aluminized polypropylene foil (0.3 to 1.2 mm^2 area, 10 μm thickness). The needle is held on an x–y positioner to aim the electrospray, and the distance between the needle and the substrate is adjustable between 2 and 4 cm. The current through the dispersed solvent varies between 0.1 and 0.3 μA, and a typical solvent consumption rate is 1 μl per minute. The surface prepared with electrospray appears to be the result of an accumulation of residues of microdroplets. The fraction of the surface covered depends on the concentration of the sample in the solvent sprayed from the needle. Monolayer coverages are *not* produced, and complete surface coverage is not attained until approximately ten times the amount nominally needed for monolayer coverage is deposited on the surface.

4.2. Matrices for Sample Preparation

Difficulties inherent in producing a surface of known and unchanging composition for analysis by particle-induced desorption ionization mass spectrometry are avoided to a large extent if a liquid matrix is used. High rates of diffusion quickly establish a surface and bulk composition consistent with the surfactant properties of the mixture. Surfactancy is a property that can be described for systems at equilibrium, and an irradiated liquid surface that is evaporating at a significant rate into the vacuum of the mass spectrometer is a system for removed from equilibrium.[406] The spectral effects of the matrix are in general an increased flux and persistence of the organic secondary ions observed in the spectrum. To this end, solid sample matrices can also be used, although questions of sample homogeneity in such a nondiffusing system remain.

4.2.1. Solid Sample Matrices

A number of solid matrices have been used in SIMS. The first to be described was ammonium chloride.[350,351] Unger[407] used this matrix to reduce the extent of intermolecular alkylation observed in the SIMS spectrum of carnitine hydrochloride, which is particularly sensitive to this complicating reaction. Ammonium chloride has also used as a matrix to obtain the SIMS spectra of organometallic and coordination compounds.[408,409] The effect of dilution in an ammonium chloride matrix on the fragmentation of an organic molecule has been investigated.[410] The matrix actually increases the absolute signal intensity for ammonium, phosphonium, and pyrylium salts admixed with these salts in a 10–10^3 weight excess.

The mechanism by which the ammonium chloride acts to enhance the secondary ion mass spectrum is based on the ease with which this salt sublimes. Ammonium chloride dissociates into gaseous ammonia and HCl upon primary particle impact. Indeed, the ammonium cation NH_4^+ is not abundant in the static SIMS spectra of ammonium chloride, although all other ammonium salts produce abundant cations. At higher primary particle fluxes, large clusters of ammonium chloride are sputtered from the surface.[411] Essentially, the matrix evaporates at the point of primary particle impact, isolating sample molecules found within the same volume in an expanding cloud of gas. The expansion into the vacuum leads to adiabatic cooling of the sample ion and thus reduces the degree of fragmentation that it undergoes. Other ammonium salts have similar thermodynamic properties that encourage their use as sublimation matrices.[412] In particular, the use of ammonium bicarbonate is particularly attractive because of its dissociation into relatively innocuous ammonia, water, and carbon dioxide. Gillen *et al.* have compared the sputter yield of ammonium chloride and glycerol,[630] and have reported similar values for secondary ion yield.

Kloppel[278] has investigated sample preparation for SIMS using a highly porous sample support matrix formed by the thermal decomposition of metal compounds. Decomposition of silver carbonate, for example, provides a highly porous silver "sponge" that sustains secondary ion emission from amino acids longer than etched silver foil supports. The porous silver targets may provide a greater surface area, but this must be counterbalanced against the fraction of sample made inaccessible to the primary beam. No diffusion mechanism for transporting sample has been suggested for this porous matrix. It was noted by the authors that further improvement in secondary ion persistence could be achieved by the addition of glycerol to the porous metal support.

McNeal[413] has described the use of a cationic surfactant (tridodecylmethyl ammonium chloride) impregnated onto an aluminized Mylar film as a solid matrix for the PD analysis of biomolecules. Absolute yield of such samples as trinucleoside diphosphates increased by a factor of 5 for the deprotonated molecular anion (M − H); the details of the fragmentation pattern were unaltered. McNeal concludes that the enhancement is due to the separation of the sample ions from one another by the creation of specific binding sites on the surface, so that the intermolecular forces are attenuated. The same argument has been invoked in explaining the effects of the ammonium chloride matrices. Additional work with cationic surfactant supports has been described.[631,632]

4.2.2. Liquid Sample Matrices

The introduction of the liquid matrix in FAB by Surman and Vickerman[414] and by Barber[415] dramatically changed the approach to sample preparation for particle-induced desorption ionization methods. The mass spectrometer user had long been accustomed to purifying the sample as much as possible before requesting a mass spectrometric analysis. In contrast, for FAB, the mass spectrometrist dissolves the sample in glycerol, adds perhaps a bit of acid or surfactant, smears an aliquot of the mixture onto a metal stage, and inserts the mix into the source of the instrument, there to be bombarded by an intense beam of high-energy particles. Objections to what seems to be an unconventional procedure are mute in the face of successful analyses of samples that could not previously be analyzed. The role of the liquid matrix in providing spectra of nonvolatile and thermally fragile molecules is central to the continued development of SIMS and FAB, and perhaps several of the other desorption ionization techniques as well. The mechanisms through which large organic secondary ions are released from a liquid matrix into the vacuum of the mass spectrometer will surely be debated over the next decade, especially as fundamental studies begin to catch up with the explosion of applications engendered by FAB.

Consider a surface bombarded by a particle beam. In the time scale involved for the first part of the argument, it is irrelevant whether the surface is liquid or solid. Some fraction of the molecules at the surface may be sputtered into the gas phase as intact molecular ions. Others may be damaged by particle impact, and may thermally degrade or may pyrolyze completely. Bursey *et al.*[416] provide a qualitative analysis of the bombardment, and conclude that only about 10 molecules are removed by a single primary particle impact. It is reasonable to assume that each impact damages or transforms many more molecules, as suggested by the results of Field[343] and Keough.[341] For a solid surface, rapid diffusional exchange of surface molecules with those from the bulk cannot occur. The detritus remains on the surface, often in the form of nonvolatile polymeric material. Under intense bombardment, it is likely that during the analysis a second particle will strike within the region previously damaged. However, the pyrolysate yields no secondary ions. Over the course of only a few seconds, the organic surface can be covered by a layer of sputter-resistant material, resulting in an immediate decay in the secondary ion signal. If the sample is removed, and scratched, for instance, to expose fresh sample to the primary ion beam, the secondary ion current will reestablish itself, but again will persist for only a few seconds. The solution to this problem suggested by Benninghoven[(19)] was to reduce the flux of the primary ion beam so that the chances of sampling a previously impacted site during the course of a normal analysis would be negligible; this is the static SIMS experiment.

The use of a liquid matrix solves the problem in a different way altogether. A primary beam impacts the surface of an organic material, and some fraction of the surface material is sputtered into the gas phase, and some fraction damaged. The distinction lies in the fact that the liquid surface is in a dynamic state of flux. A sample ion sputtered from the surface will, within a few microseconds, be replaced at the surface by a sample molecule that diffuses from within the bulk, or from another location on the surface of the liquid matrix. Pyrolysates formed at the surface also have the opportunity to diffuse *into* the bulk of the liquid, and are replaced at the surface by fresh sample molecules. This process explains the constant rejuvenation of the surface, and the persistence of the secondary ion signals that can be obtained with use of a liquid matrix. The limitation of the static SIMS experiment is removed because the renewal of the surface at any particular point occurs in a time less than that between primary particle impacts at that same site, even at equivalent bombardments of up to tens of microamps per square centimeter. Rather than a sample surface, the liquid matrix makes available a sample reservoir.

Two points are of importance in assessing the role of the liquid matrix. First, the neutral or charged nature of the primary particle is irrelevant in establishing the diffusion model. Second, for each individual primary particle impact, the solid or liquid nature of the matrix is also irrelevant. Energy dissipation takes place on the order of nanoseconds or less, and diffusion on a much slower time scale. Magee[(302)] provides the calculation that shows that the upper 10 nm of a typical liquid film can be refreshed every 3×10^{-4} s, and that under these conditions, a primary ion current of up to 10^{-2} A/cm^2 can be delivered to the surface without detectable damage.

Given a model that emphasizes the diffusivity of the sample analyte within the liquid matrix, it would seem that a wide variety of vacuum-compatible liquid matrices could be used for FAB and liquid SIMS experiments. Indeed, there have been literally dozens of matrices used, as compiled by Gower.[419] Glycerol has been used in most applications because of its ready availability, known solubilities, and the sheer momentum of its dominance in early FAB studies. The vapor pressure of glycerol at room temperature is fairly high at 10^{-3} Torr, but it is sufficiently nonvolatile that it can be placed in the source of a mass spectrometer in bulk without an excessive rise in source pressure. Glycerol is sufficiently viscous that it will remain on the sample stage as a drop (albeit with a constantly changing surface angle as it is distorted by gravity), but not so viscous that sample diffusion within it is encumbered. It dissolves a large number of organic compounds, as evidenced by previous work with glycerol in electrohydrodynamic ionization mass spectrometry. It is cheap, and readily available with a high degree of purity. Its disadvantages include the high level of background chemical noise that it generates in the FAB spectrum, and the fact that it will eventually evaporate into the source over the course of a few minutes. The "self-cleaning" attributes of the glycerol surface have been discussed by Rollgen *et al.*[564]

Other liquid matrices have been used in the SIMS/FAB experiment. The most popular of these include thioglycerol, dimethylsulfoxide, polyethylene glycols, diethanolamine, and triethanolamine. Most of these possess some degree of acidity or basicity to promote the formation of positive or negative ions of the sample, respectively. Gower[419] suggests that there are three major requirements that a matrix solvent must fulfill. The first is that it should dissolve the analyte, with the possible aid of a cosolvent or additive. The usual concentration of sample is on the order of 5 μg of sample in 5 μl of solvent, requiring solubility at the 0.1% w/w level. Although this can often be achieved, alternative solvents are sought when a FAB spectrum cannot be obtained, since this is often the sole experimental evidence for nonsolubility. The second requirement is that the solvent should be reasonably compatible with the vacuum conditions of the mass spectrometer source. The requisite properties vary from source to source, depending on the pumping conductance, but also vary with the time required to complete the desired mass spectrometric analysis, and the temperature at which the FAB sample platform is held. Most first-generation SIMS/FAB sample platforms did not include temperature control. The vapor pressure at room temperature, or at whatever temperature the heated source block imparted to the sample platform, was the controlling factor. Heated and cooled sample platforms are now available. Ackerman has measured FAB spectra as a function of the temperature of a fructose solvent matrix in one of the first studies to define the role of the matrix temperature.[420] Figure 8 compares the

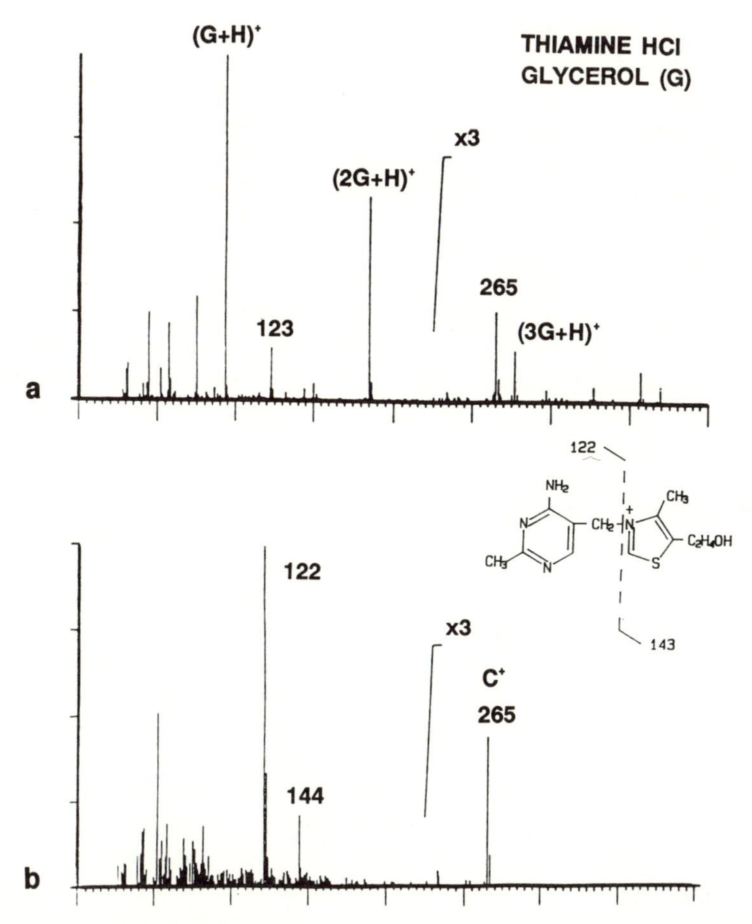

Figure 8. Comparison of the positive ion FAB mass spectra of thiamine hydrochloride obtained from a probe held (a) at room temperature and (b) at an elevated temperature (adapted from Ref. 420).

positive ion FAB mass spectrum of thiamine hydrochloride obtained with sputtering from a room temperature sample platform with the mass spectrum measured with sputtering from a heated sample platform. Control of the sample platform temperature is an empirical adjustment that controls surface activity, sample lifetimes, and diffusion rates. Signal-to-noise enhancements can be achieved if the temperature control is sufficiently well refined. In the experiments of Ackerman, temperature changes are made until the useful secondary ion current reaches a maximum, at which point the mass spectrum is recorded. One notes that establishing a best matrix temperature for complex mixtures will be a difficult process.

Finally, the third solvent property of importance is its chemical inertness and spectral transparency. It should not react with the sample of interest, or if it does, it should react in a reproducible and predictable fashion. For analyses of known compounds, sample/solvent reactivity can be assessed prior to preparation of the sample. In the analysis of unknown samples, the appearance of time- or concentration-dependent signals in the mass spectrum should be early warning signals for chemical reactivity between the sample and the solvent. This issue is addressed explicitly in Section 4.2.2.1. The solvent itself should not contribute too onerous a background mass spectrum, and those ions that do appear from the solvent should be stable so that background subtraction algorithms can be used.

Several papers report the development of a "matrixless" FAB analysis of organic compounds.[421,422] The pervasiveness of the use of the liquid matrix in FAB has led to such a description; the direct analysis of organic compounds by SIMS and FAB without the use of the matrix is a return to the early experiments of Benninghoven and Cooks. A different approach that still involves the use of an underlying liquid matrix has been advocated by Liang.[423] The sample is dissolved in a volatile solvent and then carefully deposited on the surface of a liquid viscous support (see Fig. 9). The authors suggest that the sample is dispersed on the surface of the base liquid as a finely dispersed emulsion easily sputtered by the primary particle beam. The persistence of the secondary ion signal obtained suggests that a surface diffusion of the sample occurs to replenish the depleted area. A similar mechanism was suggested by Ross to account for the persistent signal of organic secondary ions supported on a liquid metal surface.[283] This phenomenon will be discussed in Section 4.3.2.1 on surface ion management.

The benefits of a liquid surface for particle-induced desorption are

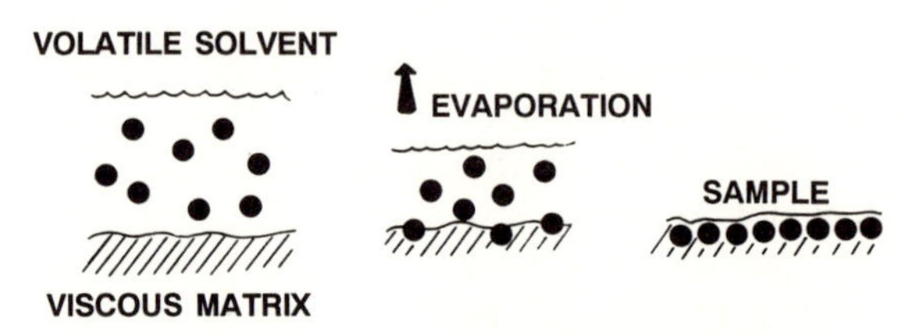

Figure 9. Surface precipitation as a method of sample preparation for particle-induced desorption ionization mass spectrometry (adapted from Ref. 423).

persuasive. Salehpour *et al.*[633] have discussed the use of liquid samples in plasma desorption mass spectrometry. There are no large differences in either the yield of secondary ions or the fragmentation patterns observed for molecules as large as bovine insulin. On the time scale of MeV particle passage through a thin film, and extraction of ions into the analyzer of a time-of-flight mass spectrometer, liquids and solids behave similarly.

4.2.2.1. Sample-Solvent Interactions. Dissolution of a sample in a liquid solvent creates the opportunity for sample–solvent interactions, especially during sample bombardment by the primary particle beam. Signals from some of the deleterious reaction products may be submerged in the chemical noise found in the spectrum, but in other cases may produce abundant signals in the mass spectrum. The keys to establishing a signal in a particle-induced desorption ionization mass spectrum as that of an artifact are its time dependence, its concentration dependence, or the impossibility of relating an ion of that particular mass to the sample or the solvent in a rational manner.[424] This section concentrates on a few instructive examples of sample-solvent interactions.

Early in the development of FAB, Dell and Morris[425] reported an *in situ* derivatization of bleomycins with acetic anhydride in glycerol. Acetyl groups were incorporated into the bleomycin skeleton to aid in the interpretation of the spectra. The strategy developed for the analysis of an unknown mixture of bleomycins was to record the FAB spectrum of the mixture, withdraw the sample, add the acetylating reagent, and then record the FAB spectrum of the derivatized mixture, pinpointing the various functional groups. When bleomycin B_2 is acetylated without the glycerol matrix, only the mono acetyl derivative is formed. Acetylation in the glycerol matrix produces the diacetyl derivative as the major product. Dissolution in glycerol apparently increases the reactivity of the substrate for the acetylation reaction. If previously established or suspected, this change presents no great problem, but can complicate structural analysis of an unknown mixture.

Buko[426] has studied the FAB spectra of peptides at higher masses, paying careful attention to the fragmentation pattern observed under different experimental parameters. The nature of the ions contributing to the molecular ion envelopes of somatostatin and insulin were established. Deuterium-labeled glycerol was used to establish that the disulfide bridges found in both of these molecules are reduced by the glycerol used in the FAB analysis. The molecular ion envelope was thus not simply a compilation of the various isotopic forms of $(M + H)^+$, but also contained contributions from $(M + 3H)^+$ and/or $(M + 5H)^+$. This is significant because in the absence of information available from exact mass measurements, the accuracy of fit between the observed molecular ion envelope and that calculated for the putative empirical formula is of great importance (see Section 5.3.1).

De Pauw[427] noted that the positive ion SIMS mass spectra of dyes dissolved in a liquid matrix exhibited enhanced abundances of $(C + 1)^+$ and $(C + 2)^+$ ions, where C^+ itself is the signal due to the intact cation of the preformed ion of the hydrochloric acid salt of the dye molecule. The propensity of a dye molecule to add hydrogen was related to the electrochemical potential for the reduction of the dye molecule in solution. The source of the incorporated hydrogen was thought to be the solvent itself. Similar results were reported by Gale,[428] who carefully measured the extent of hydrogen incorporation in a wide variety of dyes analyzed by FAB. The use of deuterium-labeled solvent established the involvement of the solvent molecules, and, through a hydrogen–deuterium exchange process, established the number of acidic hydrogens in the dye molecule. Scheifers[429] analyzed some of the same dyes in the solid state without dissolution in a matrix, and found, for instance, no evidence for the reduction of the cation of methylene blue. Fales[430] reported the plasma desorption mass spectra of these same dyes, and those spectra similarly do not provide evidence for the reduction of the intact cation. Although the solvent molecules are implicated by the labeling experiments, the mechanism is more complicated. The direct-exposure probe electron ionization and chemical ionization mass spectra of methylene blue exhibit a sample concentration dependent abundance of the $(C + 1)^+$ ion, indicating that sample/sample reactions can lead to the formation of this product as well as sample/solvent interaction.[431] Involvement of photoinduced reactions in FAB is also possible.

Unger has measured the positive ion mass spectra of antibiotics.[432] Protonated molecular ions $(M + H)^+$ were observed along with $(M + 2H)^+$ and $(M + 3H)^+$ ions, indicating that a reduction of the quinone structure to the hydroquinone had occurred. The relative abundances of these ions varied both with the particle flux and with the solvent used to dissolve the sample. Differentiation of closely related antibiotic structures was made more difficult by the occurrence of this reduction process. A complete analysis required MS/MS analysis of each of the forms of the protonated molecule.

Campana has studied the FAB spectra of synthetic porphyrins in thioglycerol matrices.[346] "Supermolecular" ions with masses greater than that of the molecular weight of the porphyrin were ascribed to reactions of sample molecules with sample fragments formed during the analysis. Thus a known mono-amide porphyrin produced abundant signals in the FAB spectrum corresponding to the diamide. The exact role of the liquid matrix in promoting this reaction was not established. However, reactive fragments formed upon bombardment can diffuse through (or across the surface of) the liquid matrix until an encounter with a suitably reactive substrate. This is an omnipresent complication associated with the use of liquid matrices.

The occurrence of sample–solvent interactions, including reductions, are particularly widespread in the FAB analysis of peptides. The oxidizing/reducing nature of the matrices is of particular interest in the analysis of peptides containing disulfide bonds that can be reduced with dissolution and/or bombardment. Reducing matrices are defined as those that cause or permit reduction of the sample molecule while molecules of the liquid matrix, or other species present in the solution, are themselves oxidized. Oxidizing matrices are those that oxidize (or preserve in an oxidized state) the sample molecules. Generally, these matrix molecules are themselves relatively easy to reduce. Reducing liquid matrices in FAB and LSIMS include thioglycolic acid, thioglycerol, glycerol, dimethylsulfoxide, and the dithiothreitol/dithioerythritol mixture. Indeed, Cleland[(634)] described the original use of dithiothreitol as a solvent designed to maintain sulfhydryl groups in their reduced form. Common oxidizing matrices include nitrobenzylalcohol and nitrophenyloctylether. The redox properties of the solvents in FAB and LSIMS depend also on the primary beam parameters, and thus no definitive ranking of the relative redox properties of these solvents can be undertaken. Finally, oxidation or reduction processes require a redox pair, and the redox properties of the sample molecules themselves determine whether the matrix can act as an oxidizer or a reducer.

Several reports investigate the production of reduced forms of a peptide sample molecule in various matrices containing a matrix modifier. For instance, although thioglycerol was found to reduce more readily oxytocin under standard LSIMS condition than glycerol, the addition of several drops of dilute HCl to the liquid matrix containing the sample was found to prevent the reduction of the sample molecule.[(635)] Once the oxytocin was reduced, however, the addition of the acid would not reoxidize the sample. Several workers have also noted reduction processes in FAB and LSIMS that result from irradiation of the sample with light, produced in conjunction with primary beam bombardment. For instance, Wirth *et al.*[(618)] have reported the reduction of azo-group-containing peptides in FAB. Irradiation of a glycerol solution of such peptides with the primary beam from a xenon discharge source produces reduction far in excess of the reduction observed with irradiation of the same sample with the primary beam from a cesium thermionic source. This difference was attributed to irradiation with the short-wavelength ultraviolet radiation and x-rays emitted from the xenon atom discharge source, which by definition must be in a line-of-sight with the sample probe. The redox properties of a solution also change as cesium ions are added to it from the primary beam bombardment, whereas an equivalent change will not be observed with the implantation of inert noble gas atoms. The presence of cesium ions in the sample solution may serve to reduce the extent of some sample molecules. Changes in the redox properties of sputtered surfaces induced by chemical effects of the primary beam itself are well known in classical SIMS experiments. These effects include increases in ion yields for some species upon bombardment with Cs^+ or O^+.[636]

Organic compounds with especially low reduction potentials, such as the quinones mentioned previously, undergo beam-induced reductions in FAB and LSIMS, and even in SIMS studies in which neat samples are analyzed. In peptides, the disulfide bond is most susceptible to reduction, and shifts in measured molecular masses in the FAB and LSIMS spectra result. Control of the matrix and instrumental conditions can be used to manipulate the redox chemistry, and to derive detailed information about the number and location of the disulfide linkages in the peptide. Interpretation of the FAB or LSIMS mass spectra of disulfide-containing peptides is complicated by the fact that the ratio of the reduced to the oxidized forms of the peptides can often change dramatically with time, in addition to changing as a function of the matrix from which the sample is sputtered. This tendency is most troublesome in matching predicted isotopic envelope patterns to those actually observed, leading to uncertainties in the derivation of the true molecular weight of the peptide of higher mass unless specific steps are taken to control the reduction chemistry.

An example of the complications introduced with disulfide reduction chemistry is given by Buko and Fraser.[637] Vasopressin (containing a single disulfide bond) was noted to undergo reduction upon bombardment of its glycerol solution, and further, the extent of the reduction was dependent on the time of the irradiation. Further, different primary beam fluxes will result in different time dependences for the reduced/nonreduced ion ratio. The use of a more-reducing matrix will result generally in a higher degree of reduction, whereas an oxidizing matrix will preserve the disulfide bond. Reports in the literature, however, must be very carefully evaluated. Oxytocin, reported by several workers to reduce with FAB sputtering from a glycerol matrix, as described above, variously does or does not appear to reduce when dissolved in a matrix of dithiothreitol/dithioerythritol, a more reducing matrix than glycerol. Such inconsistencies are due to changes in the irradiation time, sample concentration, and primary ion flux rather than to the intrinsic properties of the redox couple.

Reduction of disulfide bonds in peptides is an important reaction, but addition of hydrogen (formally a reduction) also occurs for peptides that do *not* contain disulfide bonds. Fujita *et al.*[638] discuss the formation of $(M + 2H)^+$ and $(M + 3H)^+$ for several peptides not containing disulfide bonds. Using a matrix of glycerol (acidified with aqueous trichloroacetic acid), the extent of reductions for compounds such as eledoisin, bradykinin, angiotensin I, and alpha-endorphin were about half the extent observed for oxytocin and somatostatin, but the occurrence of such reduction processes was still confirmed. One also notes that *losses* of hydrogen also contribute to the signal in the isotopic envelope in the molecular ion region, and that the actual measured signal is due to a combination of $(M - H)^+$, M^+, $(M + H)^+$, and $(M + H)^+$, requiring very accurate intensity measurements for proper deconvolution of the ion species present.

4.3. Derivatization Techniques

Particle-induced desorption ionization mass spectra typically contain abundant ions corresponding to ionic species already present in the analyte. Derivatization reactions that increase the concentration of preformed ions can thus be used to increase the sensitivity of the analysis, and since such reactions are often chemically specific, selectivity in mixture analysis is often increased as well. Derivatization reactions have most often been used in FAB and SIMS, consistent with the use of the liquid matrix. This discussion of derivatization reactions describes, first, the role of the chemical reaction in the creation of ion charge, and second, the role of physical parameters in the management of the spatial distribution of that charge.

4.3.1. Creation of Preformed Ions

4.3.1.1. Ion Addition Derivatization. Since particle-induced desorption ionization mass spectra typically contain abundant signals for ions preformed in the condensed phase, derivatization should create nonvolatile ionic species. With ions preformed in the matrix, only sputtering need occur, and the low-efficiency ionization step is avoided. Derivatization reactions can often be based on the acid/base equilibria between the sample and the matrix in which it is dissolved, resulting in a powerful but quite general method of derivatization. Specificity can be increased if the reaction proceeds only with certain functional groups. Many reactions that form ionic intermediates are used in purification procedures in organic synthesis, and can readily be adapted as derivatization procedures. Much recent research has been aimed at the identification of functional-group-specific reagents, carefully studying their reactions both with pure compounds and with complex mixtures. Derivatizations that form precursor ions by addition of charge-carrying subgroups are termed *ion-addition* reactions.

A general ion-addition reaction involves formation of a new chemical bond between the sample and a charge-carrying species. The FAB or SIMS mass spectrum then contains a new ion that can be related to the molecular weight of the sample via knowledge of the reaction sequence. Acid/base reactions constitute a general form of ion addition derivatization; acidified (HCl, oxalic, or p-toluenesulfonic acid) glycerol is used to create stable even electron $(M + H)^+$ ions. Proton abstraction with an added base produces $(M - H)^+$. Thus, ion current for the analyte can be shifted to either a positive or negative quasimolecular ion, depending on the reagent added. Since the initial description of reagents for enhanced protonation,[92] several additional reports have drawn a correlation between the ions that appear in the mass spectrum and the acid/base properties of the solvent matrix.[433–435] It should be noted that cationization reactions that lead to the formation of such ions as $(M + Na)^+$ are accommodated within the acid/base model, but are based on Lewis rather than Bronsted acid/base concepts.

Selectivity in derivatization is available with the use of more complex chemistry. New derivatization reactions developed for FAB and SIMS are based on functional group specific reactions. For instance, aldehydes and ketones can be derivatized with pyrrolidinium perchlorate to form immonium salts.[92] The sample is analyzed via the appearance of the intact cation in the FAB spectrum at a mass 54 daltons above that of the neutral molecule. Other ion-addition derivatizations for aldehydes and ketones follow classical organic reactions. The use of Girard's reagents for the transformation of aldehydes and ketones into water-soluble hydrazones is a standard organic procedure for their purification. Use of these reagents for the general derivatization of carbonyl compounds for SIMS and FAB analysis has been reported by Ross;[436] this reaction has been used to form derivatives of ketosteroids for positive ion SIMS analysis.[437,438] The steroid of mass M is analyzed via the intact cation of the ionic derivative, $(M + 104)^+$ (Girard's P) or $(M + 124)^+$ (Girard's T).

A particularly useful functional group specific reaction for biomolecules is the reaction of primary amines with pyrylium salts to form pyridinium salt derivatives.[439] Substitution of nitrogen for the ring oxygen to form the pyridinium ion is accompanied by loss of water. The pyrylium reagent reacts with aminoglycosidic antibiotics such as kanamycin, which contains a primary amino group. The FAB spectrum of kanamycin A derivatized with the trimethylpyrylium tetrafluoroborate salt contains, as expected, a signal for the intact cation of the pyridinium derivative at m/z 590. When the reagent is present in great excess, the doubly charged dipyridinium derivative may be formed. The formation and observation of doubly charged ions in the mass spectrum provide a means of extending the useful mass range of the instrument (Section 5.5.3). However, a general reaction observed is loss of a proton from the doubly charged derivative to reform the singly charged ion.[639]

The two functional groups most commonly found in biomolecules are the amino and the hydroxyl groups. Amino groups can be derivatized as described above. The reagent 2-fluoro-1-methylpyridinium *p*-toluenesulfonate reacts with hydroxyl groups to form *N*-pyridinium salt derivatives, bound through an ether linkage to the R group of the original alcohol ROH.[440] This reagent reacts, for example, with the hydroxyl group of the steroid androsterone, producing in the positive ion FAB mass spectrum a signal for the intact cation of the *N*-pyridinium salt derivative. Surface-active forms of this derivative have been described.[639]

Use of a derivatization reaction to create a preformed ion generally increases the sensitivity of the mass spectrometric experiment. The even-electron ions formed by the chemical reaction are stable species perhaps more resistant to damage from the bombardment, and less likely to participate in charge exchange processes as they traverse the selvedge. The ion current in the "quasimolecular ion" is shifted to a higher mass value where the contribution from the solvent background may be less, and the signal-to-noise ratio might therefore be increased. If an interferent is found at a particular mass, an alternative derivatization reagent can often be found. An isotopically labeled internal standard may also be used to shift the mass of the intact cation, or a 1:1 mixture of unlabeled and labeled variants of the derivatization reagent may be used to generate characteristic, easily recognized doublets in the mass spectrum. As with any derivatization reaction, care must be taken to ensure the completeness of the reaction, to minimize the formation of side-products, to determine the magnitude of matrix effects, and to preserve the integrity of the sample.

De Pauw[441] has advanced the concept of ion addition derivatization to include the formation of a surface-active derivative of the analyte. The initial goal of this work was a search for a protonation reagent that also suppressed the ion signals from the solvent matrix, thus providing an increased signal-to-noise ratio for the detection of the analyte signal. Sulfonic acids were chosen for their low vapor pressures, high acid strength, and their excellent solubility in glycerol. In addition, the surfactant properties were known to increase in the series of methane sulfonic to *para* toluenesulfonic to camphorsulfonic acids. Using leucine as the model system, the absolute intensity of the signal for the protonated molecular ion of the amino acid was shown to increase with the hydrophobicity of the sulfonic acid. At the same time, the peaks due to the glycerol solvent itself were diminished in intensity. It was concluded that the hydrophobic counter ion pumps the protonated molecular ion to the surface of the liquid matrix.

Ligon(442) also described the importance of surface activity in determining the relative yields of preformed ions in a SIMS or FAB experiment using a liquid matrix. Ligon has described surface-active derivatization reagents for the analysis of peptides(443) and for amines.(444) The reagent 2-dodecen-1-ylsuccinic anhydride was used to create a surface-active anion from primary and secondary amines.(444) Yields of 90–93% were established from simple amines derivatized at the millimolar level; the reaction fails with sterically hindered amines such as 2,2,6,6-tetramethylpiperidine. The product of the derivatization reaction is an amic acid which provides an abundant $(M - H)^-$ ion in the negative ion mass spectrum. Primary and secondary amines can be distinguished since the amic acids from primary amines are converted into imides upon treatment with acetic anhydride. Signals from the amic acids of the secondary amines remain in the mass spectrum after such treatment.

Kidwell(445) has studied the positive ion SIMS spectra of aldehydes and ketones derivatized with analogues of Girard's reagents of varying size. Since these experiments involved the direct sputtering of the ionic derivative from a solid substrate, surfactancy was not an issue. Kidwell established that higher secondary ion yields were found for the larger quaternary ammonium derivatives (despite a trend towards lesser reactivity for these compounds used in derivatizations) and attributed this to the decreased energy necessary for the separation of a large cation from its counter ion.

Kebarle(446,447) has studied the abundances of ions in FAB mass spectra as a function of conditions prevailing in the liquid matrix. Positive ion FAB mass spectra were measured for binary mixtures of neutral compounds themselves used as FAB solvents.(446) In general, the mass spectrum of the component with the lower gas-phase basicity was suppressed. The proton affinities and gas phase basicities of each of the compounds studied are available from other experiments. Complications arise if one of the compounds is surface-active in the counter solvent. Kebarle suggests that proton and charge transfer reactions in the selvedge (defined as the interface region between the surface and the vacuum, beyond which no ion/molecule reactions occur[445]) define the final distribution of protons between competing bases. In later work,[447] Kebarle described the variation in the total ion current recorded in a FAB experiment with electrolyte concentration in the liquid matrix. For alkali ions and pyridinium hydrochloride, an increase in total ion current with increasing analyte concentration was not observed. This result was interpreted in favor of a gas collision model that includes thermodynamically driven reactions occurring in the selvedge, the end products of which are essentially independent of the initial charge state of the ions in solution. This model is known as the phase-explosion model.[640,641] Lacey and Keough have shown[642] that sampling phenomena related to the surface concentration of the species in the chosen matrix complicates a simple interpretation of the data.

The gas collision model may be best applied to experiments in which the flux of the primary particle beam is sufficient to rapidly erode the surface of the liquid, creating a relatively high number density of molecules in the selvedge. That such conditions can exist was demonstrated earlier by Rollgen,[448] who showed that a "surface self-cleaning" mechanism, i.e., the rapid ablation of the surface by the primary particle beam, obviated the need to invoke a diffusion-controlled rejuvenation of the surface by exchange of molecules from the bulk of the solution. Under such conditions, the effects of ion preformation and sample surfactancy should be diminished, and the particle-induced desorption ionization mass spectrum should reduce to that of chemical ionization, since the proton transfer in the gas phase (the selvedge) determines the final ionic composition of the mass spectrum. Under less intense conditions of primary bombardment, the importance of ion preformation and surfactancy are magnified.

Ion addition derivatization reactions that have been developed for SIMS and FAB analyses are subject to the same drawbacks as any other analytical derivatization reaction. The extent of side reactions and matrix effects should be established before these reactions are used in a quantitative assay. The scale and simplicity of the reactions have to be considered in light of the small sample volumes and low concentrations routinely used in SIMS and FAB analyses, for example. Some of these derivatization reactions have already been used to probe the structure of very high mass biomolecules in other venues. For instance, the pyrylium salt reaction has been used to study the role of lysine residues in chymotrypsin.[449] Such application suggests the breadth of use that these reactions may come to encompass.

4.3.1.2. Charge-Transfer Derivatizaiton. An alternative to the ion addition derivatization chemistry described above is derivatization via formation of charge-transfer complexes. The success of nitrophenyloctyl ether as a FAB matrix is ascribed to its ability to participate in such charge-transfer reactions. The detailed chemistry of charge transfer complexes is well known.[450] Charge-transfer reactions can be used to form ionic derivatives for compounds for which no ion addition reagent is available. For instance, polynuclear aromatic hydrocarbons (PAH) form stable charge-transfer compounds with a number of electron acceptors and donors. While FAB analysis of the uncomplexed sample provides no signal for the molecular ion, the complex provides an abundant radical molecular ion.[451] Of further interest is the behavior of these complexes when exposed to a high photon flux; ion pairs can be dissociated by irradiation of the solution with light of the proper wavelength, raising the possibility that mixtures of complexes can be selectively dissociated and ionized.

The formation of the charge-transfer complex is in itself insufficient for enhancement of the signal for the radical ions in the FAB spectra. In nonpolar solvents, the complex exists as the neutral ion pair, and under these conditions, no signal enhancement would be expected. In polar solvents, solvation of the radical ions aids in their separation. Spectroscopy (electron paramagnetic resonance and UV/vis) confirms the separation of the ion pair. Optical spectroscopy in conjunction with FAB mass spectrometry has been used to investigate the charge-transfer derivatization reaction.[451,452] DiDonato[451] studied a charge-transfer system in which the donor is N,N,N,N-tetramethylphenylenediamine (TMPD) and the acceptor is a substituted benzoquinone such as 2,3-dichloro-5,6-dicyanobenzoquinone (DDQ). The presence of the $TMPD^+$ radical in solution is reflected in the UV/vis absorption spectrum, which contains a characteristic doublet at 580 and 614 nm. The UV/vis spectrum of the TMPD/DDQ complex in two common FAB solvents, glycerol and dimethylsulfoxide, reflects a complicated chemistry. In glycerol, the absorption doublet due to $TMPD^+$ increases in intensity for up to 1 h after the compounds are mixed. By contrast, in dimethylsulfoxide the intensity of this absorption band decreases in the same interval. The relative concentration of $TMPD^+$ in these two solvents will thus change explicitly with time, and this change is also reflected in the FAB spectrum, which also exhibits a time dependence. These results emphasize that simple acid/base chemistry is apt to mask the desired charge transfer effects, which may best be observed in aprotic or basic solvents.

As mentioned, polynuclear aromatic hydrocarbons (PAH) form charge-transfer complexes. Formation of these complexes may aid in the study of PAH compounds in nonvolatile matrices. Early FAB studies of samples such as coal and coal liquids[(453)] have not been particularly successful because of the nonpolar nature of the sample. Recently, it was reported that aromatic hydrocarbons dissolved in a high boiling hydrocarbon solvent could be successfully analyzed by FAB,[(454)] and thus success may be due to the formation of a charge transfer complex between the PAH and the hydrocarbon solvent. The use of other charge transfer complexes in PAH analysis has been reported.[(643)] Absorption of light in the visible and near UV range promotes the dissociation of the charge transfer complex. The wavelength of maximum absorption is directly related to the ionization potential of the donor, and the ionization potentials of the polynuclear aromatic hydrocarbons are known. It may thus be possible to selectively ionize these compounds with incident light of the appropriate wavelength. Such charge transfer experiments provide an alternative to the creation of ions from PAH with oxidation by antimony trichloride,[(455)] or the use of sulfuric acid solutions.[(644)]

4.3.1.3. Electrochemical Derivatization. The objective of the derivatization reactions described above has been the creation of preformed ions in solution by chemical means. Oxidation and reduction processes driven by electrochemistry, however, also provide a superb method for the controlled production of ions in solution. A ring disk electrode has been described by Bartmess[(456)] which is coated with a thin layer of glycerol, and potentials of +15 to −15 V applied. As the potential is swept through the oxidation or reduction potentials of appropriate organic molecules, the appropriate ions appear in the secondary ion mass spectra. This first iteration of the electrochemical direct insertion FAB probe is fitted to a sector mass spectrometer, and the electrochemical cell must be floated at the high source potential of several thousand volts. Positive or negative ion analysis can be selected, but not both. In a quadrupole-based instrument, the electrochemical cell can be referenced to ground, allowing a more accurate control and measurement of the actual potential applied. Secondly, because the mass analysis in the SIMS is accomplished with a quadrupole mass filter, both positive and negative ion analysis can proceed nearly simultaneously, taking advantage of oxidations and reductions to provide both positive and negative preformed ions.

With foreknowledge of the electrochemical properties of a target compound, electrochemical derivatization becomes straightforward. However, in a mixture, the potentials at which each sample component will be reduced or oxidized are not known. As electrochemical derivatization develops, experiments can be foreseen that step the potential on the cell by a predetermined amount, and then examine the secondary ion mass spectrum recorded for new ions. If no new ions are formed, then another equally large electrochemical step is taken and the analysis repeated. If the spectrum indicates that a new ion is being formed at that potential, the system automatically steps back to the previous potential, steps a smaller increment, and then again examines the spectrum. This interactive system is a powerful method for probing both the electrochemical and mass spectral properties of mixture components.

Although electrochemical derivatization promises to be an extremely general method of derivatization, several logistical problems still need to be solved. Poor solubility of the samples in electrochemically compatible solvents has been reported by Bartmess.[456] Diffusion in the solvents between electrodes is also a major concern, and since this is a temperature-dependent parameter, the FAB probe should be heatable and coolable. The need for a more precise control of the electrode potentials is apparent, and may be easier with a source fitted to a quadrupole-based instrument. Finally, many electrochemically generated ions may be observable only on a short time scale inconsistent with the more leisurely pace at which many mass spectra are recorded. The short-term solution is to use a selected ion monitoring experiment, but a more elegant solution may be the use of a pulsed desorption event in conjunction with a time-of-flight mass analysis.

Despite these difficulties, preliminary results are impressive. The $(M - H)^+$ ion from 1-bromohexadecane exhibits a 1000-fold increase in intensity when the potential of the cell is shifted from zero volts to +3.0 V or greater. Increases in the abundances of the M^+, $(M + H)^+$, and $(M - 3H)^+$ ions from naphthalene were also observed as the potential to the cell was applied. The latter ions result from a reaction already established in solution electrochemistry. Extension of these initial results to more complex organic and biomolecules has been reported,[645,646] and the parallels between classical solution electrochemistry and that observed with a FAB probe may aid in the interpretation of both spectra and mechanism.

4.3.2 Ion Management

The goal of derivatization for particle-induced desorption ionization mass spectrometry is not only the creation, but also the *management* of ionic charge. Simply put, the spatial distribution of ions with respect to the surface can and should be controlled. Further, the spatial distribution of the cations with respect to the anions must also be considered; neutralization by ion-pair formation undermines the sensitivity of the mass spectrometric experiment. The concept of ion management includes a description of surfactancy effects in liquid matrices, and the control of ion distribution across a solid surface (surface management). It also includes more novel approaches to control the distribution of ions in a solution through the use, for example, of host/guest complexes (bulk management). Succinctly, these approaches deal with the placement of ions in either two (surface) or three dimensions (bulk).

4.3.2.1. Surface Management. Derivatizations by ion addition and charge-transfer create preformed ions that can circulate throughout a liquid matrix, replenishing the surface as necessary to maintain a steady beam of secondary ions. Marked spectral effects can be expected if the ions are not allowed to move freely through the matrix, but rather are constrained in some manner. Imposing a structure in the matrix alters both the local electrostatic properties and the long-term order/disorder of the solution. Further, only a small percentage of the material introduced to the source as a solution is actually transformed into secondary ions. Specifically, those ions present at the surface are those that ultimately appear in the mass spectrum. Surface-ion management thus includes methods used to control the supply of sample available for sputtering, and to simplify the recovery of sample that is not consumed.

Ionic analytes that are deposited in a thin film onto an ion exchange resin support undergo strong, specific interactions with the support. In plasma desorption mass spectrometry, the mass spectra of ionic analytes deposited onto Nafion (a perfluorinated cation exchange polymer) exhibit increased abundances of characteristic positive ions, and decreased abundances of negative ions which are not strongly adsorbed at specific sites on the support.[282] On ion bombardment, the weak link between the cation and the polymer-bound anion site is cleaved, leading to an enhanced abundance of the cation. A similar effect has also been observed in FAB[457] using a finely ground ion exchange resin coated with the analyte and suspended in a glycerol solution. Pyrrolodinium perchlorate has a tendency to form abundant cluster ions CAC^+ in the positive ion FAB spectrum (C = cation and A = anion). This sample was coated onto a Dowex ion exchange resin. The FAB spectra of this mixture contained an increased relative abundance of the intact cation at m/z 72, a decreased abundance of the CAC^+ cluster ion at mass 243. Neutralization by clustering with single free anions, and cluster formation (such as the CAC^+ described) are both processes that divert ion current to species other than those directly indicative of molecular weight. The use of ion exchange resins decreases the contributions from these processes, and concentrates the ion current in an ion of interest such as the cation.

An ion exchange resin acts as a selective matrix, reducing the strong ionic interactions between the cation and the anion of the sample salt. In addition, when the resin is constructed as a film, the adsorbed ions are concentrated at the surface that can then be directly bombarded by the primary particle beam. McNeal[(413)] has described a tridodecylmethyl ammonium chloride surfactant surface onto which organic compounds can be electrosprayed. In particular, the plasma desorption mass spectra of several nucleoside phosphates were obtained with and without the cationic surfactant support. Absolute ion currents for significant quasi-molecular and fragment ion were 4–5 times as abundant when the surfactant support was used. McNeal suggests that such matrices may in the future be used to concentrate specified ionic components from a dilute aqueous mixture onto an ion exchange surface. This is an attractive experiment in that it may also reduce the degree of intermolecular interaction (including such complications as hydrogen exchange) and simplify the interpretation of the resultant mass spectrum.

Fine has observed a novel surface phenomenon produced by ion impact on the surface of bulk liquid gallium.[(458)] Organic material diffused rapidly across the surface of a liquid metal during bombardment. Since the sputtering process removes material, a mass flow of the impurities into the bombarded site was observed to occur until an atomically clean surface of gallium was obtained. Ross and Colton[(283)] have taken advantage of this phenomenon by demonstrating that polynuclear aromatic hydrocarbons can be sputtered from liquid gallium surfaces to produce a steady and persistent flux of organic secondary ions. With a liquid glycerol matrix, diffusion of organic ions from within the bulk reservoir is thought to replenish the surface in addition to the diffusion along the surface. This additional diffusion is three dimensional—both across the surface (x, y) and also from within the bulk (z). Organic compounds supported on a liquid gallium surface do *not* diffuse into the bulk, and the diffusion is limited to the metal surface. Nevertheless, this diffusion also serves to replenish the material within the area sampled by the primary beam. Since an analogous diffusion of organic material does not occur with solid gallium, but does occur with liquid gallium, the diffusion of organic materials across a metal surface can be controlled by small variations in temperature which force the solid/liquid phase transition. Gallium and gallium eutectics are promising metals for these experiments. Ga (92%)/Sn (8%) melts at 20°C; Ga (95%)/Zn(5%) melts at 25°C; Ga (67%)/In (29%)/Zn(4%) melts at

13°C. At slightly higher temperatures, but still below the temperature at which thermal degradation of organic compounds might occur, compounds of bismuth, lead, and tin might be used.

Several advantages of using solid/liquid metal surfaces for the desorption of organic compounds may be expected. In most SIMS and FAB analyses, the majority of the sample is not sputtered, and can be recovered from the probe. At this point, however, the sample is mixed with glycerol and thus an extraction is needed to remove the solvent. If the sample is supported on a gallium surface, the recoverable amount can be washed from the metal. Sensitivity may be enhanced because the sample is preconcentrated at the surface of the metal rather than diluted within the glycerol. For experiments near the detection limit, total sample depletion via surface diffusion might be completed in less time than diffusion from within the bulk, leading to a greater flux of secondary ions and lower detection limits.

4.3.2.2. Bulk Management. Crown ethers can be used to create a three-dimensional structure surrounding the analyte molecule. Crown ethers are characterized by an internal cavity, often ringed with oxygen or nitrogen atoms, that serves as a host to molecules that can both hydrogen-bond to the heteroatoms and can physically reside within the cavity. Many different forms of the crown ethers are available, with well-characterized affinities for the smaller alkali ions such as sodium or potassium. Johnstone and Rose[(459)] have investigated the complexation between alkali ions and the crown ethers, and correlated the strength of this interaction with the relative abundances of these ions as they appear in the positive ion FAB mass spectrum. Organic salts can also be incorporated into the cavity of a crown ether. Complex formation has been reported between dibenzo-30-crown-10 and the diquat dication, and it is known that the dication as guest assumes a structure similar to that in the near crystal.[(460)] The intercharge distance and the Coulombic repulsion energy within the ion will be the same as if it were in a neat crystal. Crown ethers have been used as FAB matrices for the analysis of such compounds as glutathione and free adenosine triphosphate.[(461)] The ability of the crown ether to sequester free protons in the liquid matrix led to an increase in the absolute abundances of the deprotonated molecular ions $(M - H)^+$ for these molecules. The addition of sterically hindered nitrogen bases to the liquid matrix to decrease the free proton density is an alternative experiment.[(462)] Since the initial work on crown ether matrices by Rose,[(463)] several other workers have studied the complexation of alkali ions by the crown ethers[(464,465)] using FAB mass spectrometry. A more general use of the crown ethers in the preparation of sequestered samples, or in the removal of traces of alkali impurities from solutions, can be foreseen.

5. Special Techniques

The ability to analyze organic and biological samples by particle-induced desorption ionization has catalyzed the expansion of mass spectrometry into fundamentally new areas. Avoidance of the requirements for sample vaporization has fostered the development of chromatography/mass spectrometry instruments and experiments in which the sample is analyzed directly from the chromatographic matrix. The use of the liquid solvent matrix in FAB and SIMS has made possible real-time kinetics experiments with mass spectrometric detection. Finally, the ability to create ions of very high mass from biomolecules has led not only to an order-of-magnitude improvement in the range of mass analyzers, but also to an important shift in the approach to mass spectral interpretation, and a change in emphasis in the type of information that the mass spectrum is expected to provide.

5.1. Chromatographic Interfaces

5.1.1. Dynamic Chromatography

Central to this discussion is the subdivision of chromatographic separation techniques into dynamic and static forms of chromatography. Dynamic chromatographies are defined[446] as those that measure the separation of compounds in the time domain. Examples include gas and liquid chromatography, both of which generate a characteristic retention time for the analyte. Gas chromatography coupled with mass spectrometry (GC/MS) was developed in the 1960s, and has become the method of choice for the analysis of complex mixtures. A successful interface between liquid chromatography and mass spectrometry has been the result of work in the early 1980s, and LC/MS instruments are now in place in most research laboratories. The ability of liquid chromatography to separate complex mixtures of nonvolatile species makes it very useful for biological research.

The approach to the interface between either gas or liquid chromatography and mass spectrometry has usually been to remove the carrier gas or the solvent, and to transport the sample into the source of the mass spectrometer, which operates under the usual electron ionization or chemical ionization conditions. Thermospray and electrospray ionization techniques generate the sample ions directly from the solution itself as it is nebulized through a small orifice into the vacuum. The particle-induced desorption ionization methods have been used in conjunction with liquid chromatography when the eluting samples are deposited onto a support. The intervention of this deposition restricts the use of the ionization method as a continuous, on-line detector.

Liquid chromatography has been coupled with SIMS, FAB, and PD. Benninghoven in 1980 reported[467] a combination utilizing a continuous belt interface similar to that which had earlier been used for the coupling of EI or CI sources to liquid chromatographs. The difference in operation is that the sample is not thermally desorbed from the belt and then ionized in the source as in the usual case, but rather, a beam of primary ions sputters sample ions directly from the belt as it nears the entrance aperture of a quadrupole mass spectrometer. Solutions of leucine and cytosine (0.01 *M* and 0.001 *M*, respectively) were deposited successively on the moving belt (0.1-mm-thick silver). The belt moved through three stages of pumping that removed the solvent, and then into the SIMS source. With a speed of 0.6 cm/s, selected ion monitoring of the protonated molecular ions of these two analytes revealed the distribution of sample across spots of 5 cm length. The amount of sample detected was on the order of 0.1 μg; detection limits of 0.1 ng were extrapolated using a higher primary beam flux.

In 1981 Smith[468] described a similar device, with the distinction that a triple quadrupole mass spectrometer was used to perform MS/MS studies on the sputtered ions. The moving belt interface was the same as that previously described for electron and chemical ionization studies,[469,470] but now equipped with a primary ion gun. Parameters such as belt speed, sample thickness, and primary ion flux were studied. Arginine was used as the model compound, deposited on a thin nickel belt (0.08 mm thick) in coverages ranging from 2.5 ng/cm^2 to 2.5 mg/cm^2. Full spectra were taken with the quadrupole mass spectrometer as the belt was moved at a speed of 5 cm/s through the analysis region. At lower coverages, the sample was completely consumed during the exposure, and only larger samples provided a consistent response. As in the system of Benninghoven, no continuous LC/MS analyses were reported. A residual organic background on the nickel belt was noted by Smith, and a silver overlayer deposition device was used to reduce this background. The problem is not as severe for nonvolatile samples, as the belt can be thermally cleaned after each pass through the source. For nonvolatile samples, however, simple heating will not remove the residual analyte. An alternative is sputter cleaning, possible with intense primary ion beam bombardment.

Vestal has very recently compared LD and SIMS analysis of samples deposited on a belt interface in LC/MS.[471] SIMS and LD spectra were obtained under identical chromatographic conditions. Spectral differences between LD and SIMS were observed for the amino acids and nucleosides studied, and surface coverage and dosage effects on the secondary ions observed were delineated. Figure 10 compares the SIMS and LD spectra of histidine and cytidine. Cationized species are observed to be more dominant in the LD spectra, while more fragment ions appeared in the SIMS spectra. Some differences might be expected due to the pulsed nature of the LD and the continuous bombardment of the SIMS system. Detection limits for both techniques are typically a few nanograms.

Jungclas and co-workers have published a series of papers that describe the use of PD as a detector for liquid chromatography.[472–475] A vacuum interlock system was devised in which samples eluting from the liquid chromatograph are collected for several seconds on one of several sample disks on a rotating collector. At preset intervals, the collector rotates the sample disk through a vacuum lock into the source region of a PD/TOF mass spectrometer. Fission fragments from californium initiate the sputter process, and the sample is analyzed in approximately ten minutes. Again, this is not a true on-line device, but a discrete sample collection and analysis system. The time intervals in collection and analysis depend explicitly upon the column resolution and the sample

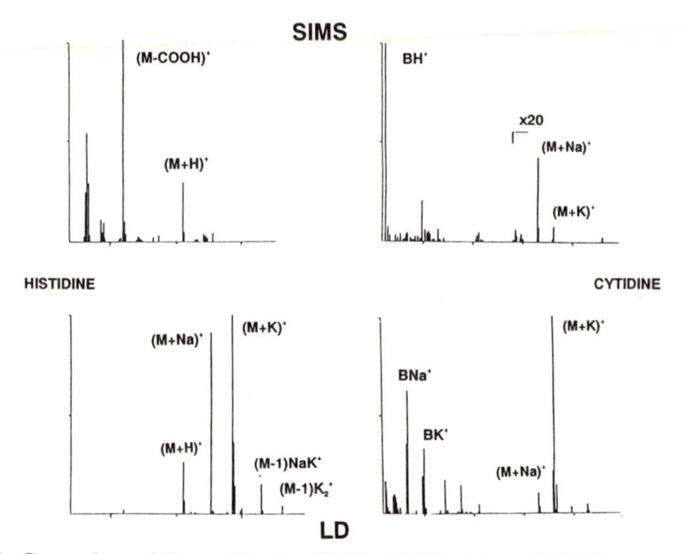

Figure 10. Comparison of the positive ion SIMS and LD spectra of histidine and cytidine. Transfer of ion current from protonated to cationized molecules is noteworthy. (Adapted from Ref. 471.)

concentration. A typical reported sample collection time is 1 min. Data acquisition time for the TOF mass analysis is about 10 s for sample concentrations of 20 μg/cm^2 and above, but as long as 1 h for samples in disk concentrations of 10 ng/cm^2. A disadvantage of this system is that in the analysis of an unknown mixture, a compromise set of conditions for sample collection and sample analysis must be used. The dynamic range of the LC/PD-TOF combination in terms of sample resolution and sample concentration is relatively low.

Flow FAB has emerged as the interface of choice in joining a microbore LC column to a mass spectrometer.[559,560] Many reviews document the development and applications of this interface. In juxtaposition to the original purpose of this device is a growing use of a continuous flow FAB probe for the introduction of discrete samples into the source of a mass spectrometer. Use of this introduction device avoids the need for constant withdrawal and reloading of the direct insertion probe. Using a continuous flow of a mixture of volatile solvents and a low (about 10%) concentration of a FAB solvent, a more reproducible surface for sputtering is obtained, and the need for retuning the FAB source for each sample is reduced. Capabilities in automated and repetitive FAB analyses are evident.

5.1.2. Static Chromatography

Static chromatography includes thin-layer and paper chromatography, as well as electrophoresis. The distinguishing characteristic of these chromatographies is that time does not explicitly enter into the separation. A retention time is not calculated, but rather, the movement of the sample relative to the movement of the solvent is determined. Furthermore, the sample is encapsulated within the chromatogram, as opposed to the dynamic forms in which the sample moves through the column and is ultimately eluted. The value of coupling these static forms of chromatography with particle-induced desorption ionization mass spectrometry lies in the simplicity, relatively low cost, flexibility, and widespread use of these separation methods. The need for an on-line, continuous detection system is avoided because the chromatogram does not develop during the course of the analysis, but is instead complete before the analysis has begun.

5.1.2.1. Special Characteristics. The advantages of the chromatography/particle-induced desorption ionization mass spectrometry combination accrue in large part because the spatial integrity of the chromatogram is retained throughout the analysis.

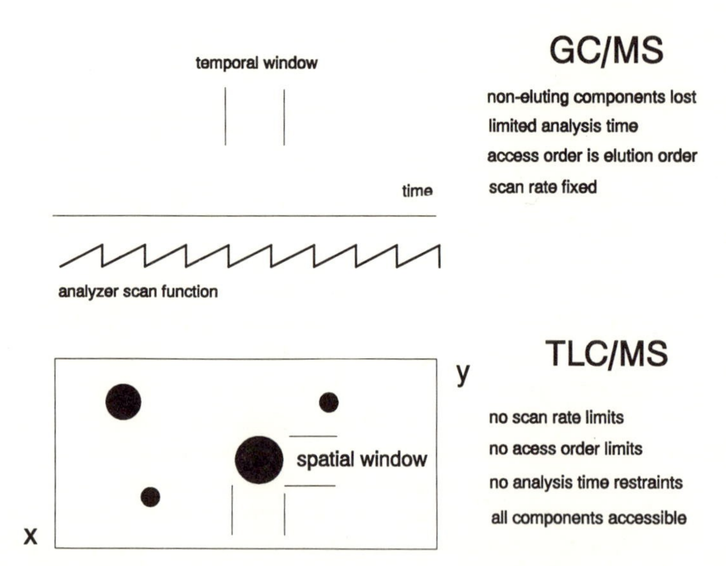

Figure 11. Characteristics of static and dynamic chromatography relevant to detection by mass spectrometry.

Accessibility: The first advantage is that of accessibility. Figure 11 compares the relevant characteristics of gas chromatography (GC) with those of thin layer chromatography (TLC) with respect to sample accessibility. The separation in GC is based in time while that in TLC is based in the x and y coordinates of the two-dimensional chromatogram. A detector at the exit of the GC column must deal with the samples as they elute from the column. A detector for TLC can be used to examine sample spots in any desired order, and repetitively if desired. This advantage is most telling when a particular compound or set of compounds in a complex mixture is targeted for analysis. In GC, the analyst must wait until the appropriate retention time window to perform the analysis. In TLC, the analyst can immediately examine the x, y coordinates where the compound or compounds should appear. Note further that in GC/MS, for example, the mass spectrometric detection is destructive, and the sample is present in the ion source for only a few seconds. The sample, once examined, cannot be reexamined unless the sample is reinjected into the GC. In TLC, on the other hand, a surface-sensitive ionization method such as SIMS leaves the bulk of the material in the chromatogram undamaged. If the initial identification of a compound present at (x, y) is drawn into question by a subsequent examination of sample at another set of coordinates, it is possible to return to the first spot for a more careful second analysis.

Informing power: A second criterion for comparison between dynamic and static chromatography is that of the information provided by each. With small-bore column technology in both GL and LC, very high separation resolutions can be achieved. The separation capabilities of TLC, for example, have also improved with the introduction of smaller and more uniform packing particles. Most of the difficult separation problems approached with TLC use the two-dimensional approach, in which development with solvent takes place in both the x and the y directions. Although the number of plates is restricted to a value far below that of high-performance GC and LC columns, resolution adequate for many complex mixture analyses can be achieved. Lack of separation power is mediated by additional informing power from the mass spectrometric detection, since the mass spectrometer itself can be operated in a more efficient manner. For instance, in GC/MS, scanning of the mass spectrometer takes some finite amount of time; each scan may require 2 s, and the concentration of the sample in the source can change significantly during that time, especially with capillary columns. The resolution of separation can be limited by constraints on the operation of the mass spectrometer. Effective resolution can be increased if a selected ion monitoring experiment is chosen, but with a decrease in the amount of information available from the mass spectrum. In TLC coupled with mass spectrometric detection, by contrast, both x and y dimensions can be independently used to establish the presence of a peak (spot) in the chromatogram. Each ion that belongs in the mass spectrum of a compound encapsulated at coordinates (x, y) must maximize at the same point in the xy plane. Furthermore, each analysis can be retraced to confirm the appropriate rise and fall of each ion. The information in GC/MS must be extracted from data in three domains—relative abundance (RA), mass-to-charge ratio (m/z), and time (t). In TLC/mass spectrometry, the information is contained in the four domains of RA, m/z, x, and y. Figure 12 shows the expected behavior for each ion that is

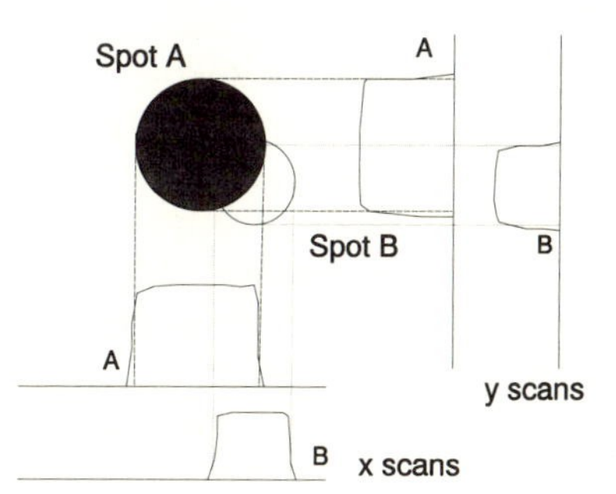

Figure 12. Spatial profiles for ions corresponding to compounds separated in x and y by chromatography.

part of the spectrum of the compound present in the chromatogram at the coordinates (x, y).

Variable resolution: Movement in the xy plane brings different compounds in a TLC to the point of analysis. The advantage of TLC over GC in this area is that the resolution of that xy movement is easily variable. Several modes of operation are possible. In the screening mode, the spot diameter is large and the point of analysis is shifted by mechanical movement in the xy plane in 0.1-mm increments. In this way, the distance between sample spots is covered as rapidly as possible. As the primary beam approaches the edge of a sample spot, ions characteristic of the sample increase in intensity, while ions characteristic of the background either stay the same or decrease in intensity. The sum of the rates of change for all ion currents rises. Above a preset threshold, a feedback loop signals the cessation of mechanical movement, and the initiation of more precise (0.01 mm) rastering of the primary ion beam. If the rate of change in ion currents is still high, the system signals a reduction in the primary beam diameter, with a further increase in effective resolution in the xy plane. If the sample spot is large, the rate of change in ion currents will decrease as the spot is traversed, and the diameter of the primary beam will increase. When analysis of the mass spectrum indicates another interface region, then the spot diameter is again decreased in order to establish thc boundary. This increase in resolution occurs on-line whenever the point of analysis crosses the background-to-sample transition, or when the composition of the spectrum acquired changes. Components that are incompletely resolved by the chromatography are automatically differentiated by mass spectrometry at the highest possible spatial resolution in both the x and the y coordinates. A further advantage is that a minimum amount of time is spent examining areas of the chromatogram in which no compounds are present, and the time establishing the exact coordinates is maximized. Since the feedback is automatic, totally unknown mixtures can be accommodated.

Sample storage: In static chromatography, the sample remains within the chromatogram as a consequence of development. This characteristic contrasts strongly with GC or LC in which the sample elutes from the column during development and is lost unless the effluent stream is routed to both a detector and a fraction collector. The matrix in which the separation occurs can be used as a storage medium for the sample, so that both the data *and* the sample can be retained. At some later time, another analytical technique can be used for the characterization of the sample mixture, or particular sample components can be excised from the chromatogram and subjected to further separation or analysis. Nonvolatile compounds analyzed by TLC are those that also seem to be suitable for storage within the chromatographic matrix.

A review of all of the interfaces developed for joining thin layer chromatography with mass spectrometry is available,[(647)] as is a more specific review detailing the use of SIMS and FAB with thin layer chromatography.[(648)]

5.1.2.2. Applications. The experiments described in this section are different in conception and execution from those in which sample and chromatographic matrix are scraped from the chromatogram, placed on the direct probe of a mass spectrometer, and the electron or chemical ionization mass spectrum obtained. This latter sequence was first suggested in 1962 by Grutzmacher,[(476)] and has been through several modifications.[(477,478)] Jamieson[(479,480)] has advanced the idea by using a high-power light source or laser to desorb thermally materials from a static chromatogram into the thermal source and bring the separated compounds into the system one after the other. Electron or chemical ionization was used to create ions from samples volatilized from the chromatogram. Jungclas[(481)] has described a combined thin layer chromatography/plasma desorption mass spectrometry experiment. Drug samples were separated by thin layer chromatography, and then the developed spots (located by another method) were extracted and a discrete sample prepared for PD analysis. These are examples of a nonintegral chromatographic analysis, that is, an analysis that removes the sample from the chromatogram before analysis, as contrasted with a direct method of analysis in which the chromatogram and all of its encapsulated components remain intact.

Nonintegral analysis: In many experiments involving particle-induced desorption ionization mass spectrometric analysis of materials separated by static chromatography, the sample spot is physically removed from the chromatogram, thus destroying the integrity of the chromatogram. Furthermore, although the intact remnant is attached to the sample stage, glycerol is added to the spot. The sample is thus extracted from the chromatogram into the liquid matrix, and the analysis is exactly that which would occur had the purified sample been added to the glycerol directly.

Chang[(482)] described such a method for this indirect analysis of chromatograms. A piece of double-stick tape on the FAB stage was pressed against the spot on the TLC plate, picking up adsorbent and sample from the plate. Glycerol or thioglycerol is then added to the adsorbent, performing the extraction. Model compounds used for the study were three coccidiostats. The sensitivity of the method was such that 0.1 μg of the sample separated on a TLC plate would produce a full, interpretable mass spectrum. No background interference from the silica gel adsorbent or from the double stick tape were noted.

Barofsky[483] has reported a similar method. Precharged compounds (arsenobetaines) were separated by TLC. The appropriate spot was excised from the TLC plate and attached with vacuum-compatible silver paint to the FAB stage. Then 1–5 μl of the mobile phase solvent and 5 μl of glycerol were added to the spot to perform the extraction. The FAB spectra obtained were identical to those obtained by direct introduction of the sample into the glycerol matrix. Kishu[484] has described the analysis of lipids separated by thin layer chromatography with SIMS. Triethanolamine was used as the extraction matrix, and most samples were excised from the plate and attached to the direct insertion probe. For a few samples, a larger portion of the chromatogram was retained, and the spatial distribution (in one dimension) only of the sample ions could be traced. The possibility that the addition of the extraction matrix would redistribute the sample within the chromatogram was not discussed.

Integral analysis: In this section, experiments are described that maintain the spatial integrity of the chromatogram, thus retaining all of the advantages of the interfaced separation and detection methods. Thus, movement of the chromatogram brings compounds into the focus of the analysis, the analysis can be repeated, and the sample can be recovered from within the chromatographic matrix for further workup. Cooks was the first to report the direct analysis of thin layer chromatograms by SIMS.[485] In the example shown in Fig. 13, a mixture of choline and butyrylcholine were separated by paper chromatography, and then the chromatogram was mounted on the manipulator of a standard SIMS instrument. Rotation of the manipulator brought the sample spots into the focus of the primary ion beam and the secondary ion extraction optics. Rotation of the manipulator could be correlated with the x-coordinate of the sample in the chromatogram. Later, samples of precharged ions were sputtered from cellulose acetate membranes of the type used in electrophoresis.[486] In both of these experiments, only one dimension of the static chromatogram was used for the separation and analysis, and the spatial resolution of the primary ion beam was on the order of 1 mm. None of these experiments involved the addition of glycerol to the chromatogram, limiting the primary ion fluxes to the static SIMS limit, but in each case, the spatial separation provided by the chromatogram was preserved.

Tantsyrev[487] has used FAB to sputter secondary ions of amino acids directly from the thin layer chromatograms in which they were separated. Glycerol was layered on top of the chromatogram, however, and

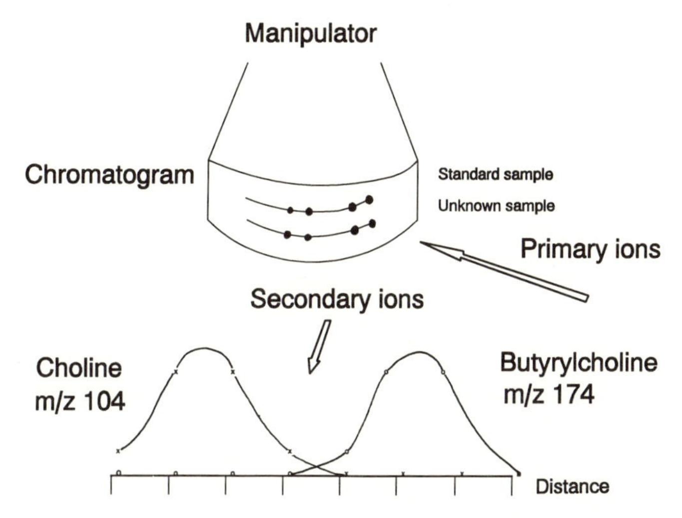

Figure 13. Separation of choline and butyrylcholine by paper chromatography and identification by secondary ion mass spectrometry (adapted from Ref. 485).

certainly extracted the amino acid into the liquid phase. The FAB spectra of the amino acids obtained in this way are identical to those using the more conventional means of sample preparation. The minimum amount of amino acid adsorbed within the chromatographic matrix that could be discerned as the full secondary ion mass spectrum was about 10^{-7} g.

Nakagawa[488–490] has described a method for TLC/SIMS that involves the intermediate extraction of the samples from the chromatographic matrix with a vacuum-compatible solvent. Sputtering from sintered glass and silica gel TLC plates was described, with applications to imidazole drugs, thermally unstable sugars, peptides, and antibiotic drugs and their metabolites. A stepper-motor direct insertion probe was constructed that allowed a scan across one dimension of the TLC plate as the probe was moved into and out of the focus of the SIMS source. Figure 14 shows the results obtained for scanning across a TLC-separated mixture of three different peptides. Commercial TLC/FAB attachments are now available from several instrument manufacturers, and increased applications can be foreseen.[649]

Hercules[491] has reported the LD spectra of triphenylmethane dyes obtained directly from a thin layer chromatogram in which they were separated. A mixture of seven dyes were separated on the plate, and the air-dried plate was mounted onto the *xy* manipulator of a LAMMA 1000 instrument. The spectra obtained directly were the same as those obtained from the pure samples with the exception of a background contribution from silica at the low-mass end of the spectrum. The spatial

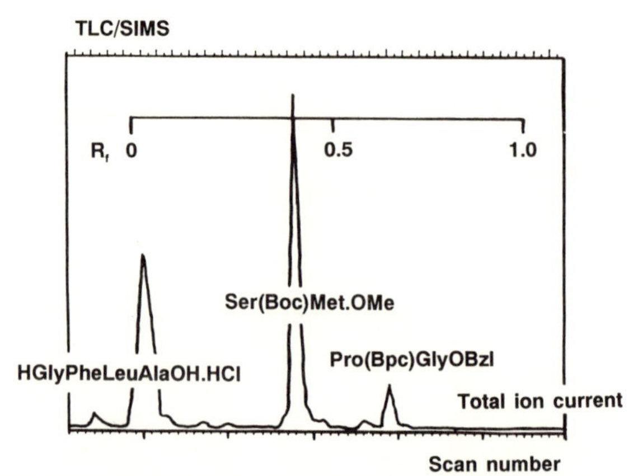

Figure 14. Separation of peptides by thin layer chromatography and identification by secondary ion mass spectrometry. Boc and Bpc are derivatizing groups. (Adapted from Ref. 488.)

resolution attained was not reported. In a recent publication, Kubis *et al.*[(650)] described the use of laser desorption to detect separated compounds directly from a polyamide TLC plate. The polyamide does not interfere with the recorded mass spectrum since the ions generated from this matrix are of low mass. The demonstrated spatial resolution of 5 μm is established by the diameter of the laser beam itself.

Fiola has described the construction of a modular SIMS system used for organic secondary ion mass spectrometry and direct chromatographic analysis.[(492)] Modifications to this instrument in a design revision have been described.[(651,652)] The instrument is based on a quadrupole mass analyzer with a range of 4000 daltons. Chromatograms are mounted on a platform attached to an *xyz* manipulator that is used to move the sample into and out of the point of instrument focus. A cesium ion beam emitted from a thermionic source is used to sputter material directly from the chromatographic matrix into the energy analyzer (a Bessel box) and the quadrupole mass analyzer. A FAB source (saddle field gun) can be fitted in place of the cesium ion gun. A gallium liquid metal ion gun is used for higher resolution on the surface. The development of the phase transition matrix[(493)] has simplified the instrumental requirements for the direct analysis of chromatograms by secondary ion mass spectrometry. A disadvantage of simple extraction of sample from the chromatographic matrix is the loss of spatial resolution. If the solvent absorbs into the chromatogram, sample can be redistributed across a much larger spot than that inherent to the chromatographic separation. Special matrices and derivatization procedures that ensure the spatial resolution in TLC/SIMS have been described.[653,654]

The rationale of the use of a phase-transition matrix is as follows. A particle-induced desorption ionization technique sputters material at the surface of the sample. Secondary ion mass spectrometry, for example, samples from perhaps the top 100 Å of the surface. The bulk of the sample residing in the chromatogram is inaccessible, unless a depth profiling experiment is used. However, at the higher primary ion current densities necessary for the profiling experiment, significant damage to the relatively fragile organic molecules also occurs. Restriction to low primary ion current densities is not necessary if a reservoir of sample, extending into the chromatographic matrix, can be established without compromise of the spatial resolution. Extraction in the z plane without diffusion in x and y is accomplished with the use of a matrix that can be controllably cycled through the solid/liquid phase transition. In the solid phase, no diffusion of samples through this matrix occurs. With liquefaction, samples in the chromatogram are extracted into a matrix reservoir that provides a stable, persistent signal of organic secondary ions. These matrices are placed on the surface of the chromatogram, and then briefly melted. When the matrix resolidifies, it generates a glassy state that is solid and limits diffusion on the chromatographic surface, but that still provides a surface (like glycerol) constantly ablated by the primary ion beam. An appropriate balance of properties between the solvent and the chromatographic matrix limits the xy diffusion to acceptable values. Threitol (1,2,3,4-butanetetrol) is used as the phase transition matrix, although other low melting point matrices can be used as well. Threitol melts at 70–72°C, a temperature easily reached with mild heating. The ability of threitol to dissolve a variety of organic samples appears to be similar to that of glycerol. Several methods have been developed for the incorporation of threitol into the chromatogram. A small amount of threitol (5%–10%) can be dissolved in the solvent system used for chromatographic development. For chromatograms that have been developed independently, the threitol is overlayered onto the surface of the chromatogram by means of spray deposition. The intact chromatogram is then placed in the SIMS instrument for analysis. The primary ion current density is adjusted to melt the threitol precisely at the point of analysis, providing a spatially discrete extraction.

Stanley has described the direct SIMS analysis of phenothiazine drugs separated by thin layer chromatography.[494] The structures of the phenothiazine drugs vary in the substituent on the nitrogen and the group (if any) substituted onto the aromatic ring at the 2-position. However, since all of the drugs are found in solution in the protonated form, a relatively abundant signal for this ion is expected to appear in the positive ion SIMS spectra. This expectation was borne out, and the intact cation is the base peak in the most of the mass spectra presented here. This result contrasts with the electron ionization mass spectra, in which the molecular ions can be of very low relative abundance, but is similar to the data obtained with chemical ionization.(495)

Spatially resolved selected ion monitoring mass spectral data measured with the chromatography/SIMS instrument is obtained by moving the chromatogram with the x and y manipulators, and monitoring the changes in the relative abundance data for the selected ion as the sample spot is brought into and out of the point of instrument focus. The x–y-mapped relative abundance of the protonated molecular ion of trimeprazine separated in a thin layer chromatogram is given in Fig. 15. The total amount of sample in this 3×4 mm oval spot is 5 μg and liquid threitol was used as the extraction solvent. The spectrum persists for well over 20 min, ample time even for the manual recording of the xy spectral intensities. The spacing of the grid lines on the axonometric plot is 0.50 mm. Each spot of a phenothiazine drug separated in a thin layer chromatogram provides similar data for the protonated molecular ion. Examination of the data shows that the initially symmetrical chromatographic spot provides data are themselves not symmetrical. The primary beam itself can promote the diffusion of the sample solution across the surface, and beam-induced tailing may distort the measured shape of the spot profile when an extended time is spent in data acquisition. Extensive applications of the imaging TLC/SIMS device have been described in the literature.(655–657) An optical detection system can be used in conjunction with the mass spectrometer to reference independent data within a

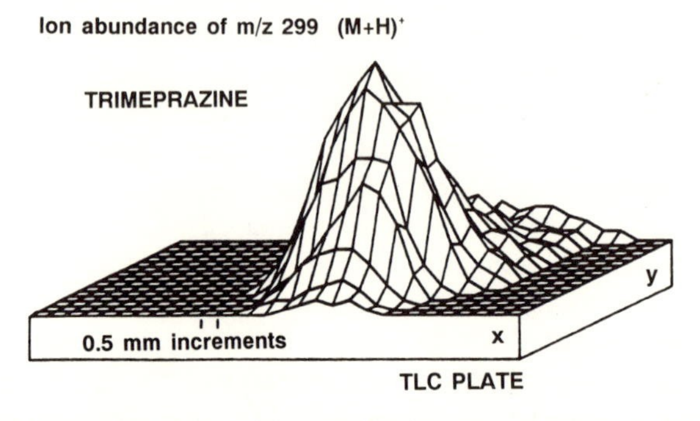

Figure 15. Spatially resolved selected ion monitoring mass spectrum for the protonated molecular ion (m/z 299) of trimeprazine separated in a thin layer chromatogram (adapted from Ref. 494).

common coordinate system.[658] The instrument has also been used in the analysis of biomolecules separated by electrophoresis.[659]

5.2. Real Time Analysis

The liquid matrix in FAB and SIMS was earlier described as a reservoir that continually replenishes its surface with fresh sample. It can also be considered as a miniature chemical reaction vial, in which new products from the reaction can be analyzed. The kinetics of simple chemical reactions can be followed, provided that an ion can be monitored that is related to the rate of disappearance of the reactant or the rate of appearance of the product. The latter is the most often reported experiment. A complication in the direct derivation of kinetic rates from these experiments is the change in concentrations of the relevant species in the solution as the liquid matrix evaporates into the vacuum of the mass spectrometer. A second experiment that has been described is the monitoring in real time of enzyme-catalyzed reactions occurring within the solution held on the probe tip in the mass spectrometer. Most often, the formation of new peptides from enzyme-cleaved larger proteins is evaluated.

5.2.1. Simple Kinetic Studies

SIMS and FAB mass spectrometry can be used to monitor reactions that occur in the solid state or the liquid solution, and further, reactions that occur in the absence of the primary ion beam and those that are induced by the bombardment. Examples include the hydrogenation of thiophene on a silver surface,[496] the formation of metal nitrides in SIMS experiments using nitrogen beams,[497] and intermolecular nitryne and methyne transfers observed in the SIMS analysis of benzotriazole and benzimidazole.[498] Such reactions are noteworthy in that they occur even under static SIMS conditions, and are reminders that even under such low irradiation conditions, the products of beam-induced reactions may complicate the interpretation of the mass spectrum. It is thought that these products form in the selvedge immediately above the surface, since the diffusion of molecules and ions in the solid itself is limited. With relatively facile diffusion through a liquid matrix, equivalent reactions might easily occur.

Figure 16 represents the change in abundance of the product of a reaction between a sulfonium salt and a triethanolamine matrix initiated by bombardment with a cesium primary ion beam.[499] Such a reaction to form the methylated triethanolamine is characteristic of the alkyl transfer

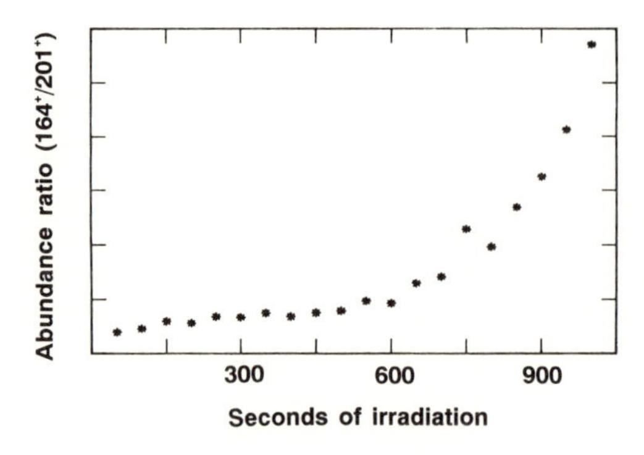

Figure 16. Change of the ratio of a beam-induced product ion (m/z 164) to the intact cation (m/z 201) of a sulfonium salt with irradiation time (adapted from Ref. 499).

reactions of the sulfonium salts, but is not observed in the absence of bombardment. The total time for this measurement is 1000 s, and data can be obtained with a time resolution of as fine as a second. Over the period of 1000 s, the total amount of triethanolamine has not changed appreciably. Detailed evaluation of the kinetics would depend on a determination of the absolute concentrations of product and reactant, and the assumption of an equivalent sampling for the two preformed ionic species. Saito[(500)] has investigated the conjugation reaction between glutathiones and aromatic nitroso compounds formed in reactions occurring on the FAB platform inside the mass spectrometer, and the exchange of hydrogens in glutathiones with deuterium from a mixture of deuterated water and deuterated glycerol.

5.2.2. Catalyzed Reactions

Caprioli[(501,502)] has described a novel experiment in which FAB is used to monitor, in real time, enzyme-catalyzed reactions that occur in the glycerol matrix as it is held on the sample stage within the mass spectrometer. The model system studied was the release of *p*-toluenesulfonyl-L-arginine from its methyl ester, used as a normal assay substrate for trypsin. The reaction produces an ion in the FAB spectrum at m/z 343, and the expected product ion at m/z 329. Over a 30-s time period, consecutive FAB mass spectra showed a consistent decrease in the former signal and a consistent increase in the latter. A more complicated analysis was the hydrolysis of (val^4) angiotensin III by yeast carboxypeptidase Y. The degradation of this peptide within the glycerol solution could be monitored by mass spectrometry. The combination of low vacuum, high glycerol content, and high energy atom bombardment produces a less-than-optimal environment for survival of even simple organisms. An experiment was performed to determine the effect of

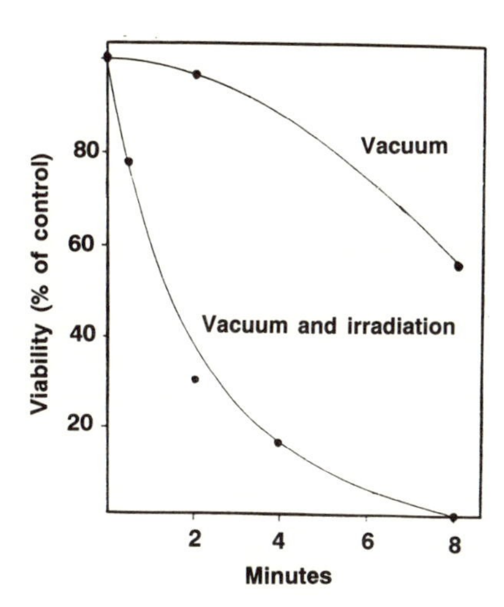

Figure 17. Comparative effect of exposure to particle beam bombardment on the viability of yeast cells (adapted from Ref. 501).

these conditions on the viability of yeast cells (*Saccharomyces cerevisiae*). Figure 17 illustrates the results, comparing the viability of the yeast cells in a glycerol solution that is exposed to vacuum with a solution exposed to both the vacuum and the high-energy atom bombardment. Although eight minutes of exposure to bombardment sterilized the system, with intermittent exposure, the survival time is lengthened. Caprioli concludes that real time measurements such as described are valuable in the identification of transient intermediates that could not otherwise be identified. This experiment is similar in concept to those in which the dosage dependence of the FAB spectra is followed. In this case, new products are actually formed as a result of a chemical process that proceeds independent of the analysis.

Caprioli[(503)] has also used FAB to determine K_m and V_{max} for tryptic peptide hydrolysis. Since the reaction occurs in solution, with the products specified by mass, the concentrations of specific species formed in the reaction can be monitored independently of other components present in the reaction mixture. The analysis takes place without the need for cleanup of the mixture or chemical derivatization. An internal standard was added to the reaction mixture to calibrate the quantitative data. Data obtained with the FAB experiment are compared with those generated by more classical GC/MS techniques. The agreement is extremely close, and suggests that this method may evolve into a general experiment for the determination of the kinetics of enzymic reactions with natural substrates. Caprioli has reviewed the quantitative aspects of enzyme kinetics using these methods.[(504,505)]

A similar real-time analysis has been carried out by Saito.[(500)]

Arylnitroso carcinogens rapidly form conjugates with reduced glutathione, but the conjugates are very labile and thus difficult to isolate and characterize. One microlilter of 10 mM reduced glutathione was mixed with an equal amount of the carcinogen and diluted about 1:1 with glycerol. The loaded FAB probe was subjected to continuous xenon atom bombardment for 100 s following an initial 140-s reaction time, and the amount of conjugate formed increased linearly with time. The power of the method is that the intermediates need never be isolated from the reaction mixture, but their empirical formulas could be confirmed by high-resolution mass spectrometry, or the structures investigated by MS/MS.

Photochemical reactions can also be followed in solutions. Schurz *et al.*[660] have described the use of FAB mass spectrometry to study polymerization reactions occurring within solutions catalyzed by ultraviolet light and the presence of iodonium salts.

5.3. High-Mass Analysis

The techniques of particle-induced desorption ionization mass spectrometry have extended mass spectrometry to ions of mass higher than those previously within reach via electron or chemical ionization. As recently as 1982, spectral determinations in the range above 4000 daltons were described as extraordinary.[506] Magnet technology has now advanced to the point where this plateau is easily reached. Quadrupoles have reached this limit for singly charged ions, and, coupled with electrospray, marched forward to over 100,000 daltons. Time-of-flight mass spectrometry with compounds in the range of 65,000–150,000 daltons is demonstrated. A different approach is necessary to deal with mass spectra in this high-mass region, as they are fundamentally different from those confined to the lower-mass ranges. This section will deal with a few of these differences.

5.3.1. Spectral Appearance

Fenselau[507,508] was among the first to note explicitly that the natural isotopic distribution of ions will inherently lead to a broad distribution of ion abundances among the isotopic cluster of high mass ions, as shown in Fig. 18. By example, consider only the ^{13}C isotope. Since the natural abundance of this isotope is about 1.1%, for the molecular ion of naphthalene, we could expect due to carbon alone an isotopic contribution at the M + 1 position of 13.2% of the relative abundance of M^+, where M is the position of the molecular ion on the mass scale. In the

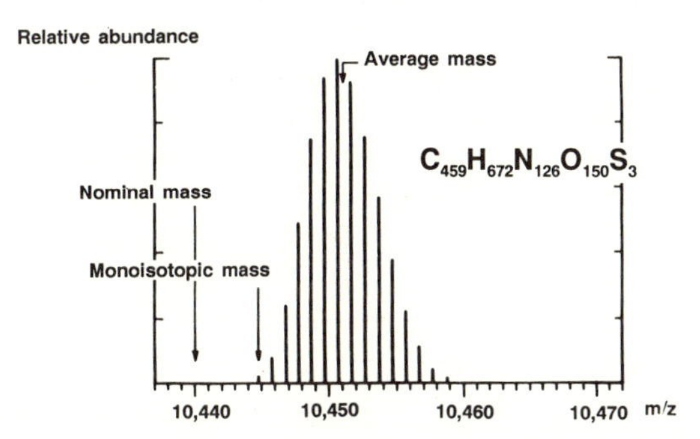

Figure 18. Broadening of the isotopic envelope for a high-mass organic ion (adapted from Ref. 507).

mass spectrum of a larger compound, an ion with 100 carbon atoms will be accompanied by a peak at the M + 1 position with a relative abundance 110% of that of M. As the number of carbons increases, the isotopic envelope broadens. Other isotopic ions contribute to this broadening as well. For very high mass ions, the envelope of the isotopic cluster can be distributed over several tens of daltons. Although the resolution of the mass spectrometer may be sufficiently great to resolve the individual integral masses in this upper mass range, the envelope for the molecular ion may overlap the envelope of fragment ions, such as loss of water or of a methyl group. Certainly it becomes difficult to distinguish between the M^+ and the $(M + H)^+$ forms of the molecular ion. In severe cases, the envelope of the molecular ion may overlap with that of the $(M + Na)^+$ ion that is only 23 daltons higher on the mass scale.

Fenselau[508] has further refined the description of the spectra of high-mass ions to include a description of the resolution necessary to separate isobars at any particular integral mass in the envelope. For instance, the ion at a mass (M + 2) in a cluster may contain contributions from $^{13}C_2$, or from sulfur, or from any number of other ions that contain a contribution for the (A + 2) isotope, where A is the mass of the low mass isotope. Using the protonated molecular ion envelope for porcine insulin as an example, Fenselau calculated that complete deconvolution of the peak at m/z 5774.6 could require a resolution of as high as 5×10^6, well beyond the capabilities of current instrumentation. It is unlikely that this level of resolution will ever be attainable. A further complication is the fact that ions such as M^+ and $(M + H)^+$ can both contribute to the molecular ion envelope. Even very high resolution, therefore, does not generate additional informational beyond that available with unit resolution of the molecular ion envelope, leading to the general recommendation that this minimum resolution be used for maximum sensitivity.

Broadening of the isotopic envelope for high-mass organic ions is expected in the mass range over 2500 daltons or so. The broadening occurs at lower masses for those ions that contain several metallic atoms, particularly those with multiple isotopes. For instance, while the contribution of a 1.1% relative abundance of ^{13}C becomes appreciable as the number of carbon atoms approaches 100, the contribution of molybdenum (seven isotopes) or tin (ten isotopes), or even magnesium (3 isotopes) becomes appreciable at much lower masses. Figure 19 is a comparison of the calculated and measured isotopic distribution for a cluster ion $Zn_7O_2(OAc)_9^+$, where (OAc) is the acetate ligand), formed by electron ionization. The ion current is distributed over an envelope 20 daltons wide even at 1000 daltons. As the particle-induced desorption ionization methods are increasingly applied to the analysis of organometallic and bio-organometallic compounds, the extent of this problem will be more widely recognized.

5.3.2. Spectral Interpretation

In general, spectra generated by particle-induced desorption ionization mass spectrometry contain mostly even-electron ions as opposed to the odd-electron ions prevalent in electron ionization. Interpretation of the spectra for samples of molecular weight with less than 1000 daltons has been relatively straightforward, with new mechanisms to explain the formation and dissociation of cluster ions. The variability of the spectra with sample preparation is a hindrance to the development of large spectral libraries, but the ability to generate these in-house is available.

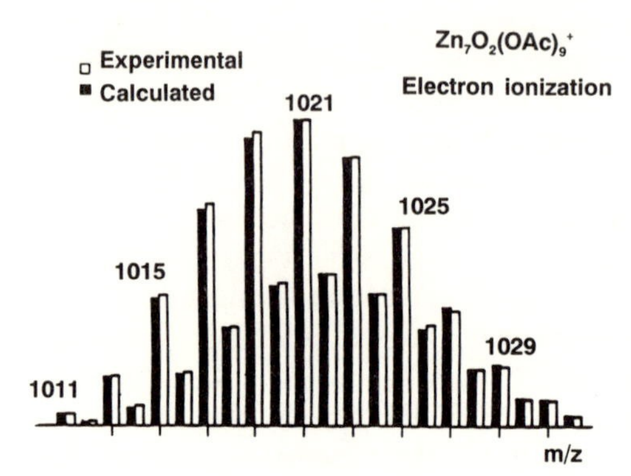

Figure 19. Comparison of the calculated and the measured isotopic envelope for the organometallic cluster ion $Zn_7O_2(acetate)_9^+$ [adapted from G. C. DiDonato and K. L. Busch, *Inorg. Chem.* **25**, 1551 (1986)].

The development of standard strategies for the interpretation of the particle-induced desorption ionization mass spectra of peptides and the deduction of the amino acid sequence is an example of how quickly the science of spectral interpretation has developed.[509]

In electron ionization, the molecular ion is generally the ion of highest mass in the spectrum. In chemical ionization, much the same formalism holds, with the complication that several adduct ions with masses higher than the molecular weight can be expected. In particle-induced desorption ionization, interpretation of a spectrum to deduce the molecular weight of the sample can become yet more complicated. For ions of moderate mass, a trio of ions at mass positions of X, X + 22, and X + 38 can be interpreted as $(M + H)^+$, $(M + Na)^+$, and $(M + K)^+$, providing confirmation of the molecular weight M. Sodium and potassium can be supplied separately, but are also common impurities especially in biological samples. In FAB, cluster ions of sample with sample or sample with matrix are also indicators of molecular weight.

At higher masses, contributions from background signals decrease in relative and absolute abundance, and ion signals assume the appearance of broad envelope distributions as discussed in the previous section. In general, the envelope of highest mass is assumed to contain the molecular ion, and it is further generally assumed that this corresponds to the protonated molecular ion. Of course, the practice of deducing the empirical formula from the exact mass measurement of a given ion in the envelope is no longer possible. The value of exact mass measurement for this purpose is diminished even at mass 1000, since the number of possibilities for the empirical formula has already grown beyond manageable proportion. At higher masses, the mass measured for any ion in the isotopic cluster (measured usually to 0.1 dalton) is compared to the mass calculated from a suspected formula based on the most abundant isotopes of the atoms in the formula. However, each integral mass can consist of several dozens of ions with different empirical formulas. Most often, the overall appearance of the envelope is compared to a calculated profile. If agreement is established, then this ion is taken to be the appropriate molecular ion, and the remainder of the spectrum is interpreted.

Plasma desorption using TOF mass analysis provides a molecular ion envelope from which the isotopically averaged molecular weight can be accurately determined. Although the individual ion contributions are not distinguished, the centroid of the signal is a reproducible measurement. The centroid mass is compared to that calculated from the proposed formula; this value is a weighted ionic mass encompassing all isotopic contributions. Even as the relative abundance of the monoisotopic peak is reduced to the level of the background, valuable chemical information can still be obtained from an accurate isotopically averaged mass centroid.

Particle-induced desorption ionization methods often provide mass spectra that contains molecular ions with only a minimal amount of fragmentation. Tandem mass spectrometry (MS/MS) (discussed in Section 5.4) is often used to derive structural information by independently examining the fragmentation of the molecular ion. The parameters of the ionization process can be chosen to encourage fragmentation, such as using higher primary particle fluxes or more energetic particles. A fundamental problem that must still be addressed is the increasing reluctance of higher-mass ions to dissociate under the conditions developed for collision-induced dissociations of lower-mass ions. In many cases, the energy added to the ion by the collision is simply accommodated within its many internal degrees of freedom, and no dissociation reactions ensue. Surface-induced dissociations or laser-based photodissociations may be possible solutions.

As the ions of interest shift to higher masses, there is increased importance attached to the determination of molecular weight. There are several practical reasons for this. Primary among them is the fact that for biomolecules of masses greater than the mass range of the instrument, well-established degradation processes can be used to reduce the molecules to smaller components. The molecular weight of all of the components is required to establish the molecular weight of the whole. The need for interpretation of the fragmentation processes, if any, from such ions is minimal. Secondly, for both time-of-flight and sector instruments, full mass spectra are not usually recorded to retain maximum sensitivity at the expected molecular weight of the analyte. To preserve sensitivity, only a relatively small mass range around the expected molecular ion is recorded. Finally, from data obtained with TOF mass analyzers and PD, it would seem that high-mass biomolecules simply may not fragment to any great extent, as previously discussed. For instance, the spectrum of bovine insulin contains the molecular ion (assumed to be the protonated molecular ion at 5734 daltons), a fragment for the alpha chain, and a second fragment ion for the beta chain. A tiger snake venom peptide at a mass of 13,309 daltons produces in PD the (protonated) molecular ion, the dimer, and the trimer, but little evidence of fragmentation. Internal energy that would cause fragmentation in a smaller ion might be comfortably accommodated within a larger ion able to distribute it among many internal degrees of freedom. The probability that such energy would be concentrated in one particular bond to cause fragmentation may be very remote.

Many high-mass biomolecules do not exhibit the diversity of

structure that might be expected from an extrapolation from lower-mass ions. For instance, peptides are composed of sequences of readily identified amino acids; polysaccharides consist of strands of sugar molecules; nucleic acids contain the bases, the sugars, and the phosphate groups; polymers are polymeric. The route to a higher-mass biomolecule is in many cases the agglomeration of many similar smaller substructures. If fragmentation of these high-mass biomolecules can be induced, the process of interpretation for many of them may ultimately be relatively simple.

5.3.3. *Strategies for Increased Mass Range*

In concept, the time-of-flight mass analyzer possesses an unlimited mass range, subject to the clearance time of the ion through the flight tube, and the ability to detect the higher-mass ions (see Section 1.3.3.). Quadrupole mass analyzers are limited to a mass range of 3000 daltons at unit mass resolution, but larger ions have been separated at less than unit mass resolution. Much of the impetus for extension of the range of mass spectrometric analysis to higher-mass biomolecules has been the result of changes in the design of magnetic sector mass spectrometers. Cottrell and Greathead have published an excellent review on extension of the mass range of a magnetic sector mass spectrometer.[510]

The relevant equation for the analysis of ions by a sector mass spectrometer is $m/z = B^2r^2/2V$, where m/z is the mass-to-charge ratio of the ion, B is the magnetic field strength, r is the radius of the magnetic sector, and V is the source accelerating voltage. Three methods to increase the mass range are to decrease the accelerating voltage, to increase the magnetic field strength, or to increase the radius of the magnetic sector. Almost all commercial instruments offer the option of operation at a decreased accelerating voltage to increase the effective mass range. As the equation indicates, at one-half of the normal accelerating voltage, the accessible mass is doubled. Conventional sources are designed for optimal ion extraction at the full accelerating voltage. Although many of the source tuning voltages track the accelerating potential, the efficiency of ion extraction is inevitably decreased. Decreased sensitivity in an experiment in which the absolute abundance of the high-mass ion is itself intrinsically low is an unfortunate combination that precludes the widespread use of the experiment. Special optics can be built into the source to compensate for the extraction problem.

A second instrumental option is to increase the strength of the magnetic field. The mass range accessible then increases as the square of the magnetic field value. Since the ion optical elements are left unchanged, there is no loss of sensitivity due to a decrease in the efficiency with which ions are extracted from the source. Practical limitations on the increase in magnetic field strength stem from the properties of the materials from which the electromagnetic is constructed, in particular the magnetic permeability. Incorporation of magnetic field inhomogeneities (a tapered field across the magnetic flux lines) provides additional focusing capabilities.

Part of the increased mass range of the new generation of commercial sector instruments also arise from an increase in the radius of the magnetic field. "Stretching" an instrument provides a higher mass range at the cost of construction of a larger instrument. In several of the newer multiple sector instruments of higher mass range, the linear flight path of the ion is well over 2 m. Additional pumping capacity must be provided, as well as added precautions for instrument support, leveling, and isolation from vibration.

Finally, the effective mass range of an instrument can be increased by study of multiply charged ions, as noted by Wood.[511] Usually, multiply charged ions are of much lower relative abundance than singly charged ions, but recent plasma desorption spectra have included abundant multiply charged ions. In highly acidic glycerol solutions used in FAB mass spectrometry, doubly and triply protonated molecular ions of peptides can be produced and sputtered into the vacuum.[512] A larger molecule can more readily accommodate multiple charges, as the charge sites can be relatively distant from each other to minimize the Coulombic repulsion. Multiple derivatization reactions offer one route to multiply charged ions, although only a few examples have been reported. The advantage of the increase in effective mass range attained through chemical derivatization is that it avoids the expense associated with an instrumental approach. The experimental difficulty of separating multiply charged ions from singly charged ions is also apparent, but can sometimes be solved with deceptively simple optical arrangements.[513] Electrospray ionization has refocused attention on the chemistry and the physics of multiply charged ions.[585]

5.4. Mass Spectrometry/Mass Spectrometry

Particle-induced desorption ionization methods often produce a simple mass spectrum in which most of the ion current is concentrated in an ion indicative of molecular weight, and the amount of fragmentation relative to electron or chemical ionization methods is reduced. While this serves well for the determination of the molecular weight of the analyte, it does not provide information about the structure of the ion. This structural information can be obtained by selecting the molecular ion individually, exciting it by collision to induce its dissociation, and analyzing the fragment ions that result, one of the experiments possible in MS/MS. Complete reviews of the basis and uses of MS/MS are found elsewhere.[514,515,661]

MS/MS has been used in the determination of ion structure and in the direct analysis of complex mixtures. Both of these applications have also seen use in conjunction with the particle-induced desorption ionization techniques. In the first place, ions generated by these methods are even-electron molecular ions, in contrast to the odd-electron ions produced in electron ionization mass spectrometry. The ion can be of a form such as $(M + Na)^+$, or $(M + Ag)^+$; while these have become familiar indicators for molecular weight determination, little is known of their actual structures. The spectra may also contain ions corresponding to clustering between the analyte and glycerol molecules, or clusters between anions and cations of the sample. The structures of these novel "supermolecular" ions are also unknown. MS/MS provides a means for such ion structure determinations, as described in the first part of this section. Secondly, several techniques, FAB and liquid SIMS in particular, make use of liquid matrices as part of the process of sample preparation. For instance, the use of the glycerol matrix in FAB leads to a high level of background chemical noise in the spectrum. The noise is removed in an MS/MS experiment that examines the dissociations of the mass-selected parent ion. More generally, MS/MS experiments can establish the relationships between ions in a very complicated mass spectrum.

MS/MS has been used most extensively with FAB and SIMS. With laser desorption and plasma desorption with TOF mass analysis, especially with a reflectron-based instrument, conditions under which MS/MS-like data can be measured have been described. Relationships between parent and daughter ions can be established, although not with the same resolution as established with the multiple sector or the multiple quadrupole instruments. The persistence of the secondary ions generated in FAB and SIMS of liquid samples makes MS/MS experiments relatively easy to complete. In LD, the transient nature of the secondary ions from a pulsed source demands a more sophisticated experimental setup.

5.4.1. Novel Ion Structures

Many of the ions observed in particle-induced desorption ionization mass spectra are even-electron species such as $(M + H)^+$ or $(M - H)^-$, as would be formed with chemical ionization mass spectrometry. Protonated molecular ion $(M + H)^+$ formed by SIMS and FAB may not be of the same structure or mixture of structures as those formed in chemical ionization. Although of exactly the same mass, the pattern of the collision induced dissociations differs. The analysis of cocaine is an example of this phenomenon.[(516)] Protonated in the gas phase by chemical ionization, cocaine produces an $(M + H)^+$ ion at m/z 304 that dissociates by loss of both benzaldehyde and benzoic acid. The $(M + H)^+$ ion of cocaine hydrochloride sampled in a SIMS or FAB experiment fragments by loss of benzoic acid, but not benzaldehyde. The difference in reactivity is a function of the process of ion formation. One possible explanation is that SIMS and FAB sputter a preformed ion in which the proton resides on the basic tertiary nitrogen; the integrity of this species is preserved as it is sputtered into the gas phase. Protonation in the gas phase by chemical ionization may yield a mixture of forms protonated on both nitrogen and oxygen, with the resultant changes in the fragmentation pattern uncovered by MS/MS. Several recent examples of the effect of ion origin on the daughter ion MS/MS spectra have been described.[(517,518)]

The distinction may seem to be of interest only in a fundamental sense, but can be of more practical importance. For instance, if a forensic analysis was based on the pattern of collision-induced dissociations, all of the variables that change the distribution must be accountable. Since peptides may also appear in a diversity of protonated forms, the use of MS/MS for sequence analysis will depend on the control of the ion mixture generated, and this will likely be an active research area in the future.

MS/MS has also been useful in establishing the nature of the novel ions formed in FAB, SIMS, PD, and LD. When the supporting sample matrix contains alkali ions, or the sample is supported on metal planchettes in SIMS, cationized forms of the sample molecule are created. The analyst often discerns molecular weight from a mixture of $(M + H)^+$, $(M + Na)^+$, or $(M + met)^+$ (met is any metal) ions, all observed simultaneously in the spectrum. Cationized ions are often more stable than the analogous protonated molecular ions, but the site of sodium ion attachment may not parallel that of hydrogen ion attachment. It is not unreasonable to expect that each metal may seek a different point of attachment to a neutral sample molecule. Silver ions may associate with alkene sites in the molecule while sodium ions may prefer other sites of attachment. There may be steric effects between metal ions of different size. Bursey[(416)] has studied the cationization of phthalic acid and of glycerol in a series of FAB experiments. Equimolar binary mixtures of alkali cations do not produce equal relative abundances of cationized glycerol ions or cationized phthalic acid upon xenon atom bombardment, and the ratios for each molecule are different. The study concluded that the alkali cation that most readily cationized one molecule may not best cationize another. Presumably this reflects the metal ion affinities of these molecules, values of which are now being determined.[519,520]

The ion chemistry involved in the interaction of alkali or metal cations with sample molecules is now being more thoroughly exploited. Examples include the sequencing of peptides as mono- or di-lithiated adducts,[662] or in the use of the specific information derived from sodium-cationized species used to determine C-terminal amino acids.[663]

A third area in which MS/MS has been used is in definition of the interrelationships between ions found in the mass spectra produced by the desorption ionization methods. A general mechanistic scheme for the creation of ions from the condensed phase, especially from liquid matrices, includes the sputtering of large cluster ions from the surface and their subsequent dissociation in the gas phase. MS/MS is used to determine the facility with which the large cluster ions dissociate to the smaller more abundant clusters. For example, MS/MS has been useful in the study of the very high mass cluster ions formed upon ion or atom bombardment of alkali halide salts. Campana[521] has reported that cluster ions of cesium iodide of mass up to 25,000 daltons can be sputtered from the surface of that salt using xenon primary ion beams. The envelope of the positive ion $Cs(CsI)_n^+$ spectrum contains irregularities that include enhanced ion intensities at values of n corresponding to particularly stable structures in the gas phase. Interpretation of this result has centered around the sputtering process itself,[522] metastable decomposition of these cluster ions in the gas phase,[523] or a combination of the two approaches.[524,525]

The contribution of gas phase dissociations in determining the relative intensities of the cluster ions is shown explicitly in the FAB spectrum of ammonium chloride. The positive ion mass spectrum contains cluster ions of the form $NH_4(NH_4Cl)_n^+$, and the relative abundances of the ions in the envelope exhibit the same irregularities as described by Campana, with enhanced abundances of ions corresponding to stable cagelike structures, defined at $n = N$, where $n = 13, 22, 31$, and 37. An MS/MS experiment has been carried out in which the cluster ions of ammonium chloride at $n = N + 1$ and 2 were chosen as parent ions, and the daughter ion MS/MS spectra recorded.[411] Losses observed in the daughter ion spectra were typically those of the neutral molecule NH_4Cl, producing clusters with successively lower values of n. However, for a parent ion at $n = N + 2$, daughter ions formed by loss of a single ammonium chloride molecule were entirely absent in the spectrum, and the preferred loss was that of *two* molecules of ammonium chloride to form the stable ion structure with $n = N$.

Cluster ions of the form $(nM + H)^+$ are often observed in the particle-induced desorption ionization mass spectra of organic molecules, corresponding to proton-bound dimers, trimers, and so on, of the sample molecule M, and of cluster ions of the forms $C_xA^+_{x-1}$ and $C_xA^-_{x+1}$ for ionic compounds. MS/MS establishes that these cluster ions dissociate to form the protonated molecular ions and the smaller cluster ions. A complication arises when the $(M + H)^+$ ion is itself examined by MS/MS, since, as discussed above, the pattern of dissociation observed in MS/MS can reflect the history of ion formation. The $(M + H)^+$ formed by dissociation may differ from that formed by protonation of the neutral sample molecule.

5.4.2. *Mixture Analysis*

When samples are mixed with glycerol or other liquid solvent, even the analysis of a pure sample becomes an exercise in mixture analysis. The simplicity of the glycerol spectrum is such that this background is easily discounted in the interpretation of the spectrum obtained. The situation is less favorable as derivatization reagents or modifiers are added to the sample mixture, when the initial sample mixture contains more than one component, or when the sample and the matrix interact with each other under the conditions of the analysis. MS/MS is often used to isolate the ion of interest, often the molecular ion, from the chemical noise that is the result of all these processes, and to examine the pattern of collision-induced dissociations of that ion to determine its identity and structure.

Striking examples of the use of MS/MS have been in the sequencing of polymeric biomolecules such as peptides or chains of nucleic acids.[526,527] The dissociation of a peptide usually occurs as a result of cleavage at the peptide bond, both in the mass spectrum and also in the daughter ion MS/MS spectrum. In both, the several series of ions observed can be used to deduce the amino acid sequence of the protein. When the chemical noise level of the original mass spectrum is high, the diagnostic sequence ions may be submerged in the background, and a link in the sequence may be lost. In MS/MS, the signal-to-noise ratio is greatly increased, and the sequence information is often more readily apparent. Figure 20 illustrates how the daughter ion MS/MS experiment is used to provide sequence information from several of the protonated

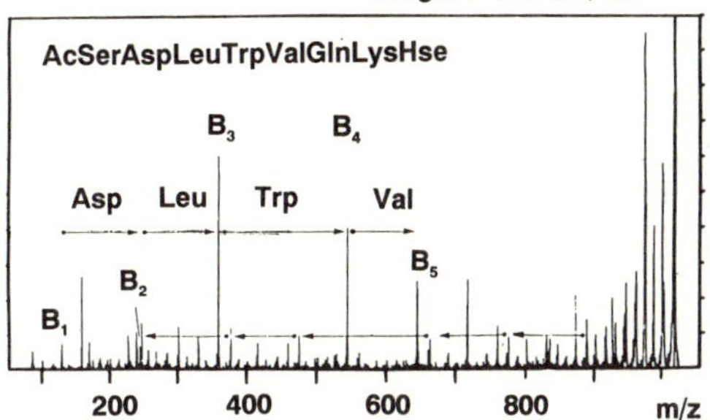

Figure 20. Daughter ion MS/MS spectrum of the protonated molecular ion of a large peptide formed by FAB ionization. The B series of ions corresponds to a standardized nomenclature for dissociations; another sequence is denoted by the unlabeled arrows. The sequence can be deduced from these ion series. (Adapted from Ref. 526).

molecular ions of peptides present in a mixture introduced into the source of the mass spectrometer.

Other MS/MS experiments available with modern instrumentation, and of particular interest for mixture analysis, are the parent ion and constant neutral loss scans. In the former, a characteristic daughter ion is set by the second mass analyzer, and the first mass analyzer is scanned to reveal all of the parent ions that dissociate to that selected daughter ion. In the latter, a characteristic neutral loss is established, and then the mass of the neutral molecule lost is used as a mass offset between the two mass analyzers, both of which scan across the entire mass range. When a parent ion/daughter ion pair related by the neutral loss is encountered, a signal appears at the detector. These two MS/MS experiments are most often used for functional group identification in mixture analysis. The daughter ion or the constant neutral loss is chosen to be characteristic of the targeted functional group. For instance, the ion 149^+ is a characteristic daughter ion for phthalate esters. Loss of 44 daltons as neutral carbon dioxide is the characteristic for dissociation of the $(M - H)^-$ ions of carboxylic acids. Constant neutral loss MS/MS scans for peptides analyzed by FAB have been reported,[664] as well as similar experiments for the acyl-carnitines.[665] The ion-addition derivatization reaction for steroids described in Section 4.3.1.1 has been used as the basis for an MS/MS experiment.[528] Since the daughter ion MS/MS spectra of steroids derivatized with Girard's T reagent all exhibit ions associated with loss of trimethylamine, a constant neutral loss (59 daltons) MS/MS spectrum highlights all of the steroids in a mixture that react with the reagent, as shown in Fig. 21.

Generally, a characteristic daughter ion can be established for an amino acid sequence as ABC^+. The rules of spectral interpretation state that any peptide containing this sequence should dissociate to an ion of that selected mass. This is then set as the characteristic daughter ion by the second mass analyzer. Scanning the first mass analyzer pinpoints *all*

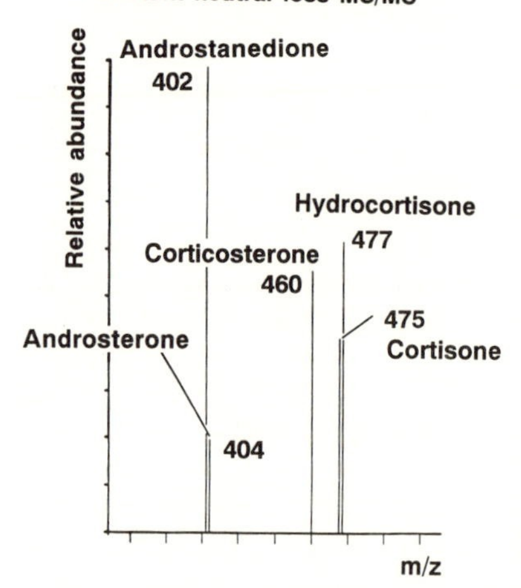

Figure 21. Constant neutral loss MS/MS spectrum for loss of trimethylamine, used to define steroids that have reacted with Girard's T reagent.

of the parent ions that dissociate to that particular daughter ion. The method also provides an inverse confirmation of sequences established via daughter ion MS/MS spectra. A sequence of MS/MS experiments can serve as an extension to the method of FAB mapping of peptides developed by Morris.[529]

More complete reviews of the applications of MS/MS in FAB and SIMS mass spectrometry are available. Biochemical applications have been most prevalent, and examples include the application of FAB MS/MS to studies of fatty acids,[530–533] acylcarnitines,[534–536] modified nucleosides and nucleotides,[537–539] marine toxins,[540] cobalamins,[541] and bile acid salts and their conjugates.[542]

6. Future Prospects

The use of particle-induced desorption ionization methods in organic mass spectrometry has been responsible for most of the growth in mass spectrometry over the past decade, and clearly there will be a continued expansion of both capabilities and applications. Much of the initial novelty has already dissipated; FAB analysis of peptides, for example, based on the excellent work of the past few years, is almost considered routine for samples of masses less than 5000 daltons. Upcoming initiatives, including the human genome project among many others, will provide a continuing source of challenge for these analytical methods. Significant work remains in the determination of mechanisms and in the optimization of abilities already demonstrated. It would be a genuine misfortune if a standard protocol developed without adequate background work to establish both the strengths and the weaknesses of the technique. The somewhat serendipitous discovery of the glycerol matrix provided a recipe that was followed for many years, with compilations of alternative matrices only now appearing, and with a systematic approach toward a more thorough understanding of the role of the liquid matrix still in the future.

Fundamental work in the processes of ion formation should result in better sensitivities and better selectivities. Ion manipulation in the condensed phase on the scale of FAB and SIMS sampling is still a young science, but the prospects of functional group specificity in the analysis of mixtures, especially in conjunction with MS/MS, are very attractive. The conjunction of electrochemistry and FAB and SIMS is being developed for the analysis of electroactive compounds. The ultimate potential of this approach is still unclear, but the tremendous information from solution organic electrochemistry may aid in an understanding of the spectra obtained with such a system. The ability of mass spectrometry to monitor very small amounts of material, or, through the use of MS/MS, to identify components in a mixture, may also make a contribution to an understanding of the electrochemical processes occurring in solution.

The cost of hardware for FAB and SIMS has already started to decline, as the inherent simplicity of these sources would suggest. In contrast to the use of accelerator sources for plasma desorption, a simple radioactive californium source, as in the original design, is small and inexpensive. An exciting prospect is the use of this source with an ion cyclotron resonance (ICR) mass spectrometer. The high-mass ions sputtered by the energetic fission fragments can be trapped in a cell for extended periods of time. Advantages of higher mass resolution and greater absolute ion intensities may result.

The large impact of the desorption ionization methods has been in the extension of mass spectrometry to new classes of molecules. Once the ions are formed, their masses can be measured (MS) or their chemical reactivity examined (MS/MS). Improvements in mass analysis technology have been catalyzed by the need to deal with ions of higher mass, but cannot be expected to increase indefinitely. The need to create ions from increasingly recalcitrant molecules is likely to be the more demanding endeavor. Understanding the processes of ionization, and the implications of sample preparation, will continue to be a central concern.

It is appropriate to close this review with the same conclusions that summarized the meeting report[666] from the first Sanibel Conference of the American Society for Mass Spectrometry, devoted to mechanisms of desorption ionization mass spectrometry:

If mass spectroscopists can agree on anything, it must certainly be that the field is moving rapidly. PD mass spectrometry is only a dozen years old, and FAB much younger. We have forgotten life before these techniques, and raised our sights even farther. Thus, we are prepared, we guess, for other techniques (either new or existing) to challenge the latest results from laser desorption and sprays. Our understanding of the mechanisms is surely less than perfect, but it is fortunate that our achievements exceed our understanding.

References

1. K. L. Busch and R. G. Cooks, *Science* **218,** 247 (1982).
2. H. R. Morris, ed., *Soft Ionization Mass Spectrometry,* Heyden, London (1981).
3. P. Hakansson, I. Kamensky, B. Sundqvist, J. Fohlman, P. Peterson, C. J. McNeal, and R. D. Macfarlane, *J. Am. Chem. Soc.* **104,** 2948 (1982).
4. P. Hakansson, I. Kamensky, J. Kjellberg, B. Sundqvist, J. Fohlman, and P. Peterson, *Biochem. Biophys. Res. Commun.*, **110,** 519 (1983).
5. A. Dell and H. R. Morris, *Biochem. Biophys. Res. Commun.*, **106,** 1456 (1982).
6. M. Barber, R. S. Bordoli, G. J. Elliott, N. J. Horoch, and B. N. Green, *Biochem. Biophys. Res. Commun.*, **110,** 753 (1983).
7. H. R. Morris, M. Panico, M. Barber, R. S. Bordoli, R. D. Sedgwick, and A. N. Tyler, *Biochem. Biophys. Res. Commun.*, **101,** 623 (1981).
8. D. M. Desiderio and I. Katakuse, *Anal. Biochem.* **129,** 425 (1983).
9. R. Self and A. Parent, *Biomed. Mass Spectrom.* **10,** 78 (1983).
10. M. E. Rose, M. C. Prescott, A. H. Wilby, and I. J. Galpin, *Biomed. Mass Spectrom.* **11,** 10 (1984).
11. R. M. Caprioli, L. A. Smith, and C. F. Beckner, *Int. J. Mass Spectrom. Ion Phys.* **46,** 419 (1983).
12. L. A. Smith and R. M. Caprioli, *Biomed. Mass Spectrom.* **10,** 98 (1983).
13. A. I. Cohen, K. A. Glavan, and J. F. Kronauge, *Biomed. Mass Spectrom.* **10,** 287 (1983).
14. J. M. Miller, *J. Organomet. Chem.* **249,** 299 (1983).
15. W. Knabe and F. R. Krueger, *Z. Naturforsch.* **37a,** 1335 (1982).
16. F. R. Krueger and W. Knabe, *Org. Mass Spectrom.* **18,** 83 (1983).
17. A. Benninghoven, D. Jaspers, and W. Sichtermann, *Appl. Phys.* **11,** 35 (1976).
18. A. Benninghoven and W. Sichtermann, *Org. Mass Spectrom.* **12,** 595 (1977).
19. A. Benninghoven, *Surf. Sci.* **35,** 427 (1973).
20. F. W. Karasek, *Res. Develop.* **25**(11), 42 (1974).
21. A. Benninghoven and W. Sichtermann, *Anal. Chem.* **50,** 1180 (1978).
22. A. Eicke, W. Sichtermann, and A. Benninghoven, *Org. Mass Spectrom.* **15,** 289 (1980).
23. H. Grade, N. Winograd, and R. G. Cooks, *J. Am. Chem. Soc.* **99,** 7725 (1977).
24. H. Grade and R. G. Cooks, *J. Am. Chem. Soc.* **100,** 5615 (1978).
25. R. J. Day, S. E. Unger, and R. G. Cooks, *J. Am. Chem. Soc.* **101,** 499 (1979).
26. R. J. Day, S. E. Unger, and R. G. Cooks, *Anal. Chem.* **52,** 557A (1980).
27. L. K. Liu, S. E. Unger, and R. G. Cooks, *Tetrahedron* **37,** 1067 (1981).

28. R. J. Colton, *J. Vac. Sci. Technol.* **18,** 737 (1981).
29. R. J. Colton, J. S. Murday, J. R. Wyatt, and J. J. DeCorpo, *Surf. Sci.* **84,** 235 (1979).
30. J. E. Campana, J. J. DeCorpo, and R. J. Colton, *Appl. Surf. Sci.* **8,** 337 (1981).
31. J. B. Westmore, W. Ens, and K. G. Standing, *Biomed. Mass Spectrom.* **9,** 119 (1982).
32. W. Ens, K. G. Standing, J. B. Westmore, K. K. Ogilvie, and M. J. Nemer, *Anal. Chem.* **54,** 960 (1982).
33. M. Barber, R. S. Bordoli, R. D. Sedgwick, and A. N. Tyler, *J. Chem. Soc. Chem. Commun.* 325 (1981).
34. F. M. Devienne and J. C. Roustan, *C. R. Hebd. Acad. Sci. Ser. C,* **276,** 923 (1973).
35. F. M. Devienne and J. C. Roustan, *Org. Mass Spectrom.* **17,** 173 (1982).
36. G. D. Tantsyrev and N. A. Kleimenov, *Dokl. Akad. Nauk. SSSR* **213,** 649 (1973).
37. G. D. Tantsyrev and M. I. Povolotskaya, *Khim. Vysokikh. Ener.* **9,** 380 (1975).
38. D. J. Surman and J. C. Vickerman, *Appl. Surf. Sci.* **9,** 108 (1981).
39. D. J. Surman, J. A. van den Berg, and J. C. Vickerman, *Surf. Int. Anal.* **4,** 160 (1982).
40. G. Borchardt, H. Scherrer, S. Weber, and S. Scherrer, *Int. J. Mass Spectrom. Ion Phys.* **34,** 361 (1980).
41. A. Iino and A. Mizuike, *Bull. Chem. Soc. Jpn.* **54,** 1975 (1981).
42. J. E. Campana, *Int. J. Mass Spectrom. Ion Phys.* **51,** 133 (1983).
43. F. W. Rollgen and U. Giessmann, *Int. J. Mass Spectrom. Ion Phys.* **56,** 229 (1984).
44. M. Barber, R. S. Bordoli, G. J. Elliott, R. D. Sedgwick, and A. N. Tyler, *Anal. Chem.* **54,** 645A (1982).
45. M. Barber, R. S. Bordoli, R. D. Sedgwick, and A. N. Tyler, *Nature* **293,** 270 (1981).
46. K. L. Rinehart, *Science* **218,** 254 (1982).
47. R. G. Cooks and K. L. Busch, *Int. J. Mass Spectrom. Ion Phys.* **53,** 111 (1983).
48. W. Aberth, K. M. Straub, and A. L. Burlingame, *Anal. Chem.* **54,** 2029 (1982).
49. M. Barber, R. S. Bordoli, and G. J. Elliott, *Gazz. Chim. Ital.* **114,** 305 (1984).
50. D. F. Torgerson, R. P. Skowronski, and R. D. Macfarlane, *Biochem. Biophys. Res. Commun.* **60,** 616 (1974).
51. R. D. Macfarlane and D. F. Torgerson, *Science* **191,** 920 (1976).
52. P. Duck, W. Treu, H. Frohlich, W. Galster, and H. Voit, *Surf. Sci.* **95,** 603 (1980).
53. P. Feigl, B. Schueler, and F. Hillenkamp, *Int. J. Mass Spectrom. Ion Phys.* **47,** 15 (1983).
54. T. Dingle, B. W. Griffiths, and J. C. Ruckman, *Vacuum* **31,** 571 (1981).
55. F. Hillenkamp, *Int. J. Mass Spectrom. Ion Phys.* **45,** 305 (1982).
56. K. Balasanmugam, T. A. Dang, R. J. Day, and D. M. Hercules, *Anal. Chem.* **53,** 2296 (1981).
57. D. M. Hercules, R. J. Day, K. Balasanmugam, T. A. Dang, and C. P. Li, *Anal. Chem.* **54,** 280A (1982).
58. M. L. Vestal, in *Ion Formation from Organic Solids,* (A. Benninghoven, ed.), Springer-Verlag, New York (1983).
59. C. A. Evans, Jr., and C. D. Hendricks, *Rev. Sci. Instrum.* **43,** 1527 (1972).
60. B. N. Colby and C. A. Evans, Jr., *Anal. Chem.* **45,** 1884 (1973).
61. B. S. Simons, B. N. Colby, and C. A. Evans, Jr., *Int. J. Mass Spectrom. Ion Phys.* **15,** 291 (1974).
62. S. T. F. Lai and C. A. Evans, Jr., *Org. Mass Spectrom.* **13,** 733 (1978).
63. B. P. Stimpson and C. A. Evans, Jr., *Biomed. Mass Spectrom.* **5,** 52 (1978).
64. B. P. Stimpson, D. S. Simons, and C. A. Evans, Jr., *J. Phys. Chem.* **82,** 660 (1978).
65. K. W. S. Chan and K. D. Cook, *J. Am. Chem. Soc.* **104,** 5031 (1982).
66. K. D. Cook and J. H. Callahan, *Adv. Mass Spectrom.* **10,** 1291 (1986).

67. M. Dole, L. L. Mack, R. L. Hines, R. C. Mobley, L. D. Ferguson, and M. B. Alice, *J. Chem. Phys.* **49,** 2240 (1968).
68. J. Gieniec, L. L. Mack, K. Nakamae, C. Gupta, V. Kumar, and M. Dole, *Biomed. Mass Spectrom.* **11,** 259 (1984).
69. M. Yamashita and J. B. Fenn, *J. Phys. Chem.* **88,** 4671 (1984).
70. C. M. Whitehouse, R. N. Dreyer, M. Yamashita, and J. B. Fenn, *Anal. Chem.* **57,** 675 (1985).
71. A. Bruins, T. Covey, and J. Henion, Presented at the 34th Annual Conference on Mass Spectrometry and Allied Topics, June 1986, Cincinnati, Ohio, p. 585.
72. M. L. Vestal, *Science* **226,** 275 (1984).
73. C. R. Blakley and M. L. Vestal, *Anal. Chem.* **55,** 750 (1983).
74. M. L. Vestal, *Int. J. Mass Spectrom. Ion Phys.* **46,** 193 (1983).
75. M. L. Vestal, *Mass Spectrom. Rev.* **2,** 447 (1983).
76. D. J. Liberato, C. C. Fenselau, M. L. Vestal, and A. L. Yergey, *Anal. Chem.* **55,** 1741 (1983).
77. D. J. Liberato and A. L. Yergey, *Anal. Chem.* **58,** 6 (1986).
78. C. E. Parker, R. W. Smith, S. J. Gaskell, and M. M. Bursey, *Anal. Chem.* **58,** 1661 (1986).
79. J. V. Iribarne and B. A. Thomson, *J. Chem. Phys.* **64,** 2287 (1976).
80. B. A. Thomson and J. V. Iribarne, *J. Chem. Phys.* **71,** 4451 (1979).
81. M. Tsuchiya and T. Taira, *Int. J. Mass Spectrom. Ion Phys.* **34**, 351 (1980).
82. H. Kuwabara and M. Tsuchiya, *Mass Spectrosc.* **30,** 313 (1982).
83. M. Tsuchiya, H. Kuwabara, and K. Musha, *Anal. Chem.* **58,** 695 (1986).
84. R. E. Honig, *J. Appl. Phys.* **29,** 549 (1958).
85. W. Sichtermann, M. Junack, A. Eicke, and A. Benninghoven, *Fres. Z. Anal. Chem.* **301,** 115 (1980).
86. A. Benninghoven, A. Eicke, M. Junack, W. Sichtermann, J. Krizek, and H. Peters, *Org. Mass Spectrom.* **15,** 459 (1980).
87. A. Benninghoven and W. Sichtermann, *Int. J. Mass Spectrom. Ion Phys.* **38,** 351 (1981).
88. W. Sichtermann and A. Benninghoven, *Int. J. Mass Spectrom. Ion Phys.* **40,** 177 (1981).
89. A. Benninghoven, *J. Vac. Sci. Technol.* **A3,** 451 (1985).
90. R. J. Day, S. E. Unger, and R. G. Cooks, *Anal. Chem.* **52,** 353 (1980).
91. R. J. Day, S. E. Unger, and R. G. Cooks, *J. Am. Chem. Soc.* **101,** 501 (1979).
92. K. L. Busch, S. E. Unger, A. Vincze, R. G. Cooks, and T. Keough, *J. Am. Chem. Soc.* **104,** 1597 (1982).
93. R. E. Honig, in R. E. Finnigan, ed., *Retrospective Lectures—32nd ASMS,* American Society for Mass Spectrometry, East Lansing, Michigan (1984).
94. F. J. Vastola and A. J. Pirone, *Adv. Mass Spectrom.* **4,** 107 (1968).
95. F. J. Vastola, R. O. Mumma, and A. J. Pirone, *Org. Mass Spectrom.* **3,** 101 (1979).
96. R. O. Mumma and F. J. Vastola, *Org. Mass Spectrom.* **6,** 1373 (1972).
97. M. A. Posthumus, P. G. Kistemaker, H. L. C. Meuzelaar, and M. C. Ten Noever de Brauw, *Anal. Chem.* **50,** 985 (1978).
98. R. J. Cother and J.-C. Tabet, *Int. J. Mass Spectrom. Ion Phys.* **53,** 151 (1983).
99. R. Wechsung, F. Hillenkamp, R. Kaufmann, R. Nitsche, E. Unsold, and H. Vogt, *Microchim. Acta* **2,** 282 (1978).
100. F. Hillenkamp, R. Unsold, R. Kaufmann, and R. Nitsche, *Appl. Phys.* **8,** 341 (1975).
101. H. J. Heinen, S. Meier, H. Vogt, and R. Wechsung, *Int. J. Mass Spectrom. Ion Phys.* **47,** 19 (1983).

102. R. Stoll and F. W. Rollgen, *Org. Mass Spectrom.* **14,** 642 (1979).
103. V. S. Antonov, V. S. Letokhov, and A. N. Shibanov, *Appl. Phys.* **25,** 71 (1981).
104. J. E. Hunt, R. D. Macfarlane, J. J. Katz, and R. C. Dougherty, *Proc. Natl. Acad. Sci. USA* **77,** 1745 (1980).
105. R. D. Macfarlane, D. Uemura, K. Ueda, and Y. Hirata, *J. Am. Chem. Soc.* **102,** 875 (1980).
106. T. Ogita, N. Otake, Y. Miyazaki, H. Yonehara, R. D. Macfarlane, and C. J. McNeal, *Tetrahedron Lett.* **21,** 3203 (1980).
107. N. Furstenau, W. Knippelberg, F. R. Krueger, G. Weiss, and K. Wien, *Z. Naturforsch* **32a,** 711 (1977).
108. N. Furstenau, *Z. Naturforsch.* **33a,** 563 (1970).
109. Y. Le Beyec, S. Della Negra, C. Deprun, P. Vigny, and Y. M. Ginot, *Rev. Phys. Appl.* **15,** 1631 (1980).
110. H. Danigel, H. Jungclas, and L. Schmidt, *Int. J. Mass Spectrom. Ion Phys.* **52,** 223 (1983).
111. D. F. Torgerson, *Int. J. Mass Spectrom. Ion Phys.* **52,** 223 (1983).
112. P. Roepstorff, P. Hojrup, P. F. Nielsen, B. Sundqvist, G. Jonsson, P. Hakansson, I. Kamensky, and M. Lindberg, *Adv. Mass Spectrom.* **10,** 1617 (1986).
113. M. Barber, R. S. Bordoli, and R. D. Sedgwick, in *Soft Ionization Biological Mass Spectrometry* (H. R. Morris, ed.), Heyden, London (1981).
114. R. S. Lehrle, J. C. Robb, and D. W. Thomas, *J. Sci. Instrum.* **39,** 458 (1962).
115. H. D. Beckey, *Int. J. Mass Spectrom. Ion Phys.* **2,** 500 (1969).
116. G. W. Wood, *Tetrahedron* **39,** 1125 (1982).
117. H.-R. Schulten, *Int. J. Mass Spectrom. Ion Phys.* **32,** 97 (1979).
118. K. H. Ott, F. W. Rollgen, J. J. Zwinselman, R. H. Fokkens, and N. M. M. Nibbering, *Angew. Chem. Int. Ed. Engl.* **20,** 111 (1981).
119. H. Wollnick, *Spectra* **9,** 37 (1983).
120. R. P. Morgan, J. H. Beynon, R. H. Bateman, and B. N. Green, *Int. J. Mass Spectrom. Ion Phys.* **28,** 171 (1978).
121. P. H. Dawson, ed., *Quadrupole Mass Spectrometry and its Applications,* Elsevier, New York (1976).
122. K. Wittmaack, *Vacuum* **32,** 65 (1982).
123. T. A. Lehmann and M. M. Bursey, *Ion Cyclotron Resonance Spectrometry,* Wiley, New York (1976).
124. A. G. Marshall, *Acc. Chem. Res.* **18,** 316 (1985).
125. M. v. Ardenne, *Z. Tech. Phys.* **20,** 344 (1939).
126. H. J. Liebl and R. F. K. Herzog, *J. Appl. Phys.* **34,** 2893 (1963).
127. H. A. Storms, K. F. Brown, and J. D. Stein, *Anal. Chem.* **49,** 2023 (1977).
128. P. Williams, R. K. Lewis, C. A. Evans, Jr., and P. R. Hanley, *Anal. Chem.* **49,** 1399 (1977).
129. P. J. Todd, G. L. Glish, and W. H. Christie, *Int. J. Mass Spectrom. Ion Phys.* **61,** 215 (1984).
130. J. L. Hundley, *Phys. Rev.* **30,** 864 (1927).
131. J. P. Blewett and E. J. Jones, *Phys. Rev.* **50,** 464 (1936).
132. S. K. Allison and M. Kamegai, *Rev. Sci. Instrum.* **32,** 1090 (1961).
133. W. Aberth and A. L. Burlingame, in *Ion Formation from Organic Solids* (A. Benninghoven, ed.), Springer-Verlag, New York (1983).
134. R. Stoll, U. Schade, F. W. Rollgen, U. Giessmann, and D. F. Barofsky, *Int. J. Mass Spectrom. Ion Phys* **43,** 227 (1982).
135. G. K. Wehner, *Phys. Rev.* **102,** 690 (1956).

136. S. S. Wong, R. Stoll, and F. W. Rollgen, *Z. Naturforsch.* **37a,** 718 (1982).
137. R. Clampitt, K. L. Aitken, and D. K. Jeffries, *J. Vac. Sci. Technol.* **12,** 1208 (1975).
138. R. Clampitt and D. K. Jeffries, *Nucl. Instrum. Methods* **149,** 739 (1978).
139. L. W. Swanson, G. A. Schwind, A. E. Bell, and J. E. Brady, *J. Vac. Sci. Technol.* **16,** 1864 (1979).
140. A. R. Waugh, A. R. Bayly, and K. Anderson, *Vacuum* **34,** 103 (1984).
141. D. R. Kingham and L. W. Swanson, *Vacuum* **34,** 941 (1984).
142. D. F. Barofsky, U. Giessmann, L. W. Swanson, and A. E. Bell, *Int. J. Mass Spectrom. Ion Phys.* **46,** 495 (1983).
143. D. F. Barofsky, U. Giessmann, A. E. Bell, and L. W. Swanson, *Anal. Chem.* **55,** 1318 (1983).
144. R. G. Stoll, D. J. Harvan, and J. R. Hass, *Int. J. Mass Spectrom. Ion Proc.* **61,** 71 (1984).
145. R. G. Stoll, R. B. Cole, D. J. Harvan, and J. R. Hass, *Int. J. Mass Spectrom. Ion Proc.* **69,** 239 (1986).
146. H. Liebl, *Anal. Chem.* **46,** 22A (1974).
147. H. Liebl, *Scanning* **3,** 79 (1980).
148. R. Castaing and G. Slodzian, *J. Microsc.* **1,** 395 (1962).
149. S. L. Koontz and M. B. Denton, *Int. J. Mass Spectrom. Ion Phys.* **37,** 241 (1981).
150. M. A. LaPack, S. J. Pachuta, K. L. Busch, and R. G. Cooks, *Int. J. Mass Spectrom. Ion Phys.* **53,** 323 (1983).
151. V. Franchetti, B. H. Solka, W. E. Baitinger, J. W. Amy, and R. G. Cooks, *Int. J. Mass Spectrom. Ion Phys.* **23,** 29 (1977).
152. G. Dearnaley, J. H. Freeman, K. S. Nelson, and J. H. Stephen, *Ion Implantation,* North-Holland, Amsterdam (1973).
153. G. Dearnaley, *J. Metals* **34,** 18 (1982).
154. P. D. Prewett, *Vacuum* **34,** 931 (1984).
155. D. W. Breck, W. G. Eversole, R. M. Milton, J. B. Reed, and T. L. Thomas, *J. Am. Chem. Soc.* **78,** 5963 (1956).
156. O. Heinz and R. T. Reaves, *Rev. Sci. Instrum.* **39**, 1229 (1968).
157. G. I. Taylor, *Proc. R. Soc. London, Ser. A* **280,** 383 (1964).
158. J. F. Mahoney, A. Y. Yahiku, H. L. Daley, R. D. Moore, and J. Perel, *J. Appl. Phys.* **40,** 5101 (1969).
159. P. D. Prewett, D. K. Jeffries, and D. J. McMillan, *Vacuum* **34,** 107 (1984).
160. H. W. Werner and P. R. Boudewijn, *Vacuum* **34,** 83 (1984).
161. D. Broughton and R. Clampitt, *Vacuum* **34,** 275 (1984).
162. S. P. Thompson, *Vacuum* **34,** 223 (1984).
163. F. G. Rudenauer, W. Steiger, E. Wieser, R. Grotzschel, and F. Nahring, *Vacuum* **35,** 315 (1985).
164. F. G. Rudenauer, *Springer Ser. Chem. Phys. SIMS IV* **36,** 133 (1984).
165. A. R. Waugh, A. R. Bayly, and K. Anderson, *Springer Ser. Chem. Phys. SIMS IV* **36,** 138 (1984).
166. D. F. Barofsky, J. H. Murphy, A. M. Ilias, and E. Barofsky, *Springer Ser. Chem. Phys. SIMS IV,* **36,** 377 (1984).
167. M. A. Rudat, *Anal. Chem.* **54,** 1917 (1982).
168. W. Aberth, R. Reginato, and A. L. Burlingame, *Springer Ser. Chem. Phys. SIMS IV* **36,** 380 (1984).
169. N. Klaus, *Vacuum* **35,** 131 (1985).
170. C.-S. Su, *Vacuum* **34,** 649 (1984).
171. O. J. Orient, A. Chutjian, and S. H. Alajajian, *Rev. Sci. Instrum.* **56,** 69 (1985).

172. I. Brodie and C. A. Spindt, *Appl. Surf. Sci.* **2,** 149 (1979).
173. I. G. Brown, J. E. Galvin, B. F. Gavin, and R. A. MacGill, *Rev. Sci. Instrum.* **57,** 1254 (1986).
174. W. F. DiVergillo, H. Goede, and V. F. Fosnight, *Rev. Sci. Instrum.* **57,** 1254 (1986).
175. S. R. Walther, K. N. Leung, and W. B. Kunkel, *Rev. Sci. Instrum.* **57,** 1531 (1986).
176. The Construction and Operation of Fast Atom Bombardment Ion Sources, Sponsored by the American Society for Mass Spectrometry, Baltimore, Maryland, December 11, 1981.
177. H. Kobayashi, K. Susuki, Y. Yanagisawa, and K. Yukawa, *Mass Spectros.* **25,** 315 (1977).
178. A. H. McIlraith, *J. Vac. Sci. Technol.* **9,** 207 (1971).
179. A. M. Ghander and R. K. Fitch, *J. Vac. Sci. Technol.* **24,** 489 (1974).
180. J. Franks and A. M. Ghander, *Vacuum* **24,** 489 (1974).
181. D. J. Surman, J. A. van den Berg, and J. V. Vickerman, *Surf. Int. Anal.* **4,** 160 (1982).
182. J. Franks, *Int. J. Mass Spectrom. Ion Phys.* **46,** 343 (1983).
183. W. V. Ligon, *Int. J. Mass Spectrom. Ion Phys.* **41,** 205 (1982).
184. R. A. McDowell and H. R. Morris, *Int. J. Mass Spectrom. Ion Phys.* **46,** 443 (1983).
185. J. F. Mahoney, J. Perel, and A. T. Forrester, *Appl. Phys. Lett.* **38,** 320 (1981).
186. J. F. Mahoney, D. M. Goebel, J. Perel, and A. T. Forrester, *Biomed. Mass Spectrom.* **10,** 61 (1983).
187. K. F. Faull, J. D. Barchas, C. N. Kenyon, and P. C. Goodley, *Int. J. Mass Spectrom. Ion Phys.,* **46,** 347 (1983).
188. K. F. Faull, A. N. Tyler, H. Sim, J. D. Barchas, I. J. Massey, C. N. Kenyon, P. C. Goodley, J. F. Mahoney, and J. Perel, *Anal. Chem.* **56,** 308 (1984).
189. J. Perel, K. Faull, J. F. Mahoney, A. N. Tyler, and J. D. Barchas, *Am. Lab.* **16**(11), 94 (1984).
190. G. C. Stafford, D. C. Bradford, and D. R. Stephens, Proceedings of the 29th Annual Conference on Mass Spectrometry and Allied Topics, Minneapolis, Minnesota, May, 1981, p. 163.
191. G. C. Stafford and D. C. Bradford, Proceedings of the 30th Annual Conference on Mass Spectrometry and Allied Topics, Honolulu, Hawaii, June 1982, p. 539.
192. P. E. Filpus-Luyckx and E. A. Schweikert, *Anal. Chem.* **58,** 1686 (1986).
193. F. Hillenkamp, in: *Ion Formation from Organic Solids* (A. Benninghoven, ed.), Springer-Verlag, New York (1983).
194. R. J. Conzemius and J. M. Capellen, *Int. J. Mass Spectrom. Ion Phys.* **34,** 197 (1980).
195. P. G. Kistemaker, G. J. Q. van der Peyl, and J. Haverkamp, in: *Soft Ionization Biological Mass Spectrometry* (H. R. Morris, ed.) Heyden, London (1981).
196. R. Stoll and F. W. Rollgen, *Z. Naturforsch.* **37a,** 9 (1982).
197. G. J. Q. van der Peyl, K. Isa, J. Haverkamp, and P. G. Kistemaker, *Org. Mass Spectrom.* **16,** 416 (1981).
198. P. G. Kistemaker, M. M. J. Lens, G. J. Q. van der Peyl, and A. H. J. Boerboom, *Adv. Mass Spectrom.* **8,** 928 (1980).
199. G. J. Q. van der Peyl, J. Haverkamp, and P. G. Kistemaker, *Int. J. Mass Spectrom. Ion Phys.* **42,** 125 (1982).
200. G. J. Q. van der Peyl, W. J. van der Zande, and P. G. Kistemaker, *Int. J. Mass Spectrom. Ion Proc.* **62,** 51 (1984).
201. D. A. McCrery, D. A. Peake, and M. L. Gross, *Anal. Chem.* **57,** 1181 (1985).
202. D. A. McCrery, E. B. Ledford, and M. L. Gross, *Anal. Chem.* **54,** 1435 (1982).
203. J. Dennemont and J.-Cl. Landry, *Adv. Mass Spectrom.* **10,** 993 (1986).

204. M. J. Southton, M. C. Witt, A. Harris, E. R. Wallach, and J. Myatt, *Vacuum* **34,** 903 (1984).
205. C. Schiller, K.-D. Kupka, and F. Hillenkamp, *Fres. Z. Anal. Chem.* **308,** 308 (1981).
206. K.-D. Kupka, F. Hillenkamp, and C. Schiller, *Adv. Mass Spectrom.* **8,** 935 (1980).
207. E. D. Hardin and M. L. Vestal, *Anal. Chem.* **53,** 1492 (1981).
208. G. D. Daves, T. D. Lee, W. R. Anderson, D. F. Barofsky, G. A. Massey, J. C. Johnson, and P. A. Pincus, *Adv. Mass Spectrom.* **8,** 1012 (1980).
209. G. D. Daves, *Acc. Chem. Res.* **12,** 359 (1979).
210. F. Hillenkamp, M. Karas, D. Holtkamp, and P. Klusener, *Int. J. Mass Spectrom. Ion Proc.* **69,** 265 (1986).
211. P. K. Dutta and Y. Talmi, *Anal. Chim. Acta* **132,** 111 (1981).
212. H. J. Heinen, *Int. J. Mass Spectrom. Ion Phys.* **38,** 309 (1981).
213. L. Edelman, *Fres. Z. Anal. Chem.* **308,** 218 (1981).
214. A. Orsulakova, R. Kaufmann, C. Morgenstern, and M. D'Haese, *Fres. Z. Anal. Chem.* **308,** 221 (1981).
215. D. W. Lorch and H. Schafer, *Fres. Z. Anal. Chem.* **308,** 246 (1981).
216. A. Mathey, *Fres. Z. Anal. Chem.* **308,** 249 (1981).
217. U. Seydel and B. Lindner, *Fres. Z. Anal. Chem.* **308,** 253 (1981).
218. A. H. Verbueken, F. L. Van de Vyver, W. J. Visser, M. E. De Broe, and R. E. Van Grieken, *Adv. Mass Spectrom.* **10,** 987 (1986).
219. R. Bohm, *Fres. Z. Anal. Chem.* **308,** 258 (1981).
220. J. T. Dickinson, L. B. Brix, and L. C. Jensen, *J. Phys. Chem.* **88,** 1698 (1984).
221. J. T. Dickinson, L. C. Jensen, and M. K. Park, *Appl. Phys. Lett.* **41,** 827 (1982).
222. B. T. Chait and K. G. Standing, *Int. J. Mass Spectrom. Ion Phys.* **40,** 185 (1981).
223. F. Heresch, E. R. Schmid, and J. F. Huber, *Anal. Chem.* **52,** 1803 (1980).
224. D. Zakett, A. E. Schoen, R. G. Cooks, and P. H. Hemberger, *J. Am. Chem. Soc.* **103,** 1295 (1981).
225. P. Feigl, B. Schueler, and F. Hillenkamp, *Int. J. Mass Spectrom. Ion Phys.* **47,** 15 (1983).
226. R. J. Cotter, *Anal. Chem.* **52,** 1767 (1980).
227. D. V. Davis, R. G. Cooks, B. N. Meyer, and J. L. McLaughlin, *Anal. Chem.* **55,** 1302 (1983).
228. R. B. van Breemen, M. Snow, and R. J. Cotter, *Int. J. Mass Spectrom. Ion Phys.* **49,** 35 (1983).
229. J.-C. Tabet and R. J. Cotter, *Int. J. Mass Spectrom. Ion Phys.* **54,** 151 (1983).
230. J.-C. Tabet, M. Jablonski, R. J. Cotter, and J. E. Hunt, *Int. J. Mass Spectrom. Ion Proc.* **65,** 105 (1985).
231. F. G. Rudenauer, *Vacuum* **22,** 609 (1972).
232. P. H. Dawson, *Int. J. Mass Spectrom. Ion Phys.* **17,** 447 (1975).
233. P. H. Dawson and P. A. Redhead, *Rev. Sci. Instrum.* **48,** 159 (1977).
234. R. J. Beuhler and L. Friedman, *Int. J. Mass Spectrom. Ion Phys.* **23,** 81 (1977).
235. R. J. Beuhler and L. Friedman, *Nucl. Instrum. Methods* **170,** 309 (1980).
236. A. F. Dillon, R. S. Lehrle, J. C. Robb, and D. W. Thomas, *Adv. Mass Spectrom.* **4,** 477 (1968).
237. Y. Yang, E. A. Sokoloski, H. M. Fales, and L. K. Pannell, *Biomed. Environ. Mass Spectrom.* **13,** 489 (1986).
238. J. B. Westmore, W. Ens, and K. G. Standing, *Biomed. Mass Spectrom.* **9,** 119 (1982).
239. B. T. Chait and F. H. Field, *Int. J. Mass Spectrom. Ion Phys.* **41,** 17 (1981).
240. P. Steffens, E. Niehus, T. Friese, and A. Benninghoven, in *Ion Formation from Organic Solids,* (A. Benninghoven, ed.) Springer-Verlag, New York (1983).

241. E. Niehus, T. Heller, H. Feld, and A. Benninghoven, *Springer Ser. Chem. Phys. SIMS V* **44,** 188 (1986).
242. W. P. Poschenrieder, *Int. J. Mass Spectrom. Ion Phys.* **9,** 357 (1972).
243. G. H. Oetjen and W. P. Poschenrieder, *Int. J. Mass Spectrom. Ion Phys.* **16,** 353 (1975).
244. B. A. Mamyrin, V. I. Karataev, D. C. Snmikk, and V. A. Zagulin, *Sov. Phys. JETP* **37,** 45 (1973).
245. W. Gohl, R. Kutsches, H. J. Laue, and H. Wollnick, *Int. J. Mass Spectrom. Ion Phys.* **48,** 411 (1983).
246. E. W. Muller and S. V. Krishnaswamy, *Rev. Sci. Instrum.* **45,** 1053 (1974).
247. D. A. Laude, C. L. Johlman, R. S. Brown, D. A. Weil, and C. L. Wilkins, *Mass Spectrom. Rev.,* **5,** 107 (1986).
248. M. L. Gross and D. L. Rempel, *Science* **226,** 261 (1984).
249. D. H. Russell, *Mass Spectrom. Rev.* **5,** 167 (1986).
250. I. J. Amster, F. W. McLafferty, M. E. Castro, D. H. Russell, R. B. Cody, and S. Ghaderi, *Anal. Chem.* **58,** 485 (1986).
251. R. E. Shomo, A. G. Marshall, and C. R. Weisenberger, *Anal. Chem.* **57,** 2940 (1985).
252. R. B. Cody, I. J. Amster, and F. W. McLafferty, *Proc. Natl. Acad. Sci. USA* **82,** 6367 (1985).
253. M. E. Castro and D. H. Russell, *Anal. Chem.* **56,** 578 (1984).
254. S. K. Viswanadham, D. M. Hercules, R. R. Weller, and C. S. Giam, *Biomed. Environ. Mass Spectrom.* **14,** 43 (1987).
255. J. Shabanowitz, D. F. Hunt, R. T. McIver, R. L. Hunter, and J. E. P. Syka, *Adv. Mass Spectrom.* **10,** 933 (1986).
256. R. B. Cody, J. A. Kinsinger, S. Ghaderi, I. J. Amster, F. W. McLafferty, and C. E. Brown, *Anal. Chim. Acta* **178,** 43 (1985).
257. S. Ghaderi and D. Littlejohn, *Adv. Mass Spectrom.* **10,** 875 (1986).
258. C. la Lau, Mass discrimination caused by electron multiplier detectors, Topics in Organic Mass Spectrometry, *Adv. Anal. Chem. Instrum.* **8,** (1970).
259. G. C. Stafford, *Environ. Health Perspect.* **36,** 85 (1980).
260. P. G. Cullis, G. M. Neumann, D. E. Rogers, and P. J. Derrick, *Adv. Mass Spectrom.* **8,** 1729 (1980).
261. T. Sakurai and T. Hashizume, *Rev. Sci. Instrum.* **57,** 236 (1986).
262. B. Sundqvist and R. D. Macfarlane, *Mass Spectrom. Rev.* **4,** 421 (1985).
263. R. J. Day, Ph.D. dissertation, Purdue University, 1980.
264. J. H. Beynon, D. Cameron, and J. F. J. Todd, *Org. Mass Spectrom.* **17,** 346 (1982).
265. W. Aberth and A. L. Burlingame, *Anal. Chem.* **56,** 2915 (1984).
266. N. Furstenau, *Z. Naturforsch.* **33a,** 563 (1978).
267. J. E. Hunt, T. J. Michalski, and J. J. Katz, *Int. J. Mass Spectrom. Ion Phys.* **53,** 335 (1983).
268. R. Wurster, U. Haas, and P. Wieser, *Fres. Z. Anal. Chem.* **308,** 206 (1981).
269. F. R. Krueger, *Chromatographia* **101,** 151 (1970).
270. C. J. McNeal, R. D. Macfarlane, and E. L. Thurston, *Anal. chem.* **41,** 2036 (1979).
271. A. Benninghoven and W. Sichtermann, *Int. J. Mass Spectrom. ion Phys.* **38,** 351 (1981).
272. R. A. Gibbs and N. Winograd, *Rev. Sci. Instrum.* **52,** 1148 (1981).
273. D. W. Moon, R. J. Bleiler, C. C. Chang, and N. Winograd, *Springer Ser. Chem. Phys. SIMS V* **44,** 825 (1986).
274. R. A. Gibbs, S. P. Holland, K. E. Foley, B. J. Garrison, and N. Winograd, *J. Chem. Phys.* **76,** 684 (1982).

275. D. W. Moon and N. Winograd, *Int. J. Mass Spectrom. Ion Phys.* **52,** 217 (1983).
276. A. Benninghoven, W. Lange, M. Jirikowsky, and D. Holtkamp, *Surf. Sci.* **123,** L721 (1982).
277. D. Holtkamp, W. Lange, M. Jirikowski, and A. Benninghoven, *Appl. Surf. Sci.* **17,** 296 (1984).
278. K. D. Kloppel, K. Weyer, and G. von Bunau, *Int. J. Mass Spectrom. Ion Phys.* **51,** 47 (1983).
279. H. Grade and R. G. Cooks, *J. Am. Chem. Soc.* **100,** 5615 (1978).
280. D. F. Klemperer, in *Chemisorption and Reactions on Metallic Films,* (J. R. Anderson, ed.) Academic Press, New York (1971).
281. J. A. Gardella, J. H. Wandass, P. A. Cornelio, and R. L. Schmitt, *Springer Ser. Chem. Phys. SIMS V* **44,** 534 (1986).
282. E. A. Jordan, R. D. Macfarlane, C. R. Martin, and C. J. McNeal, *Int. J. Mass Spectrom. Ion Phys.* **53,** 345 (1983).
283. M. M. Ross and R. J. Colton, *Anal. Chem.* **55,** 1170 (1983).
284. S. A. Martin, C. E. Costello, and K. Biemann, *Anal. Chem.* **54,** 2362 (1982).
285. H. Danigel and R. D. Macfarlane, *Int. J. Mass Spectrom. Ion Phys.* **39,** 157 (1981).
286. L. Schmidt, H. Danigel, and H. Jungclas, *Nucl. Instrum. Methods* **198,** 165 (1982).
287. A. Albers, K. Wien, P. Duck, W. Treu, and H. Voit, *Nucl. Instrum. Methods* **198,** 69 (1982).
288. F. R. Krueger and K. Wien, *Z. Naturforsch.* **33a,** 638 (1978).
289. N. Furstenau, W. Knippelberg, F. R. Krueger, G. Weiss, and K. Wien, *Z. Naturforsch.* **32a,** 711 (1977).
290. Y. Le Beyec, S. Della Negra, C. V. Deprun, P. Vigny, and Y. M. Ginot, *Rev. Phys. Appl.* **15,** 1631 (1980).
291. P. Hakansson, I. Kamensky, and B. Sundqvist, *Surf. Sci.* **116,** 302 (1982).
292. B. Sundqvist, P. Hakansson, I. Kamensky, and J. Kjellberg, in *Ion Formation from Organic Solids* (A. Benninghoven, ed.) Springer-Verlag, New York (1983).
293. E. Nieschler, B. Nees, N. Bischof, H. Frohlich, W. Tiereth, and H. Voit, *Surf. Sci.* **145,** 294 (1984).
294. C. A. Anderson, H. J. Roden, and C. F. Robinson, *J. Appl. Phys.* **40,** 3419 (1969).
295. H. W. Werner and A. E. Morgan, *J. Appl. Phys.* **47,** 1232 (1976).
296. K. Wittmaack, *J. Appl. Phys.* **50,** 493 (1979).
297. G. Muller, *Appl. Phys.* **10,** 217 (1976).
298. C. W. Magee and W. L. Harrington, *Appl. Phys. Lett.* **33,** 193 (1978).
299. G. Slodzian, *Ann. Phys. (Paris)* **9,** 591 (1964).
300. D. Briggs, A. Brown, J. A. van den Berg, and J. C. Vickerman, in *Ion Formation from Organic Solids* (A. Benninghoven, ed.) Springer-Verlag, New York (1983).
301. S. J. Doherty, Ph.D. thesis, Indiana University, 1989.
302. C. W. Magee, *Int. J. Mass Spectrom. Ion Phys.* **49,** 211 (1983).
303. D. V. McCaughan, R. A. Kushner, and V. T. Murphy, *Phys. Rev. Lett.* **30,** 614 (1973).
304. P. Hakansson, E. Jayasinghe, A. Johanssen, I. Kamensky, and B. Sundqvist, *Phys. Rev. Lett.* **47,** 1227 (1981).
305. P. Duck, H. Frohlich, N. Bischof, and H. Voit, *Nucl. Instrum. Methods* **198,** 39 (1982).
306. E. Nieschler, B. Nees, N. Bischof, H. Frohlich, W. Tiereth, and H. Voit, *Rad. Effects* **83,** 121 (1984).
307. R. Ambros and A. M. Kleinfeld, *Nucl. Instrum. Methods* **99,** 173 (1972).

308. L. E. Sieberling, J. E. Griffith, and T. A. Tombrello, *Radiat. Eff.* **52,** 201 (1980).
309. W. L. Brown, W. M. Augustiniak, L. J. Lanzerotti, R. E. Johnson, and R. Evatt, *Phys. Rev. Lett.* **45,** 1632 (1980).
310. B. R. Karlsson, *Nucl. Instrum. Methods* **198,** 121 (1982).
311. R. G. Ulbrich, N. Narayanamurti, and M. Chin, *Phys. Rev. Lett.* **45,** 1432 (1980).
312. V. S. Antonov, V. S. Letokhov, and A. N. Shibanov, *Appl. Phys. B* **28,** 245 (1982).
313. V. S. Antonov, S. E. Egorov, V. S. Letokhov, Y. A. Matveets, and A. N. Shibanov, *JETP Lett.* **36,** 33 (1982).
314. W. B. Martin and R. M. O'Malley, *Int. J. Mass Spectrom. Ion Proc.* **59,** 227 (1984).
315. R. J. Cotter, *Anal. Chem.* **56,** 1257A (1984).
316. R. E. Krailler and D. H. Russell, *Anal. Chem.* **57,** 1211 (1985).
317. W. D. Bowers, S.-S. Delbert, R. L. Hunter, and R. T. McIver, *J. Am. Chem. Soc.* **10,** 7288 (1984).
318. W. Reuter, *Springer Ser. Chem. Phys. SIMS V* **44,** 94 (1986).
319. G. A. Schick, J. P. Baxter, J. Subbia-Singh, P. H. Kobrin, and N. Winograd, *Springer Ser. Chem. Phys. SIMS V* **44,** 90 (1986).
320. D. L. Donohue, W. H. Christie, D. E. Goeringer, and H. S. McKown, *Anal. Chem.* **57,** 1193 (1985).
321. M. Karas, D. Bachmann, and F. Hillenkamp, *Adv. Mass Spectrom.* **10,** 969 (1986).
322. M. Karas, D. Bachmann, and F. Hillenkamp, *Anal. Chem.* **57,** 2935 (1985).
323. V. Walther and H. Hintenberger, *Z. Naturforsch.* **18a,** 843 (1963).
324. Y. M. Fogel, *Sov. Phys. Usp.* **10,** 17 (1967).
325. R. C. Bradley, *J. Appl. Phys.* **30,** 1 (1959).
326. Z. Sroubeck, *Surf. Sci.* **44,** 47 (1984).
327. H. W. Werner, *Vacuum* **24,** 493 (1973).
328. K. Wittmaack, *Surf. Sci.* **89,** 668 (1979).
329. H. Kambara, *Org. Mass Spectrom.* **17,** 29 (1982).
330. D. F. Hunt, W. M. Bone, J. Shabanowitz, J. Rhodes, and J. M. Ballard, *Anal. Chem.* **53,** 1704 (1981).
331. K. G. Standing, B. T. Chait, W. Ens, G. McIntosh, and R. Beavis, *Nucl. Instrum. Methods* **198,** 33 (1982).
332. R. G. Orth, H. T. Jonkman, D. H. Powell, and J. Michl, *J. Am. Chem. Soc.* **103,** 6026 (1983).
333. R. G. Orth, H. T. Jonkman, D. H. Powell, and J. Michl, *J. Am. Chem. Soc.* **104,** 1834 (1982).
334. T. Tagaki, *Vacuum* **36,** 27 (1986).
335. L. Holland and W. Steckelmacher, *Vacuum* **2,** 346 (1952).
336. A. Albers, K. Wien, P. Duck, W. Treu, and H. Voit, *Nucl. Instrum. Methods* **198,** 69 (1982).
337. P. Hakansson and B. Sundqvist, *Rad. Effects* **61,** 179 (1982).
338. B. Nees, E. Nieschler, N. Bischof, H. Frohlich, K. Riemer, W. Tiereth, and H. Voit, *Surf. Sci.* **145,** 197 (1984).
339. I. Kamensky, P. Hakansson, B. Sundqvist, C. J. McNeal, and R. Macfarlane, *Nucl. Instrum. Methods* **198,** 65 (1982).
340. S. E. Unger, T. M. Ryan, and R. G. Cooks, *Anal. Chim. Acta* **118,** 169 (1980).
341. T. Keough, F. S. Ezra, A. F. Russell, and J. D. Pryne, Proceedings of the 33rd Annual Conference on Mass Spectrometry and Allied Topics, San Diego, California, May 1985, p. 964.
342. W. V. Ligon, *Int. J. Mass Spectrom. Ion Phys.* **52,** 189 (1983).

343. F. H. Field, *J. Phys. Chem.* **86,** 5115 (1982).
344. H. Kambara, in: *Ion Formation from Organic Solids* (A. Benninghoven, ed.) Springer-Verlag, New York (1983).
345. J. E. Campana, M. M. Ross, S. L. Rose, J. R. Wyatt, and R. J. Colton, in: *Ion Formation from Organic Solids* (A. Benninghoven, ed.), Springer-Verlag, New York (1983).
346. L. Kurlansik, T. J. Williams, J. E. Campana, B. N. Green, L. W. Anderson, and J. M. Strong, *Biochem. Biophys. Res. Commun.* **111,** 478 (1983).
347. J. Meili and J. Seibl, Proceedings of the 31st Annual Conference on Mass Spectrometry and Allied Topics, Boston, MA, May 1983, p. 294.
348. W. D. Lehmann, M. Kessler, and W. A. Konig, *Biomed. Mass Spectrom.* **11,** 217 (1984).
349. D. F. Barofsky, U. Giessmann, and E. Barofsky, *Int. J. Mass Spectrom. Ion. Phys.* **53,** 319 (1983).
350. L. K. Liu, K. L. Busch, and R. G. Cooks, *Anal. Chem.* **53,** 109 (1981).
351. K. L. Busch, B.-H. Hsu, Y.-X. Xie, and R. G. Cooks, *Anal. Chem.* **55,** 1157 (1983).
352. R. Michael and D. Stulik, *J. Vac. Sci. Technol.* **A4,** 1861 (1986).
353. F. Okuyama and Y. Fujimoto, *J. Vac. Sci. Technol.* **A4,** 237 (1986).
354. J. S. Sovey, *J. Vac. Sci. Technol.* **16,** 813 (1979).
355. M. Maurette, A. Banifatemi, S. Della-Negra, and Y. Le Beyec, *Nature* **303,** 159 (1983).
356. K. Wittmaack, *Nucl. Instrum. Methods* **168,** 343 (1980).
357. G. Staudenmaier, *Rad. Eff.* **13,** 87 (1972).
358. C.-E. Richter and M. Trapp, *Int. J. Mass Spectrom. Ion Phys.* **38,** 21 (1981).
359. F. Honda, G. M. Lancaster, Y. Fukuda, and J. W. Rabalais, *J. Chem. Phys.* **69,** 4931 (1978).
360. A. Benninghoven and A. Mueller, *Phys. Lett.* **40A,** 169 (1972).
361. L. Kelner and S. P. Markey, *Int. J. Mass Spectrom. Ion Phys.* **59,** 157 (1984).
362. L. Kelner and T. C. Patel, *Springer Ser. Chem. Phys. SIMS V* **44,** 494 (1986).
363. G. J. Q. van der Peyl, W. J. van der Zande, R. Hoogerbrugge, and P. G. Kistemaker, *Adv. Mass Spectrom.* **10,** 1511 (1986).
364. R. P. Opsal, K. G. Owens, and J. P. Reilly, *Anal. Chem.* **57,** 1884 (1985).
365. P. Hakansson, I. Kamensky, and B. Sundqvist, *Nucl. Instrum. Methods* **198,** 43 (1982).
366. G. J. Q. van der Peyl, W. J. van der Zande, K. Bederski, A. J. H. Boerboom, and P. G. Kistemaker, *Int. J. Mass Spectrom. Ion Phys.* **47,** 7 (1983).
367. G. J. Q. van der Peyl, W. J. van der Zande, and P. G. Kistmaker, *Int. J. Mass Spectrom. Ion Phys.* **62,** 51 (1984).
368. P. Sigmund, in: *Sputtering by Particle Bombardment I* (R. Behrisch, ed.), Springer-Verlag, New York (1981).
369. G. K. Wehner, *Phys. Rev.* **102,** 690 (1956).
370. D. Onderdelinden, *Can. J. Phys.* **46,** 139 (1968).
371. J. E. Demuth, K. Christmann, and P. N. Sando, *Chem. Phys. Lett.* **76,** 201 (1980).
372. R. J. Cotter, M. Snow, and M. Colvin, in: *Ion Formation from Organic Solids* (A. Benninghoven, ed.), Springer-Verlag, New York (1983).
373. F. R. Krueger, *Z. Naturforsch.* **32a,** 1089 (1977).
374. D. Zakett, A. E. Schoen, R. G. Cooks, and P. H. Hemberger, Presented at the 29th Annual Conference on Mass Spectrometry and Allied Topics, Minneapolis, Minnesota, 1981.
375. R. J. Cotter, *Anal. Chem.* **53,** 719 (1981).

376. D. M. Hercules, *Pure Appl. Chem.* **55,** 1869 (1983).
377. A. E. Schoen, Ph.D. dissertation, Purdue University, 1982.
378. R. J. Cotter and A. L. Yergey, *J. Am. Chem. Soc.* **103,** 1596 (1981).
379. U. Giessmann and F. W. Rollgen, *Org. Mass. Spectrom.* **11,** 1094 (1976).
380. G. J. Q. van der Peyl, K. Isa, J. Haverkamp, and P. G. Kistemaker, *Int. J. Mass Spectrom. Ion Phys.* **47,** 11 (1983).
381. G. J. Q. van der Peyl, K. Isa, J. Haverkamp, and P. G. Kistemaker, *Org. Mass Spectrom.* **16,** 416 (1981).
382. T. M. Ryan, R. J. Day, and R. G. Cooks, *Anal. Chem.* **52,** 2054 (1980).
383. D. F. Barofsky and U. Giessmann, *Int. J. Mass Spectrom. Ion Phys.* **46,** 359 (1983).
384. D. N. Heller, J. Yergey, and R. J. Cotter, *Anal. Chem.* **55,** 1310 (1983).
385. T. A. Dang, R. J. Day, and D. M. Hercules, *Anal. Chem.* **56,** 866 (1984).
386. B. Sundqvist, P. Roepstorff, J. Fohlman, A. Hedin, P. Hakansson, I. Kamensky, M. Lindberg, M. Salehpour, and G. Sawe, *Science* **226,** 696 (1984).
387. B. Sundqvist, A. Hedin, P. Hakansson, i. Kamensky, J. Kjellberg, M. Salehpour, G. Save, and S. Widdiyasekera, *Int. J. Mass Spectrom. Ion Phys.* **53,** 167 (1983).
388. G. P. Jonsson, A. B. Hedin, P. L. Hakansson, B. U. R. Sundqvist, B. G. S. Save, P. F. Nielsen, P. Roepstorff, K.-E. Johansson, I. Kamensky and M. S. L. Lindberg, *Anal. Chem.* **58,** 1084 (1986).
389. G. Schmelzeisen-Redecker, U. Giessmann, and F. W. Rollgen, *Angew. Chem. Int. Ed. Engl.* **23,** 892 (1984).
390. L. D. Detter, R. G. Cooks, and R. A. Walton, *Inorg. Chim. Acta* **115,** 55 (1986).
391. L. Laxhuber, H. Mohwald, and M. Hashmi, *Int. J. Mass Spectrom. Ion Phys.* **51,** 93 (1983).
392. K. D. Kloppel and W. Seidel, *Int. J. Mass Spectrom. Ion Phys.* **31,** 151 (1979).
393. E. De Pauw and J. Marien, *Int. J. Mass Spectrom. Ion Phys.* **38,** 11 (1981).
394. K. L. Busch and R. G. Cooks, *Spectra* **8,** 22 (1982).
395. G. M. Lancaster, F. Honda, Y. Fukuda, and J. W. Rabalais, *J. Am. Chem. Soc.* **101,** 1951 (1979).
396. P. T. Murray and J. W. Rabalais, *J. Am. Chem. Soc.* **103,** 1007 (1981).
397. H. T. Jonkman, J. Michl, R. N. King, and J. D. Andrade, *Anal. Chem.* **50,** 2078 (1978).
398. R. G. Orth, H. T. Jonkman, and J. Michl, *Int. J. Mass Spectrom. Ion. Phys.* **43,** 41 (1982).
399. M. Barber, J. C. Vickerman, and J. Wolstenholme, *J. Chem. Soc. Faraday Trans. I* **76,** 549 (1980).
400. J. Marien and E. De Pauw, *J. Chem. Soc. Chem. Commun.* 949 (1982).
401. R. N. Katz, T. Chaudhury, and F. H. Field, *J. Am. Chem. Soc.* **108,** 3897 (1986).
402. R. J. Day, J. Zimmerman, and D. M. Hercules, *Spectrosc. Lett.* **14,** 773 (1981).
403. K. Balasanmugam and D. M. Hercules, *Anal. Chem.* **55,** 146 (1983).
404. S. W. Graham and D. M. Hercules, *Spectrosc. Lett.* **15,** 1 (1982).
405. E. Brunnix and G. Rudstam, *Nucl. Instrum. Methods* **13,** 131 (1961).
406. P. J. Todd and G. S. Groenewold, *Anal. Chem.* **58,** 895 (1986).
407. S. E. Unger, R. J. Day, and R. G. Cooks, *Int. J. Mass Spectrom. Ion Phys.* **39,** 231 (1981).
408. J. L. Pierce, K. L. Busch, R. A. Walton, and R. G. Cooks, *J. Am. Chem. Soc.* **103,** 2583 (1981).
409. J. L. Pierce, K. L. Busch, R. G. Cooks, and R. A. Walton, *Inorg. Chem.* **21,** 2597 (1982).

410. B.-H. Hsu, Y.-X. Xie, K. L. Busch, and R. G. Cooks, *Int. J. Mass Spectrom. Ion Phys.* **51,** 225 (1983).
411. S. J. Pachuta and R. G. Cooks, in: *Desorption Mass Spectrometry* (P. A. Lyon, ed.) ACS Symposium Series 291, Washington, D.C. (1985).
412. R. F. Chaiken, D. J. Sibbett, J. E. Sutherland, D. K. Van de Mark, and A. Wheeler, *J. Chem. Phys.* **37,** 2311 (1962).
413. C. J. McNeal and R. D. Macfarlane, *J. Am. Chem. Soc.* **108,** 2132 (1986).
414. D. J. Surman and J. C. Vickerman, *J. Chem. Res. Synop.*, 170 (1981).
415. M. Barber, R. S. Bordoli, G. J. Elliott, R. D. Sedgwick, and A. N. Tyler, *J. Chem. Soc. Faraday Trans. I* **79,** 1249 (1983).
416. M. M. Bursey, G. D. Marbury, and J. R. Hass, *Biomed. Mass Spectrom.* **11,** 522 (1984).
417. P. J. Todd and C. P. Leibman, *Springer. Ser. Chem. Phys. SIMS V* **44,** 500 (1986).
418. E. De Pauw, *Mass Spectrom. Rev.* **5,** 191 (1986).
419. J. L. Gower, *Biomed. Mass Spectrom.* **12,** 191 (1985).
420. B. L. Ackerman, J. T. Watson, and J. F. Holland, *Anal. Chem.* **57,** 2656 (1985).
421. J. M. Miller and R. Therberge, *Org. Mass Spectrom.* **20,** 600 (1985).
422. C. G. Sanders, T. R. Sharp, and E. L. Allred, *Tetrahedron Lett.* **27,** 3231 (1986).
423. M.-Y. Zhang, X.-Y. Liang, Y.-Y. Chen, and X.-G. Liang, *Anal. Chem.* **56,** 2288 (1984).
424. M. A. Baldwin and K. J. Welham, *Org. Mass Spectrom.* **21,** 235 (1986).
425. A. Dell, H. R. Morris, M. D. Levin, and S. M. Hecht, *Biochem. Biophys. Res. Commun.* **102,** 730 (1981).
426. A. M. Buko, L. R. Philips, and B. A. Fraser, *Biomed. Mass Spectrom.* **10,** 408 (1983).
427. G. Pelzer, E. De Pauw, D. V. Dung, and J. Marien, *J. Phys. Chem.* **88,** 5065 (1984).
428. P. J. Gale, B. L. Bentz, B. T. Chait, F. H. Field, and R. J. Cotter, *Anal. Chem.* **58,** 1070 (1986).
429. S. M. Scheifers, S. Verma, and R. G. Cooks, *Anal. Chem.* **55,** 2260 (1983).
430. L. K. Pannel, E. A. Sokoloski, H. M. Fales, and R. L. Tate, *Anal. Chem.* **57,** 106 (1985).
431. G. C. DiDonato, Ph.D. thesis, Indiana University, 1987.
432. R. Copper and S. Unger, *J. Antibiotics* **38,** 24 (1985).
433. J. Inchaouh, J. C. Blais, G. Bolbach, and A. Brunot, *Int. J. Mass Spectrom. Ion Proc.* **61,** 153 (1984).
434. Q. W. Huang, G. L. Wu, and H. T. Tanj, *Int. J. Mass Spectrom. Ion Proc.* **70,** 145 (1986).
435. B. D. Musselman, J. T. Watson, and C. K. Chang, *Org. Mass Spectrom.* **21,** 215 (1986).
436. M. M. Ross, D. A. Kidwell, and R. J. Colton, *Int. J. Mass Spectrom. Ion Proc.* **63,** 141 (1985).
437. G. C. DiDonato and K. L. Busch, *Biomed. Mass Spectrom.* **12,** 354 (1985).
438. D. A. Kidwell, M. M. Ross, and R. J. Colton, *Biomed. Mass Spectrom.* **12,** 254 (1985).
439. K. L. Busch, B.-H. Hsu, K. V. Wood, R. G. Cooks, C. G. Schwartz, and A. R. Katritzky, *J. Org. Chem.* **49,** 764 (1984).
440. K. L. Busch, K. J. Kroha, R. A. Flurer, and G. C. DiDonato, *Springer Ser. Chem. Phys. SIMS V* **44,** 512 (1986).
441. E. De Pauw, G. Pelzer, D. V. Dung, and J. Marien, *Biochem. Biophys. Res. Commun.* **123,** 27 (1984).
442. W. V. Ligon and S. B. Dorn, *Int. J. Mass Spectrom. Ion Proc.* **62,** 315 (1985).

443. W. V. Ligon, *Anal. Chem.* **58,** 485 (1986).
444. W. V. Ligon and S. B. Dorn, *Anal. Chem.* **58,** 1889 (1986).
445. D. A. Kidwell, M. M. Ross, and R. J. Colton, *Adv. Mass Spectrom.* **10,** 1607 (1986).
446. J. A. Sunner, R. Kulatunga, and P. Kebarle, *Anal. Chem.* **58,** 1312 (1986).
447. J. A. Sunner, R. Kulatanga, and P. Kebarle, *Anal. Chem.* **58,** 2009 (1986).
448. S. S. Wong, F. W. Rollgen, I. Manz, and B. Przybylski, *Biomed. Mass Spectrom.* **12,** 43 (1985).
449. M. H. O'Leary and G. A. Samberg, *J. Am. Chem. Soc.* **93,** 3530 (1971).
450. R. Foster, *Organic Charge Transfer Complexes,* Academic Press, New York (1969).
451. G. C. DiDonato and K. L. Busch, *Biomed. Mass Spectrom.* **12,** 354 (1985).
452. E. De Pauw, *Anal. Chem.* **55,** 2196 (1983).
453. A. F. Gaines and F. M. Page, *Fuels* **62,** 1041 (1983).
454. G. Dube, *Org. Mass Spectrom.* **19,** 242 (1984).
455. G. S. Groenewold, P. J. Todd, and M. V. Buchanan, *Anal. Chem.* **56,** 2251 (1984).
456. J. E. Bartmess and L. R. Philips, *Adv. Mass Spectrom.* **10,** 1581 (1986).
457. K. L. Busch, G. C. DiDonato, K. J. Kroha, and L. R. Hittle, Presented at the 1985 Pittsburgh Conference on Analytical Chemistry and Applied Spectroscopy, New Orleans, Louisiana, March 1985.
458. J. Fine, S. C. Hardy, and T. D. Andreadis, *J. Vac. Sci. Technol.* **18,** 1310 (1981).
459. R. A. W. Johnstone, I. A. S. Lewis, and M. E. Rose, *Tetrahedron* **39,** 1597 (1983).
460. H. M. Coloquhon, J. F. Stoddart, D. J. Williams, J. B. Wolstenholme, and R. Zarzycki, *Angew. Chem. Int. Ed. Engl.* **20,** 1051 (1981).
461. I. Fujii, R. Isoba, and K. Kanematsu, *J. Chem. Soc., Chem. Commun.,* 405 (1985).
462. G. C. DiDonato and K. L. Busch, unpublished results.
463. R. A. W. Johnstone and M. E. Rose, *J. Chem. Soc. Chem. Commun.* 1268 (1983).
464. P. D. Beer, *J. Chem. Soc. Chem. Commun.* 1115 (1985).
465. I. Fujii, R. Isobe, and K. Kanematsu, *J. Chem. Soc. Chem. Commun.* 405 (1985).
466. D. C. Fenimore and C. M. Davis, *Anal. Chem.* **53,** 253A (1981).
467. A. Benninghoven, A. Eicke, M. Junack, W. Sichtermann, J. Krizek, and H. Peters, *Org. Mass Spectrom.* **15,** 459 (1980).
468. R. D. Smith, J. E. Burger, and A. L. Johnson, *Anal. Chem.* **53,** 1603 (1981).
469. W. H. McFadden, H. L. Schwartz, and S. Evans, *J. Chromatogr.* **122,** 389 (1976).
470. N. J. Alcock, C. Eckers, D. E. Games, M. P. L. Games, M. S. Lant, M. A. McDowall, M. Rossiter, R. W. Smith, S. A. Westwood, and H.-Y. Yong, *J. Chromatogr.* **251,** 165 (1982).
471. T. P. Fan, E. D. Hardin, and M. L. Vestal, *Anal. Chem.* **56,** 1870 (1984).
472. H. Jungclas, H. Danigel, and L. Schmidt, *Org. Mass Spectrom.* **17,** 86 (1982).
473. H. Jungclas, H. Danigel, L. Schmidt, and J. Dellbrugge, *Org. Mass Spectrom.* **17,** 499 (1982).
474. L. Schmidt, H. Danigel, and H. Jungclas, *Nucl. Instrum. Methods* **198,** 165 (1982).
475. H. Jungclas, H. Danigel, and L. Schmidt, *J. Chromatogr.* **271,** 35 (1983).
476. K. Heyns and H. F. Grutzmacher, *Angew. Chem. Int. Ed. Engl.* **1,** 400 (1962).
477. G. J. Down and S. A. Gwyn, *J. Chromatogr.* **103,** 208 (1975).
478. R. Kaiser, *Chem. Br.* **5,** 54 (1969).
479. L. Ramaley, M.-A. Vaughan, and W. D. Jamieson, *Anal. Chem.* **57,** 353 (1985).
480. L. Ramaley, M. E. Nearing, M.-A. Vaughan, R. G. Ackman, and W. D. Jamieson, *Anal. Chem.* **55,** 2285 (1983).
481. H. Danigel, L. Schmidt, H. Jungclas, and K.-H. Pfluger, *Biomed. Mass Spectrom.* **12,** 542 (1985).
482. T. D. Chang, J. O. Lay, and R. J. Francel, *Anal. Chem.* **56,** 109 (1984).

483. M. L. Zimmermann and D. F. Barofsky, *Adv. Mass Spectrom.* **10,** 1257 (1986).
484. Y. Kushi and S. Handa, *J. Biochem.* **98,** 265 (1985).
485. S. E. Unger, A. Vincze, R. G. Cooks, R. Chrisman, and L. D. Rothman, *Anal. Chem.* **53,** 976 (1981).
486. A. Ba-Isa, K. L. Busch, R. G. Cooks, A. Vincze, and I. Granoth, *Tetrahedron* **39,** 591 (1983).
487. G. D. Tantsyrev, M. I. Povolotskaya, and V. A. Saraev, *Bioorgan. Khim.* **10,** 848 (1984).
488. Y. Nakagawa, *Proc. Soc. Med. Mass Spectrom.* **9,** 39 (1984): CA **102:** 137616z.
489. K. Iwatani, T. Kadono, and Y. Nakagawa, *Mass Spectrosc.* **34,** 181 (1986).
490. K. Iwatani and Y. Nakagawa, *Mass Spectrosc.* (*Jpn*) **34,** 189 (1986).
491. D. M. Hercules, *Pure Appl. Chem.* **55,** 1869 (1983).
492. J. W. Fiola, G. C. DiDonato, and K. L. Busch, *Rev. Sci. Instrum.* **57,** 2294 (1986).
493. G. C. DiDonato and K. L. Busch, *Anal. Chem.* **58,** 3231 (1986).
494. M. S. Stanley and K. L. Busch, *Anal. Chim. Acta* **194,** 199 (1987).
495. A. P. Melikian, N. W. Flynn, F. Petty, and J. D. Wander, *J. Pharm. Sci.* **66,** 228 (1977).
496. S. E. Unger, R. G. Cooks, B. J. Steinmetz, and W. N. Delgass, *Surf. Sci.* **116,** L211 (1982).
497. D. A. Baldwin, N. Shamir, and J. W. Rabalais, *Surf. Sci.* **141,** 617 (1984).
498. B.-H. Hsu and R. G. Cooks, *Anal. Chem.* **57,** 2925 (1985).
499. K. L. Duffin and K. L. Busch, *Int. J. Mass Spectrom. Ion Proc.* **74,** 141 (1986).
500. K. Saito and R. Kato, *Proc. Soc. Med. Mass Spectrom.* **9,** 45 (1984).
501. R. M. Caprioli, L. A. Smith, and C. F. Beckner, *Int. J. Mass Spectrom. Ion Phys.* **46,** 419 (1983).
502. L. A. Smith and R. M. Caprioli, *Biomed. Mass Spectrom.* **10,** 98 (1983).
503. R. M. Caprioli and L. Smith, *Anal. Chem.* **58,** 1080 (1986).
504. R. M. Caprioli, *Mass Spectrom.* **8,** Specialist Periodical Report.
505. R. M. Caprioli, in *Desorption Mass Spectrometry* (P. A. Lyon, ed.), ACS Symposium Series 291, Washington, D.C. (1985).
506. C. Fenselau, *Anal. Chem.* **54,** 105A (1982).
507. J. Yergey, D. Heller, G. Hansen, R. J. Cotter, and C. Fenselau, *Anal. Chem.* **55,** 353 (1983).
508. J. A. Yergey, R. J. Cotter, D. Heller, and C. Fenselau, *Anal. Chem.* **56,** 2262 (1984).
509. S. Naylor, A. F. Findeis, B. W. Gibson, and D. H. Williams, *J. Am. Chem. Soc.* **108,** 6359 (1986).
510. J. S. Cottrell and R. J. Greathead, *Mass Spectrom. Rev.* **5,** 215 (1986).
511. G. W. Wood and W. F. Sun, *Biomed. Mass Spectrom.* **7,** 399 (1980).
512. L. R. Shronk and R. J. Cotter, *Biomed. Environ. Mass Spectrom.* **13,** 395 (1986).
513. H. I. Kenttamaa, K. L. Busch, R. G. Cooks, and K. V. Wood, *Org. Mass Spectrom.* **18,** 561 (1983).
514. F. W. McLafferty, ed., *Tandem Mass Spectrometry,* John Wiley and Sons, New York (1983).
515. R. G. Cooks and G. L. Glish, *Chem. Eng. News* **59** (30 Nov), 40 (1981).
516. G. L. Glish, P. J. Todd, K. L. Busch, and R. G. Cooks, *Int. J. Mass Spectrom. Ion Phys.* **56,** 177 (1984).
517. S. Nacson, A. G. Harrison, and W. R. Davidson, *Org. Mass Spectrom.* **21,** 317 (1986).
518. D. A. McCrery, D. A. Peake, and M. L. Gross, *Anal. Chem.* **57,** 1181 (1985).
519. S. A. McLuckey, A. E. Schoen, and R. G. Cooks, *J. Am. Chem. Soc.* **104,** 848 (1982).

520. J. C. Tabet, D. Fraisse, and G. Geminet, *Adv. Mass Spectrom.* **10,** 1535 (1986).
521. J. E. Campana, T. M. Barlak, R. J. Colton, J. J. DeCorpo, J. R. Wyatt, and B. I. Dunlap, *Phys. Rev. Lett.,* **47,** 1046 (1981).
522. T. M. Barlak, J. E. Campana, R. J. Colton, and J. J. DeCorpo, *J. Am. Chem. Soc.* **104,** 1212 (1982).
523. W. Ens, R. Beavis, and K. G. Standing, *Phys. Rev. Lett.* **50,** 27 (1983).
524. B. I. Dunlap, J. E. Campana, B. N. Green, and R. H. Bateman, *J. Vac. Sci. Technol. A* **1,** 432 (1983).
525. B. I. Dunlap and J. E. Campana, *Org. Mass Spectrom.* **21,** 221 (1986).
526. C. V. Bradley, I. Howe, and J. H. Beynon, *Biomed. Mass Spectrom.* **8,** 85 (1981).
527. K. Eckart, H. Schwarz, K. B. Tomer, and M. L. Gross, *J. Am. Chem. Soc.* **107,** 6765 (1985).
528. R. A. Flurer and K. L. Busch, Presented at the 1986 Pittsburgh Conference on Analytical Chemistry and Applied Spectroscopy, Atlantic City, New Jersey, March 1986, p. 864.
529. H. R. Morris, M. Panico, and G. W. Taylor, *Biochem. Biophys. Res. Commun.* **117,** 299 (1983).
530. N. J. Jensen, K. B. Tomer, and M. L. Gross, *Anal. Chem.* **57,** 2018 (1985).
531. N. J. Jensen, K. B. Tomer, and M. L. Gross, *J. Am. Chem. Soc.* **107,** 1863 (1985).
532. M. Cervilla and G. Puzo, *Anal. Chem.* **55,** 2100 (1983).
533. M. Riviere, M. Cervilla, and G. Puzo, *Anal. Chem.* **57,** 2444 (1985).
534. D. S. Millington, C. R. Roe, and D. A. Maltby, *Biomed. Mass Spectrom.* **11,** 236 (1984).
535. D. S. Millington, T. P. Bohan, C. R. Roe, A. L. Yergey, and D. J. Liberato, *Clin. Chim. Acta* **145,** 69 (1985).
536. S. J. Gaskell, C. Guenat, D. S. Millington, D. A. Maltby, and C. R. Roe, *Anal. Chem.* **58,** 2801 (1986).
537. S. E. Unger, A. E. Schoen, R. G. Cooks, D. J. Ashworth, J. D. Gomes, and C.-j. Chang, *J. Org. Chem.* **46,** 4765 (1981).
538. P. P. Wickramanayake, B. L. Arbogast, D. R. Buhler, M. L. Deinzer, and A. L. Burlingame, *J. Am. Chem. Soc.* **107,** 2485 (1985).
539. K. B. Tomer, M. L. Gross, and M. Deinzer, *Anal. Chem.* **58,** 2527 (1986).
540. K. D. White, J. A. Sphon, and S. Hall, *Anal. Chem.* **58,** 562 (1986).
541. I. J. Amster and F. W. McLafferty, *Anal. Chem.* **57,** 1208 (1985).
542. K. B. Tomer, N. J. Jensen, M. L. Gross, and J. Whitney, *Biomed. Environ. Mass Spectrom.* **13,** 265 (1986).
543. H. M. Scheibel and H.-R. Schulten, *Mass Spectrom. Rev.* **5,** 249 (1986).
544. B. U. R. Sundqvist, *Adv. Mass Spectrom.* **11A,** 363 (1989).
545. M. Karas and F. Hillenkamp, *Adv. Mass Spectrom.* **11A,** 416 (1989).
546. M. Karas and F. Hillenkamp, *Anal. Chem.* **60,** 2299 (1988).
547. M. Karas, A. Igendoh, U. Bahr, and F. Hillenkamp, *Biomed. Environ. Mass Spectrom.* **18,** 841 (1989).
548. G. D. Roberts, S. A. Carr, C. A. Meyers, K. O. Johanson, and L. M. Miles, *Adv. Mass Spectrom.* **11B,** 1534 (1989).
549. T. Sakurai, T. Matsuo, M. Morris, Y. Fujita, H. Matsuda, and I. Katakuse, *Adv. Mass Spectrom.* **11A,** 220 (1989).
550. K. Biemann and S. Martin, *Mass Spectrom. Rev.* **6,** 1 (1987).
551. D. M. Desiderio, ed., *Mass Spectrometry of Peptides,* CRC Press, Boca Raton, Florida (1990).
552. R. M. Caprioli, *Mass Spectrom. Rev.* **6,** 237 (1987).
553. R. M. Caprioli, *Adv. Mass Spectrom.* **11A,** 422 (1989).

554. K. Mock, J. Firth, and J. S. Cottrell, *Org. Mass Spectrom.* **24,** 591 (1989).
555. G. Cerveau, C. Chuit, R. J. P. Corriu, L. Gerbier, C. Reye, J. L. Aubagnac, and B. El Amrani, *Int. J. Mass Spectrom. Ion Proc.* **82,** 259 (1988).
556. G. Bonas, C. Bosso, and M. R. Vignon, *Adv. Mass Spectrom.* **11A,** 546 (1989).
557. E. De Pauw, *Adv. Mass Spectrom.* **11A,** 383 (1989).
558. K. D. Cook, P. J. Todd, and H. J. Friar, *Biomed. Environ. Mass Spectrom.* **18,** 492 (1989).
559. Y. Ito, T. Takeuchi, D. Ishii, and M. Goto, *J. Chromatogr.* **346,** 161 (1985).
560. R. Caprioli, T. Fan, and J. Cottrell, *Anal. Chem.* **58,** 2949 (1986).
561. K.-H. Chen and R. J. Cotter, *Adv. Mass Spectrom.* **11A,** 174 (1989).
562. M. J. Bertrand, H. Perreault, and Q. Zha, Presented at the 37th ASMS Conference on Mass Spectrometry and Allied Topics, May 21–25, 1989, Miami Beach, Florida, p. 395.
563. K. P. Wirth, E. Junker, F. W. Rollgen, P. Fonrobert, and M. Przybylski, *Adv. Mass Spectrom.* **11A,** 430 (1989).
564. S. S. Wong, K. P. Wirth, and F. W. Rollgen, *Adv. Mass Spectrom.* **10B,** 1589 (1986).
565. J. B. Pallix, U. Schuhle, C. H. Becker, and D. L. Huestis, *Anal. Chem.* **61,** 805 (1989).
566. L. Li and D. Lubman, *Anal. Chem.* **60,** 2591 (1988).
567. L. M. Nuwaysir and C. L. Wilkins, *Anal. Chem.* **61,** 689 (1989).
568. C. H. Watson, G. Baykut, and J. R. Eyler, *Anal. Chem.* **59,** 1133 (1987).
569. L. Van Vaeck, J. Bennett, P. Van Espen, E. Schweikert, R. Gijbels, F. Adams, and W. Lauwers, *Org. Mass Spectrom.* **24,** 782 (1989).
570. L. Van Vaeck, J. Bennett, P. Van Espen, E. Schweikert, R. Gibjels, F. Adams, and W. Lauwers, *Org. Mass Spectrom.* **24,** 797 (1989).
571. L. Van Vaeck, P. Van Espen, F. Adams, R. Gijbels, W. Lauwers, and E. Esmans, *Biomed. Environ. Mass Spectrom.* **18,** 581 (1989).
572. S. Della-Negra, Y. Le Beyec, and P. Hakansson, *Nucl. Instrum. Methods* **B9,** 103 (1985).
573. S. Della-Negra, C. Deprun, Y. le Beyec, F. W. Rollgen, K. Standing, B. Monart, and G. Bolbach, *Int. J. Mass Spectrom. Ion Proc.* **75,** 319 (1987).
574. M. Salehpour and J. Hunt, *Int. J. Mass Spectrom. Ion Proc.* **85,** 99 (1988).
575. M. Salehpour and J. E. Hunt, *Adv. Mass Spectrom.* **11,** 218 (1989).
576. B. Nees, R. Schmidt, Ch. Schoppmann, and H. Voit, *Int. J. Mass Spectrom. Ion Proc.* **94,** 305 (1989).
577. B. Nees, R. Schmidt, Ch. Schoppmann, and H. Voit, *Int. J. Mass Spectrom. Ion Proc.* **94,** 205 (1989).
578. J. H. Callahan, K. Hool, and K. D. Cook, *Anal. Chem.* **60,** 714 (1988).
579. K. D. Cook, J. H. Callahan, and V. F. Man, *Anal. Chem.* **60,** 706 (1988).
580. N. B. Zolotoy and G. V. Karpov, *Adv. Mass Spectrom.* **11A,** 498 (1989).
581. T. D. Lee, K. Legesse, J. F. Mahoney, and J. Perel, Presented at the 37th ASMS Conference on Mass Spectrometry and Allied Topics, May 21–26. 1989, Miami Beach, Florida, p. 582.
582. A. P. Bruins, T. R. Covey, and J. D. Henion, *Anal. Chem.* **59,** 2642 (1987).
583. R. D. Smith, C. J. Barinaga, and J. D. Henion, *Anal. Chem.* **60,** 1948 (1988).
584. S. F. Wong, C. K. Meng, and J. B. Fenn, *J. Phys. Chem.* **92,** 546 (1988).
585. K. W. M. Siu, G. J. Gardner, and S. S. Berman, *Org. Mass Spectrom.* **24,** 931 (1989).
586. P. J. Rudewicz, *Biomed. Environ. Mass Spectrom.* **15,** 461 (1988).
587. I. G. Beattie and T. J. A. Blake, *Biomed. Environ. Mass Spectrom.* **18,** 872 (1989).
588. E. Gelpi, *Adv. Mass Spectrom.* **11B,** 1152 (1989).

589. K. Tanaka, H. Waki, Y. Ido, S. Akita, Y. Yoshida, and T. Yoshida, *Rapid Commun. Mass Spectrom.* **2,** 151 (1988).
590. R. C. Beavis and B. T. Chait, *Rapid Commun. Mass Spectrom.* **3,** 233 (1989).
591. R. P. Lattimer and H.-R. Schulten, *Anal. Chem.* **61,** 1201A (1989).
592. M. Salehpour, D. L. Fishel, and J. E. Hunt, *Adv. Mass Spectrom.* **11A,** 428 (1989).
593. J. E. Hunt, M. Salehpour, A. E. Ruthenberg, R. Amrein, B. Zabrancky, E. Kanter, and D. L. Fishel, *Adv. Mass Spectrom.* **11A,** 468 (1989).
594. M. Salehpour, D. L. Fishel, and J. E. Hunt, *Int. J. Mass Spectrom. Ion Proc.* **84,** R7 (1988).
595. W. Aberth and A. Burlingame, *Anal. Chem.* **60,** 1426 (1988).
596. M. Barber and B. N. Green, *Rapid Commun. Mass Spectrom.* **1,** 80 (1987).
597. F. G. Rudenauer, W. Stieger, H. Studnicka, and P. Pollinger, *Int. J. Mass Spectrom. Ion Proc.* **77,** 63 (1987).
598. A. E. Bell, G. A. Schwind, S. Rao, and L. W. Swanson, *Int. J. Mass Spectrom. Ion Proc.* **88,** 59 (1989).
599. Y. Saito and T. Noda, *Z. Phys. D. At. Mol. Clusters* **12,** 225 (1989).
600. L. F. Jiang, T. Y. Yen, and D. F. Barofsky, Presented at the 37th ASMS Conference on Mass Spectrometry and Allied Topics, May 21–25, 1989, Miami Beach, Florida, p. 1071.
601. L. F. Jiang, E. Barofsky, and D. F. Barofsky, in *Secondary Ion Mass Spectrometry,* SIMS VI, A. Benninghoven, A. M. Huber, and H. W. Werner, eds., John Wiley, New York (1988), p. 683.
602. S. J. Doherty and K. L. Busch, 1988 Pittsburgh Conference on Analytical Chemistry and Applied Spectroscopy, March 1988, New Orleans, abstract 1089.
603. R. Levi-Setti, G. Crow, and Y. L. Wang, in *Secondary Ion Mass Spectrometry,* SIMS V, A. Benninghoven, R. J. Colton, D. S. Simons, and H. W. Werner, eds., Springer-Verlag, New York (1986).
604. E. Niehuis, T. Heller, H. Feld, and A. Benninghoven, *J. Vac. Sci. Technol. A* **5,** 1243 (1987).
605. P. Steffans, E. Niehuis, T. Friese, D. Greifendorf, and A. Benninghoven, *J. Vac. Sci. Technol A* **3,** 1322 (1985).
606. J. A. Eccles, J. A. van den Berg, and J. C. Vickerman, *J. Vac. Sci. Technol. A* **4,** 1888 (1986).
607. A. Brown, J. A. van den Berg, and J. C. Vickerman, *Spectrochim. Acta* **40B,** 871 (1985).
608. J. E. Delmore and A. D. Appelhans, *J. Chem. Phys.* **84,** 6238 (1986).
609. A. D. Appelhans, J. E. Delmore, and D. A. Dahl, *Anal. Chem.* **59,** 1685 (1987).
610. R. E. Shomo III, A. G. Marshall, J. E. Delmore, A. D. Appelhans, and D. A. Dahl, *Adv. Mass Spectrom.* **11A,** 476 (1989).
611. R. C. Beavis and B. T. Chait, *Rapid Commun. Mass Spectrom.* **3,** 432 (1989).
612. R. C. Beavis and B. T. Chait, *Rapid Commun. Mass Spectrom.* **3,** 436 (1989).
613. S. Ghaderi, *Adv. Mass Spectrom.* **11B,** 1724 (1989).
614. G. Krier, J. F. Muller, and D. Deparis, *Adv. Mass Spectrom.* **11B,** 1736 (1989).
615. R. J. Cotter, *Biomed. Environ. Mass Spectrom.* **18,** 513 (1989).
616. R. J. Cotter, J. Honovich, N. Qureshi, and K. Takayama, *Biomed. Environ. Mass Spectrom.* **14,** 591 (1987).
617. J. Perel, J. F. Mahoney, T. D. Lee, and K. Legesse, Presented at the 37th ASMS Conference on Mass Spectrometry and Allied Topics, May 21–25, 1989, Miami Beach, Florida, p. 1075.

618. K. P. Wirth, E. Junker, F. W. Rollgen, P. Fonrobert, and M. Przybylski, *J. Chem. Soc. Chem. Commun.* 1387 (1987).
619. D. J. Burinsky, R. L. Dilliplane, G. C. DiDonato, and K. L. Busch, *Org. Mass Spectrom.* **23,** 231 (1988).
620. C. W. Kazakoff, P. H. Bird, and R. T. B. Rye, *Org. Mass Spectrom.* **24,** 703 (1989).
621. J. P. Thomas, P. E. Filpus-Luyckx, M. Fallavier, and E. A. Schweikert, *Phys. Rev. Lett.* **55,** 103 (1985).
622. W. Brandt, A. Ratkowski, and R. H. Ritchie, *Phys. Rev. Lett.* **33,** 1325 (1974).
623. T. Keough, F. S. Ezra, A. F. Russell, and J. D. Pryne, *Org. Mass Spectrom.* **22,** 241 (1987).
624. L. E. Scriven and C. V. Sternling, *Nature* **187,** 186 (1960).
625. M. W. Collins, M.S. thesis, Indiana University, 1989.
626. J. C. Blais, R. B. Cole, and J. C. Tabet, *Adv. Mass Spectrom.* **11A,** 446 (1989).
627. J. C. Blais, R. Galera, G. Bohlbach, and M. Saint-Pierre Chazalet, *Adv. Mass Spectrom.* **11A,** 448 (1989).
628. K. Heckles, R. A. W. Johnstone, and A. H. Wilby, *Tetrahedron Lett.* **28,** 103 (1987).
629. R. A. W. Johnstone and A. H. Wilby, *Int. J. Mass Spectrom. Ion Proc.* **89,** 249 (1989).
630. G. Gillen, J. W. Christiansen, I. S. T. Tsong, B. Kimball, and P. Williams, *Rapid Commun. Mass Spectrom.* **2,** 67 (1988).
631. C. J. McNeal and R. D. Macfarlane, *Adv. Mass Spectrom.* **11A,** 410 (1989).
632. J. Ai and R. D. Macfarlane, Presented at the 37th ASMS Conference on Mass Spectrometry and Allied Topics, May 21–25, Miami Beach, Florida, p. 289.
633. M. Salehpour, P. Hakansson, B. U. R. Sundqvist, and A. G. Craig, *Int. J. Mass Spectrom. Ion Proc.* **77,** 173 (1987).
634. W. W. Cleland, *Biochemistry* **3,** 480 (1964).
635. S. Mohara and M. Tanimoto, *Mass Spectrosc.* **35,** 248 (1987).
636. A. Benninghoven, F. F. Rudenauer, and H. W. Werner, *Secondary Ion Mass Spectrometry. Basic Concepts, Instrumental Aspects, Applications, and Trends,* Wiley-Interscience, New York (1987).
637. Y. Yazdanparast, P. Andrews, D. L. Smith, and J. E. Dixon, *Anal. Biochem.* **153,** 348 (1986).
638. Y. Fujita, T. Matsuo, T. Sakurai, H. Matsuda, and I. Katakuse, *Int. J. Mass Spectrom. Ion Proc.* **63,** 231 (1985).
639. K. L. Busch, K. L. Duffin, and S. M. Brown, *Adv. Mass Spectrom.* **11A,** 456 (1989).
640. J. Sunner, M. G. Ikonomou, and P. Kebarle, *Int. J. Mass Spectrom. Ion Proc.* **82,** 221 (1989).
641. J. Sunner, A. Morales, and P. Kebarle, *Int. J. Mass Spectrom. Ion Phys.* **86,** 169 (1988).
642. M. P. Lacey and T. Keough, *Rapid Commun. Mass Spectrom.* **3,** 46 (1989).
643. L. K. L. Dean and K. L. Busch, *Adv. Mass Spectrom.* **11B,** 1646 (1989).
644. C. P. Leibman, P. J. Todd, and G. Mamantov, *Org. Mass Spectrom.* **23,** 634 (1988).
645. J. E. Bartmess, *Anal. Chem.* **59,** 2012 (1987).
646. L. R. Phillips and J. E. Bartmess, *Biomed. Environ. Mass Spectrom.* **18,** 878 (1989).
647. K. L. Busch, *Thin Layer Chromatography Coupled with Mass Spectrometry* (J. Sherma and B. Fried, eds.), Handbook of Thin Layer Chromatography, Marcel Dekker, New York (1990).
648. K. L. Busch, *J. Planar Chromatog.* **2,** 355 (1989).
649. K. Masuda, K.-I. Harada, M. Suzuki, H. Oka, N. Kawamura, and M. Yamada, *Org. Mass Spectrom.* **24,** 74 (1989).

650. A. J. Kubis, K. V. Somayajula, A. G. Sharkey, and D. M. Hercules, *Anal. Chem.* **61,** 2516 (1989).
651. K. L. Duffin, R. A. Flurer, K. L. Busch, L. A. Sexton, and J. W. Dorsett, *Rev. Scient. Instrum.* **60,** 1071 (1989).
652. M. S. Stanley and K. L. Busch, *Anal. Instrum.* **18,** 243 (1989).
653. S. J. Doherty and K. L. Busch, *Anal. Chim. Acta* **218,** 217 (1989).
654. S. M. Brown and K. L. Busch, *Anal. Chim. Acta* **218,** 231 (1989).
655. K. L. Duffin and K. L. Busch, *J. Planar Chrom.* **1,** 249 (1988).
656. J. C. Dunphy and K. L. Busch, *Biomed. Environ. Mass Spectrom.* **17,** 405 (1988).
657. J. C. Dunphy and K. L. Busch, *Talanta* **37,** 471 (1990).
658. S. J. Doherty and K. L. Busch, *J. Planar Chrom.* **2,** 149 (1989).
659. M. S. Stanley and K. L. Busch, *J. Planar Chrom.* **1,** 138 (1988).
660. H. H. Schurz and K. L. Busch, Presented at the 37th ASMS Conference on Mass Spectrometry and Allied Topics, May 21–25, 1989, Miami Beach, Florida, p. 268.
661. K. L. Busch, G. L. Glish, and S. A. McLuckey, *Mass Spectrometry/Mass Spectrometry: Techniques and Application of Tandem Mass Spectrometry,* VCH Publishers, New York (1988).
662. J. A. Leary, T. D. Williams, and G. Bott, *Rapid Commun. Mass Spectrom.* **3,** 192 (1989).
663. W. Kulick, W. Heerma, and J. K. Terlouw, *Rapid Commun. Mass Spectrom.* **3,** 276 (1989).
664. E. De Pauw, S. Fachetti, G. Pelzer, and K. Nasar, *Adv. Mass Spectrom.* **11B,** 1372 (1989).
665. K. N. Cheng, J. Rosankiewicz, B. M. Tracey, and R. A. Chalmers, *Biomed. Environ. Mass Spectrom.* **18,** 668 (1989).
666. K. L. Busch, R. J. Cotter, and M. M. Ross, *Rapid Commun. Mass Spectrom.* **3,** 165 (1989).

4

Laser Resonant and Nonresonant Photoionization of Sputtered Neutrals

Christopher H. Becker

Glossary of Symbols and Acronyms

A	Electronically excited neutral molecular state
A^+	Electronically excited ionic molecular state resulting in rapid fragmentation
F	Molecular photofragments neutral in charge
FWHM	Full width at half-maximum
g	Ground electronic state of neutral molecule
g^+	Ground electronic state of ionized molecule
IP	Ionization potential
MPI	Multiphoton ionization
MS	Mass spectrometry
NRMPI	Nonresonant multiphoton ionization
PMMA	Poly(methyl methacrylate)
Q	Unimolecular decomposition rate of state A
SALI	Surface analysis by laser ionization
REMPI	Resonantly enhanced multiphoton ionization

Christopher H. Becker • Molecular Physics Laboratory, SRI International, Menlo Park, California 94025.

Ion Spectroscopies for Surface Analysis (Methods of Surface Characterization series, Volume 2), edited by Alvin W. Czanderna and David M. Hercules. Plenum Press, New York, 1991.

SIMS	Secondary ion mass spectrometry
SPI	Single-photon ionization
TOF	Time-of-flight
UHV	Ultrahigh vacuum
UV	Ultraviolet
VUV	Vacuum ultraviolet

1. Introduction

This chapter deals with photoionization of atoms and molecules sputtered from a surface under vacuum. In order to understand the origin and motivation of the work discussed here, first note the initial experiments on sputtering performed in 1910 by J. J. Thomson.[1] In that early work Thomson sputtered a cathode and noted that some of the sputtered atoms were in a charged state while the majority were neutral in charge. After many years of general inactivity in the field, beginning in the late 1950's Honig[2,3] began developing the sputtering of solids as a means of surface and materials chemical characterization using the charged particles driven from the surface, so-called secondary ions. The use of secondary ions as a means of analysis has flourished in the intervening years.[4–6] The secondary ion mass spectrometry (SIMS) technique became popular because of its sensitivity, and also, while not experimentally trivial, it was relatively straightforward to bombard a sample with a (primary) beam of ions and examine those secondary ions directly with a mass spectrometer. As the SIMS technique grew in popularity, Thomson's original observation was amplified over the years, that is, it was commonly observed that the vast majority of sputtered species normally were neutral in charge, often by factors of orders of magnitude. Furthermore, it became appreciated that the small fraction of charged particles was a quantity that has large relative variations (orders of magnitude) depending on the local chemical composition.[7] This is the origin of the so-called matrix effects where the sensitivity to a given chemical moiety (the secondary ion yield) is strongly dependent on the local chemical matrix. Thus, as the general composition varies, the sensitivity varies and the quantitation is jeopardized unless this matrix effect is well characterized. Characterization of such matrix effects is particularly difficult for surfaces, interfaces, and extremely thin films where standard reference materials are not available. These are exactly the situations of ever increasing technological importance.

Thus the motivation was set by the early 1960s to begin examining the neutral sputtered component that is relatively invariant as a function of chemical matrix because it dominates the emission. The question has been how to approach this technologically. Because of the additional step of ionizing the sputtered neutrals, a practical system becomes more difficult, especially when one wants to maintain the very high sensitivity often found with SIMS. In principle, one can imagine various ways of converting a sputtered neutral to an ion for mass spectral analysis. These would prominently include some form of electron impact ionization (electrons generated by a filament or low density plasma) and photoionization (any one of several schemes).

The first experiment to examine the sputtered neutrals mass spectrometrically was performed in the late 1950s again by Honig, using electron bombardment ionization with electron emission from a hot filament.[2] Subsequently, Honig and others began to use a low-pressure plasma as a source of hot electrons to cause ionization.[8–10] While some improvements in ionization efficiency are obtained with the plasma ionization, there is typically an associated loss compared to standard electron bombardment in the transmission of ions into the mass spectrometer. Using state-of-the-art designs it is now possible with the electron bombardment or plasma ionization techniques to obtain overall efficiencies (sometimes called useful yields) of approximately 1 detected ion per 10^9 surface atoms.[11] While the overall efficiency for SIMS varies by about five orders of magnitude depending on the particular system, SIMS typically is much more sensitive compared to a 10^{-9} useful yield, with a range in useful yield commonly of about 10^{-2}–10^{-6}. These "postionization" (ionization taking place above the surface after sputtering) techniques developed using these forms of electron impact did perform a useful scientific role, but they were technologically deficient.

A breakthrough came in 1982 when Winograd and co-workers successfully used a resonant photoionization scheme as a means of detecting sputtered neutral atoms.[12] This photoionization scheme, discussed below, had been developed earlier especially by Hurst and co-workers for very sensitive detection of gaseous atoms.[13,14] Not long after Winograd's success, Becker and Gillen reported[15] on another laser-based approach using a nonresonant photoionization scheme. A single-photon ionization scheme recently has been developed as well, especially for organic surface analysis.[16,17] These approaches, the subject matter of this chapter, form a powerful new approach to surface and materials characterization, an approach that undoubtedly will grow in importance as the techniques are commercialized and also as the capabilities of lasers continually improve.

In the last few years since the initial work was completed, these techniques have already shown the power of laser technology applied to surface analysis. It is appropriate that we discuss briefly the basic photoionization concepts, and then discuss some of the experimental details that must be considered for successful implementation of the photoionization approaches. Examples then will follow and the chapter will conclude with some remarks about future directions.

2. Photoionization

Three types of photoionization schemes will be considered. They are resonantly enhanced multiphoton ionization (REMPI), nonresonant multiphoton ionization (NRMPI), and single-photon ionization (SPI). For most atoms and with standard lasers now used, the ionization potential is greater than the energy per photon. Thus to ionize, a multiphoton process is required. Figure 1 is a simple schematic electronic energy level diagram for an idealized atom showing the four different

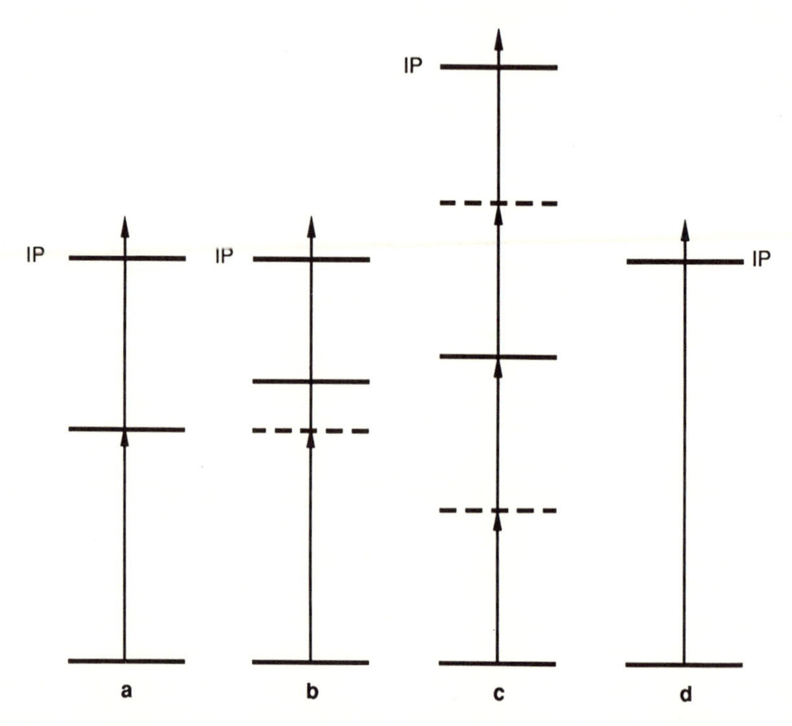

Figure 1. Schematic energy level diagram for (a) two-photon resonantly enhanced MPI through a real intermediate energy level (solid line); (b) two-photon nonresonant MPI through a virtual energy level (dashed line); (c) four-photon resonantly enhanced MPI through one real and two virtual energy levels; and (d) single-photon ionization, where IP is the ionization potential for a particular particle in the ground state.

photoionization schemes of concern for a single frequency (multifrequency irradiation is also possible). A similar figure would apply for molecules. In resonantly enhanced multiphoton ionization[13,14,18] the laser is tuned to excite an electron to a "real" intermediate electronic level, i.e., where the lifetime is $\geq 10^{-9}$ s (Fig. 1a). In a nonresonant process (Fig. 1b), the laser wavelength is not tuned to an intermediate level and the process can proceed only through a virtual state; or another point of view is to say that more than one photon must be near the atom during an optical cycle which is on the order of 10^{-15} s.[19,20] While the cross sections for resonant and nonresonant processes of course do vary with species and wavelengths, to give some feeling for typical pulsed laser power densities required, commonly 10^5 W/cm^2 will saturate a resonant bound–bound transition, 10^7 W/cm^2 efficiently drives a one-photon bound–free atomic transition, and 10^9 W/cm^2 or more will be required for nonresonant multiphoton processes. These differences in required power densities become key differences in experimental geometry in terms of the laser beam diameters, the interaction volume, and consequently the collection efficiency for the photoionization. Finally this can translate into a sensitivity difference for common lasers in use now. Lasers that easily saturate a transition need not be tightly focused, unlike those required to efficiently drive a nonresonant multiphoton process. Lasers used to drive a nonresonant process commonly are focused to cross sectional areas of 10^{-3} cm^2 or less whereas beam diameters of several millimeters sometimes are used for species that are readily ionized in a linear process.

Figure 1c shows that there are also situations where, in a sense, there is a combination of both the resonant and nonresonant process though these are still usually referred to as resonant processes. Figure 1c illustrates the situation often encountered with atoms of high ionization potential. In this case, two photons are required to reach the intermediate level and two photons from the intermediate level are needed to reach the ionization continuum. This is referred to in shorthand notation as a 2 + 2 process. Further variations are possible.[14,18] Much higher intensities are needed to efficiently drive the process of Fig. 1c versus Fig. 1a. Thus lasers must typically be tightly focused for the Fig. 1c process.

Finally, in Fig. 1d the conceptually simple one-photon ionization process is represented. This scheme is used primarily for molecules. The reason this method has not been commonly used until now is that the wavelength required to ionize most chemical species by a single photon is shorter than those from ordinarily available laser systems. Nevertheless it is becoming appreciated that vacuum ultraviolet (VUV) radiation is readily generated by nonlinear techniques in sufficient quantities to make practical the use of such short-wavelength radiation.[21–24] This method historically has not been considered as resonant or nonresonant, though we can perhaps categorize it as the latter because no special wavelength selection is needed and a distribution of electron kinetic energies and molecular internal eneriges results. This one-photon ionization technique is particularly useful for molecular photoionization.[16,25–27]

Cross sections for photoionization are larger for molecules with, say, >10 atoms ($\gtrsim 10^{-16}\,\mathrm{cm}^2$) versus atomic and small molecule cross sections (commonly $\sim 10^{-18}$–$10^{-17}\,\mathrm{cm}^2$).[28] An important point in favor of one-photon ionization for general (nonselective) analysis of large molecules is that as the molecules grow in size, photoionization cross sections in the VUV become relatively uniform even for very different molecular types (e.g., benzene and n-hexane).[28] On the other hand, for experiments where selectivity is desired (see, e.g., Refs. 29, 30, and 31), REMPI is advantageous, though some spectral selectivity can diminish as molecular complexity increases.[32]

In a MPI process, because more than one photon is required, fairly high intensities are associated with efficient ionization. One can readily reach a situation where a molecule continues to absorb photons to bring the total internal energy of the molecule to the point where fragmentation becomes extensive. Whereas for single-photon ionization, by working at intensities low enough to avoid multiple absorption and yet high enough to be a few percent efficient, the amount of energy available to the molecule is less than or equal to the photon energy minus the ionization potential; the exact amount depends on how much (kinetic) energy the ejected electron possesses. By limiting the internal energy, a "softer" ionization, i.e., ionization with low fragmentation will result especially as the molecule grows in size and can more readily hold internal energy without fragmentation.[25,27,33]

To understand in more detail how molecular fragmentation can arise in MPI, consider the following simple but general model for molecular multiphoton ionization in the presence of a single laser field. The model molecular system has four electronics states, the ground state (g), the first excited state (A), the ground state of the molecular ion (g^+), and the excited state of the ion (A^+), which rapidly decomposes into fragments. Three single-photon transitions (first-order processes) connect these states with associated absorption cross sections: the first resonance absorption from g to A with cross section σ_1, the ionization transition from A to g^+ with σ_2, and the fragmenting absorption from g^+ to A^+ with σ_3. Unimolecular dissociation of the excited neutral state, A (when it is a "real" state reached by resonant absorption) gives neutral fragments (F), which might be subsequently ionized, but cannot contribute to the parent ion signal. The rate coefficient for unimolecular decomposition of the excited neutral state A is represented by Q (s^{-1}). In equation form, the processes are (ignoring stimulated and spontaneous emission of radiation):

$$g \xrightarrow[(\sigma_1)]{h\nu} A \tag{1}$$

$$A \xrightarrow[(\sigma_2)]{h\nu} g^+ + e^- \tag{2}$$

$$A \xrightarrow[(Q)]{} F \tag{3}$$

$$g^+ \xrightarrow[(\sigma_3)]{h\nu} A^+ \tag{4}$$

See the inset in the top of Fig. 2.

The photoabsorption cross section between two states is usually well approximated as proportional to the product of the square of the electronic transition moment times the square of the vibrational wave function overlap integral (the Franck–Condon factor, $F_{v,v'}$).[34] The electronic transition moment for an electronic bound-to-bound transition (such as g to A and g^+ to A^+) is typically significantly larger than for a bound-to-free transition due principally to the oscillatory nature of the continuum wave function of the departing electron. This basic physical fact is a key to the ionic fragmentation difficulty for MPI. Thus for Franck–Condon factors of comparable size, the ionizing cross section σ_2 is usually the smallest.

Clearly if one wishes to avoid fragmentation while using an ionizing scheme as described above, a logical approach is to find a small ratio for σ_3/σ_2. While this may be accomplished in some cases, there are two common situations regarding electronic transition moments and Franck–Condon factors that arise making this difficult. First, there usually is at least one electronic state energetically accessible from g^+ when considering ionization of a stable, electronically closed shell molecule. Radical (ionic) molecules tend to have electronic transitions lying lower in energy than for stable molecules; this is because the energy differences between the highest few occupied orbitals are generally less than between the highest occupied and lowest unoccupied orbital. Second, we can generally expect significant changes in equilibrium geometry in the photoionization process. This means that even if the initial ground state molecules (g) are vibrationally cold, potentially resulting in the population of only a few vibrational levels in the intermediate state (A), photoionization should produce a distribution of vibrational levels in g^+,

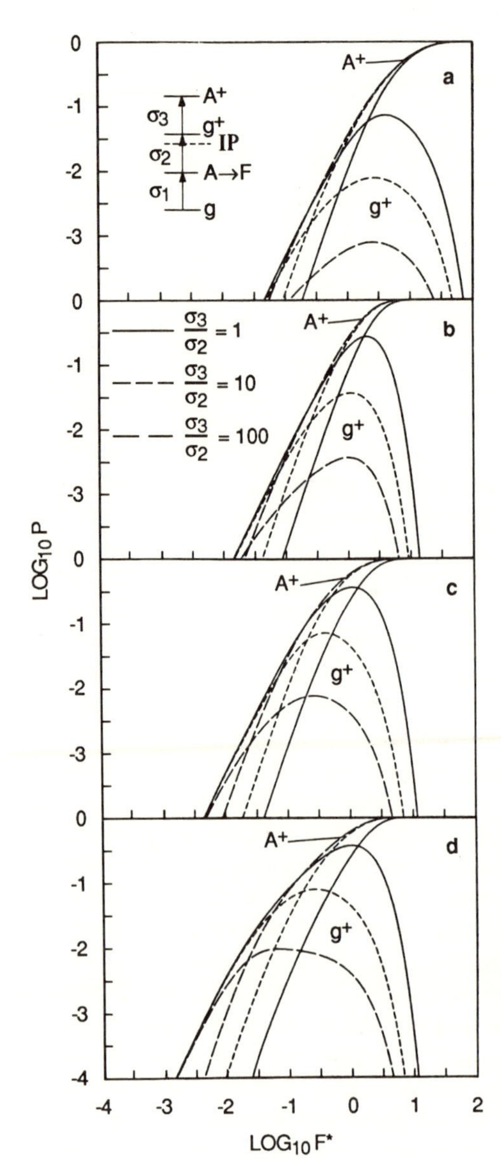

Figure 2. Plots of the probabilities P of producing the ground state of the ion g^+ (curves with distinct maxima) and the fragmenting state of the ion A^+ (curves asymptotically approaching $P = 1$ at high laser fluences). F^* is the normalized ionization laser fluence $\int \sigma_2 I(t)\, dt$, that is, $\log_{10} F^* = 0$ ($F^* = 1$) corresponds to the saturation fluence for the ionization step. Part (a) $\sigma_1/\sigma_2 = 0.1$; (b) $\sigma_2/\sigma_1 = 1$; (c) $\sigma_1/\sigma_2 = 10$; and (d) $\sigma_1/\sigma_2 = 100$.

according to the values of $F_{v,v'}$.[34,35] Thus it is difficult to avoid populating vibrational levels of g^+ that have significant Franck–Condon factors with energetically accessible vibrational levels of the A^+ state.

The simple kinetic model then for this system represented by Eqs. (1)–(4) is given by the following rate equations:

$$\dot{n}_g = -n_g\sigma_1 I \tag{5}$$

$$\dot{n}_A = n_g\sigma_1 I - n_A(\sigma_2 I + Q) \tag{6}$$

$$\dot{n}_F = Qn_A \tag{7}$$

$$\dot{n}_{g^+} = n_A\sigma_2 I - n_{g^+}\sigma_3 I \tag{8}$$

$$\dot{n}_{A^+} = n_{g^+}\sigma_3 I \tag{9}$$

where I is the light intensity in photons/cm^2 s, and n_g, n_A, n_F, n_{g^+}, and n_{A^+} are the concentrations of the molecules in the g, A, F, g^+, and A^+ states, respectively. The solutions to these equations for comparison to experiment are the time-integrated concentrations. We assume all the population begins in state g, normalized so that initially $n_g = 1$, and that the laser is of constant intensity of duration t. The final values of these normalized concentrations correspond to the probabilities of formation. The solution can be written in closed form.[17]

The solutions are plotted in Fig. 2 for various realistic ratios of the three cross sections, under the assumption that there is no neutral fragmentation ($Q = 0$). The plots basically show there is a limit to the amount of ground state ions that can be formed, after which further increases in the laser fluence lead to a decrease in the ground state ion signal due to an increased degree of fragmentation. As the ratio σ_3/σ_2 becomes larger, it is increasingly difficult to retain much of the stable ground state ion as "runaway" absorption takes place. Also the smaller the value of σ_1/σ_2, the more fragmentation dominates stable ion production at fluences required for highest ground state ion yield. As an example, note that for benzene at 248 nm,[36] $\sigma_1/\sigma_2 \cong 0.1$ and substantial fragmentation is observed (from the amount of fragmentation we estimate $\sigma_3/\sigma_2 \sim 2$).

The preceding has ignored unimolecular decomposition of the excited neutral (A) but it is known that this too can play a significant role in the multiphoton dissociation of organic molecules.[37] Neutral fragmentation does not directly change the mass spectrum, it mainly tends to reduce its intensity. For larger values of Qt, this reduction is proportional to $1/(Qt)$. On the other hand, the neutral fragments are also likely to be multiphoton ionized by the strong laser field. As a result, neutral fragmentation followed by ionization can present similar mass spectra as that of ionization followed by fragmentation. Fragmentation can be minimized (with lower ionization probability) by lowering the laser power, only if all fragmentation occurs through the ion ($Q = 0$). If dissociation of the excited neutral state plays an important role, a reduction in laser power will allow neutral dissociation to compete more effectively with ionization of the parent molecule. Neutral fragmentation has been shown to be significantly reduced by shortening the laser pulse length, for example using picosecond pulses.[37] In this case more parent ions will be formed and ionic fragmentation dominates once again. There appear to be no ideal laser conditions that are generally acceptable for studying organic species by MPI without prior knowledge of the excited state spectroscopy. Note, however, that with knowledge of a particular molecule's spectroscopy and the use of two lasers for resonantly enhanced MPI, fragmentation can be largely diminished with efficient ionization, as demonstrated for some volatile aromatic compounds;[38] this is noteworthy for selective applications.

3. Experimental Details

3.1. Lasers

Pulsed dye lasers primarily are used for resonantly enhanced multiphoton ionization. Frequently the dye lasers are pumped by either Nd:YAG lasers or excimer lasers, though there are variations possible. For NRMPI, powerful pulsed lasers are needed. Commonly an excimer or also amplified Nd:YAG laser is used especially with frequenty conversions into the ultraviolet (UV) for the latter. Common pulse widths are on the order of 10 ns; however, short pulsewidths are possible including picoseconds. Picosecond lasers have been shown to reduce molecular fragmentation, especially where there is a unimolecular decay channel of the excited neutral into fragments with rate Q as discussed in the last section. The fast laser pulse can effectively bypass this decay channel.

A frequency conversion scheme is needed for SPI because most species have ionization potentials above common laser photon energies. Because the electron can assume a continuum of energies in the single photon process, generally there is no need to tune this photon. A very convenient scheme used to generate VUV radiation at 10.5 eV is to make the ninth harmonic from a Nd:YAG laser system. This is accomplished by successive tripling first in potassium dihydrogen phosphate crystal, then a static gas cell of xenon phase matched with argon.[16,21–23] It is generally appropriate to disperse the more intense UV beam from the VUV beam; while a prism can do this, it is possible to use the necessary refocusing LiF or MgF_2 lens in an off-axis manner to minimize the number of optical elements that are sources of loss.[16]

One of the very bright aspects of these approaches to surface analysis is the fact that they rely on laser technology, which is continually evolving, providing more and more powerful, versatile, and convenient tools. One such tool that is worth noting in particular for NRMPI is the chirped pulse amplification scheme now becoming popular.[39] In this method a solid state laser is amplified many times after broadening the pulse in time so that the laser power density does not become so high as to damage the amplifying medium. After extensive amplification the pulse is then compressed in time $\sim 10^{-12}$ s. This method is being used to produce extremely high laser power densities for very basic physics experiments, even particle physics experiments, with estimated limits of $>10^{20}$ W/cm^2. Without such extensive amplification these tools still can easily provide power densities on the order of 10^{14} W/cm^2, which generally is enough to saturate every element in the periodic table. Species requiring more than two photons in a nonresonant process typically require at least 10^{12} W/cm^2 or higher power densities to become efficient, whereas two-photon processes typically are efficient at power densities of $\sim 10^{10}$ W/cm^2.

3.2. Ion Beam Systems and Mass Spectrometers

As in all surface analysis systems, UHV is required for system cleanliness. With the photoionization approaches there are two reasons for this. First, common to all surface analyses, is that contaminating adsorbates do not become prevalent during the analysis. Second, because the ionization occurs in the gas phase above the sample, gas phase impurities can also cause a background signal. These backgrounds are usually mass specific, but can contribute to a background spanning many masses if metastable ion decomposition occurs.

Pulsed ion beams are often used for analysis of the surface monolayer in order to maintain so-called static conditions, i.e., erosion of much less than 1 monolayer. When depth profiling materials, dc operation of the ion beam is appropriate and high-quality ion beams are needed, with typical current densities on the order of microamps per 100 μm^2. Furthermore, the ion beam should be rastered during depth profiling (e.g., see Werner and Boudweijn, Vol. 3 in this series) to obtain a flat-bottomed crater as depth profiling proceeds and the laser pulse should be gated so that atoms and molecules originating from near the center of the crater are probed. This will give the best depth resolution.

Because of the pulsed nature of the measurement, when performing a static (pulsed ion beam) or dynamic (dc ion beam) surface analysis, as much beam current as possible in temporal proximity to the laser pulse is of benefit to sensitivity. Therefore, an intense ion beam typically originating from a duoplasmatron ion gun is needed. However, small spot analysis and mapping, which requires very finely focused ion beams, is a rapidly developing technology. Typically duoplasmatrons focus only to about 5–10 μm FWHM diameter spots. To achieve smaller dimensions, liquid metal field emission ion guns are needed. These have lower overall currents but higher current densities than duoplasmatron guns; they provide spatial resolution to at least the 100-nm range.

The pulsed nature of most of these measurements is a natural fit to the time-of-flight (TOF) type of mass spectrometer (MS). In the TOF-MS mass separation occurs by the simple kinetic energy equation $K = mv^2/2$ with all ions having the same electrostatically determined kinetic energy. There are two features of the TOF-MS that make it an ideal choice in a pulsed ionization measurement. First, there is an inherent multiplex advantage that all masses arrive at the detector for each laser pulse. This is valuable from a sensitivity point of view when numerous species must be monitored and also when unknowns may be present. The second key feature is that they have very high efficiencies, where the transmission frequently is in the range of 0.2–0.8.

A popular version of TOF-MS used for photoionization of sputtered neutrals is the reflecting TOF-MS.[40] This mass spectrometer has an energy compensating device (to first and second order), the reflector, in order to achieve high mass resolution. This device works by having higher kinetic energy ions penetrate deeper into the reflecting electrostatic field and so take a longer time to turn around.[40] The faster particles' shorter drift time is exactly matched by tuning the instrument to the longer turnaround time required. The instrument is set so that fast and slow ions bunch at the point of detection, typically a channelplate particle multiplier. Furthermore, the reflector can electrostatically discriminate against (i.e., reject by transmission) any secondary ions thereby eliminating a potential source of background.[15] Alternatively, with the laser off, TOF SIMS measurements can be made after a small adjustment in spectrometer voltages. Another type of energy compensation device for TOF-MS is based on Poschenreider's design.[41] This method is similar in nature, though it has only first-order focusing; its main drawback, however, is that it is much more difficult to construct than the reflecting TOF mass spectrometer. A general schematic for the apparatus with reflecting TOF-MS is shown in Fig. 3.

The need for an energy compensation device in TOF-MS is clear

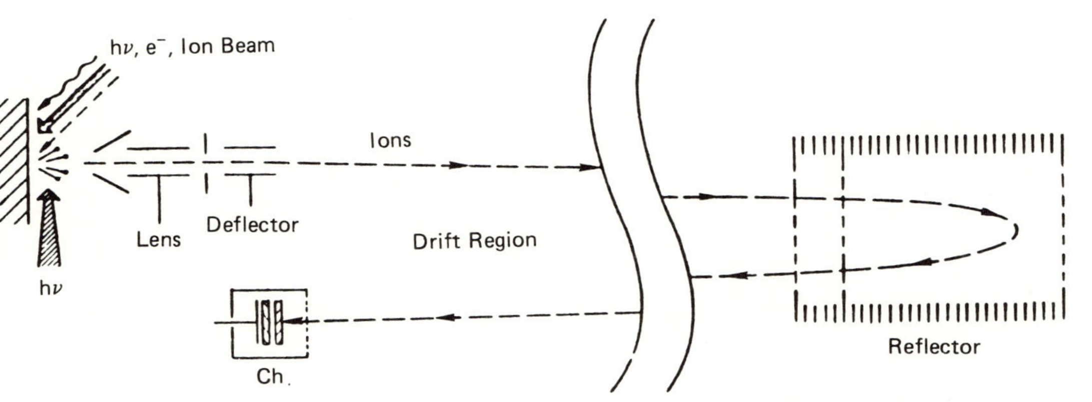

Figure 3. Schematic of the surface analysis by laser ionization (SALI) instrument. Shown are three alternative probe beams (upper left), the ionizing laser beam ($h\nu$, lower left), and the path of the photoions, focused to compensate for transverse velocities, and deflected slightly into the reflecting TOF mass analyzer leading to a microchannel plate detector (Ch).

especially for applications to surface analysis. There are two energy spreads that typically occur and energy compensation is necessary for acceptable mass resolution. The first spread in energy is due to the kinetic energy distribution of the particles themselves due to sputtering. This can range from fairly low energies for heavy molecules to tens of electron volts for sputtered atoms. The other source of energy spread results because photoionization occurs in a region of electric field for extraction into the TOF-MS and because the laser beam has an appreciable width, even if it is focused through the field region. Thus, energy distribution results from the range of electric potential sampled. Unfortunately, extremely high mass resolution is probably not achievable at least as it appears now for situations of sputtering atomic species with photoionization. This is because the virtual source of the particles is different for these two sources of energy spreads. For the kinetic energy spread from sputtering, the virtual source is behind the sample, whereas for the energy spread due to extraction the virtual source is downstream in the MS. Thus only a compromise condition can be obtained. It is known for conditions of single energy spread (SIMS, or photoionization of slow molecules) that TOF mass spectrometers can achieve $m/\Delta m \geq 10{,}000$.[42,43] However, for appreciable energy spreads from both sputtering and photoionization, to date mass resolutions of only about 2000 have been measured with a focused laser.[44] However, this is sufficient for many applications and even separating some isotopes at nominally the same integer mass. The mass resolution with a larger laser beam will typically be lower than with a focused laser beam either because a greater potential difference is experienced across the laser beam if the extraction fields are comparable, or because a greater time spread develops at low initial energies at lower extraction fields. Nevertheless, with compensating energy devices, TOF mass resolutions of approximately 200 can be achieved with laser beams that are even several millimeters in diameter.[45,46]

3.3. Detection Electronics

Focusing now on TOF mass recording from a particle multiplier, the detection may be performed in either an analog or digital form. When large numbers of ions arrive at a given mass, as commonly occurs for high concentrations, then pulse pile-up occurs making digital detection impossible. Analog detection is accomplished with high-speed multichannel analog-to-digital converters, commonly called transient digitizers or fast waveform recorders. The output from the channelplate is passed through a 50 Ω resistor and the voltage swing is recorded in real time, typically with time resolution of at least 10 ns. When performing survey analyses where large numbers of ions occur from each laser pulse, it is not only necessary to use analog detection, but necessary to use either a logarithmic amplifier prior to waveform detection or a variable attenuator in concert with a linear amplifier.

When less than 1 photoion per pulse arrives for a given mass, that is, in a particular time window, then digital detection is possible and is the method of choice from a signal-to-noise consideration. Time-to-digital converters now are common with 1 ns or better time resolution and increasingly they have multistop capabilities, meaning they can detect more than one ion per laser pulse with minimal dead time between arrival times. Signal-to-noise improvements of ~5 have been observed when comparing digital with analog signals.[44]

All of these detection schemes can be implemented with signal averaging over many laser pulses. Either local buffer memory can be used or the data can be transferred to the computer after each pulse for accumulation/signal averaging. To push the repetition rate to its limits, local buffer memories are preferable.

4. Artifacts, Quantitation, Capabilities, and Limitations

4.1. Resonantly Enhanced Multiphoton Ionization (REMPI)

Soon after REMPI was developed for sensitive and selective gaseous atom detection,[13,14] Winograd and co-workers applied the technique to the detection of sputtered surface atoms, initially In atoms.[12] This method was quickly adopted by other groups.[47–49] The use of REMPI for material analysis has also been implemented using laser ablation[50,51] and thermal desorption;[52] work is continuing with these other desorption probes but will not be further discussed here as they generally do not satisfy the lateral and depth resolutions specifically associated with surface analyses.

4.1.1. *Inorganic Analyses*

Because a specific resonance is chosen in REMPI (concentrating now on atoms and small molecules; large organic molecules will be discussed separately), a hallmark of the technique is selectivity. This selectivity requires foreknowledge of the energy levels of the species and a modest number of populated quantum states. The degree of selectivity often determines the background level in a measurement, which in turn can control the detection limit obtained. The degree of selectivity varies depending on the ionization scheme, associated laser power(s) at the laser wavelength(s), and the other species present, in particular, sputtered molecules often have broad absorption features and therefore accidental resonances. In general, a 1 + 1 ionization scheme will provide the greatest selectivity under sensitive (efficient) conditions because very high laser power densities can be avoided. The ionization scheme used for In is an optimal example[(45)]. In that case, the resonance transition is driven with a relatively short wavelength pulse (304 nm), but less than 1 mJ was needed with a laser beam of several millimeters diameter. Greater intensity is generally required to efficiently drive the bound–free transition, and in the In study[(45)] it was done with a lower energy photon at 608 nm with 7 mJ per 10-ns pulse. Avoiding the greater intensity short-wavelength pulse, even in a 1 + 1 scheme, is advantageous in reducing nonresonant transitions, which can lead to isobaric interferences and also mass-nonspecific (broad distributions of arrival times) background associated with metastable decay of photoionized molecules present from the sputtering or in the gas phase. As an example[(53)] of how selectivity can diminish at higher intensities, Fig. 4 shows the effect of increasing the power density for the photoionization of Mo, showing the effects of unidentified resonance lines and the beginnings of a nonresonant background increase.

Isobaric interferences can be greatly suppressed using REMPI, as discussed for the example of $^{28}Si_2$ and ^{56}Fe in Section 5.1. REMPI also

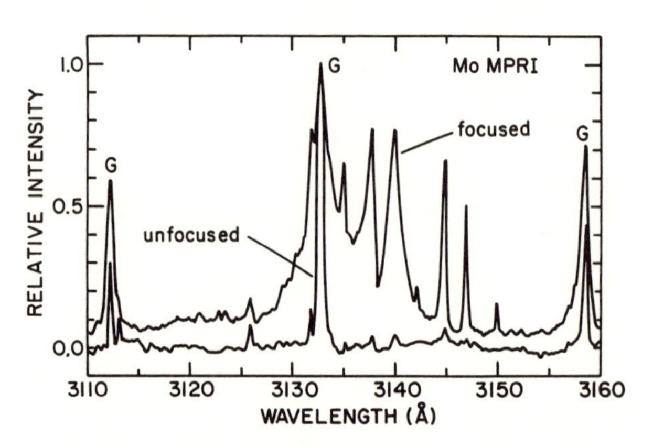

Figure 4. Ion intensity versus excitation wavelength for one-color, REMPI of Mo atoms sputtered from Mo foil by 5 keV Ar^+ (2 μA, 5-μs pulses, 30 Hz): (upper spectrum) laser tightly focused, >1500 mJ/cm^2; (lower spectrum) laser unfocused, 11 mJ/cm^2. Each spectrum is normalized to the 3133-Å peak.

has been used to determine isotope ratios for small quantities of bulk materials with pulsed thermal heating,[52] but at the present time the author knows of no example combining ion bombardment (giving spatial resolution) with REMPI for isotope ratio determinations. Care must be taken to consider subtle mass effects in sputtering, i.e., the effect of differing velocity distributions in detection or preferential sputtering, a subject of study in geochemical research using SIMS.[54]

Consideration of the velocity distributions of particles leads naturally to two other subjects: density versus flux measurements, and factors determining overall efficiencies ("useful yields"). Let us consider first that the measurements (for REMPI and also for other pulsed photoionization approaches!) are of the densities of the particles, for most situations. Generally the particles are well approximated as stationary during the laser pulse, resulting in a density detection. When considering the relative concentrations of different isotopes (or of different species in survey analyses, more relevant for NRMPI) what is really desired is the relative *flux* of material, related to the density by a Jacobian equal to the velocity (density × velocity = flux). It is noteworthy that all of the photoionization approaches can make *in situ* measurements of the velocity distributions (to be discussed shortly). For a first approximation for the conversion from density to flux, however, one can assume similar kinetic energy distributions during sputtering and simply use a mass correction of $m^{-1/2}$; this assumes equal mass spectrometer transmission and detector response for the particles.

The overall efficiency of a surface analysis measurement using a postionization technique can be factored into the product of the mass spectrometer transmission (including detector response) times the overall laser sampling efficiency. For TOF mass spectrometers, the transmission can be very high including the microchannel plate efficiency (~30%); this is aside from the significant multiplex advantage in mass. The overall laser sampling efficiency is the product of the photoionization efficiency for those particles within the laser beam times the probability that the particles fall within the laser beam. This latter probability of intersection is itself dependent on factors such as ion beam temporal pulse width, the geometry of the laser beam and its position relative to the surface, the particles' velocity distributions, and the viewing area of the mass spectrometer's extraction system. This intersection probability is discussed in some detail in Refs. 53 and 55, to which the interested reader is referred. Simply, one can make a partition between the solid angle of the laser beam viewing the sputter area times the fraction of the particles' velocities which will lead to their being within the laser beam at the time of the light pulse for pulsed sputtering. Clearly, the cross-sectional area of the laser beam and its distance to the surface are critical parameters; one wishes to maximize the laser beam cross sectional area and minimize the distance to the surface without laser–surface interactions for maximum sensitivity. In principle it should be possible to reach useful yields of ~0.1 using REMPI for several millimeter diameter laser beams.[53,55]

The distance that the laser beam, traveling parallel to the surface, passes above the surface may be purposely increased when making velocity and/or angular distribution measurements of sputtered particles for the sake of resolution. By varying the time delay between a short desorbing pulsed and the ionizing laser pulse (for any photoionization scheme) with a fixed path length that the neutral atoms or molecules must transmit, a time-of-flight from surface to photoionization is determined yielding a velocity/kinetic energy distribution for the sputtered neutrals. Spatially resolved photoionization can also yield desorption angular distributions. Winograd and co-workers have used a ribbon-shaped laser beam to measure velocity and angular distributions simultaneously from sputtering.[56,57] They have investigated in particular Rh atom sputtering as a function of Rh crystal structure and adsorbate coverage. This very powerful technique can reveal much of the basic physics involved in the sputtering technique, and it has significant analytical implications; the work is discussed in greater detail in this volume by Winograd and Garrison (Chapter 2). While these experiments are integral in the sense that nearly all angles are detected simultaneously with the modest intensity ribbon-shaped laser beam, this author's group, using a focused laser for NRMPI, has begun related measurements in a differential angular mode; first results have been obtained for electron-stimulated desorption of CH_3OH and CO from Ni(110).[58]

Considering limitations of the REMPI method for surface analysis, it is already clear that the specific and selective detection of the technique, while providing many opportunities, also renders it inappropriate for survey analyses, and that quantitative comparison even between two species is difficult unless, again, standard reference materials are available for each species. A somewhat more subtle but important difficulty exists which may sometimes limit applications of the technique to bulk analyses rather than the analyses specifically of surfaces, interfaces, and extremely thin films: matrix effects (i.e., sensitivities that vary with the local chemical composition).

Because with REMPI the laser(s) is tuned to a specific energy level, if the relative population of this electronic state (and rovibronic level as well for molecules) changes with local composition, then detection sensitivities will vary. While matrix effects, principally for ground state atoms, will generally not vary so widely as the case for the secondary ions in SIMS, variations of as much as factors of ~50 have been observed for the yields for sputtered ground state U atoms[59,60] and Cr atoms[61] as a function of matrix composition. For In[53] and Fe[47] much smaller matrix effects were observed. The case of oxidizing an In surface is shown in Fig. 5.[53] This shows the relatively minor matrix effect for the ground state In atom versus the secondary ion yield which changes more than a factor of 200. The loss of ground state atoms is in the form of excited state atoms, ions, and molecules. While essentially all channels can be probed in principle, this matrix effect can be a practical difficulty except for depth profiling dilute systems in the bulk of some other compound. However, it is important to emphasize again that these matrix effects will generally be small compared to SIMS.

4.1.2. Organic Analyses

Some different issues need to be considered for the analyses of complex molecular systems characteristic of organic compounds such as biochemicals and polymers. Even some fairly small molecules, with molecular weights of only ~100–200 amu, have broad absorption features even if the internal energy of the molecule is quite low, as demonstrated with molecular beam spectroscopy of the pyrimidine bases uracil and thymine.[32] While selectivity has been found in supersonic jet cooled organic compounds,[29] the results of Ref. 32 show that spectral selectivity between similar compounds may be difficult to obtain. On the other hand, even with broad absorption features, selectivity between, say, aromatic and nonaromatic systems may be very high. Such selec-

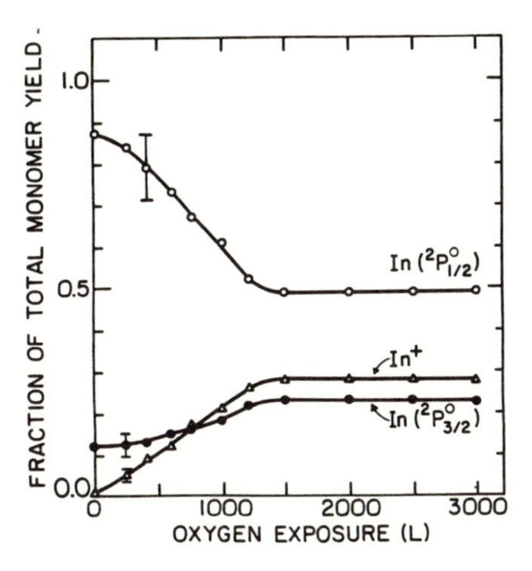

Figure 5. Fraction of the total monomer yield for (○) ground state ($^2P_{1/2}$) In atoms, (●) metastable ($^2P_{3/2}$) In atoms, and (△) In^+ secondary ions versus oxygen exposure. Approximately one monolayer of In_2O_3 is formed after 800-L exposure. The range bars present the relative standard deviation of the data. 1 L = 10^{-6} torr s.

tivity, of course, may be desired, but it is unlikely that the REMPI approach will generally be appropriate as a survey technique for organic compounds.

The discussion in Section 2 about REMPI showed that molecular fragmentation can dominate even at fractional photoionization probabilities of $\sim 10^{-2}$ or less, depending on the cross sections involved. Unless the fragmentation situation is extremely unfavorable, however, limiting the laser power to minimize fragmentation and reduce efficiency may still leave relatively high sensitivity. This is just now being observed with ion-stimulated desorption,[62] but such analyses are more extensive so far for laser-induced desorption, with supersonic jet entrainment[29,30] and without jet entrainment.[31] In particular, there have been cases where some fragmentation persists even at the lowest laser powers used to obtain a signal.[29,31] On the other hand, more and more internal energy resulting from additional photon absorption can be accommodated as the size of the molecule increases without fragmentation over the duration of the measurement, i.e., the transit time in the mass spectrometer.[33]

4.2. Nonresonant Multiphoton Ionization (NRMPI)

Nonresonant multiphoton ionization has been well known to the physics community at least since the 1960s (for reviews see Refs. 19 and 20), but there was no analytical use of the process until its application to surface analysis.[15] The NRMPI technique, like the REMPI technique, can be applied with other desorption probes such as laser ablation[63,64] and thermal desorption.[65] However, we will focus on ion beam desorption, which consistently gives the lateral and depth resolution required for surface analysis, even though in some special cases other desorption probes will be valuable for surface analysis. The use of untuned lasers for nonselective photoionization as tools for surface analysis has been termed "surface analysis by laser ionization" (SALI) by the author; this encompasses the NRMPI work of this subsection as well as that for untuned SPI discussed in Section 4.3.

NRMPI is a tool appropriate for elemental and limited molecular information. Large molecules are typically largely fragmented using NRMPI because the formed photoion will more readily absorb photons than the neutral molecule, and in the intense laser field required to drive the nonlinear process, continued photon absorption inevitably leads to molecular destruction. Some compositional information is obtained, however, from diatomics and small sputtered clusters, though there can be a fundamental question as to whether the small molecule (especially a diatomic) was formed during sputtering from near neighbors or was actually bound on the surface;[66] undoubtedly both pathways exist, though their relative contributions are generally not characterized. Examining larger molecules, triatomics on up, makes formation during the sputtering process itself less likely and has more potential for giving compositional phase information.[67]

During the course of analyses, especially where many species are being monitored, it is not uncommon for some atomic or small molecular species to have an accidental resonance with the untuned laser, which typically has a photon energy of 5–6 eV for lasers of common use today. However, even under such circumstances the goal of the NRMPI approach is for nonselective and uniform ionization. The mass spectrometer alone serves the purpose of discrimination. Another way of stating this goal is to say that we want to remove the photoionization cross section as a variable by driving the ionization to completion (saturation). Thus, one does not need to rely on the knowledge of cross sections (as a function of wavelength), the electronic state distribution of the sputtered neutrals, or even the partition of the neutrals into atoms and various sorts of molecules. This can be accomplished to a large degree by the use of very intense laser radiation ($\sim 10^{10}$ W/cm^2 for two-photon ionization).

While the ionization is not absolutely uniform from species to species, it has been found that for the lasers commonly in use now (see also comments on future directions), for those species with IPs less than the energy of two photons, the elemental detection sensitivities fall within about factors of 3 of each other.[15,63] This means that without standards (including through interfaces and on outer surfaces) that rapid semiquantitative elemental survey analyses are available (sensitivies to be discussed below). For such analyses, the elemental constituents of molecular signals are added to that measured for the atomic photon. With generally unknown degrees of molecular formation during sputtering and photodissociation, as long as all the sputtered components are ionized, then this elemental analysis can be performed. It is clear, however, that for molecular analysis, quantitation must require some knowledge/calibration of photodissociation. Changing distributions of electronic states of the neutrals are not important. Only sputtering into the ion channel represents a loss mechanism.

The reasons for modest deviation from complete ionization uniformity for elemental analysis are principally due to differing effective ionization volumes, and velocity and angular distributions of the sputtered particles. The differing effective ionization volume occurs because there are differences in MPI cross sections for different species and therefore the wings of the focused laser beam are more or less effective for ionization depending on these cross sections. This effect can be minimized by using well-collimated laser beams. The extent to which differences in effective ionization volumes can be eliminated has not yet been established, but very high power solid state lasers may be effective in reducing this discrimination below effects due to velocity and angular distributions. The latter can in principle be measured as indicated in Section 4.1.1, though this is not a desirable situation for routine analytical work. By subtending a significant solid angle (commonly about 10% of the possible 2π steradians) and by working close to the sample, angular distribution effects are probably secondary to velocity effects more often than the reverse.

To improve quantitation of relative abundances of different species beyond the level of direct interpretation of ratios of photoion intensities in the raw data, some use of standards is necessary. The key points here are that the reference materials do not have to be very closely matched to the sample, and that because neutral emission dominates nearly all sputtering situations, then surfaces, interfaces, and extremely thin films can be confidently examined.

The overall efficiency, or useful yield, of the NRMPI technique is of course a strong function of the laser beam size and position. For the common situation of the beam passing about 1 mm above and parallel to the surface with a focused diameter of approximately 0.3 mm and extracting over 2–3 mm length, useful yields of 10^{-3} are obtained. This useful yield is determined by knowing the ion beam intensity and pulse width, and then determining the number of photoions arriving at the detector for the various isotopes/concentrations. With more intense lasers than now commonly used (now, e.g., a standard condition is 10^{10} W/cm^2 for 100 mJ, 10 ns pulses) enlarging the laser beam cross section can allow for increases in the useful yields. Most of the overall loss comes from the fraction of intersected sputtered neutrals, with only a small component of the loss from the TOF-MS. While a value of 10^{-3} is below that in optimal REMPI cases using 1 + 1 ionization schemes, the 10^{-3} value compares very well with other nonselective postionization methods, namely, electron impact ionization and low-pressure plasma ionization of sputtered neutrals; these last two methods have useful yields of approximately 10^{-9} plus enjoy no multiplex advantage in mass.[11] With a useful yield of 10^{-3}, sensitivities of 1 ppm for 1% of a monolayer removal have been obtained.[15] Minimum detection levels of 100 ppb concentrations have been performed in modest accumulation times, such as ~2 min.

One must be aware of sputtering artifacts whenever using ion beam techniques, especially preferential sputtering (discussed in some detail by Werner in Vol. 3 of this series). However, it is very important to note for depth profiling that under conditions of preferential sputtering it is very advantageous to examine the removed component directly rather than to probe the resulting surface itself as done by Auger electron spectroscopy (AES) or x-ray photoelectron spectroscopy (XPS). The pathological behavior of Hg preferential sputtering in oxides of HgCdTe presents a classic example[68,69] and also raises the issue of quantitation in depth profiling with AES or XPS in any multicomponent system. In the HgCdTe work it was shown that with steady-state conditions the sputtered atom signal gave the true composition (corroborated with Rutherford backscattering analysis for these high concentrations),[68] due simply to conservation of mass, whereas the AES measurements are made on an always Hg-depleted and Cd- and Te-enriched surface. Furthermore, for interfaces and extremely thin films, it can be expected that when the ion mixing depth is much larger than the depth needed to reach quasisteady-state sputtering, the SALI depth profiles should closely resemble the true concentration profiles of the elements. The downside of this is that for simply measuring the composition of the outer surface itself, AES and XPS can be used without concern for preferential sputtering (for high concentrations) whereas any ion-beam-based analysis must be concerned for preferential sputtering. Most commonly, preferential sputtering effects are usually thought to be at most factors of 2,[70] the example of Hg being, as said, extreme. Another sputtering difficulty during depth profiling is the overall change in erosion rate as a function of composition; this can lead to a nonconstant relationship between sputter time and depth (a well-known problem); a solution to this, however, lies in integrating the intensity for all mass peaks as a function of depth to estimate the change in overall erosion rate as a function of depth.

While the common laser systems used now do not efficiently drive three- and four-photon NRMPI, detection of some of these high IP atoms can still be sensitively performed. More powerful lasers becoming available will soon change this whole issue (high IP atoms such as O and Ar have been saturated in the authors laboratory with an intense, 1×10^{14} W/cm^2, 1 ps pulse width laser operating in the visible region of the spectrum by NRMPI). For 193-nm (6.42-eV) excimer radiation, the atoms needing three or four photons are H, N, O, F, He, Ne, Ar, and Kr. By using sputtered diatomics, at least H, N, O, and F often can be detected very sensitively. The oxygen atomic ion has been detected readily at 100 ppm concentrations using the $^{16}O^+$ signal directly with 10^{10} W/cm^2 at 193 nm;[44] in contrast, F is more difficult to see atomically. The case of SiF discussed in the next section provides an example of F detection as well as depth profiling through an interface.

4.3. Single-Photon Ionization (SPI)

Single-photon ionization was first applied to mass spectrometry (of volatile compounds) with discharge lamps by Tanaka and co-workers,[25] with important early contributions also from Inghram's group.[26] It was rapidly realized by many in the mass spectrometry community that the characteristics of SPI (generally uniform and relatively soft ionization) were well suited to chemical analysis. SPI can in a sense be considered a "universal detector." A review of the field, sometimes referred to as photoionization mass spectrometry, was made by Reid.[27] For surface analysis, however, the power density of light from a discharge lamp is simply too low for the small quantities of material to be examined (high sensitivities needed).

It was not until a sufficiently intense coherent VUV beam could be easily produced that there was a potential for application of SPI to surface analysis. The scheme was for 118-nm (10.5-eV) light produced by the ninth harmonic of the Nd:YAG laser, which was initially developed in physics laboratories with picosecond lasers in the 1970s.[21,22] With a proper choice of conditions, a 10^{-4} conversion efficiency of 355-nm photons to 118-nm photons was realized for a common 6-ns pulsewidth laser,[16] yielding 10^{12} photons per pulse; this is sufficient to make the technique practical. The limit in detection sensitivities has not yet been firmly established or pushed, but it is probably $\sim 10^{-17}$ mole (about 10^{-6} of a monolayer over $1\ mm^2$).[16]

5. Applications

5.1. Depth Profiling of Bulk Material Using REMPI

At present, the greatest sensitivity demonstrated for (pulsed) sputtering followed by REMPI is a (bulk) detection of 10^{-10} concentration of ^{113}In in Si with an estimated removal of about 1/3 of a monolayer over $0.07\ cm^2$ in the 5-min measurement.[45] This measurement can be extrapolated to a 10^{-11} concentration detection limit. For several Si samples with dilute concentrations of uniformly doped In, a linear relationship between signal and concentration was established, shown in Fig. 6.[45] This correlation forms the basis for quantitative determination of concentration; note, however, that a reference standard is needed to convert relative to absolute measurements, and that all the measurements are of dilute concentrations within the bulk, taken after sputter precleaning, reaching steady state.

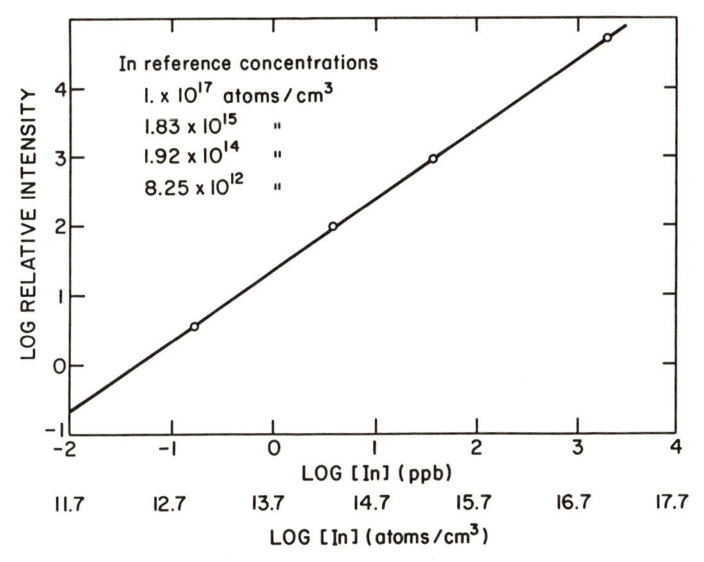

Figure 6. In^+ REMPI intensity versus bulk In concentration.

While detection of In is indeed a favorable case, other heavy metals are generally facile as well, usually ionizable with a 1 + 1 REMPI scheme.[(18)] The depth profile by Ar^+ sputtering of implanted Fe in Si is another notable example, reaching a minimum concentration for ^{56}Fe of about 10^{-8}, with extrapolated limits of 2×10^{-9} for ^{56}Fe (limited by nonresonant MPI of Si_2) and 5×10^{-10} for ^{54}Fe.[(55)] The surmounting of the ^{56}Fe–Si_2 isobaric interference with only modest mass resolution is an excellent example of the power of the REMPI method.

5.2. Multielement Analysis by NRMPI

An example[(63)] of the power of NRMPI for multielement comparisons is shown in Figs. 7 and 8. A SALI TOF mass spectrum for GaAs under steady-state Ar^+ sputtering and 248-nm ionization is shown in Fig. 7. Note that the stoichiometry of the raw data is largely due to saturation of the ionization even though there is a 4-eV difference in ionization potentials of Ga and As. Steady-state sputtering ensures stoichiometric removal. These data were recorded with the channel plate output attenuated by 100 to avoid nonlinear response of the fast amplifier in the detection system.

Following the measurement shown in Fig. 7, experiments were conducted using the 1.06-μm laser beam as the probe beam instead of an

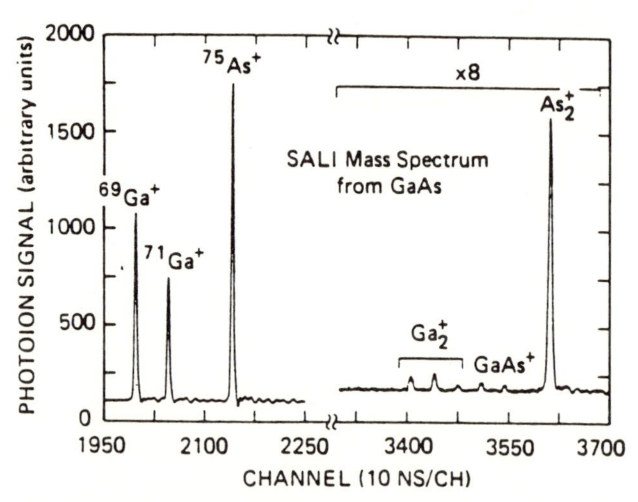

Figure 7. SALI time-of-flight mass spectrum from GaAs using a steady-state Ar^+ beam sputtering and a 248-nm ionizing laser beam.

Ar^+ beam. During this work the surface underwent substantial heating and a plasma was (unintentionally) initiated at the surface, which resulted in surface contamination from sputtering of the ion extractor (stainless steel covered with remnants of previous samples). This contamination allowed, however, a useful many-element comparison of SALI and SIMS. The static SALI spectrum[63] of the contaminated surface (Fig. 8(b) shows by comparison with Fig. 7 that the surface is arsenic rich from the previous laser irradiation. It also shows significant coverages of Cu, Fe, Ni, and other previous sample material from the ion extractor. Figure 8a is a SIMS TOF positive ion spectrum using the same pulsed Ar^+ beam as the primary ion beam (static mode). Note that the SIMS spectrum is dominated by low ionization potential elements not really representative of the true surface composition. In particular, the As is nearly undetectable, which is the usual case for a positive SIMS spectrum of GaAs. Such spectra provide direct information on relative neutral-to-ion yields.

5.3. Depth Profiling of Bulk Material Using NRMPI

Depth profiling is now routinely performed using NRMPI of sputtered species. Despite the pulsed nature of the ionization/measurement, sensitive depth profiles to micrometer depths with a high density of depth points can be routinely obtained in modest times (~10–20 min. with 50–100-Hz lasers) using dc ion beams or switching between dc and pulsed ion beam modes. A "conventional"

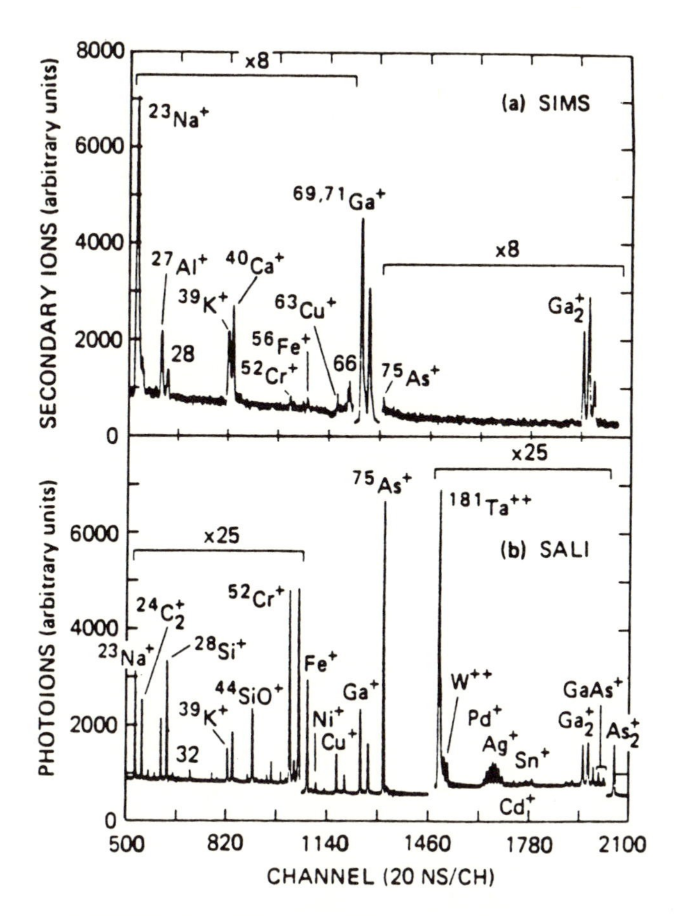

Figure 8. (a) TOF positive ion SIMS pectrum from a contaminated surface of GaAs using Ar^+ static mode bombardment (see the text). (b) SALI spectrum from the same surface with the same Ar^+ probe beam as in (a). The ionizing laser wavelength was 248 nm with a power density of approximately 10^9 W/cm^2.

depth profile of an implant (B in Si) has been demonstrated to the 10^{-6} concentration level,[71] though SIMS can usually handle such cases of dilute concentrations throughout the uniform bulk without difficulty. To repeat, a key application of postionization with NRMPI is depth profiling through interfaces and extremely thin films as exemplified below.

5.4. Interface Analysis with NRMPI

Selective chemical vapor deposition of W using WF_6 is currently being used for interconnections in VLSI circuitry. The effect of fluorine from this reactive process on long-term gate dielectric reliability is not well understood. Fluorine-implanted gate structures were made to better understand F migration, in correlation with electrical measurements.[71] The test devices had three layers: the top layer is a (~200-nm-thick gate) polycrystalline Si, the next layer is a 41-nm-thick gate oxide, and the substrate is single-crystal Si. After the oxide was grown on the Si wafer, polysilicon was deposited at 950°C for 25 min. Then F was implanted into the samples at 93 keV and 10^{15} atoms/cm^2. An example of a sputter depth profile through the polycrystalline Si–SiO_2 interface is shown in Fig. 9. The mobility of F under annealing conditions was the processing issue highlighted in this work.[71] But in Fig. 9, the F depth distributions are shown of unannealed samples where a smooth F distribution is expected through the interface due to similar nuclear stopping powers for Si and SiO_2. A comparison is made of SALI and a SIMS profile, the latter taken on a Cameca instrument. The SIMS data show the common

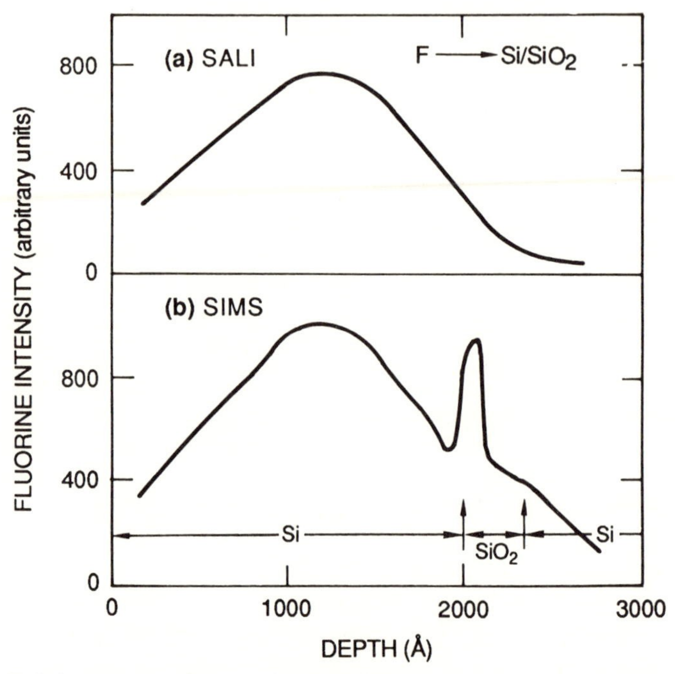

Figure 9. Relative concentrations as a function of depth (depth profiles) of F implanted into Si on SiO_2 measured by (a) SALI with Ar^+ sputtering and 248 nm photoionization, and by (b) positive ion SIMS with O_2^+ sputtering. The arrow marks the Si/SiO_2 interface. A smooth distribution of F through the interface is expected.

influence of matrix effects at an interface, actually being a fairly mild matrix effect here. The neutral sputtered diatomic molecule SiF still provides a sensitive and useful monitor of the F concentration. Because the SiF bond is strong and the F is dilute compared to the Si concentration, the probability of SiF formation is proportional to the F concentration rather than the product of the Si and F concentrations. With confidence in the unannealed results, interpretation of the diffused F profiles[71] after various anneals could be considered trustworthy.

Sometimes diatomic molecular photoions are readily formed, as in the case of SiF^+, but compared to SIMS, the abundance of molecular ions compared to atomic ions is considerably reduced as a general rule. Probably the reason for this is due principally to photodissociation by the intense laser field. Not all molecules will be equally susceptible to photodissociation, of course, and the probability of fragmentation will also have some wavelength dependence. Thus there exists the distinct possibility that for cases of isobaric interferences of an atomic and molecular ion, this interference can be reduced by the process of photodissociation without need of extremely high mass resolution. The case of the P–SiH interference is just such an example, with low yields of SiH^+ relative to P^+ for 193-nm radiation.[44]

Even though the primary ion beam is usually several keV, high depth resolution is still achievable. Typically an ~80-μm-diam. ion beam is rastered over a 1-mm^2 and the laser is triggered to fire only when the ion beam is sampling from the center of the flat-bottom crater. Figure 10 displays the depth profile recorded in 10 min of a sample consisting of a 280-nm-thick GaAs layer on top of 20 layers of InGaAs (5.8 nm)/GaAs (21.5 nm) where the In^+ signal intensity if plotted as a function of depth (sputter time).[72] A depth resolution of 5 nm is measured at the onset of the layered structure at ~280 nm from the surface. After this point the depth resolution degrades slightly, which is probably due to surface topography changes from the extensive sputtering, but each layer is still easily resolvable throughout the entire layered structure between 280 and 820 nm. Of course this depth resolution is also available with any technique that uses a raster/gated well-collimated ion beam for sputtering.

With a high useful yield it is possible to perform combined high lateral and depth resolution measurements. The author has recently installed a Ga liquid metal ion gun in his apparatus and performed both static and dynamic selected area analyses.[44] Chemical imaging has been performed with beam spot sizes of 0.2 and 1.0 μm. With a 0.2 μm spot size, a dynamic range of 100 was observed. Multielement detection with TOF-MS can be particularly valuable for these applications. Sample

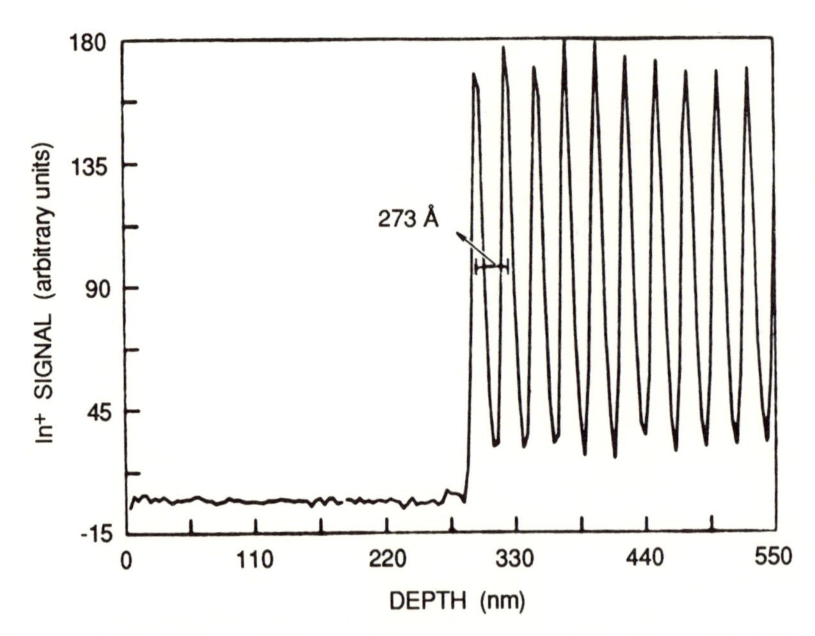

Figure 10. In^+ signal versus depth in a layered structure containing 280-nm-thick GaAs on top of 20 layers of InGaAs (5.8 nm)/GaAs (21.5 nm) taken with SALI using a 7-keV Ar ion beam rastered over 1 mm^2.

positioning before chemical analysis was accomplished with the use of imaging by secondary electron detection.

5.5. Analysis of Organic Compounds Using SPI and Ion Beam Desorption

For analysis of compounds with negligible vapor pressure, ion beam desorption offers an attractive and convenient means of detection. An example[16] is shown in Fig. 11 of the environmental toxin 2,3,7,8-tetrachlorodibenzo-*p*-dioxin (TCDD), recorded at room temperature after depositing the compound on a Si wafer and allowing the solvent (nonane) to dry before introduction into the UHV chamber; small crystallites were observed. The submonolayer (pulsed) Ar^+ beam dose at 7 keV was kept below $3 \times 10^9/cm^2$ per pulse over 0.02 cm^2. While characteristic fragments are seen in the low-mass region, notably m/e = 50 amu (CH_3Cl) and 84 amu (CH_2Cl_2), the high-mass part of the spectrum shows parent molecules (m/e = 320–328) in a clear fashion with the characteristic pattern for the Cl isotopes of 35 and 37. Decomposition by loss of Cl and COCl is seen to be a minor process. The

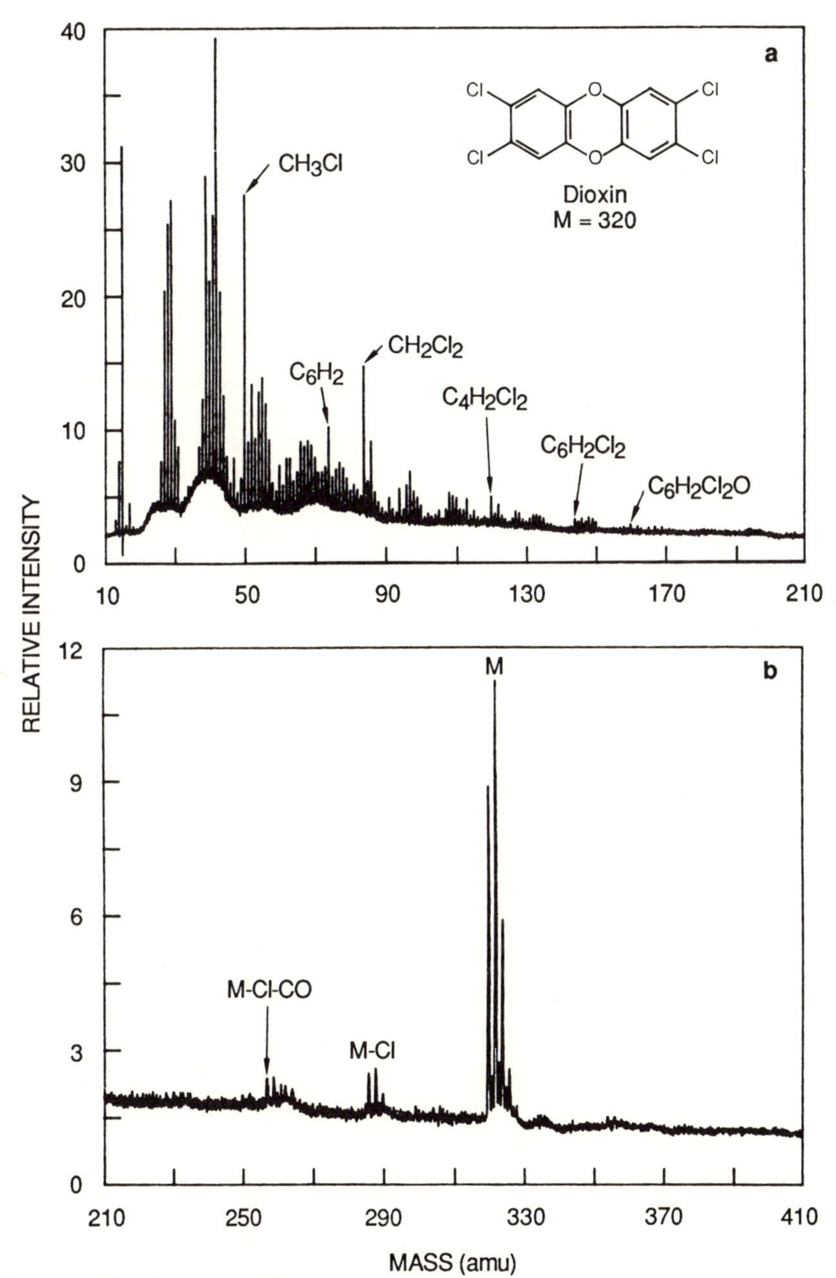

Figure 11. Photoionization TOF mass spectrum of 2,3,7,8-tetra-chloro-dibenzo-*p*-dioxin. The spectrum was obtained from 200 pulses of Ar^+ bombardment for desorption and 118 nm radiation for ionization.

parent molecule peak at 322 amu for one single pulse had S/N of 10. Assuming a desorption yield of 1–10 molecules per incident ion, a single desorption pulse corresponds to 3×10^{-17} to 3×10^{-16} mole; a larger amount of material was on the sample mount for these initial experiments. The sensitivity can be improved with better spatial dispersion of the sample on the substrate and with a higher intensity VUV source. For the current 1-mm-diam focus of the light (passing about 1 mm above the sample), the estimated ionization efficiency is about 1%; while some light intensity increase would be desirable, an efficiency too high would lead to significant multiple absorption and excessive fragmentation. We also note that with multiphoton ionization at 248 nm and 308 nm, no parent molecular ion or even structurally significant fragments were observed.

In Fig. 11 there is significant fragmentation at lower masses. Given the limited amount of energy available from a single 10.5-eV photon, it is very likely that the Ar^+ beam is causing this fragmentation. Similar observations of significant (though not overwhelming) fragmentation have been made with other compounds including biochemicals.[16] That the ion beam is primarily responsible for most of this fragmentation was proven conclusively in the analysis of polymers by comparing different means of desorption.[17] The use of laser- and electron-induced desorption with SPI has not yet been applied for compounds such as biochemicals, but this situation will undoubtedly change probably before this chapter is printed.

Though SIMS has been valuable for biochemical analysis and has received much attention (see Chapter 3), it is reasonable to anticipate that ultimately the most powerful approach to the mass spectrometry of nonvolatile compounds will be through the decoupling of the desorption and ionization, yielding the greatest flexibility and control for desorption as well as the opportunity to observe the majority neutral channel. With a soft ionization, when some fragmentation is desired for structural analysis, then additional more powerful laser light can be added in a controlled fashion. This concept is already being applied with variable intensity laser light for MPI after laser desorption and supersonic jet entrainment,[30] and multiphoton dissociation of trapped secondary ions.[73] Spatial resolution laterally and in depth, as well as ease of application, will ensure that ion beam desorption will play an important role in this field.

5.6. Bulk Polymer Analysis Using SPI and Ion Beam Desorption

SPI is also valuable for the surface analysis of *bulk* polymers[17,74] where many different chemistries may be present, including nonaromatic

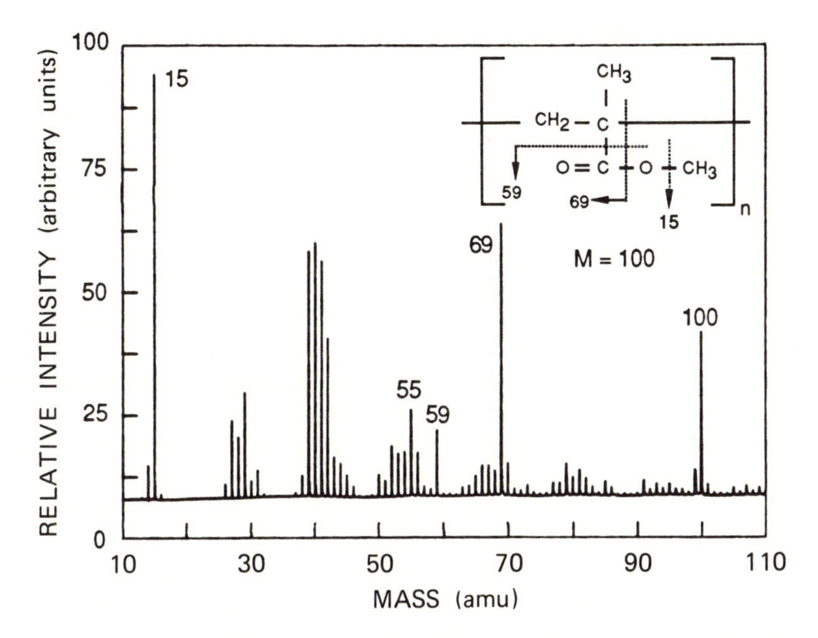

Figure 12. SALI mass spectrum of poly(methyl-methacrylate) (PMMA) obtained with pulsed 7-keV Ar^+ sputtering (submonolayer analysis) with single-photon ionization of sputtered neutral molecules (118 nm at 3×10^5 W/cm^2).

systems not readily accessible with REMPI. An example[17] is shown in Fig. 12 for pulsed Ar^+ desorption (dose $< 10^{13}$ Ar^+/cm^2) of poly(methyl-methacrylate) (PMMA) with 118-nm photoionization at an intensity of 3×10^5 W/cm^2. While there are some minor mass peaks extending beyond mass 110 amu, the major peaks are displayed. Note especially the structurally important peaks at $m/e = 100$ (monomer of PMMA), 69 (monomer minus OCH_3), 59 ($OCOCH_3$), and 15 (CH_3). Comparison with SIMS studies[75] of the positive and negative spectra from PMMA show that while there is some similarity, the SALI spectra can be analyzed more straightforwardly including a strong monomer peak which is absent in the SIMS spectra. Generally, comparisons[17] between the SALI–SPI spectra with SIMS spectra show that the SALI–SPI spectra are simpler and easier to analyze, although for some spectra the SALI and SIMS data are quite similar. Furthermore, for bulk polymer systems the mass spectra (for the neutrals or secondary ions) generally do not extend beyond several hundred amu, unlike the case for dispersed oligomers, where parent mass distributions can be observed.[76,77]

6. Future Directions

The outlook for change in the current practice of laser resonant and nonresonant photoionization of sputtered neutrals can be classified perhaps into two areas: technological improvements and new types of applications.

The key area for technological improvement is in lasers. The laser industry has a history of constantly improving performance in terms of pulse repetition rate, pulse energy, spatial characteristics, and reliability. We can expect continued improvements in these general areas. In particular, untuned powerful Nd:YAG picosecond pulsed lasers may become important for nonresonant MPI at power densities $\geq 10^{13}$ W/cm^2,[39] and for generation of relatively intense ninth harmonic pulses at 118 nm;[22] such lasers allow saturation of ionization of high IP atoms, with possible enlarging of the laser beam focal diameter for greater collection efficiency. As mentioned in subsection 4.2, high IP atoms such as atomic O are now being saturated with very intense, 10^{14} W/cm^2, ps pulsed lasers in a surface analysis setting. Other wavelengths[24] can be valuable as well, and may play a prominent role. There are occasions where wavelengths shorter than the LiF cutoff (approximately 105 nm) would be desirable, requiring differential pumping and curved reflective grating optics for focusing and wavelength separation. There are occasions where longer wavelengths are acceptable or desirable, and larger photon intensities are usually available above 130 nm.

Considering new types of applications, commercial surface analysis instruments based on the technology described in this chapter are only now about to become available. As the use grows from custom machines in a few laboratories to optimized and standardized commercial equipment, many new types of applications will develop, and the technology will mature. In the next few years it is likely that this surface analysis technology will be applied to three-dimensional chemical mapping using a finely focused liquid metal ion gun with beam diameter of ~100 nm. By using a finely focused electron gun, very high lateral resolution imaging (~10–100 nm) may be possible for some organic systems. The use of photoionization of sputtered neutrals also should be incorporated into processing equipment for the chemical and electronic industries, especially for integrated circuit development and production. Equipment for these specialized applications would also probably require a high degree of automation.

7. Summary

The application of laser technology to mass spectrometric surface analysis rests primarily on the physical fact that under nearly all conditions, the secondary neutral emission channel from ion impact dominates the abundance of ejected particles and therefore is more important than the secondary ion channel. With the current sophisticated laser technology and prospects for continued improvement, lasers can contribute substantially to surface analysis through the photoionization of sputtered neutral atoms and molecules.

For elemental analysis, the lowest detection limit demonstrated to date is a concentration of 10^{-10} for a heavy metal using a 1 + 1 REMPI scheme. Though this is an optimum situation, the REMPI technique has many potential applications to address, relying often on selectivity as well as sensitivity. Two limitations must also be noted for the REMPI approach. Only one element is detected at any time, and a specific electronic state(s) is probed and not other populated atomic states and molecules. Together these features mean that REMPI can be more difficult to quantify and is more susceptible to matrix effects compared to NRMPI, though in general matrix effects for REMPI are expected to be mild compared to SIMS. (REMPI can examine small sputtered molecules, but this generally is not practical owing to the large number of populated levels.) While similar sensitivities for NRMPI and REMPI for the higher ionization potential atoms are expected, REMPI generally has lower detection limits for heavy atoms because the laser beam often can have a larger diameter and still achieve efficient photoionization. NRMPI is better suited for general analytical purposes, such as analyzing unknowns and simultaneously detecting many species including sputtered molecules. A particular strength of the NRMPI approach lies in quantitation while depth profiling through interfaces. There are now numerous examples of both thin film depth profiling and bulk analyses for NRMPI and REMPI of inorganic materials.

For organic chemical analysis, SPI has been shown to be a promising tool for nonselective and sensitive detection by avoiding substantial photofragmentation. Bulk polymer surfaces have been analyzed in a static mode (doses $<10^{13}$ Ar^{+}/cm^{2}), as have deposited organic compounds including biochemicals. A good deal of exploration remains in the area of biochemical analysis.

The most rapid change in the technology can be expected by the influence of the availability of commercial equipment and continued improvements in laser technology. Many aspects of the performance of photoionization of sputtered neutral species lies in the laser light characteristics. Improvements in specifications such as repetition rates, pulse energy, spatial characteristics, wavelength availability, ease of use, and reliability will directly translate into improved performances for mass spectrometric surface analysis by resonant and nonresonant photoionization of sputtered neutrals.

Acknowledgments

The author thanks J. B. Pallix, U. Schühle, T. N. Tingle, K. T. Gillen, and D. L. Huestis for their many contributions. The author also thanks N. Winograd for stimulating discussions. Financial support from the National Science Foundation, Gas Research Institute, and Perkin-Elmer Corporation is gratefully acknowledged.

References

1. J. J. Thomson, *Phil. Mag.* **20,** 752 (1910).
2. R. E. Honig, *J. Appl. Phys.* **29,** 549 (1958).
3. R. E. Honig, in *Secondary Ion Mass Spectrometry SIMS V,* (A. Benninghoven, R. J. Colton, D. S. Simons, and H. W. Verner, eds.), Springer-Verlag, New York (1986), p. 2.
4. See, e.g., *Secondary Ion Mass Spectrometry SIMS VI,* (A. Benninghoven, A. M. Huber, and H. W. Werner, eds.), John Wiley and Sons, New York, (1988).
5. A. W. Czanderna, Chap. 1, this volume, Refs. 40–49.
6. N. Winograd and B. Garrison, Chap. 2, and K. Busch, Chap. 3, this volume, references therein.
7. A. Benninghoven, *Surf. Sci.* **53,** 596 (1975).
8. R. E. Honig, in *Advances in Mass Spectrometry* (J. C. Waldron, ed.), Pergamon Press, London (1959), p. 162.
9. J. R. Woodyard and C. B. Cooper, *J. Appl. Phys.* **35,** 1107 (1964).
10. H. Oechsner and W. Gerhard, *Surf. Sci.* **44,** 480 (1974).
11. W. Reuter, in: *Secondary Ion Mass Spectrometry SIMS V* (A. Benninghoven, R. J. Colton, D. S. Simons, and H. W. Werner, eds.), Springer-Verlag, New York (1986), p. 94.
12. N. Winograd, J. P. Baxter, and F. M. Kimock, *Chem. Phys. Lett.* **88,** 581 (1982).
13. G. S. Hurst, M. G. Payne, M. H. Nayfeh, J. P. Pudish, and E. B. Wagner, *Phys. Rev. Lett.* **35,** 82 (1975).
14. G. S. Hurst, M. G. Payne, S. D. Kramer, and J. P. Young, *Rev. Mod. Phys.* **51,** 767 (1979).
15. C. H. Becker and K. T. Gillen, *Anal. Chem.* **56,** 1671 (1984).
16. U. Schühle, J. B. Pallix, and C. H. Becker, *J. Am. Chem. Soc.* **110,** 2323 (1988).
17. J. B. Pallix, U. Schühle, C. H. Becker, and D. L. Huestis, *Anal. Chem.* **61,** 805 (1989).
18. G. S. Hurst and M. G. Payne, *Spectrochim. Acta* **43B,** 715 (1988).

19. P. Lambropoulos, *Adv. At. Mol. Phys.* **12,** 87 (1976).
20. J. Morellec, D. Normand, and G. Petite, *Adv. At. Mol. Phys.* **18,** 97 (1982).
21. A. H. Kung, J. F. Young, and S. E. Harris, *Appl. Phys. Lett.* **22,** 301 (1973); **28,** 239 (erratum).
22. L. J. Zych and J. F. Young, *IEEE J. Quantum Electron.* **QE-14,** 147 (1978).
23. A. H. Kung, *Opt. Lett.* **8,** 24 (1983).
24. R. Hilbig, G. Hilber, A. Lago, B. Wolff, and R. Wallenstein, *Comments At. Mol. Phys.* **18,** 157 (1986).
25. F. P. Lossing and I. Tanaka, *J. Chem. Phys.* **25,** 1031 (1956).
26. H. Hurzeler, M. G. Inghram, and J. D. Morrison, *J. Chem. Phys.* **28,** 76 (1958).
27. N. W. Reid, *Int. J. Mass Spectrom. Ion Phys.* **6,** 1 (1971).
28. J. Berkowitz, *Photoabsorption, Photoionization, and Photoelectron Spectroscopy,* Academic Press, New York (1979), references therein.
29. R. Tembreull and D. M. Lubman, *Anal. Chem.* **59,** 1082 (1987).
30. J. Grotemeyer, U. Boesl, K. Walter, and E. W. Schlag, *Org. Mass Spectrom.* **21,** 645 (1986).
31. F. Engelke, J. H. Hahn, W. Henke, and R. N. Zare, *Anal. Chem.* **59,** 909 (1987).
32. B. B. Brady, L. A. Peteanu, and D. H. Levy, *Chem. Phys. Lett.* **147,** 538 (1988).
33. P. J. Robinson and K. A. Holbrook, *Unimolecular Reactions,* Wiley-Interscience, New York (1972).
34. D. W. Turner, C. Baker, A. D. Baker, and C. R. Brundle, *Molecular Photoelectron Spectroscopy,* Wiley, New York (1970).
35. S. R. Long, J. T. Meek, and J. P. Reilly, *J. Chem. Phys.* **79,** 3206 (1983).
36. W. K. Bischel, L. E. Jusinski, M. N. Spencer, and D. J. Eckstrom, *J. Opt. Soc. Am.* **B2,** 877 (1985).
37. S. D. Colson, *Nucl. Instrum. Methods Phys. Res.* **B27,** 130 (1987), references therein.
38. J. W. Hager and S. C. Wallace, *Anal. Chem.* **60,** 5 (1988).
39. P. Maine and G. Mourou, *Opt. Lett.* **13,** 467 (1988).
40. B. A. Mamyrin, V. I. Karataev, D. V. Shmikk, and V. A. Zagulin, *Sov. Phys.—JETP* (Engl. Transl.) **37,** 45 (1973).
41. W. P. Poschenreider, *Int. J. Mass Spectrom. Ion Phys.* **9,** 357 (1972).
42. E. Niehuis, T. Heller, H. Feld, and A. Benninghoven, in: *Secondary Ion Mass Spectrometry SIMS V* (A. Benninghoven, R. J. Colton, D. S. Simons, and H. W. Werner, eds.), Springer-Verlag, New York (1986), p. 188.
43. K. Walter, U. Boesl, and E. W. Schlag, *Int. J. Mass Spectrom. Ion Proc.* **71,** 309 (1986).
44. J. B. Pallix and C. H. Becker, unpublished results, 1988.
45. D. L. Pappas, D. M. Hrubowchak, M. H. Ervin, and N. Winograd, *Science* **243,** 64 (1989).
46. C. E. Young, M. J. Pellin, W. F. Calaway, B. Jorgensen, E. L. Schweitzer, and D. M. Gruen, in: *Resonance Ionization Spectroscopy 1986,* (G. S. Hurst and C. G. Morgan, eds.), Institute of Physics, Bristol, UK (1987), p. 163.
47. M. J. Pellin, C. E. Young, W. F. Callaway, and D. M. Gruen, *Surf. Sci.* **144,** 619 (1984).
48. D. L. Donohue, W. H. Christie, D. E. Goeringer, and H. S. McKown, *Anal. Chem.* **57,** 1193 (1985).
49. J. E. Parks, H. W. Schmitt, G. S. Hurst, and W. M. Fairbank, Jr., *Thin Solid Films* **108,** 69 (1983).
50. S. Mayo, T. B. Lucatorto, and G. G. Luther, *Anal. Chem.* **54,** 553 (1982).

51. R. C. Estler and N. S. Nogar, *Appl. Phys. Lett.* **52,** 2205 (1988).
52. J. D. Fassett, J. C. Travis, L. J. Moore, and F. E. Lytle, *Anal. Chem.* **55,** 765 (1983).
53. F. M. Kimock, J. P. Baxter, D. L. Pappas, P. H. Kobrin, and N. Winograd, *Anal. Chem.* **56,** 2782 (1984).
54. H. Gnaser and I. D. Hutcheon, *Surf. Sci.* **195,** 499 (1988).
55. C. E. Young, M. J. Pellin, W. F. Callaway, B. Jorgensen, E. L. Schweitzer, and D. M. Gruen, *Nucl. Instrum. Methods Phys. Res. B* **27,** 119 (1987).
56. N. Winograd, P. H. Kobrin, G. A. Schick, J. Singh, J. P. Baxter, and B. J. Garrison, *Surf. Sci.* **176,** L817 (1986).
57. C. T. Reimann, M. El-Maazawi, K. Walzl, B. J. Garrison, N. Winograd, and D. M. Deaven, *J. Chem. Phys.* **90,** 2027 (1989).
58. J. J. Vajo, J. H. Campbell, and C. H. Becker, to be published; see also J. J. Vajo, J. H. Campbell, and C. H. Becker, *J. Vac. Sci. Technol. A* **7,** 1949 (1989).
59. D. E. Goeringer, W. H. Christie, and R. E. Valiga, *Anal. Chem.* **60,** 345 (1988).
60. J. M. R. Hutchinson, K. G. W. Inn, J. E. Parks, D. W. Beekman, M. T. Spaar, and W. F. Fairbank, Jr., *Nucl. Instrum. Methods Phys. Res. B* **26,** 578 (1987).
61. W. Husinsky, P. Wurz, B. Strehl, and G. Betz, *Nucl. Instrum. Methods Phys. Res. B* **18,** 452 (1987).
62. D. M. Hrubowchak and N. Winograd, private communication, 1988.
63. C. H. Becker and K. T. Gillen, *J. Vac. Sci. Technol. A* **3,** 1347 (1985).
64. B. Schueler and R. W. Odom, *J. Appl. Phys.* **61,** 4652 (1987).
65. C. H. Becker and K. T. Gillen, *Appl. Phys. Lett.* **45,** 1063 (1984).
66. N. Winograd, D. E. Harrison, Jr., and B. J. Garrison, *Surf. Sci.* **78,** 467 (1978).
67. J. B. Pallix, C. H. Becker, N. Missert, K. Char, and R. H. Hammond, in: *Thin Film Processing and Characterization of High-Temperature Superconductors* (J. M. E. Harper, R. J. Colton, and L. C. Feldman, eds.), American Institute of Physics Conf. Proc. No. 165, New York (1988), p. 413.
68. C. M. Stahle, D. J. Thomson, C. R. Helms, C. H. Becker, and A. Simmons, *Appl. Phys. Let.* **47,** 521 (1985).
69. C. M. Stahle, C. R. Helms, H. F. Schaake, R. L. Strong, A. Simmons, J. B. Pallix, and C. H. Becker, *J. Vac. Sci. Technol. A* **7,** 474 (1989); **8,** 3373 (erratum).
70. P. Williams, in *Applied Atomic Collision Physics,* Vol. 4 (S. Datz, ed.) Academic, Orlando, Florida (1983), p. 327.
71. J. B. Pallix, C. H. Becker, and K. T. Gillen, *Appl. Surf. Sci.* **32,** 1 (1988).
72. J. B. Pallix and C. H. Becker, in *Advanced Characterization Techniques for Ceramics,* (W. S. Young, G. L. McVay, and G. E. Pike, eds.), American Ceramic Society, Westerville, Ohio (1989), p. 62.
73. D. F. Hunt, J. Shabanowitz, and J. R. Yates, III, *J. Chem. Soc., Chem. Commun.* 548 (1987).
74. U. Schühle, J. B. Pallix, and C. H. Becker, *J. Vac. Sci. Technol. A* **6,** 936 (1988).
75. A. Brown and J. C. Vickerman, *Surf. Interface Anal.* **8,** 75 (1986).
76. I. V. Bletsos, D. M. Hercules, D. van Leyen, and A. Benninghoven, *Macromolecules* **20,** 407 (1987).
77. L. D. Nuwaysir and C. L. Wilkins, *Anal. Chem.* **60,** 279 (1988).

5

Rutherford Backscattering and Nuclear Reaction Analysis

L. C. Feldman

1. Introduction

This chapter deals with the use of high-energy ion beams in the characterization of solids. In this context "high-energy" refers to positive ions with incident kinetic energies of greater than 25 keV and extends to the region of many MeV. Historically this regime is associated with nuclear physics and the basic processes are of a nuclear type. For surface analysis one picks out the subset of well-understood nuclear processes to apply to materials problems. The primary advantage of these ion scattering techniques is quantitative analysis, which, in turn, results from the successful understanding of these processes by the nuclear physics community. A second main advantage is in-depth profiling in a nondestructive manner.

The fundamental relations that govern the particle–solid interaction and their use in surface analysis will be emphasized throughout the chapter. The second section describes the physical principles underlying Rutherford backscattering and nuclear reaction analysis. Section 3 considers the apparatus used, while Section 4 describes in detail the sensitivities and limitations of the techniques set by the physics and apparatus. In Section 5 there is a discussion of ion scattering as a

L. C. Feldman • AT & T Bell Laboratories, Murray Hill, New Jersey 07974.

Ion Spectroscopies for Surface Analysis (Methods of Surface Characterization series, Volume 2), edited by Alvin W. Czanderna and David M. Hercules. Plenum Press, New York, 1991.

structural tool. This subject is an entire field unto itself; while it is not the main point of this chapter, these structural considerations do play an important role in elemental analysis. The concluding section compares the ion scattering technique to other surface probes. Nuclear physicists were using ion scattering for the analysis of the targets as early as 1951.[1] However, it is in the last 20 years or so that these techniques have been carefully honed to a high level of sophistication as materials technology requirements became ever more stringent. A number of fine research treatises have been written on these subjects.[2,3] The recent books by Chu, Mayer, and Nicolet[4] and Feldman and Mayer[5] approach the subject in a more fundamental way.

2. Principles of the Methods

2.1. Rutherford Backscattering

2.1.1. Impact Parameters

The basis of MeV ion scattering for materials analysis rests on the fundamental relations that govern ion scattering in solids. Atomic and nuclear collisions can be characterized by the impact parameter of the scattering process (Fig. 1). The largest impact parameters possible in a solid are the order of 1 Å. For such a distant collision, the fast ions

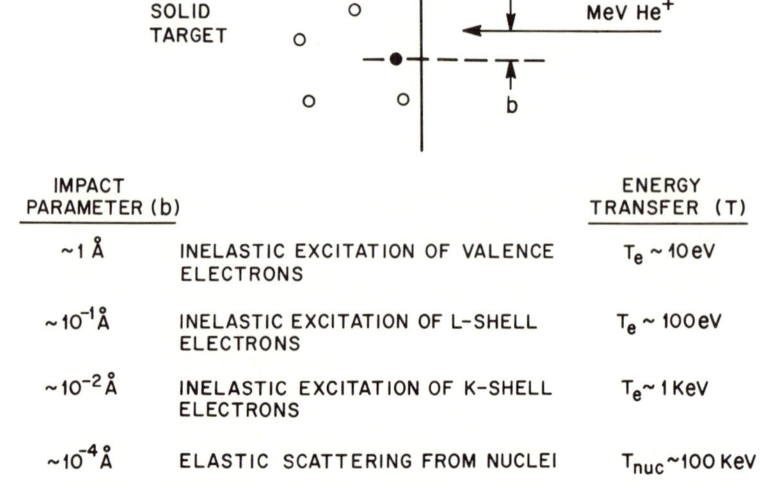

Figure 1. Schematic of an energetic ion approaching an atom within a solid target at impact parameter b. The impact parameter defines the type of atomic interactions as classified in the figure. The energy transfer T is either to an electron (T_e) or another nucleus (T_{nuc}).

interact primarily through excitation of valence electrons, with energy transfers of the order of 10 eV/collision. These processes have cross sections of the order of atomic dimensions, $\sim 10^{-16}\,\text{cm}^2$. At smaller impact parameters, the particles make "harder" collisions, inner shell excitation for example, with larger energy transfer and a smaller cross section. At the smallest impact parameter, the order of nuclear dimensions $10^{-12}\,\text{cm}$, the incident ion scatters from the atomic nucleus in a billiard ball type of collision with a large energy transfer, $\sim 100\,\text{keV}$. The cross sections for these processes are $\sim 10^{-24}\,\text{cm}^2$. The picture of an energetic ion traversing a solid consists of an ion losing energy gradually to electrons as it passes each monolayer, coming to rest tens of microns below the surface. A small fraction of the particles will undergo energetic nuclear encounters before coming to rest. It is these nuclear backscattering events that represent the signal in ion scattering studies. The final energy of the particle determines the depth at which the scattering event occurred, since the particle will lose energy to electrons in penetrating the solid. Electron encounters are generally called inelastic events while the nuclear backscattering event is usually denoted an elastic event.

Close collsions can initiate a number of atomic or nuclear processes which are useful in analysis. Small impact parameter processes may give rise to inner shell electron excitation with the subsequent emission of a characteristic x-ray; for ions this analysis process is known as PIXE, particle induced x-ray emission. If the impact parameter is of the order of the nuclear size, a nuclear reaction may occur; this process is part of ion beam analysis and denoted by the abbreviation NRA, nuclear reaction analysis. The most probable process at small impact parameters is elastic scattering; in analysis this is primarily known as RBS, Rutherford backscattering spectrometry. This latter process is by far the most used for materials characterization, and this chapter will concentrate primarily on RBS. In surface characterization NRA is used for light atom detection (hydrogen, deuterium, carbon, oxygen) and is important for surface and thin film analysis involving these elements. PIXE has had relatively little use in ion beam/surface work.

The use of ion backscattering and nuclear reactions as a quantitative material analysis tool depends on an accurate knowledge of the nuclear and atomic scattering processes. In the rest of this section, we shall consider in detail the relations that describe these scattering mechanisms.

2.1.2. Kinematics

Consider the geometry in Fig. 2, which illustrates a projectile of mass M_1, atomic number Z_1, and energy E_0 scattering from a surface atom of a solid composed of atoms of mass M_2 and atomic number Z_2

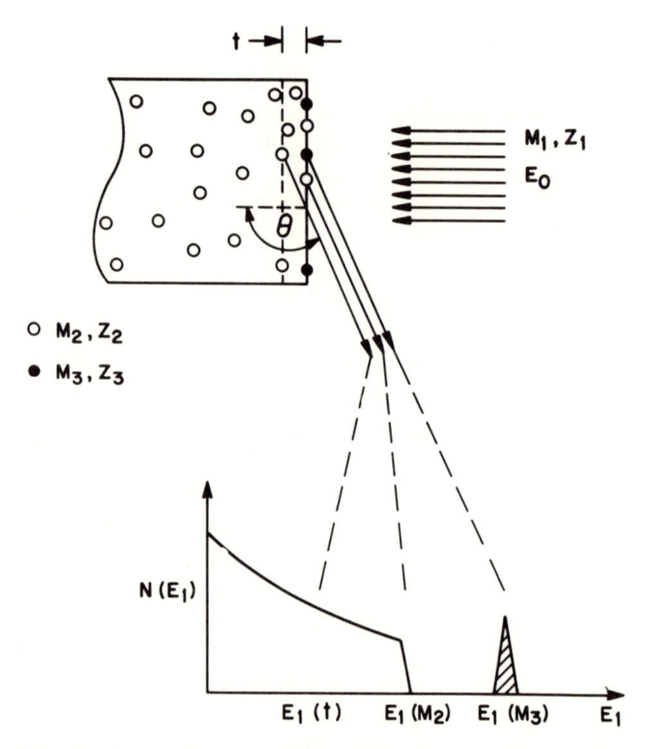

Figure 2. Geometry for particles of incident energy E_0, mass M_1, and atomic number Z_1 scattering from substrate atoms of a solid of mass M_2, atomic number Z_2, and a surface impurity of mass M_3 ($M_3 > M_2$) and atomic number Z_3. Shown below is the energy spectrum E_1 of scattered particles corresponding to scattering from the surface atoms M_2 and M_3 or from various depths (t) of the substrate atoms.

and surface impurity, M_3, Z_3. The scattering angle is θ. The energy E_1 of particles backscattering from a surface atom M_2 is related to the incident energy through the kinematic factor K, where

$$K_{M_2}(\theta) = E_1/E_0 \tag{1}$$

and

$$K_{M_2}(\theta) = \left[\frac{(M_2^2 - M_1^2 \sin^2\theta)^{1/2} + M_1\cos\theta}{M_2 + M_1}\right]^2 \tag{2}$$

This formula acquires a particularly simple form at $\theta = 90°$ and 180°:

$$K_{M_2}(90°) = (M_2 - M_1)/(M_2 + M_1)$$

and

$$K_{M_2}(180°) = (M_2 - M_1)^2/(M_2 + M_1)^2$$

The mass of the scattering atom determines the backscattered energy, $E_1 = KE_0$.

2.1.3. Cross Sections

One of the main advantages of ion scattering in any application is the ability for quantitative analysis. This is a result of the scattering being determined by a process with a known cross section. In the large majority of cases the differential scattering cross section, $d\sigma/d\Omega$, is given by the familiar Rutherford formula:

$$(d\sigma/d\Omega)_c = \left(\frac{Z_1 Z_2 e^2}{4E_c \sin^2 \theta_c/2}\right)^2 \tag{3}$$

where the subscript c refers to the center of mass system. In the lab system, we have

$$\frac{d\sigma}{d\Omega} = \left(\frac{Z_1 Z_2 e^2}{2E_0 \sin^2 \theta}\right)^2 \frac{\{\{1 - [(M_1/M_2)\sin\theta]^2\}^{1/2} + \cos\theta\}^2}{\{1 - [(M_1/M_2)\sin\theta]^2\}^{1/2}}$$

The total number of particles of impurity mass $M_3(>M_2)$, atomic number Z_3, and surface density N_3 (atoms/cm^2) may be extracted from the measured yield, Y_3, by

$$Y_3 = N_3 \frac{d\sigma}{d\Omega} \Delta\Omega Q \tag{4}$$

where Q is the measured number of incident projectiles and $\Delta\Omega$ is the solid angle of the detector. Y_3 would be the shaded area in Fig. 2. Measurement of N_3 by this method is straightforward and is usually accurate to better than 10%. In the best cases impurity layers of 10^{12} atoms/cm^2 can be determined.

Deviations from pure Rutherford scattering can be found at low energies due to atomic screening effects and at high energies due to nuclear scattering, i.e., when the distance of closest approach is about equal to the nuclear radius. These deviations are discussed more completely in Refs. 4 and 5.

2.1.4. Scattering from the Bulk: Stopping Power

Energetic light ions can penetrate microns into a solid losing energy primarily through electronic interactions. Understanding the rate of energy loss, dE/dx, is an important component in the design of backscattering experiments. The rate of energy loss dE/dx, or stopping power as it is sometimes called, determines the depth sensitivity and depth resolution in a given experiment.

Understanding the rate of the energy loss of ions traversing matter is one of the classic problems in modern day physics. During the development of atomic and nuclear physics, the evaluation of energy loss rates has been addressed by Bohr, Bethe, Fermi, Bloch, and other prominent physicists. The calculation of this quantity is usually formulated in terms of the impulse approximation, the calculation of energy transfer to an electron by a fast ion. Complexities arise in properly treating the different bound electrons in a solid and the "resonant" energy transfers to the valence electrons of the solid. A concise formula for the stopping power is given by the Bethe–Bloch formula:[6]

$$\frac{dE}{dx} = \frac{4\pi Z_1^2 e^4 N Z_2}{mV^2} \ln\left(\frac{2mV^2}{Z_2 I_0}\right) \tag{5}$$

where m is the mass of the electron, NZ_2 the number of electrons/volume in the material, V the velocity of the ion, and $I_0 \cong 10$ eV.

In practice formulas of the form of Eq. (5) are used as a guide in semiempirical fits to stopping power data to produce the most accurate stopping powers. Extensive tables of dE/dx are now available. Figure 3 shows a graphical compilation from the tables of Ziegler[7] illustrating the stopping power of He^+ in Si over a broad energy range from 1 to 10^5 keV. Note the broad maximum in the stopping power at ~1.0 MeV. Since depth resolution is inversely proportional to the stopping power this energy region is the optimum for ion scattering analyses. Furthermore the scattering cross section is almost pure Rutherford at ~1.0 MeV for He scattering from all elements in the periodic table (except hydrogen isotopes). This combination of stopping power optimization and Rutherford scattering have made the 0.5–2.5 MeV range the dominant one for RBS analysis with He.

In surface and thin film analyses it may be sufficient to assume a constant stopping power (independent of energy). This assumption allows a simple illustration of many of the features of ion scattering; more complete analyses are not conceptually difficult but may require numerical techniques. In the constant dE/dx approximation the final energy, $E_1(t)$, of a particle at normal incidence that backscatters from depth t is given by

$$E_1(t) = K_{M_2}(\theta)[E_0 - t(dE/dx)_{E_0}] - (t/|\cos\theta|)(dE/dx)_{E_1}$$

where the subscripts on dE/dx refer to the energy at which dE/dx is evaluated. This equation is simply an extension of Eq. (1) and uses the

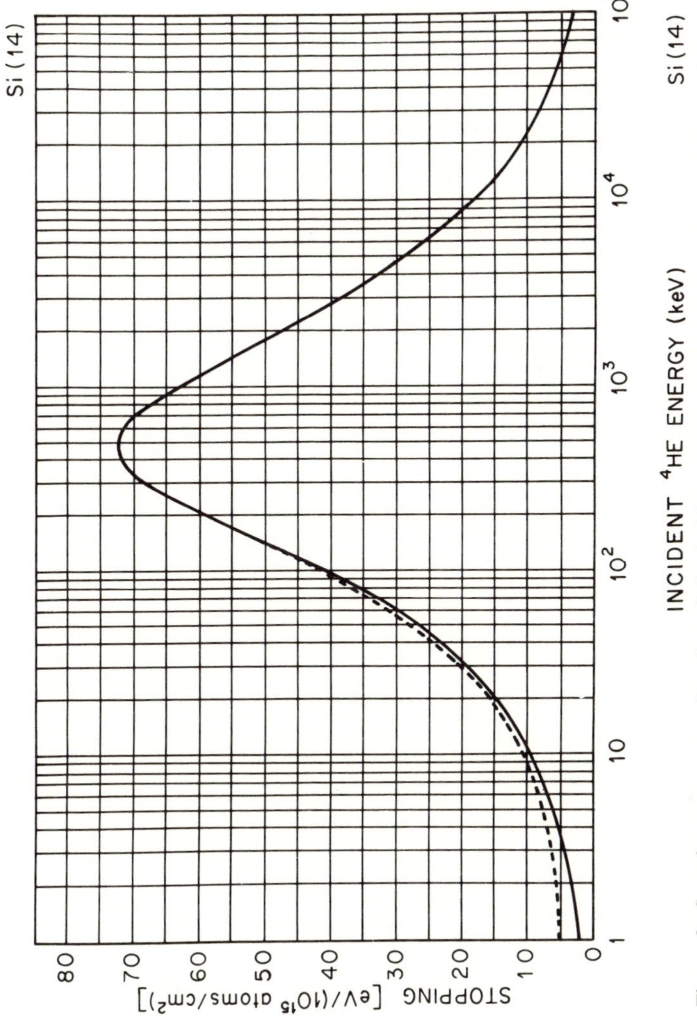

Figure 3. Stopping power as a function of the incident energy for He ions in Si (from Ref. 7). The solid curve is for electronic stopping and the dashed curve is for electronic plus nuclear stopping.

fact that at depth t the incident particle energy, E_t is

$$E_t = E_0 - t\left(\frac{dE}{dx}\right)_{E_0}$$

due to energy loss in the entrance path. The second term in the equation for $E_1(t)$ arises from energy loss on the outward path, $(t/|\cos\theta|)$, at energy E_1. A variation in thickness δt corresponds to a variation in energy δE_1 through

$$\delta E_1 = \delta t\left(K_{M_2}\left(\frac{dE}{dx}\right)_{E_0} + \left(\frac{dE}{dx}\right)_{E_1}\Big/|\cos\theta|\right)$$

so that the depth resolution in an experiment is determined by the energy resolution of the detector δE, dE/dx and scattering angle, θ. Typical depth resolution is on the order of 30–100 Å.

2.1.5. The Energy Spectrum

Projectiles can scatter from any depth t, resulting in a continuous energy spectrum. The yield from a slice of width Δt at depth t is given by (for $\theta = 180°$)

$$Y(t)\Delta t = \left(\frac{Z_1 Z_2 e^2}{4E_t}\right)^2 NQ\Delta\Omega\Delta t$$

where N is the volume density of atoms/cm^3 in the solid and $\Delta\Omega$ is the solid angle subtended by a detector of 100% efficiency. To convert this to a spectrum, $N(E_1)dE_1$, of the measured energy E_1 we note that (for $\theta = 180°$)

$$dt = dE_1\Big/\left[K\left(\frac{dE}{dx}\right)_{E_0} + \left(\frac{dE}{dx}\right)_{E_1}\right]$$

and

$$E_t = (E_1 + AE_0)/(K + A)$$

where A is the ratio $(dE/dx)_{E_0}/(dE/dx)_{E_1}$. If $K \cong 1$, then $A \simeq 1$ and

$$N(E_1) \propto \frac{1}{(E_0 + E_1)^2}$$

This yield spectrum of helium with $E_0 = 1.4$ MeV incident onto gold is indicated in Fig. 4.

As a final illustration Fig. 5 shows a schematic of the ion scattering spectrum of a thin film on a substrate of lower mass material. The energy

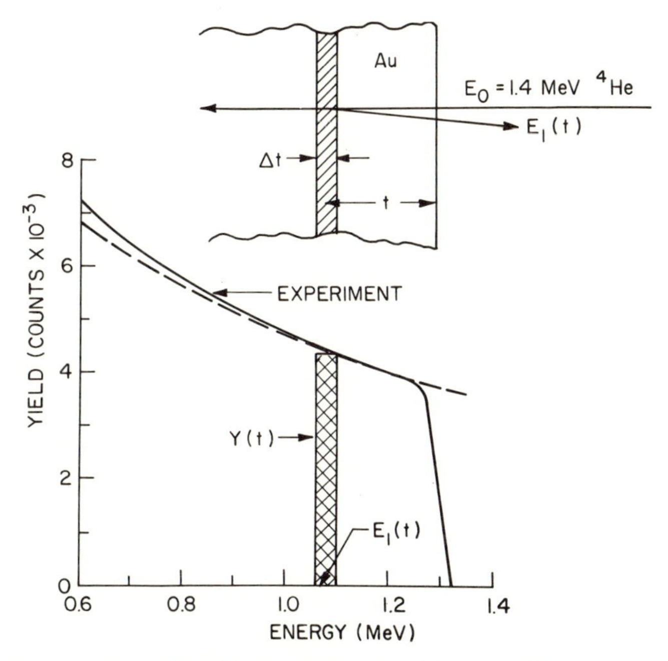

Figure 4. Spectrum for 1.4-MeV He ions backscattered from a solid gold target (solid line) and the theoretical shape of the spectrum (dotted line) as given by the relationship $N(E_1) \propto (E_0 + E_1)^{-2}$.

spectrum can be thought of as a depth profile of each of the masses, independently. It is this thin film application that has been the major use of ion scattering.

2.2. Nuclear Reaction Analysis

2.2.1. Kinematics

Symbolically, a nuclear reaction may be written as

$$M_1 + M_2 \rightarrow M_3 + M_4 + Q_W \tag{6}$$

where M_1 is the incident nucleus, M_2 is the target mass, M_3 is the emitted radiation, M_4 is the residual nucleus, and Q_W is the energy released or absorbed in the reaction. In general M_3 may refer to a particle or gamma ray.

The kinematic formulas associated with nuclear reactions are not as simple as those for elastic scattering. The relationship between the

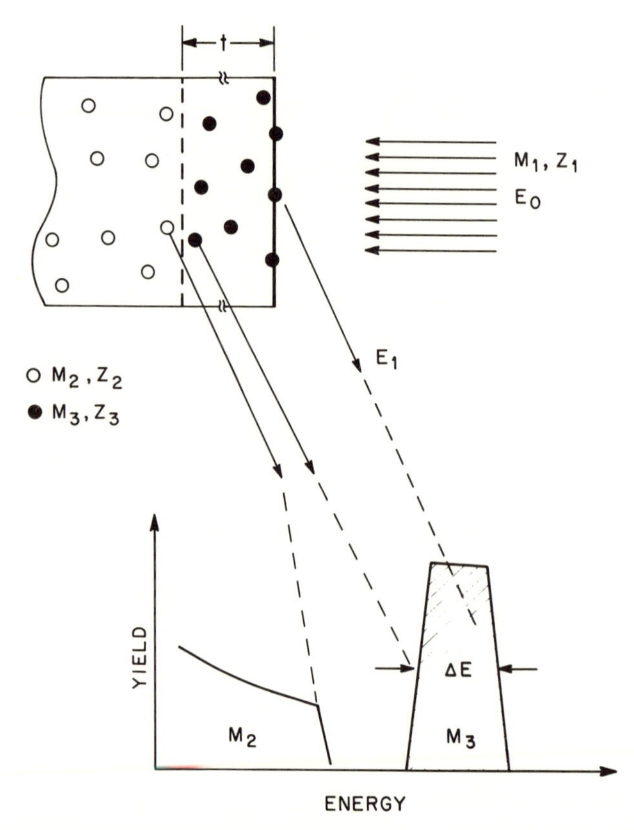

Figure 5. Schematic of the ion scattering spectrum from a solid consisting of a thin film on a lighter mass substrate.

masses of the constituents, the energy available, and the Q_W value of the reaction have been summarized in Ref. 5. The general idea is that a large positive Q_W value implies an excess of energy available to the outgoing reaction product. Nuclear reaction analysis is useful when the outgoing energy is greater than the energy of elastically scattered incident particles. This determination can be made by using well-established Q_W values and the kinematic formulas.

For a reaction induced by an incident particle M_1, and energy E_0 the energy of the emitted particle, E_3 (at angle θ), is given by

$$E_3^{1/2} = A \pm (A^2 + B)^{1/2}$$

where

$$A = \frac{(M_1 M_3 E_0)^{1/2}}{M_3 + M_4} \cos\theta \quad \text{and} \quad B = \frac{M_2 Q_W + E_0(M_4 - M_1)}{M_3 + M_4}$$

As illustrated in the preceding discussion accurate surface coverage determination at the submonolayer level is possible using RBS for the case of an adsorbate heavier than substrate. However, many interesting surface problems are with light adsorbates, i.e., hydrogen, carbon or oxygen. Quantitative analysis on the monolayer scale is then possible through the use of a selected nuclear reaction with a large, positive "Q_W value". Q_W is the energy balance for a nuclear reaction and exoergic reactions have positive Q_W values.[5] For example, deuterium coverage is measured via the $d(^3\text{He}, p)^4\text{He}$ reaction with a Q_W value of 18.35 MeV. An incident 0.7 MeV ^{3}He beam impinges on a deuterium-covered surface, undergoing a nuclear reaction with the deuterium. The emerging reaction products are a 13-MeV proton and a 6-MeV ^{4}He. The energetic outgoing proton appears as a signal in the spectrum at an energy far in excess of the abundant, backscattered ^{3}He. Knowledge of the absolute cross section permits quantitative coverage determination. The use of nuclear reactions for analysis has been tabulated by Feldman and Picraux[8] and has recently been reexamined by Davies *et al.*[9] Quantitative coverage determination for light elements, particularly hydrogen and its isotopes, is critical in surface science.

2.2.2. Cross Section

It is not possible to write down a simple formula for the cross section for nuclear reactions. The cross sections are governed by nuclear scattering phenomena and are not necessarily monotonic with energy or angle. Nevertheless nuclear physicists have measured such cross sections over the years and there are many convenient tabulations. As in the case of Rutherford scattering, the nuclear reaction yield from a surface impurity is given by

$$Y = N_s \frac{d\sigma}{d\Omega} \Delta\Omega Q$$

where N_s is the surface atom density (atoms/cm^2), $d\sigma/d\Omega$ is the nuclear reaction cross section, $\Delta\Omega$ is the solid angle subtended by the detector, and Q is the number of incident ions.

2.2.3. Tables of Nuclear Reactions

There has now been sufficient experience with nuclear reaction techniques so that the useful reactions are readily tabulated and the operating parameters specified. Table 1 lists commonly detected elements by nuclear reactions. The element to be detected is listed along with the

Table 1. Most Used Charged Particle Reactions for Light-Atom Detection[a]

Nucleus	Reaction	Incident energy (E_0) (MeV)	Emitted energy[b] (MeV)	σ_{LAB} (E_0) (mb/sr)	Yield[c] (counts/μC)
^{2}H	$^{2}H(d, p)^{3}H$	1.0	2.3	5.2	30
^{2}H	$^{2}H(^{3}He, p)^{4}He$	0.7	13.0	61	380
^{3}He	$^{3}He(d, p)^{4}He$	0.45	13.6	64	400
^{6}Li	$^{6}Li(d, \alpha)^{4}He$	0.7	9.7	6	35
^{7}Li	$^{7}Li(p, \alpha)^{4}He$	1.5	7.7	1.5	9
^{9}Be	$^{9}Be(d, \alpha)^{7}Li$	0.6	4.1	~1	6
^{11}B	$^{11}B(p, \alpha)^{8}Be$	0.65	5.57(α_0)	0.12(α_0)	0.7
		0.65	3.70(α_1)	90 (α_1)	550
^{12}C	$^{12}C(d, p)^{13}C$	1.20	3.1	35	210
^{13}C	$^{13}C(d, p)^{14}C$	0.64	5.8	0.4	2
^{14}N	$^{14}N(d, \alpha)^{12}C$	1.5	9.9(α_0)	0.6(α_0)	3.6
		1.2	6.7(α_1)	1.3(α_1)	7.0
^{15}N	$^{15}N(p, \alpha)^{12}C$	0.8	3.9	~15	90
^{16}O	$^{16}O(d, p)^{17}O$	0.90	2.4(p_0)	0.74(p_0)	5
		0.90	1.6(p_1)	4.5(p_1)	28
^{18}O	$^{18}O(p, \alpha)^{15}N$	0.730	3.4	15	90
^{19}F	$^{19}F(p, \alpha)^{16}O$	1.25	6.9	0.5	3
^{23}Na	$^{23}Na(p, \alpha)^{20}Ne$	0.592	2.238	4	25
^{31}P	$^{31}P(p, \alpha)^{28}Si$	1.514	2.734	16	100

[a] From Refs. 8 and 5.
[b] Laboratory emission angle of 150° with recoil nucleus in ground stae (excited state).
[c] For a $1 \times 10^{16}/cm^2$ surface layer and a solid angle of 0.1 sr at 150°.

suggested nuclear reaction and the appropriate incident energy. The notation $^{12}C(d, p)^{13}C$ refers to the detection of ^{12}C. The nuclear reaction is a 1.2-MeV incident deuteron and ^{12}C, forming ^{13}C and a 3.1-MeV energetic proton. The incident beam energy is the suggested incident energy to maximize the efficiency of detection. The "emitted energy" refers to the kinetic energy of the outgoing particle and $\sigma_{LAB}(E_0)$ refers to the cross section for the reaction (in mb/sr = 10^{-27} cm^2/sr) at the suggested incident energy. The yield (counts/μC) gives the expected count rate per microcoulomb of beam under a set of standard operating procedures specified as follows: N_s, the quantity of impurity is $1 \times 10^{16}/cm^2$ or about 10 monolayers, and $\Delta\Omega$, the solid angle of the detector is 0.1 steradian at a scattering angle of 150°.

An example of nuclear reactions in surface analysis is shown in Fig. 6, which indicates the absolute deuterium coverage on a W(001) surface due to a finite deuterium gas exposure in a UHV system. The nuclear reaction is $d(^{3}He, p)^{4}He$ using 700-keV ^{3}He. The saturation coverage is almost exactly 2.0×10^{15} atoms/cm^2 which corresponds to two hydro-

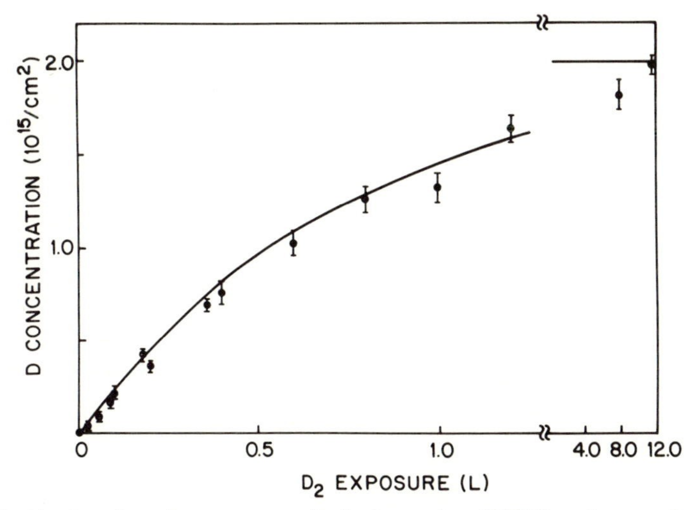

Figure 6. Absolute deuterium coverage adsorbed on a clean W(001) surface as a function of deuterium exposure in langmuirs (L). Redrawn from Ref. 25.

gen atoms/tungsten surface atom. The calculated line is the expected dose dependence for Langmuir kinetics.

2.3. Hydrogen Detection

The detection of hydrogen and its isotopes plays a special role in thin film analysis. Hydrogen cannot be detected by Auger electron spectroscopy (AES) or x-ray photoelectron spectroscopy (XPS) and is detected with difficulty in secondary ion mass spectroscopy (SIMS). Furthermore, hydrogen is ubiquitous. It is the dominant partial pressure of gas in most ultrahigh vacuum (UHV) deposition systems and may readily be incorporated into thin films. In the following section we describe two basic methods of using ion beams for hydrogen detection: forward recoil scattering and nuclear reaction.

2.3.1. Forward Recoil Scattering

Forward recoil scattering (FRS) makes use of the detection of the energetic recoil from an ion–atom collision. It is used primarily for the detection of light impurities (principally hydrogen and deuterium). Similar to backscattering, FRS yields a depth profile of the light impurity, with a depth resolution of ~600 Å and a sensitivity of ~1%. The forward recoil technique is similar to backscattering analysis but instead of measuring the energy of the scattered helium ion, the energies of the recoiling ^1H or ^2H nuclei are measured (Fig. 7). Hydrogen is lighter than helium, and both particles are emitted in the forward direction after a close collision. A Mylar foil ($\simeq$10 μm) is placed in front of the detector to block the penetration of the abundantly scattered helium ions while permitting the passage of the H ions. The stopping power is such that a 1.6-MeV ^1H ion only loses 300 keV in penetrating a Mylar film that completely stops 3-MeV ^4He ions. The Mylar absorber does introduce energy straggling (see Section 4.2), which combines with the energy resolution of the detector to give a total energy resolution of about 40 keV for scattering at the sample surface. Depth profiles are determined by the energy loss of the incident He ion along the inward path and the energy loss of the recoiling ^1H or ^2H along the outward path.

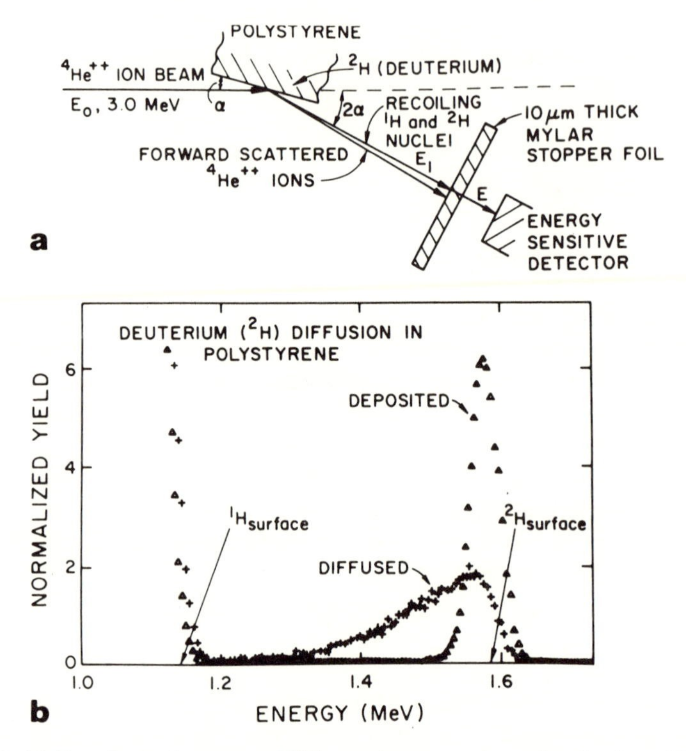

Figure 7. (a) Experimental geometry FRS experiments to determine depth profiles of ^1H and ^2H in solids. (b) Recoil spectrum of ^2H diffused in a sample of polystyrene for 1 h at 170°C. The sample consisted of a bilayer film of 120 Å of deuterated polystyrene on a large molecular weight (MW = 2×10^7) film of polystyrene. [From Mills *et al.*, *Appl. Phys. Lett.* **45,** 957 (1984), and Ref. 5.]

As an example the diffusion of deuterium (^2H) in polystyrene is shown in Fig. 7b. In that case a ^2H ion, detected at an energy of 1.4 MeV, corresponds to a collision that originated via a ^2H recoil from a depth of about 4000 Å below the surface. The use of FRS allows the determination of hydrogen and deuterium diffusion coefficients in the range of 10^{-12}–10^{-14} cm^2/s, a range that is difficult to determine by conventional techniques.

2.3.2. Nuclear Reactions

Many nuclear reactions have the property that the reaction yield exhibits one or more sharp peaks or "resonances" as a function of bombarding energy. Such a resonance is measured experimentally by varying the incident beam energy in small increments and measuring the quantity of radiation emitted per unit beam fluence at each energy. Resonance methods in depth profiling of trace elements takes advantage of the sharp peak in the nuclear reaction cross section as a function of energy. Consider the ideal case shown in Fig. 8a, where only one resonance exists in the stopping cross-section $C(x)$ curve and where off-resonance cross-section values can be neglected. The method consists of measuring the reaction yield (most often γ rays) due to the interaction between the incident beam and the impurity atoms as a function of incident beam energy. Incident ions having an energy E_0 (i.e., larger than E_R, the resonance energy) are slowed down until E_R is reached at depth x, where the nuclear reaction will then occur at a rate proportional to the impurity concentration. The depth x and the incident-beam energy E_0 are related through the equation

$$E_0 = E_R + \left(\frac{dE}{dx}\right)_{\text{in}} \frac{x}{\cos\theta_1} \tag{7}$$

where $\theta_1$is the angle between the incident beam and the surface normal. The stopping power $(dE/dx)_{\text{in}}$ for the incident beam is assumed to be a constant. A more elaborate analysis can be carried out by taking into account the detailed cross-section function, energy straggling, and other factors.[2]

Neglecting the finite experimental depth resolution, it is seen that the yield curve can be converted into the desired concentration profile by simply changing scales of yield and energy to corresponding scales of concentration and depth, respectively. An example of the use of nuclear resonance is shown in Fig. 8b, which gives the gamma-ray yield as a function of beam energy for a hydrogen-implanted target. The reaction between fluorine and hydrogen has a strong resonance at about 16.4 MeV

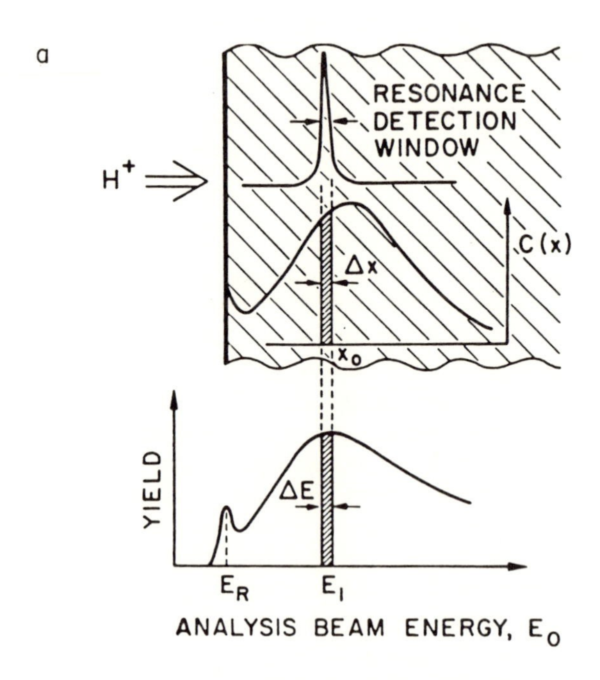

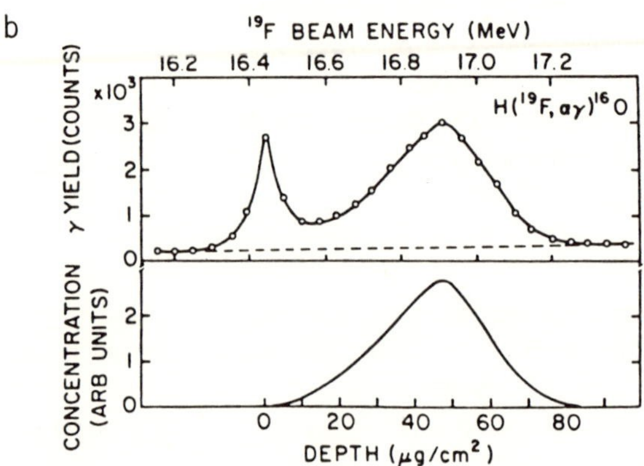

Figure 8. (a) Principle of concentration profile measurements using resonant nuclear reactions. (b) Range profile of 12-keV H implanted into Al_2O_3 to a fluence of $4 \times 10^{16}/cm^2$ measured by use of the nuclear reaction $^1H(^{19}F, \alpha\gamma)\ ^{16}O$. The upper part of the figure is the raw experimental data and the lower part is the extracted depth profile. [Redrawn from J. Bøttiger, S. T. Picraux, and N. Rud, in *Ion Beam Surface Layer Analysis,* Vol. 2, O. Meyer, G. Linker, and F. Käppeler, eds. Plenum Press, New York (1976), p. 811.]

so that the hydrogen concentration profile can be obtained directly. The extracted hydrogen concentration profile indicates only the hydrogen within the sample and does not include the surface hydrogen present due to contamination.

Nuclear reaction analysis (NRA) is a method of determining the absolute concentration (atoms/cm^2) of not only hydrogen but other light impurities in and on a solid. It thus provides an absolute calibration for other surface sensitive techniques—particularly AES and SIMS. In a typical application, a light particle of interest is implanted into a heavier substrate. SIMS provides a sensitive depth profile while NRA determines an absolute concentration. Reaction analysis is particularly useful for hydrogen detection and absolute hydrogen surfaces coverages. The most useful reactions for ^{1}H detection are ^{1}H(^{19}F, $\alpha\gamma$)^{16}O with an incident F energy of ~17 MeV and ^{1}H(^{15}N, γ)^{16}O with an incident N energy of ~6.5 MeV. Deuterium detection is usually carried out via the ^{2}H(^{3}He, p)^{4}He reaction at an incident ^{3}He energy of 0.7 MeV.

3. Apparatus

3.1. General Setup

Apparatus for ion scattering analysis of solid surfaces consists of three main components: (1) An accelerator for the production of energetic charged particle beams; (2) a beam line for the transport and shaping of the beam; and (3) a scattering chamber, which houses the samples, charged particle detectors, and other apparatus in a suitable vacuum.

Accelerator expertise has developed from the past work of nuclear physicists, although there are now commercial accelerators specifically designed for surface analysis. The Van de Graaff electrostatic generator, invented in 1931 by R. J. Van de Graaff, is the basis for early MeV accelerators. A "charging belt" carries electrons away from a terminal housed within an insulating gas-filled tank. Within the positively charged terminal an ion source creates positive ions which are then accelerated to ground to form the ion beam. The terminal potential is determined by the charging rates and charging potentials, values up to 30 MV exist in the largest machines. A 2-MV accelerator, ideal for surface analysis, is relatively easily accomplished. To maintain such a potential requires apparatus of the order of a few meters. It is the rather large size of this apparatus that has been a major drawback in the proliferation of ion scattering.

There have been two interesting variations in basic accelerator design for ion scattering/surface analysis. In one case the "charging belt" has been replaced by a "charging chain," which provides a more uniform means of maintaining the potential and hence less "voltage ripple." A second variation is the "tandem accelerator." Here the ion source is at ground and creates a negative ion. Acceleration of the negative ion from ground to a terminal at a positive potential V yields a particle of kinetic energy Ve. At the terminal a two-electron stripping process results in the creation of a positive ion with further acceleration to ground and a final kinetic energy of $2Ve$. This design allows some compactness in accelerator size.

The ion beam from a charged particle accelerator is transported through an evacuated beam line eventually impinging on a sample held on a sample manipulator within a vacuum chamber. The exact configuration of the system is dependent on the users applications; however, some common features exist. The beam line is essentially an evacuated drift tube with components for shaping and steering the beam. Common beam line components include collimators, steerers, and focusing elements such as quadrupoles. Collimators are usually chosen to define the angular divergence of the beam and define the size of the beam spot on the sample. For channeling applications one is usually interested in a divergence that is much less than the channeling critical angle, about 1°. A convenient size beam spot is often 1 mm on sample. These requirements are easily met by placing two collimators, about 1 mm in diameter, two meters apart. Of course, it may be desirable to still have much smaller beam spots and divergences. The collimator size, however, is limited by the fluctuation in the spot position of the beam. This fluctuation arises from an inherent energy jitter in the accelerator, which transforms to a position fluctuation after beam transport through a magnetic or electrostatic beam-switching component. There is a substantial loss in incident beam current if a collimator is considerably smaller than the extent of beam jitter or if the divergence is substantially smaller than the output divergence of the beam from the accelerator.

Simple electrostatic steerers provide a way of deflecting the beam over small distances. In the MeV regime it is sufficient to use electrostatic deflection in a simple "parallel plate capacitor" like arrangement. Focusing elements may be used for microbeam applications where one desires spatial information on the micrometer scale, as in a scanning electron microscope. The focusing element may be an electrostatic or magnetic quadrupole. The disadvantage of focused ion beams is that ion beam damage effects scale as the number of incident ions/cm^2. Nevertheless, the ion microprobe has proved extremely useful in specialized applications. The reader is referred to the work of Doyle *et al.* for further discussions of the point.[10]

Sample chambers range from a very simple variety for straightforward RBS analysis to rather complex ones for UHV analysis with specialized spectrometers for enhanced energy resolution in the detection of backscattered ions. A description of a UHV chamber apparatus used at Bell Labs follows.

3.2. A UHV Ion Scattering Chamber

The scattering chamber is a 45-cm-diam stainless steel UHV chamber coupled via a differentially pumped beamline to an ion accelerator. Ion and titanium sublimation pumps maintain the chamber in the low 10^{-10}-Torr range. The beamline, connected to the chamber by a 1-mm-diam aperture, is held to 10^{-8} Torr by a turbomolecular pump. At the upstream end of the beamline (2.5 m away), a variable collimator provides further isolation from the accelerator and beam switching magnet, which are typically in the 10^{-6}-Torr range. The two collimators define the divergence of the beam which must be $\leq 0.06°$ for channeling applications. The accelerator routinely produces microamps of beam of any gas from 0.1 to 2.0 MeV.

As shown in Fig. 9, the chamber contains LEED optics (also used

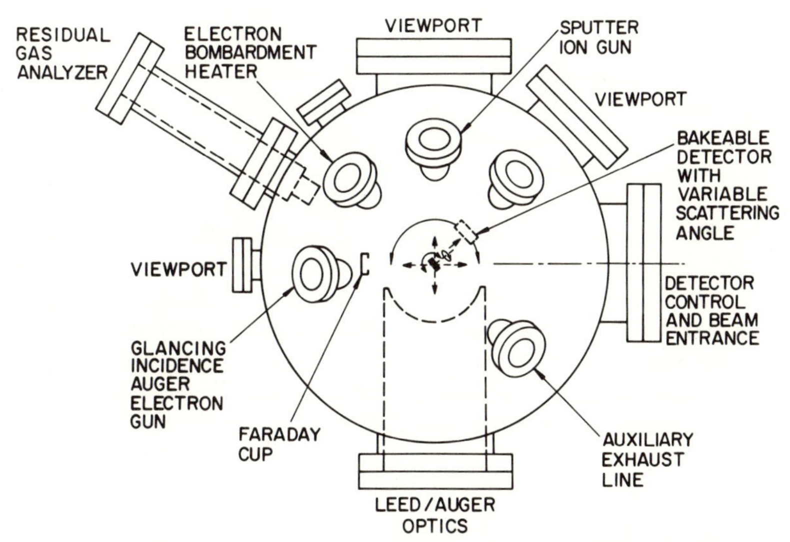

Figure 9. Schematic of the experimental UHV ion scattering chamber described in the text.

with a glancing incidence electron gun for retarding field Auger spectroscopy) for monitoring the condition of the sample surface, sputter ion and electron bombardment guns for sample cleaning, and a residual gas analyzer. A leak valve manifold permits the admission of a variety of gases for sputtering or adsorption studies; this relatively high-pressure gas may then be removed rapidly by the turbopump through an auxiliary exhaust line which bypasses the chamber's beam entrance aperture. Other ports contain *e*-beam sources or Knudsen cells for thin film and epitaxy studies.

The sample manipulator used in these experiments is based on a UHV concentric rotary and linear motion feedthrough mounted on a movable stage. It permits three orthogonal translations, a tilt, and two rotations: one about an axis in the sample surface (±180°) and one about an axis normal to the sample (±90°). Both rotations have ~0.1° precision, making crystal alignment for ion channeling relatively easy. The compact design of the sample holder allows a clear view of the LEED screen, while the open center makes transmission experiments through thin crystals possible. Samples may be heated by a resistive heating element or by electron bombardment from either side, and temperature is usually monitored externally with infrared and optical pyrometers. For accurate ion beam charge integrations, the sample is biased at +300 V to suppress secondary electron emission.

Also held on the manipulator, above the sample position, is a scattering "standard" consisting of a calibrated concentration of a heavy element implanted in Si at low energy. By comparing the scattering from the sample to the standard, an absolute determination of the areal atomic density in the sample can be made without specific knowledge of the solid angle subtended by the detector or of the exact scattering angle.

Energy of scattered ions is measured with a surface barrier detector made to withstand the 200°C chamber bakeout without deterioration. Cooling is provided for improving the energy resolution. The detector is mounted on an arm which can be rotated around the chamber's central axis over a range of 0–160° using an external control. In this way, a grazing exit-angle geometry can be obtained for all incident directions, resulting in improved depth resolution. A slit over the detector limits its acceptance angle to ~1°.

3.3. Charged Particle Spectrometers

Most RBS analysis involves the use of silicon solid state charged particle detectors. The operation of such a device is illustrated in Fig. 10.

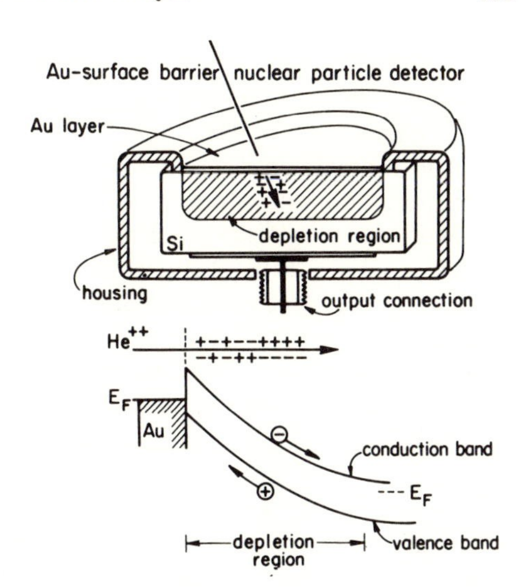

Figure 10. Schematic of a surface barrier solid state detector.

An incoming charged particle of energy E creates electron–hole pairs losing energy at the rate of 3.6 eV/pair. In the presence of the applied field (reverse bias) the charges are swept to the electrodes, creating a voltage pulse of magnitude $V = ne/C$, where C is the capacitance of the device and $n = E/3.6$, the number of pairs created. This voltage signal is proportional to the energy of the incoming radiation. Intrinsic fluctuations in the energy loss process result in a finite resolution to the detector of about 10 keV. This in turn sets a limit on the depth resolution and mass resolution of the RBS technique. A solid state detector is an extremely convenient, inexpensive device with 100% efficiency; as a result it has played a significant role in the development of the ion scattering technique.

Substantial improvement in absolute resolution can be achieved with the use of electrostatic or magnetic analyzers. In this case the resolution is correctly given as a fractional quantity, $\Delta E/E$ and a value of $\sim 4 \times 10^{-3}$ is achievable. Such analyzers represent an increase in cost, and a decrease in convenience and efficiency. However, a new, relatively small, electrostatic analyzer has recently become available, useful for operation in the 50–400-keV range.[11] With such an instrument there is a substantial gain in depth resolution to ~10 Å.

4. Quantitative Analysis and Sensitivity

4.1. Mass Resolution

The basic equation that governs mass resolution is given by the kinematic factor, K in Section 2.1.2. Mass resolution, the ability to distinguish different masses, is a function of energy, detector energy resolution, the scattering angle, and the mass of the incident particle.

Consideration of the simple equations in Section 2.1.2 demonstrates that the ultimate in mass resolution can be achieved with higher-mass projectiles and detection at an angle of 180°. For practical reasons, however, most analysis is carried out with He projectiles at scattering angles of ~170°. Table 2 shows the backscattering factor for 2.0-MeV He impinging on different atoms. Since the energy resolution of a solid state detector is typically 15 keV one can clearly see that light elements are easily distinguished, but heavy elements may not be resolved. In particular note that the important species Ga and As can just be resolved and the scattering from tungsten and gold fall within the resolution of the detector. In fact RBS is not the most useful way to survey unknown samples for impurities; rather, it is most useful for controlled experiments in which one seeks a depth profile or quantification of a known species. These limits in mass resolution are also shown in Fig. 11 in terms of discrimination of adjacent elements in the periodic chart.

A good example of the use of mass resolution and quantification in RBS is shown in Fig. 12, illustrating the composition of the new "123" superconducting material and the clear separation of the three principal elements Ba, Y, and Cu.

4.2. Depth Resolution

As discussed in Section 2 the basic equation which relates the depth of the scattering event to the emergent energy is

$$E_1(t) = K\left[E_0 - t\left(\frac{dE}{dx}\right)_{E_0}\right] + \frac{t}{|\cos\theta|}\left(\frac{dE}{dx}\right)_{E_1} \tag{8}$$

Table 2. Mass Resolution for E_0 = 2.0 MeV $^4He^+$ and θ = 180°

Element	K (180°)	Energy (eV)
Carbon	0.25	500
Oxygen	0.36	720
Aluminum	0.55	1101
Silicon	0.56	1125
Gallium	0.79	1586
Arsenic	0.81	1615
Tungsten	0.92	1833
Gold	0.92	1844

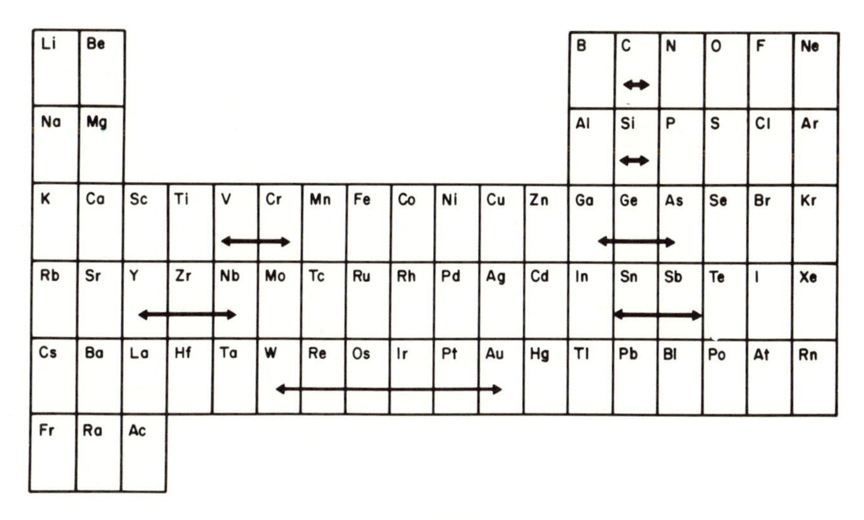

Figure 11. Mass resolution, under typical conditions, shown as a spread on a periodic chart.

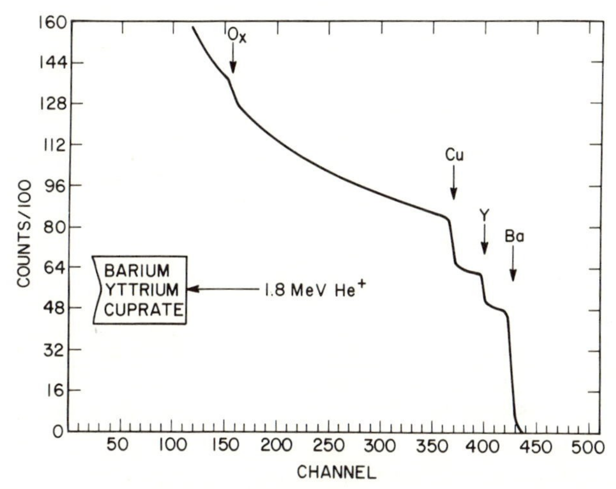

Figure 12. RBS spectrum from the 123 superconducting material $YBa_2Cu_3O_7$ illustrating mass resolution from a complex bulk target.

In this analysis we have assumed the "surface approximation," that is, the stopping power does not change with penetration. Using this form we can write the relationship between depth resolution, Δt, and energy resolution, ΔE_1, as

$$\Delta t = \frac{\Delta E_1}{K\left(\frac{dE}{dx}\right)_{E_0} + \frac{1}{|\cos\theta|}\left(\frac{dE}{dx}\right)_{E_1}} \tag{9}$$

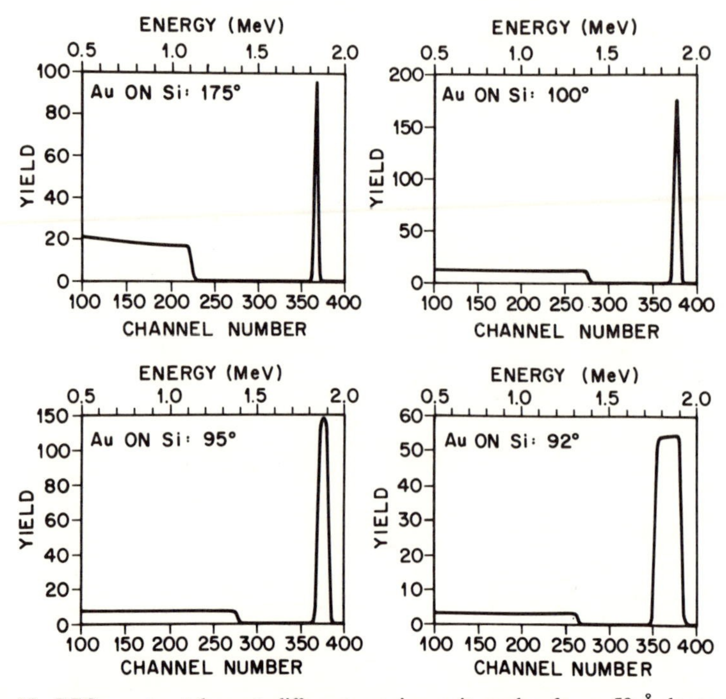

Figure 13. RBS spectra taken at different grazing exit angles for a 50-Å layer of Au deposited on Si. Note that the Au peak becomes "flat topped" at $\theta = 95°$ indicating a depth resolution of ~50 Å. At 92° the 50 Å film is clearly resolved and the depth resolution is considerably less than 50 Å.

Standard conditions may be taken as $\Delta E_1 = 15\,\text{keV}$, typical of a solid state detector, and $\theta = 180°$ for standard backscattering. Then $\Delta t \sim 250\,\text{Å}$ for He in silicon. As shown in Fig. 13 the depth resolution may be substantially improved by using a grazing exit angle geometry, for example $\Delta t \sim 45\,\text{Å}$ at $\theta = 95°$.

In addition to energy resolution and exit angle the major factor that limits depth resolution is the stopping power. Figure 14 shows a typical plot of the stopping power for He in a number of elemental materials over a range of energies indicating that the 1.0-MeV regime is where the stopping power is a maximum and the depth resolution a minimum. Note that the stopping power is larger for higher-Z elements so that better resolution is expected for heavier materials.

Optimization of detector energy resolution and take-off angle can only influence depth resolution near the surface. At greater depth, the resolution is limited by energy straggling. Straggling is associated with the statistical fluctuation in the energy loss process and yields a small difference between the energy loss of two particles that have penetrated to the same depth.

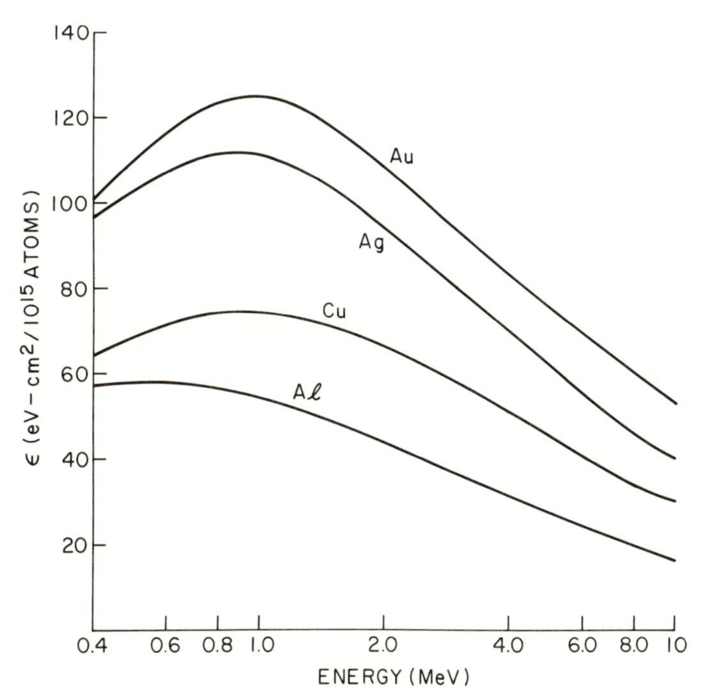

Figure 14. Stopping power as a function of energy for various elements. (Adapted from Ref. 7.)

The Bohr value for the straggling of a particle, Z_1, backscattering from a depth t of a sample with atomic number Z_2 and volume concentration N is

$$\Omega_B^2 = 4\pi Z_1^2 e^4 N Z_2 t(1 + 1/|\cos\theta|) \quad (10)$$

where Ω_B^2 is the mean square value of the straggling. In this equation, θ is the scattering angle (>90°) and normal incidence is assumed. The value of $\sqrt{\Omega_B^2}$ for He scattering from Si at a depth of 1000 Å and to an angle of 95° (normal incidence) is ~10 keV. This root-mean-square value would

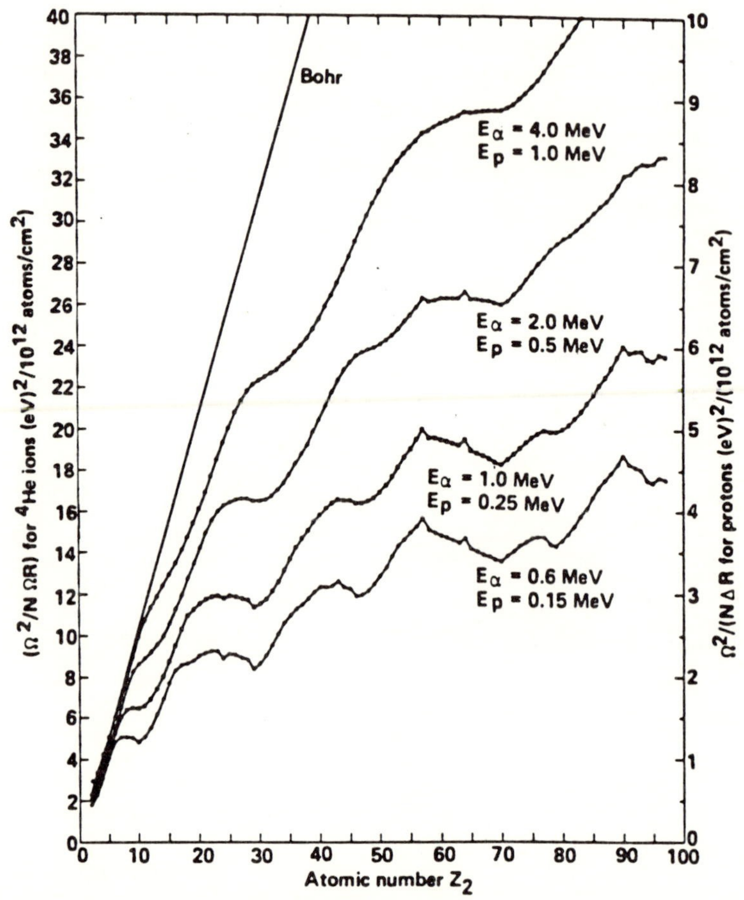

Figure 15. Energy straggling per unit target thickness versus target atomic number Z_2 calculated for ^{4}He projectiles (left-hand scale) and protons (right-hand scale). Bohr's theory is independent of energy. (From Ref. 12.)

then correspond to an effective resolution of $2.35\Omega_B$ or 23 keV. Since this value is substantially greater than the detector resolution (~15 keV) the system resolution is governed by straggling. The Bohr value of straggling is only an approximation; more accurate calculations have been carried out by Chu *et al.*[12] and are indicated in Fig. 15.

4.3. Quantitative Analysis

The basic relationship for quantitative analysis may be written as

$$Y = N_s \frac{d\sigma}{d\Omega} \Delta\Omega Q \tag{4}$$

where Y is the measured yield of scattered particles, N_s is the areal density of particles (atom/cm^2), $\Delta\Omega$ is the detector solid angle (steradians), Q is the number of incident particles, and $d\sigma/d\Omega$ is the differential scattering cross section (cm^2/steradian). This form is applicable for the case of an impurity on or near the surface of a solid.

Figure 16 shows the spectrum associated with the use of Eq. (4) in

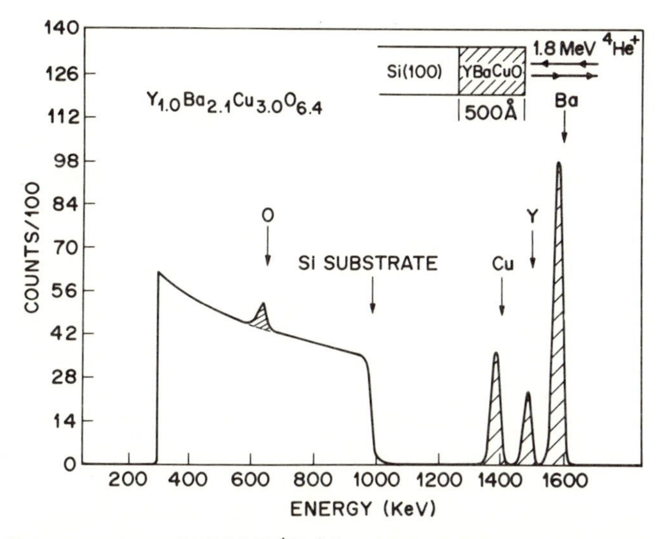

Figure 16. Energy spectrum of 1.8-MeV $^4He^+$ ions backscattered from a thin film (~500 Å) of YBaCuO on the surface of a silicon substrate. The vertical arrows indicate the energies of particles backscattered from the various atomic constituents. The absolute yield of atoms/cm^2 for each individual element is given straightforwardly by using the integrated area of the peaks (cross-hatched) in Eq. (4). In the case of oxygen there is a background subtraction for scattering from the "infinite" Si substrate. The ratio of the peak intensities is given as $Y_{1.0}Ba_{2.1}Cu_{3.0}O_{6.4}$ close to the ideal value for the superconducting material $Y_{1.0}Ba_{2.0}Cu_{3.0}O_7$. (From B. A. Davidson, private communication.)

establishing the known number of atom/cm^2 in a complex thin film on a silicon wafer. The yield is the integrated area of the peaks in RBS spectrum. All factors in the basic yield equation are known except for N_s, the desired quantity. In this case of thin film analysis, the deposited film is calculated to be close to the desired ratio of $YBa_2Cu_3O_7$, for the new superconducting material.

In other cases one is interested in the stoichiometry of a thicker film such as illustrated in Fig. 17. Here one uses the height of the backscattering spectrum for qualitative analysis. For example the ratio of the heights of the leading edge of the spectrum is given by[4]

$$\frac{H_A}{H_B} = \frac{N_A \sigma_A / [\varepsilon]_A}{N_B \sigma_B / [\varepsilon]_B}$$

where σ_A and σ_B are the appropriate cross sections, $[\varepsilon]_{A,B}$ refers to the backscattering factor for scattering from A, B, and N_A/N_B is the desired stoichiometry ratio. For Rutherford scattering $\sigma_A/\sigma_B = (Z_A/Z_B)^2$.

In the constant dE/dx approximation the backscattering factor, $[\varepsilon]_A$ may be written as

$$[\varepsilon]_A = K_A \left(\frac{dE}{dx}\right)_{in} + \left(\frac{dE}{dx}\right)_{out} \tag{11}$$

for $\theta = 180°$. In this equation $(dE/dx)_{in}$ and $(dE/dx)_{out}$ are the appropriate stopping powers for the compound composed of elements A and B.

As an example we consider a material of the form A_xB_y and determine the ratio x/y. The stopping power formula makes use of Bragg's law so that the total stopping power ε is given by

$$\varepsilon = x\varepsilon_A + y\varepsilon_B$$

In this equation ε_A, ε_B represent the stopping power for a material made up solely of A, B atoms and ε is the total stopping power of the compound.

Then $(dE/dx)_{in} = N_m \varepsilon^{in}$ where N_m is the number of molecules/unit volume and ε^{in} means the stopping power evaluated at the incident energy. For simplicity we write

$$R \equiv H_A/H_B = \frac{N_A \sigma_a / [\varepsilon]_A}{N_B \sigma_B / [\varepsilon]_B}$$

where $N_A = xN_M$, $N_B = yN_M$, H_A/H_B is the ratio of the heights of the

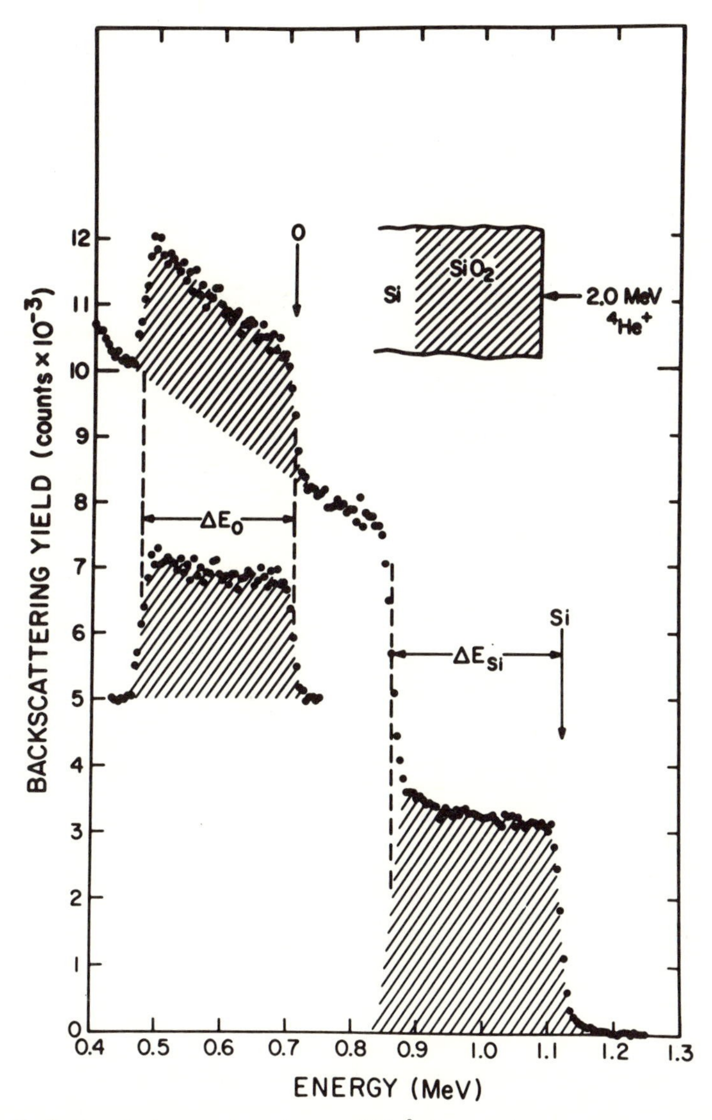

Figure 17. Backscattering spectrum from a 5000-Å film of SiO_2 thermally grown on a Si substrate. The ratio of the heights of the spectra near the Si and O edges can be used in a "height analysis" to give the stoichiometry of the film. [Redrawn from W. K. Chu, J. W. Mayer, M.-A. Nicolet, T. M. Buck, G. Amsel, and F. Eisen, *Thin Solid Films* **17,** 1 (1973).]

signals derived from the experimental spectrum, and

$$q \equiv \frac{x}{y} = \left(\frac{Z_B}{Z_A}\right)^2 R \frac{[\varepsilon]_A}{[\varepsilon]_B} \tag{12}$$

This equation may yield a crude approximation to the stoichiometry of the material by taking $\{[\varepsilon]_A/[\varepsilon]_B\} = 1$.

A more careful analysis considers the x/y dependence in the backscattering factors. We write

$$\frac{[\varepsilon]_A}{[\varepsilon]_B} = \frac{K_A(x\varepsilon_A^{in} + y\varepsilon_B^{in}) + (x\varepsilon_A^{out} + y\varepsilon_B^{out})}{K_B(x\varepsilon_A^{in} + y\varepsilon_B^{in}) + (x\varepsilon_A^{out} + y\varepsilon_B^{out})} \tag{13}$$

or

$$\frac{[\varepsilon]_A}{[\varepsilon]_B} = \frac{q(K_A\varepsilon_A^{in} + \varepsilon_A^{out}) + K_A\varepsilon_B^{in} + \varepsilon_B^{out}}{q(K_B\varepsilon_A^{in} + \varepsilon_A^{out}) + K_B\varepsilon_B^{in} + \varepsilon_B^{out}} \tag{14}$$

Using (14) we can rewrite Eq. (12) in the form,

$$q = A_R\left(\frac{qS_1 + S_2}{qS_3 + S_4}\right)$$

where $q = x/y$, $A_R = R(Z_B/Z_A)^2$ and S_1, S_2, S_3, S_4 follow from the last equation for $[\varepsilon]_A/[\varepsilon]_B$. Solving for q we obtain

$$q = \frac{-(S_4 - A_RS_1) + [(S_4 - A_RS_1)^2 + 4A_RS_2S_3]^{1/2}}{2S_3}$$

Since q must be positive only the positive sign in the quadratic solution is valid. Note also that if $S_1 = S_2 = S_3 = S_4$ we find

$$\frac{x}{y} = R\left(\frac{Z_B}{Z_A}\right)^2$$

i.e., simply the ratio of the heights scaled by the Rutherford cross section. This is the equivalent of taking the ratio of the backscattering factors as unity. This form for q is a convenient way to explicitly solve for the stoichiometry in the constant stopping power approximation.

In practice quantitative analysis is rarely carried out "by hand." Instead simulation programs are used that take into account essentially all aspects of the particle solid interaction and the experimental setup. The simulation can then generate an RBS spectrum for comparison to the data.[13]

4.4. Lateral Resolution

Lateral resolution in most analytical instrumentation is set by the beam spot size and the interaction between the incident radiation with the solid. Electron beams can be focused to very small spots (<100 Å) although they quickly "plume" as the penetrate, owing to multiple scattering. MeV ion beams have the advantage of reduced multiple scattering although they cannot be readily focused. Nevertheless, a number of laboratories have produced specialized ion microbeams which operate close to $\sim 1\ \mu m$ resolution. The reader is referred to Ref. 10 for further discussion.

4.5. Beam Damage and Desorption

The Z^2 dependence of the Rutherford scattering cross section clearly indicates a large sensitivity to heavy elements. It is of interest to ask for the ultimate sensitivity, i.e., the smallest amount of material detectable.

The limit for ion scattering is set by sputtering. Sputtering is a process in which an energetic ion impinging on a solid creates a solid cascade due to small angle nuclear scattering. As a result of multiple scattering some fraction of the secondary ions acquire the correct momentum to escape the solid, giving rise to an erosion process. The basic question is under what conditions will the erosion of the material occur before a measurement is complete. Sputtering is defined in terms of a sputtering yield S, which is the number of atoms ejected from the solid per incident ion. In the following we calculate the limit to sensitivity set by the sputtering process.[5]

We consider a thin layer of material (possibly submonolayer) containing N_s atoms/cm^2. The yield of scattering ions, Q_D, is given by the usual equation in a Rutherford backscattering measurement:

$$Q_D = \sigma(\theta)\Omega Q N_s \tag{15}$$

where $\sigma(\theta)$ is the scattering cross section, Ω is the solid angle of the detector, and Q is the number of incident ions.

For the same number of incident ions the loss of atoms from the layer, ΔN_s, due to sputtering by the incident ions is

$$\Delta N_s = SQ/a$$

where a is the area of the probing beam spot. We require that the amount of erosion be less than the original film thickness, i.e.,

$$\Delta N_s < N_s$$

which can be expressed as a limit on the value of Q given by

$$Q < \left(\frac{Q_D a}{S\sigma(\theta)\Omega}\right)^{1/2}$$

and a corresponding minimum value of N_s given by

$$N_s > \left(\frac{Q_D S}{\sigma(\theta) a \Omega}\right)^{1/2} \tag{16}$$

In evaluating this quantity, we use the following standard numbers for the case of a layer of gold ($Z = 79$): $\sigma(\theta)$, the cross section for 2-MeV He scattering at 170° is 10^{-23} cm^2/sr; Ω, the solid angle of the detector corresponding to a 1 cm^2 detector 5 cm from the target, is 4×10^{-2} sr; S, the sputtering yield, is 10^{-3}; a, the area of the probing beam, is 10^{-2} cm^2; and Q_D is arbitrarily taken as 10^2, the minimum number of counts required for a statistically significant measurement. With these values we find a minimum layer thickness of 5×10^{12} Au atoms/cm^2 or about 1/200 of a monolayer. From the standard scattering expressions, we see that the scattering parameters and geometry could be optimized in a number of ways to further increase the sensitivity. Experience suggests that the absolute best sensitivity realizable for this favorable case of a heavy scatterer is 5×10^{11} atoms/cm^2. Note that the incident charge required is not prohibitive; Q corresponds to an incident particle dose of 5×10^{13} ions or about 10 μC.

5. Ion Scattering as a Structural Tool

The scattering spectrum from a single crystal aligned with a major symmetry direction parallel to the beam is drastically modified from a disordered sample. In the noncrystalline material each atom senses a uniform distribution of impact parameters and obeys scattering laws as previously discussed. The beam will undergo small-angle multiple scattering, in an uncorrelated fashion, which does not affect this impact parameter distribution. In an aligned single crystal the impact parameter distribution is also uniform at the first monolayer. However the small angle scattering events determine the impact parameter for the second atom in a correlated fashion, and so on throughout the crystal. This correlation produces a unique (nonuniform) flux distribution inside the crystal which modifies the scattering probability. Deep inside the crystal this phenomenon is known as channeling[(14)] and is useful in a variety of crystallographic studies.

Correlated scattering also plays a role at the surface or within the first few monolayers of a single crystal. The scattering from a surface atom determines the impact parameter distribution at the second atom and so on along the atomic row. In the simplest view the surface atom shadows the underlying atoms reducing their scattering yield. In backscattering this surface interaction gives rise to the surface peak, SP, a measure of the scattering yield from the surface. The description of channeling is most elegantly described in the early work of Lindhard.[14] The process is more generally described in the book by Morgan[15] while materials and surface applications are emphasized in Ref. 16.

5.1. Shadowing

The simplest example of this correlated small angle scattering is illustrated in Fig. 18. Ions incident at the smallest impact parameter undergo large-angle scattering; those at larger impact parameter suffer small deflections which determine the flux distribution of ions near the second atom. In a static representation the ion beam has a distance of closest approach to the second atom, R, within which there are no particles. For Coulomb scattering

$$R = R_c = 2(Z_1 Z_2 e^2 d/E)^{1/2} \tag{17}$$

where d is the atomic spacing, E is the incident ion energy, and Z_1, Z_2 the atomic numbers of the incident and target atom. For 1.0-MeV He^+

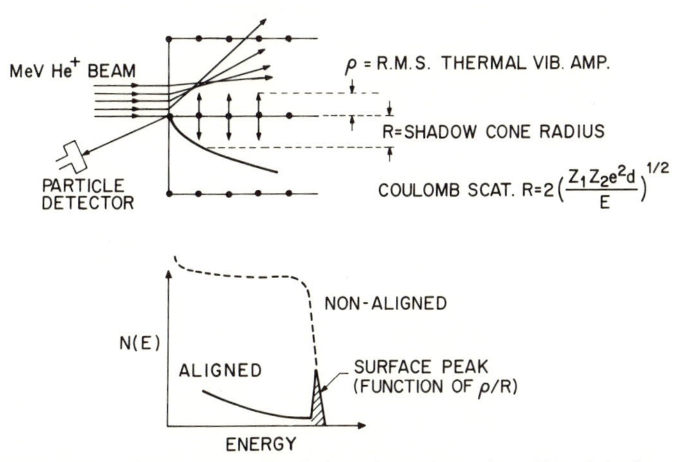

Figure 18. Schematic showing the interactions at the surface of an aligned single crystal and the formation of the shadow cone. The energy spectra for the aligned and nonaligned case are shown below.

incident along the $\langle 100 \rangle$ direction of tungsten, $R_c = 0.16$ Å. Since the bulk value of ρ, the two-dimensional thermal vibrational amplitude, is 0.06 Å (at room temperature), the shadowing is still substantial, even in a nonstatic case.

5.2. Channeling

The energetic ion beams can penetrate micrometers into the bulk. When the beam is parallel to a major symmetry direction of the crystal the particles acquire channeling trajectories and, eventually, the channeling flux distribution. Channeling effects have been of considerable interest in recent years both as a solid state tool and for a variety of other purposes.[16]

For most of the surface applications described here the details of the bulk effects are not essential and we give primarily a qualitative description of the evolution of the channeling process. The uniform incident ion beam encounters the first monolayer of the solid and undergoes small angle scattering to acquire the "shadow cone" flux distribution. A correlated sequence of scattering can continue for hundreds of atoms, until an incident particle is gently steered away from the closest string of atoms. Interaction with an adjacent row is similar, so the initial motion is oscillatory in nature, bouncing from string to string, with a wavelength hundreds of angstroms long. The important fact is that particles cannot get close enough to the atoms of the solid to undergo small impact parameter processes such as Rutherford scattering. Eventually the phases of the different trajectories are mixed and the flux distribution can be described by statistical equilibrium concepts.[14] In the case of incidence precisely along the channeling direction, the flux distribution, $f_B(r)$, well inside the crystal is of the form

$$f_B(r) = \ln\left(\frac{Nd}{Nd - \pi r^2}\right) \tag{18}$$

where r is the perpendicular distance from the atomic row and N is the volume density of atoms inside the solid, so that Nd is the area of a unit cell. This is a distribution that is sharply peaked at the center of the channel (i.e., far from the string of atoms) and decreases as r^2 to zero intensity at the atomic sites.

The scattering yield from bulk atoms is then the fold of the Gaussian positional distribution for the atoms with the bulk flux distribution and is roughly given by $Nd\pi\rho^2$. Since the distribution is normalized to unity for nonchanneling conditions, the channeling bulk yield is the order of 1% of the nonchanneling yield.

These results allow us to develop a description of the backscattering energy distribution in the channeling case: at energies corresponding to the surface of the crystal there is a surface peak corresponding to at least one monolayer. At lower backscattered energies (i.e., deeper depths) the scattering yield undergoes some oscillations and then reaches a steady state value of ~1% of the non-single-crystal yield. At even greater depths dechanneling processes occur. Since the basic process is Rutherford scattering, RBS yield analysis can be used to give the number of monolayers contributing to the surface peak.

5.3. The Surface Peak

Some of the most sophisticated uses of ion beams in surface analysis involve measurements of the surface peak (SP). In the following we consider the SP in more detail.

In cases of practical interest the SP corresponds to scattering from a few atoms in the atomic row. The most direct approach to a SP prediction is computer simulation in which a large number of incident ion trajectories are computed and the large angle scattering probability calculated for each layer, as outlined by Barrett[(17)] and Stensgaard *et al.*[(18)] An additional advantage of the simulation technique is that particular surface structure models are easily included for comparison to experiments. In the following we discuss the basic assumptions that are used in the simulation and the parameterization of the results.

The assumptions in the simulation center on two aspects of the problem: (1) the scattering of the projectile and (2) the model of thermal vibrations used to govern the relative positions of the atoms in the crystal. The Moliere approximation to the Thomas–Fermi potential is assumed to be the best available analytical potential for the impact parameter range of interest here and is used in most calculations. Results obtained using other reasonable potentials deviate insignificantly from the Moliere potential results.[(18)] Once the potential is specified, the scattering problem is reduced to the simplest and most accurate method of solving the scattering integral.

While the scattering properties of energetic ions by atoms in a solid are well understood, the least established part of the simulation concerns the model of thermal vibrations of the atoms. An MeV ion passes a lattice spacing in a time which is short compared to the characteristic phonon times in a solid. Thus the solid can be viewed as an ordered arrangement of atoms, with the atoms statically displaced from their equilibrium positions in accord with the chosen model of thermal vibrations. In most cases a harmonic approximation is used in which the thermal position distribution of atoms is a Gaussian whose width is related to a measured Debye temperature. Correlations in thermal vibrations and enhanced surface vibration may be included.

The results of the SP calculation for a large number of cases show that the SP can be expressed as a function of one parameter ρ/R_M where R_M corresponds to the shadow cone radius at the second atom using the Thomas–Fermi potential. Over the range of interest R_M/R_C varies from 0.7 to 1.0.

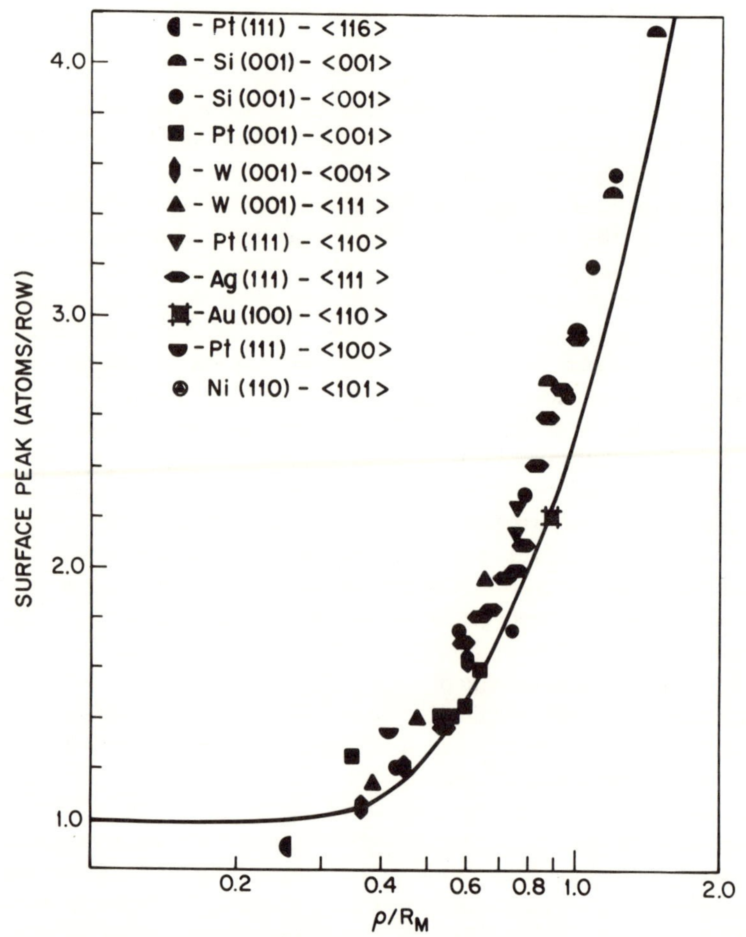

Figure 19. Comparison of the "theoretical" curve for the surface peak, based on computer simulations, with experimental values for a number of different surfaces. The experimental values were determined from backscattering measurements. The notation Ni(110)–⟨101⟩ indicates a Ni crystal with a (110) surface plane and a backscattering measurement in a ⟨101⟩ axial direction (from Ref. 19.)

A comparison of SP measurements with theory is shown in Fig. 19. This curve, originally published in 1981, represents all SP measurements known to that time. (By now there has been a substantial increase in the number of measurements and the reader is referred to Ref. 20 for a recent update.) In each measurement the sample consisted of an AES clean surface and displayed a (1 × 1) LEED pattern. The (1 × 1) LEED pattern is taken as an indication of a bulklike surface so that comparison with a bulklike SP calculation is valid. One can see that the overall trend is well established and the systematics of the surface peak are well explained by the ρ/r scaling.

In establishing this curve, bulk thermal vibrations are used throughout—i.e., for calculation and experimental plotting. This assumption is almost certainly wrong at some level, as surface thermal vibrations are not expected to be the same as bulk values. In general surface vibration amplitudes are not known for a specific material and surface. Theoretical estimates suggest that the root-mean-square thermal vibration amplitude can be up to a factor of $\sqrt{2}$ larger than the bulk value. The overall deviation in Fig. 19 between theory and experiment can be explained by an overall enhancement in amplitude of a factor of ~1.2.

One of the main advantages of ion scattering in surface studies is that simple geometrical pictures can be employed in experimental design. In Fig. 20 we show schematically the three most common methods in which SP measurements are used to probe surface structures. Figure 20a shows the ideal SP from a "bulklike" structure. Surface reconstruction (atomic translations in the plane of the surface) is indicated by a SP increase for a beam incident normal to the surface (Fig. 20b). Such measurements have been extensively applied to elemental semiconductor surfaces establishing that surface displacements of 0.1 Å can persist a few monolayers into the surface.[21–24] Reconstruction in the topmost layer of W(001) has also been established.[25]

Surface relaxation is shown in Fig. 20c. In this case reordering of the surface occurs by a change of the spacing between the first and second atomic layers; atomic spacings in the plane of the surface remain the same. A measurement of the SP in a nonnormal direction yields an increase above the expected value. Relaxation is usually explored by measurement of the SP as a function of small angular variations about the off-normal direction. Asymmetry in this angular scan reveals the sign of the lattice constant change, contraction or expansion, and the magnitude. The largest relaxation measured to date is for Pb(110), ~15%.[26] Relaxation measurements have also been reported in Ni, Cu, and Ag.[27–29]

Adsorbate site determination can be explored by SP measurements as a function of adsorbate coverage (Fig. 20d). An adsorbate will form a

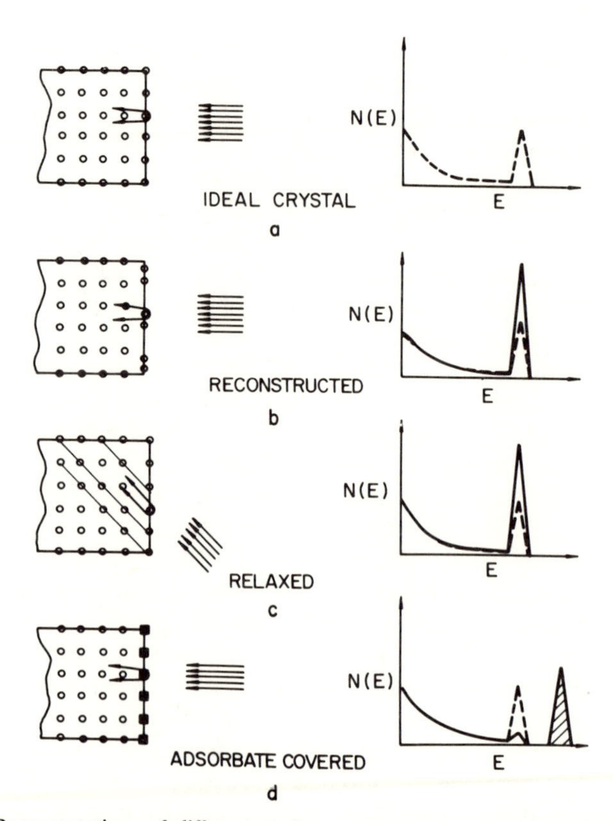

Figure 20. Representations of different surface structures on a simple cubic crystal. The backscattering spectra shown on the right-hand side represent the expected signal from the different structures. The dashed line represents the spectrum from a bulklike crystal. (From Ref. 16.)

shadow cone which will reduce the substrate SP in certain cases of registry. Obviously a positive effect can only be expected if the adsorbate is aligned to an atomic string of the substrate. This limitation makes this technique most useful in cases where the LEED pattern suggests registry or in epitaxial crystal growth investigations. An elegant and general method for adsorbate site determination is given by the channeling/blocking technique to be described later. Epitaxy studies are described in Refs. 30–32.

Surface peak analysis is also useful in studies of thin film formation and interface structure. The basic scheme is shown in Fig. 21, which illustrates: the ideal SP, the (unchanged) SP associated with a nonreactive overlayer, the increased SP formed by a reacted layer, and the reacted, thick film spectrum.

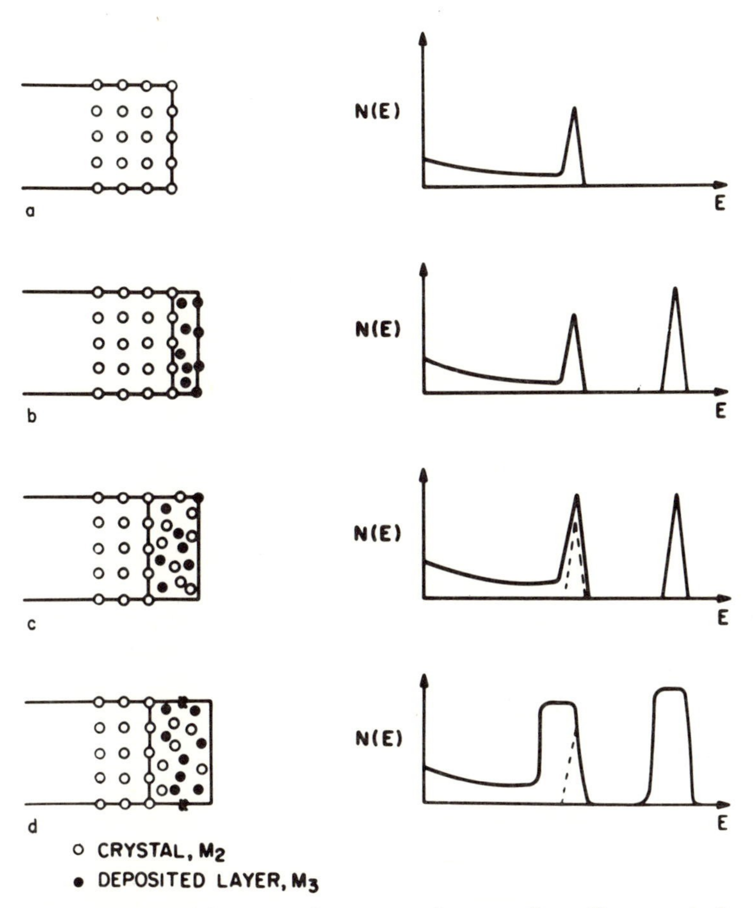

Figure 21. Schematic of ion scattering spectra for amorphous films on single-crystal substrates. (a) The uncovered crsytal with a bulklike surface; (b) a passive (nonreacting) overlayer; (c) a thin reactive film wth an increase of atoms off lattice sites; and (d) a thick reactive film. (From Ref. 16.)

For metal/semiconductor applications the object is to understand the initial phases of reactive thin film formation under UHV conditions. The experiment usually consists of the measurement of a substrate SP as a function of a deposited overlayer as illustrated schematically in Fig. 21. For example, the deposition of Ti on clean Si results in the formation of a silicide, Ti_xSi_y, which is assumed nonepitaxial. In this case one measures the reaction rate of the deposited material with the clean substrate. An accurate measure of the SP with coverage yields the stoichiometry of the initial phase that is formed (Fig. 22).[33]

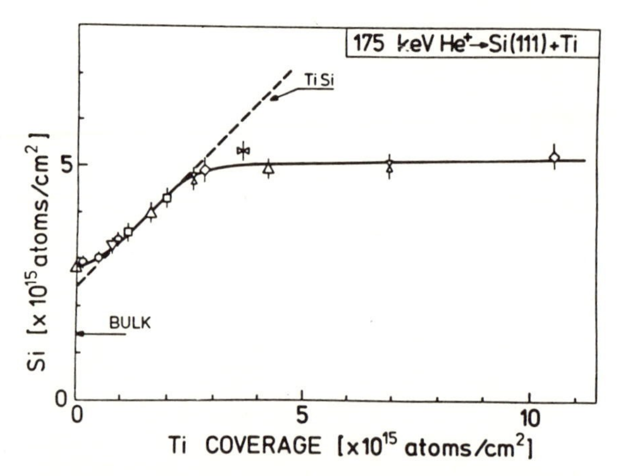

Figure 22. Increase of the Si substrate SP as a function of Ti deposition. The slope of the line indicates the initial stoichiometry, TiSi (from Ref. 33).

Another major interest is characterizing the crystal structure at the interface between a disordered layer and a single crystal. Such interfaces are common in technology as in the Si/SiO_2 interface or metal–oxide–semiconductor multilayer stacks. The principal measurement is the substrate SP as a function of overlayer thickness, similar to the thin film formation studies. An example of an aligned energy spectrum for a thin

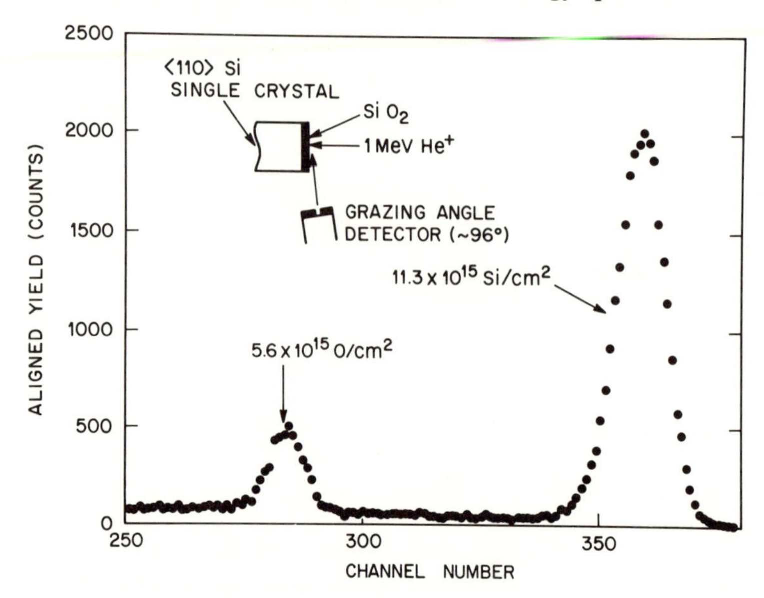

Figure 23. Backscattering spectrum from a 1.0-MeV $^4He^+$ ion incident on an aligned Si single crystal containing a ~12-Å layer of SiO_2.

layer of SiO_2 on Si is shown in Fig. 23. The Si and oxygen peaks are easily converted to a known number of atoms/cm^2 using the basic RBS analysis equations. In this case the Si peak contains contributions from Si in the oxide, the Si SP and any disordered Si at the interface. Analysis of samples with various oxide thicknesses yield the compilation shown in Fig. 24, which shows the oxide is stoichiometric SiO_2. The intercept is somewhat higher than that expected for the ideal single crystal indicating one to two monolayers of disordered Si at the interface.[34]

It is important to note from these previous two applications that the SP must be properly taken into account when analyzing thin film stoichiometry in a channeling direction. At relatively high energy (~2.0 MeV) the intrinsic (substrate) surface peak can correspond to scattering from ~20 to 30 Å of the substrate material; thus the intrinsic SP must be subtracted from the thin film yield for accurate stoichiometry determination.

5.4. Double Alignment and Transmission

The main emphasis in Section 5 has been on understanding of the surface peak. Since the surface peak is a general feature of all backscattering/channeling spectra and critical in thin film analysis it is

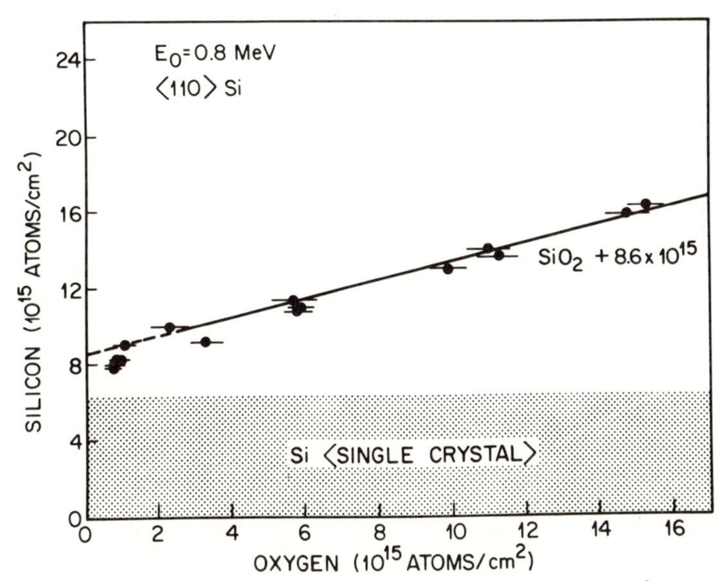

Figure 24. Si versus oxygen areal densities for a range of oxides up to ~40 Å. [From L. C. Feldman, I. Stensgaard, P. J. Silverman and T. E. Jackman, in: *The Physics of SiO_2 and Its Interfaces,* S. T. Pantelides, ed., Pergamon, New York (1978).]

important to demonstrate that the SP is well understood. In this section we describe two other surface sensitive ion scattering geometries, i.e., double alignment and thin crystal transmission.

In the double alignment geometry one makes use of channeling and blocking to ascertain surface atom positions. The geometry is shown in Fig. 25. The principal idea is to use single alignment to confine backscattering to the first few exposed layers. Backscattered particles may be blocked on their outward path. The blocking pattern is then used to yield surface coordinates. This double alignment technique has been pioneered by van der Veen, Saris, and co-workers.[11] The double

a

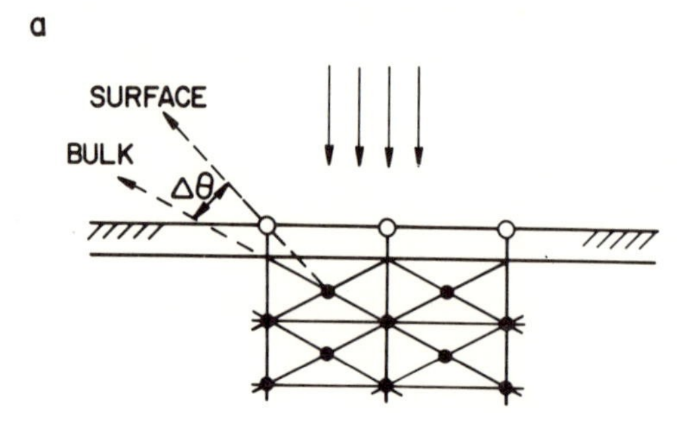

b

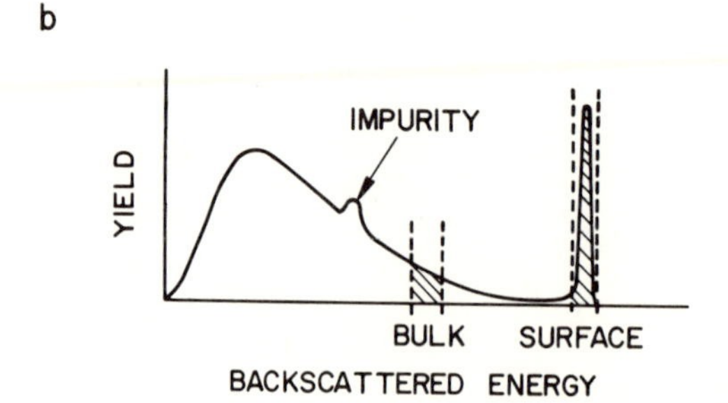

c

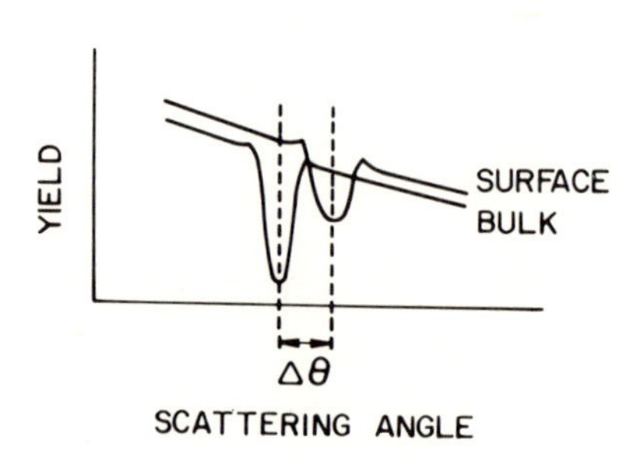

Figure 25. (a) Geometry for a double alignment experiment. (b) The backscattered energy spectrum for a light impurity on a heavier adsorbate. (c) Angular scan for scattering from the bulk and from the surface. The shift, $\Delta\theta$, indicates the relative lattice spacing between the surface layer and the bulk. [Redrawn from W. C. Turkenburg, W. Soszka, F. W. Saris, H. H. Kersten, and B. G. Colenbrander, *Nucl. Instrum. Methods* **132,** 587 (1976).]

alignment technique has been particularly fruitful, providing information over and above a single alignment experiment.

The advantage of double alignment over single alignment measurements can be realized by regarding the former as a double differential measurement in angle, while the latter is single differential, i.e., only the incident channeling direction is specified. Usually the price paid for a double differential measurement is one of efficiency; however, the

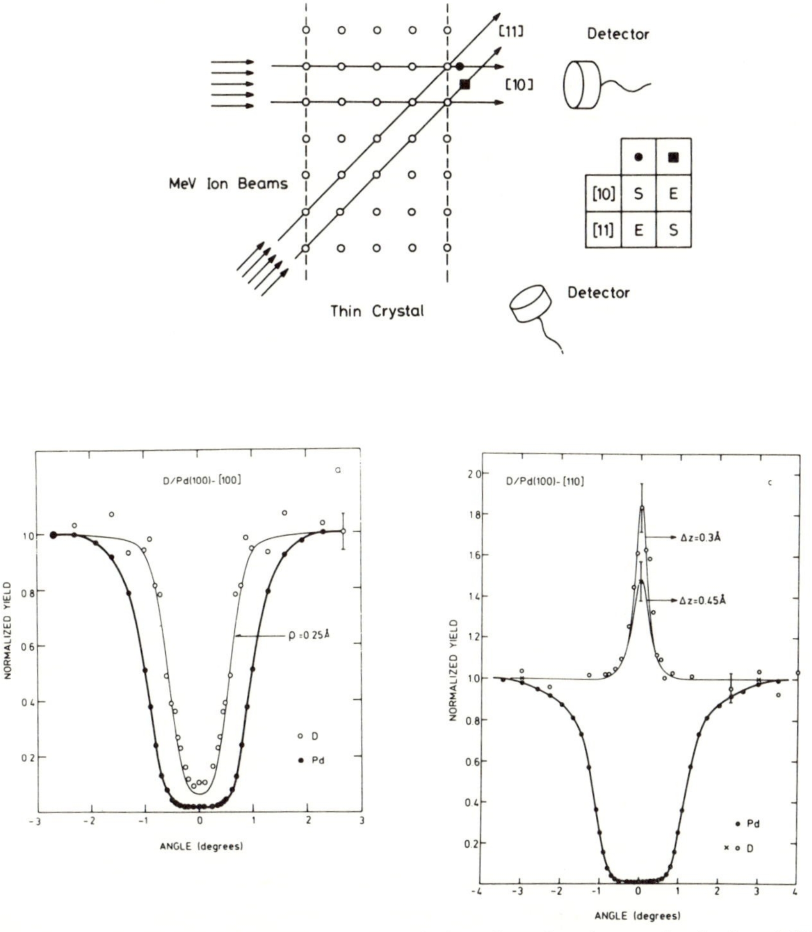

Figure 26. (a) Schematic illustrating how a particular adsorption site may be shadowed (S) or exposed (E) for different incident-beam directions in transmission-channeling experiments. Angular scans for deuterium in Pd(100) showing the normalized yield for Pd (filled circles) and deuterium (open circles) for measurements along the (b) [100] and (c) [110] directions. (From Ref. 37.)

development of an angular dispersive, high-resolution spectrometer has overcome this potential difficulty. The double alignment geometry and its role in surface crystallography has been reviewed recently by van der Veen.[35]

In the transmission geometry adsorbate surface location and thin films are studied on the back side of a thin single-crystal substrate. The major advantage in such thin film studies is that the substrate surface peak makes essentially no contribution to the scattering yield. Interpretation of ultrathin film stoichiometry is then simplified, since the substrate SP need not be considered.[36]

More recently thin crystal transmission has been used for the determination of deuterium location on single-crystal metal surfaces. The experiments make use of the bulk channeling concept, that the flux of the channel beam is sharply focused at the channel center. The basic interaction probe is the ^{3}He(d, p) ^{4}He reaction discussed earlier. Deuterium on surface sites that are in between atomic strings give rise to an enhanced yield for channeling, while deuterium on substitutional sites result in a decreased yield. Some very beautiful results using this technique are shown in Fig. 26.[37]

6. Applications

High-energy ion scattering has been used for a diversity of applications in the investigation of the near surface properties of solids. Many specific applications have been discussed in the earlier parts of this chapter as illustrations of the techniques. A survey of the broader range of applications is given in Table 3. The applications are listed in a generic format with specific materials examples given in the second column and a single reference to a particular example. A few comments are appropriate in interpreting this table.

The fundamental principles underlying an analysis technique determines the limits of applicability. These ion-scattering techniques are extremely well understood and it is straightforward to estimate the viability of a particular application. Clearly if the experiment is not going to work "on paper" it will not work in practice, given that the fundamentals are well established. The next factor that determines the applicability of an analysis technique is convenience and accuracy. For example, both Auger analysis and ion scattering yield a thin film profile; however, through experience an analyst learns the most appropriate technique. Factors under consideration concern the required quantitative accuracy, the depth resolution necessary to provide a useful answer and

Table 3. Applications

Process or structure	Typical material	Ref.
Film composition		
Insulating layers (oxides and nitrides)	Si/SiO_2	4, 38
Metal–semiconductor interactions	Ni/Si(111)	4, 39
Metal–polymer interactions	Au/polystyrene	40
Superconducting films	$Y_1Ba_2Cu_3O_7$	41
Epitaxial layers	AlGaAs/GaAs	42
Silicide formation	NiSi	43
Corrosion	Aluminium	4
Hydride analysis	PdH; NbD	44, 45
Interdiffusion	Sn/Si	4
Metallurgy	Cr/Cu; W/Cu	4
Implantation		
Dose calibration	As in Si	4
Damage and annealing studies	Si; III–V; Al	16
Impurity profiles	As in Si; Ni/Si	4
Impurity lattice location	C in Fe; B in Si	46
Formation of metastable phases	WCu	47
Buried oxides and nitrides	$Si/SiO_2/Si$	48
Implanted silicodes	Ni in Si	49
Epitaxial regrowth	Si; GaAs	16
Impurity segregation	As in Si	50
Laser annealing	Si; Ge	50
Defect analysis	Si; Al	16
Bulk properties of materials		
Composition	Superconductors	41
Diffusion of impurities	As in amorphous Si	50
Solubility studies	Dopants in Si	50
Crystallinity	Oxides	41
Thermal vibrations	Superconductors	51
Hydrogen analysis	Glasses; hydrides	5
Polymer–polymer diffusion	Polystyrenes	5
Precipitation phenomena	Dopants in Si	50
Phase transitions	Ferroelectrics	52
Impurity defect interactions	Mn in Al	16
Structure	Na β-alumina	53
Structural properties of surfaces		
Surface relaxation	Ag(110); FCCs	54
Surface reconstruction	W(100); Si(111)	16
Adsorbate coverage	Hydrogen on Ni	37
Adsorbate surface location	Sulfur on Ni	55
Hydrogen coverage	W(100)	16
Adsorbate induced reconstruction	Pt(110)	56

(continued)

Table 3. Continued

Process or structure	Typical material	Ref.
Surface melting	Pb(110)	57
Surface phase transitions	W(100)	58
Epitaxial layers		
Crystalline quality	CaF_2/Si(111)	59
Critical thickness measurements	Ge_xSi_{1-x}/Si(100)	60
Strained layers	GaAs/Si; GeSi/Si	61
Superlattices—composition and strain	GeSi . . . GeSi	61
	Si/Si Growth	32
Epitaxial insulating films	CaF_2 on Si	59
Epitaxial silicide formation	$CoSi_2$/Si(111)	62
Metal/metal epitaxy	Au/Ag; Pd/Ag(111)	16
Initial stages of epitaxy	Ge/Si; Au/Ag	63, 16
Ion beam initiated epitaxy	Si regrowth	64
Superconducting epitaxial layers	$Ba_2Y_1Cu_3O_7$/SrTiO	41
Epitaxial regrowth of implanted layers	Si; Ge; GaAs	50
Beam annealing	Si	64
Impurity segregation	Dopants in Si	50
Interfaces		
Crystalline–amorphous interface	SiO_2/Si(100)	16
Defects at crystal–crystal interfaces	Si/Sapphire	65
Interfacial reactions	Pt, W, Au/GaAs	66
Phase formation	BeAlFe Alloy	67
Grain boundary diffusion	Sb in Fe	68
Metal–ceramic	Au/silica	69
Surface processing		
Sputtering processes	Damage—InP, Si	70
Preferential sputtering	PtSi	71
Low-energy ion beam damage	Si	5
Reactive ion etching—damage	Si	72
Reactive ion etching—surface chemistry	CF/Si	72
Plasma processing	Si, InP	73, 74
Sputter deposition and damage	AlO_2 on InP	74
Plasma deposition	SiO_2 on InP	74
Surface cleaning techniques	HF/Si	75
Surface analysis		
Hydrogen analysis	Hydrogen/Ni(110)	37
Calibration for SIMS, Auger	As/Si	5
Thin films (not included above)		
Clustering and morphology	Sn, Ga on Si	76
Stoichiometry	High Tc superconductors	41
Polycrystallinity	Metal/Si	16

convenience and speed. Generally, unless a given probe does one aspect of analysis better than competing techniques it will not survive. Through trial, error, and practice the range of applications is learned by the relevant technological community. Table 3 lists those classes of applications that have been most usefully and extensively approached by ion scattering.

Specific references to ion scattering studies may be found in the extensive bibliographies listed in Refs. 4 and 16. Areas that have come into some prominence since publication of these original works are the use of ion scattering in composition determination of the new superconducting oxides (Ref. 41); the use of ion scattering/channeling for strain analysis of lattice mismatched epitaxial layers (Ref. 60) and ion scattering as a probe of thin film morphology and clustering (Ref. 76).

7. Outstanding Strengths of RBS in Relation to AES, XPS, and SIMS

Table 4 summarizes the basic properties of RBS as a surface and thin film analysis probe. Most aspects of the summary have been dealt with in the main text of the chapter.

It is of interest to compare ion scattering to the three other principal, thin film analysis techniques: AES, SIMS, and XPS. Each of these techniques, and RBS, has risen into prominence because it does (at least) one aspect of analysis well.

AES combined with sputtering is the most useful, overall thin film analysis technique. Compared to RBS it has better mass resolution, lateral resolution, and depth resolution; however, it is not sufficiently quantitative and suffers from possible sputtering artifacts.

SIMS is particularly useful in the investigation of low concentrations of dopants and impurities. Compared to RBS its sensitivity can be better by a factor of 10^2 or more for heavy Z elements (where RBS is best) and a factor of 10^4 better for low-Z elements in a high-Z matrix. Difficulties with SIMS center on quantitative analysis and matrix affects, although the post-ionized secondary neutral mass spectroscopies (treated elsewhere in this volume) are minimizing the latter. RBS can often provide a quantitative calibration for SIMS.

XPS is sensitive to chemical shifts and provides detailed chemical binding information. RBS provides no chemical information. Depth profiling with XPS is also sensitive to sputtering artifacts and the accompanying difficulties of analysis. XPS has no capability for hydrogen detection.

Table 4. Summary of Ion Scattering Analysis

Basic interactions	
Depth sampled	Few monolayers to ~1 μm
Damage	Minimal, rarely interferes with measurements
Sample parameters	
Vacuum	Must be vacuum worthy
Geometry	Wafers generally, and are particularly necessary for grazing exit studies
Type	Metals, semiconductors—OK Thick insulators—can have charging problems
Z_2 Dependence	Scattering $\propto Z_2^2$
Isotope effects	Yes—particularly for nuclear reactions
Matrix effects	No
Surface structure	Yes
Near surface structure	Yes
Experimental details	
Vacuum requirements	10^{-4}–10^{-10} Torr
Time to obtain spectrum	~1–5 min
Depth profiling	Implicit in the measurement
Lateral resolution	Typically 1 mm, but probes as small as 1 μm reported
Data analysis	
Mass identification	Yes, but resolution is difficult at high Z
Quantitaive coverage	Yes
Sensitivity	$10^{12}/cm^2$ (high Z); $10^{14}/cm^2$ (low Z)
H Detection	Yes, with nuclear reactions
Depth resolution	30–200 Å
Chemical effects	No
Simplicity of interpretation	Yes

The outstanding strength of RBS is in quantitative analysis. As described in the chapter it can measure a surface coverage accurately and easily to 5%. For thin films RBS can also yield the stoichiometry and depth distribution of the stoichiometry at a level of a few percent. In a channeling geometry, ion scattering is a near surface structural tool. Ion beam techniques can also be used for quantitative hydrogen analysis through NRA.

References

1. A. B. Brown, C. W. Snyder, W. A. Fowler, and C. C. Lauritsen, *Phys. Rev.* **82,** 159 (1951).
2. J. A. Davies, in: *Material Characterization Using Ion Beams* (J. P. Thomas and A. Cachard, eds.), p. 405, Plenum Press, London (1978).
3. L. C. Feldman and J. M. Poate, *Ann. Rev. Mater. Sci.* **12,** 149 (1982).

4. W. K. Chu, J. W. Mayer, and M. A. Nicolet, *Backscattering Spectrometry,* Academic Press, New York (1978).
5. L. C. Feldman and J. W. Mayer, *Fundamentals of Surface and Thin Film Analysis,* Elsevier, New York (1986).
6. F. Bloch, *Ann. Phys. (Leipzig)* **16,** 287 (1933).
7. J. F. Ziegler, *Helium-Stopping Powers and Ranges in All Elemental Matter,* Vol. 4, Pergamon, New York (1977).
8. L. C. Feldman and S. T. Picraux, in *Ion Beam Handbook for Materials Analysis* (J. W. Mayer and E. Rimini, eds.), Academic Press, New York (1977).
9. J. A. Davies and P. R. Norton, *Nucl. Instrum Methods* **168,** 611 (1980).
10. B. L. Doyle, *J. Vac. Sci. Technol.* **A3,** 1374 (1985); and *Nucl. Instrum Methods* **B15,** 654 (1986).
11. J. F. van der Veen, R. G. Smeenk, R. M. Tromp, and F. W. Saris, *Surf. Sci.* **79,** 212 (1979).
12. W. K. Chu, in: *Ion Beam Handbook for Materials Analysis* (J. W. Mayer and E. Rimini, eds.), Academic Press, New York (1977).
13. L. R. Doolittle, *Nuc. Instrum. Methods* **B9,** 344 (1985).
14. J. Lindhard, *Mat. Fys. Medd. Dan. Vid. Selsk.* **34,** 1 (1965).
15. D. V. Morgan, *Channeling,* Wiley, New York (1973).
16. L. C. Feldman, J. M. Mayer, and S. T. Picraux, *Materials Analysis by Ion Channeling,* Academic Press, New York (1982).
17. J. H. Barrett, *Phys. Rev. B* **3,** 1527 (1971).
18. I. Stensgaard, L. C. Feldman, and P. J. Silverman, *Surf. Sci.* **77,** 513 (1978).
19. L. C. Feldman, *Nucl. Instrum Methods* **191,** 211 (1981).
20. L. C. Feldman, in: *Ion Beams for Materials Analysis* (J. R. Bird and J. S. Williams, eds.), Academic Press, Australia (1989).
21. I. Stensgaard, L. C. Feldman, and P. J. Silverman, *Surf. Sci.* **102,** 1 (1981).
22. R. M. Tromp, R. G. Smeenk, and F. W. Saris, *Surf. Sci.* **104,** 13 (1981).
23. R. J. Culbertson, L. C. Feldman, and P. J. Silverman, *Phys. Rev. Lett.* **45,** 2043 (1980).
24. R. M. Tromp, E. J. van Loenen, M. Iwami, and F. W. Saris, *Solid State Commun.* **44,** 971 (1983).
25. I. Stensgaard, L. C. Feldman, and P. J. Silverman, *Phys. Rev. Lett.* **42,** 247 (1979); and L. C. Feldman, P. J. Silverman, and I. Stensgaard, *Surf. Sci.* **87,** 410 (1979).
26. J. W. M. Frenken and J. F. van der Veen, *Phys. Rev. Lett.* **54,** 134 (1985).
27. D. L. Adams, H. B. Nielsen, J. N. Andersen, I. Stensgaard, R. Feidenhans'l, and J. E. Sorensen, *Phys. Rev. Lett.* **49,** 669 (1982).
28. R. Feidenhans'l and I. Stensgaard, *Surf. Sci.* **133,** 453 (1983).
29. Y. Kuk and L. C. Feldman, *Phys. Rev. B* **30,** 5811 (1984).
30. R. J. Culbertson, L. C. Feldman, P. J. Silverman, and H. Boehm, *Phys. Rev. Lett.* **47,** 657 (1981).
31. Y. Kuk, L. C. Feldman, and P. J. Silverman, *J. Vac. Sci. Technol.* **A1,** 1060 (1983).
32. H.-J. Gossmann and L. C. Feldman, *Phys. Rev. B* **32,** 6 (1985).
33. E. J. van Loenen, A. E. M. J. Fischer and J. F. van der Veen, *Surf. Sci.* **155,** 65 (1985).
34. R. Haight and L. C. Feldman, *J. Appl. Phys.* **53,** 4884 (1982) and references therein.
35. J. F. van der Veen, *Surf. Sci. Rep.* **5,** 1 (1985).
36. N. W. Cheung, L. C. Feldman, P. J. Silverman, and I. Stensgaard, *Appl. Phys. Lett.* **35,** 859 (1979).
37. F. Besenbacher, I. Stensgaard, and K. Mortensen, *Surf. Sci.* **191,** 288 (1987).
38. W. K. Chu, J. W. Mayer, M.-A. Nicolet, T. M. Buck, G. Amsel, and F. Eisen, *Thin Solid Films* **17,** 1 (1973).

39. K. N. Tu, W. K. Chu, and J. W. Mayer, *Thin Solid Films* **25,** 403 (1975).
40. R. M. Tromp, F. Legoues, and P. S. Ho, *J. Vac. Sci. Technol.* **A3,** 782 (1985).
41. J. M. E. Harper, R. J. Colton, and L. C. Feldman, eds, *Thin Film Processing and Characterization of High-Temperature Superconductors,* American Institute of Physics, New York (1988).
42. F. W. Saris, W. K. Chu, C. A. Cheng, R. Ludeke, and L. Esaki, *Appl. Phys. Lett.* **37,** 931 (1980).
43. R. T. Tung, J. C. Bean, J. M. Gibson, J. M. Poate, and D. C. Jacobson, *Appl. Phys. Lett* **40,** 684 (1982).
44. R. S. Blewer, *Appl. Phys. Lett.* **23,** 593 (1973).
45. W. A. Lanford, H. P. Troutvelter, J. F. Ziegler, and J. Keller, *Appl. Phys. Lett.* **28,** 566 (1976).
46. L. C. Feldman, E. N. Kaufman, J. M. Poate, and W. M. Augustyniak, in *Ion Implantation in Semiconductors and Other Materials* (B. L. Crowder, ed.), Plenum Press, New York (1973).
47. J. M. Poate and A. G. Cullis, in *Treatise on Materials Science and Technology,* (J. Hivonen, ed.), Academic Press, New York (1980).
48. G. K. Celler, P. L. F. Hemment, K. W. West, and J. M. Gibson, *Appl. Phys. Lett.* **48,** 532 (1986).
49. A. E. White, K. T. Short, R. C. Dynes, J. P. Garno, and J. M. Gibson, *Appl. Phys. Lett.* **50,** 95 (1987).
50. J. M. Poate and J. W. Mayer, eds., *Laser Annealing of Semiconductors,* Academic Press, New York (1982).
51. R. P. Sharma, L. E. Rehn, P. M. Baldo, and J. Z. Liu, *Phys. Rev. B* **38,** 9287 (1988).
52. D. S. Gemmell, *Rev. Mod. Phys.* **46,** 129 (1974).
53. S. J. Allen, L. C. Feldman, D. B. McWhan, J. P. Remeika, and R. E. Walstedt, in *Superionic Conductors* (G. D. Mahon and W. L. Roth, eds.), Plenum Press, New York (1976).
54. Y. Kuk and L. C. Feldman, *Phys. Rev. B* **30,** 5811 (1984); and S. M. Yalisove, W. R. Graham, E. D. Adams, M. Copel, and T. Gustafsson, *Surf. Sci.* **171,** 400 (1986).
55. J. F. van der Veen, R. M. Tromp, R. G. Smeenk, and F. W. Saris, *Surf. Sci.* **82,** 468 (1979).
56. T. E. Jackman, K. Griffiths, J. A. Davies, and P. R. Norton, *J. Chem. Phys.* **79,** 3529 (1983).
57. Joost W. M. Frenken and J. F. van der Veen, *Phys. Rev. Lett.* **54,** 134 (1985).
58. I. Stensgaard, K. G. Purcell, and D. A. King, *Phys. Rev. B* **39,** 897 (1989).
59. J. M. Phillips, L. C. Feldman, J. M. Gibson, and M. L. McDonald, *Thin Solid Films* **107,** 217 (1983).
60. A. T. Fiory, J. C. Bean, L. C. Feldman, and I. K. Robinson *J. Appl. Phys.* **56,** 1227 (1984).
61. L. C. Feldman, J. Bevk, B. A. Davidson, H.-J. Gossmann, and J. P. Mannaerts, *Phys. Rev. Lett.* **59,** 664 (1987); and W. K. Chu, F. W. Saris, C. A. Chang, R. Ludeke, and L. Esaki, *Phys. Rev. B* **26,** 1999 (1982).
62. R. T. Tung, J. M. Gibson, and J. M. Poate, *Phys. Rev. Lett.* **50,** 429 (1983).
63. H.-J. Gossmann, L. C. Feldman, and W. M. Gibson, *Surf. Sci.* **155,** 413 (1985).
64. J. Linnros, G. Holmen, and B. Svensson, *Phys. Rev. B* **32,** 2770 (1985).
65. S. T. Picraux and P. Rai-Chadbury, in *Semiconductor Characterization Techniques* (P. A. Barnes and G. A. Rozgonyi, eds.), Electrochemical Society, Princeton (1978).
66. A. K. Sinha and J. M. Poate, *Appl. Phys. Lett.* **23,** 666 (1973).
67. S. M. Mayers and J. E. Smugeresky, *Metal Trans.* **7A,** 795 (1976).

68. M. Guttmann, P. R. Krake, F. Abel, G. Amsel, M. Bruneaux, and C. Cohen, *Scr. Metall.* **5,** 479 (1971).
69. D. V. Morgan, M. J. Howes, and C. J. Madoms, *J. Electrochem Soc.* **123,** 295 (1976).
70. R. S. Williams, *Solid State Commun.* **41,** 153 (1982).
71. Z. L. Liau and J. W. Mayer, in *Ion Implantation* (J. K. Hirvonen, ed.), Academic Press, New York (1980).
72. G. S. Oeherlein, R. M. Tromp, J. C. Tsang, Y. H. Lee, and E. J. Petrillo, *J. Electrochem. Soc.* **132,** 1441 (1985).
73. B. Robinson, T. N. Nguyen, and M. Copel, in *Deposition and Growth: Limits for Microelectronics* (G. W. Rubloff, ed.), American Institute of Physics, New York (1988).
74. W. C. Dautremont-Smith and L. C. Feldman, *Thin Solid Films* **105,** 187 (1983).
75. See references, in *Silicon-Molecular Beam Epitaxy* (E. Kasper and J. C. Bean, eds.), C.R.C. Press, Boca Raton, Florida (1988).
76. M. Zinke-Allmang, L. C. Feldman, and S. Nakahara, *Appl. Phys. Lett.* **51,** 975 (1987).

6

Ion Scattering Spectroscopy

E. Taglauer

1. Introduction

Surface characterization by ion scattering spectroscopy means the determination of the atomic masses and their geometric arrangement on a solid surface. A low-energy ion beam is a well-suited probe for such investigations because of the strong interaction between the ions in the considered energy regime and the surface atoms. "Low-energy" here refers to a range from a few hundred electron volts up to several keV. There are some essential features of ion scattering spectroscopy (ISS) or low-energy ion scattering (LEIS) which define its usefulness as a surface analytical method. (1) It is extremely surface sensitive, to the extent that the scattering signal can arise exclusively from the outermost atomic layer. (2) The signal is mass sensitive such that the elemental composition and mass selective structural information are obtained from the surface. (3) The information is obtained in real space and therefore the interpretation of the data, at least in a first approach, can rely on fairly simple concepts. (4) By the sputtering effect of the ion beam, compositional depth profiles of near surface layers can be gained. (5) Ion scattering equipment can be incorporated into an ultrahigh vacuum apparatus without too much expense in terms of investment and construction.

The method of ISS was introduced in 1967 by D. P. Smith[1] and

E. Taglauer • Surface Physics Department, Max-Planck-Institut für Plasmaphysik, EURATOM Association, D-8046 Garching bei München, Germany.

Ion Spectroscopies for Surface Analysis (Methods of Surface Characterization series, Volume 2), edited by Alvin W. Czanderna and David M. Hercules. Plenum Press, New York, 1991.

since then numerous applications have been published. The technique was further developed in recent years by adding the use of alkali ions to the originally used noble gas ions, by detecting scattered neutral particles and also direct recoil ions, and by using large scattering angles (close to 180°) for structural studies.

Naturally there are also limitations in the ion scattering method, the most serious lying in the fact that it is not a priori quantitative due to the neutralization of the ions at the surface. The neutralization probability in noble gas ion scattering is not always sufficiently well known. The use of alkali ions or neutral detection can also encounter problems because of multiple scattering effects. Nevertheless, there are many examples in the literature of very successful uses of the unique features of ISS for surface characterization.

In the following sections the basic principles of ISS are described, followed by a description of the essential experimental features and an outline of theoretical methods of calculating intensity distributions. Examples of polycrystalline and amorphous samples show the potential of ISS for elemental compositional analysis of surfaces and near-surface depth profiles. The last section refers to characterization of single crystalline surfaces and ordered adsorption structures. This includes deviations from a rigid lattice due to defects or thermal vibrations.

2. Basic Principles

2.1. Parameter Range

For the discussion of the physical processes relevant to the interaction between the energetic ions and a surface let us consider a typical

Table 1. Physical Parameters Relevant to Low-Energy Ion Scattering

		Ion (He^+, Li^+)	Solid (Ni)
Energy (eV)	Kinetic	1000	Phonon 0.03
	Ionization potential	24.5 (He); 5.5 (Li)	Work function 5
Velocity ($cm\ s^{-1}$)		2×10^7	3×10^4
Length (Å)	λ De Broglie	10^{-2}	Lattice parameter 2–3
	Distance of closest approach r_0	0.5	
Time (s)	Collision time	5×10^{-16}	Vibrational period 10^{-13}

case such as a 1-keV He^+ or Li^+ ion* impinging on a metal surface, such as nickel. Some of the pertinent physical parameters for such a case are listed in Table 1, giving the basis for the description. The interaction between the ions and the surface atoms can be treated by classical mechanics because quantum effects are negligible if we consider scattering angles larger than Bohr's critical angle[2,3] θ_c. $\theta_c \approx \lambda/r_0$ is indeed very small compared to all practically used scattering angles. Similarly, diffraction effects are not relevant since $\lambda \ll d$. It is also obvious from the values in Table 1 that the scattering process can be considered as one collision or a series of two-body collisions between the projectile and individual target atoms. Thermal motions of these atoms are slow compared to the ion velocities, so that owing to the short interaction time the projectile interacts with a lattice of thermally displaced atoms essentially at rest. For the description of the scattering kinematics it is therefore sufficient to consider two-body collisions for which the scattered ion energies and the scattering cross sections can be calculated.

2.2. Binary Collisions

2.2.1. Energy Spectrum

For the collision of two masses M_1 and M_2 interacting through a centrosymmetric potential, the energies after the encounter can be

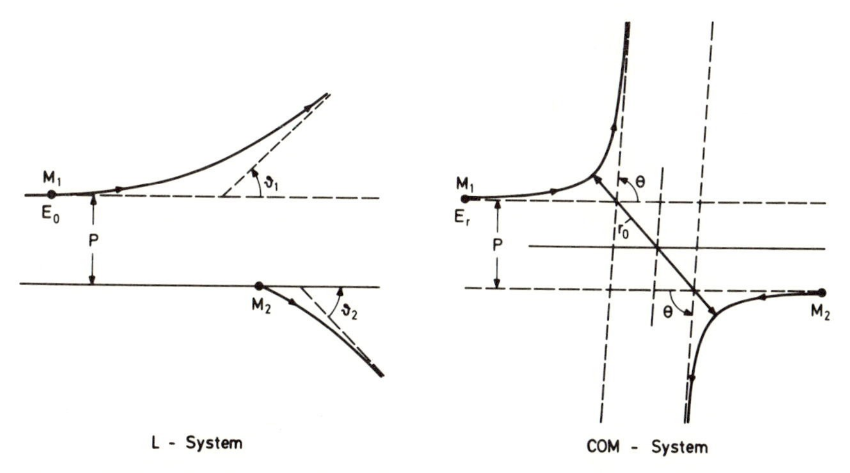

Figure 1. Trajectories for the elastic collision of two masses M_1 and M_2 with scattering angles of ϑ in the laboratory system (left) and θ in the center-of-mass system (right). The impact parameter is p; distance of closest approach is r_0.

* In this chapter the most abundant isotopic species are always taken for the projectiles without specification, i.e., ^4He, ^{20}Ne, ^7Li, ^{39}K, ^{40}Ar.

calculated from the conservation of energy and momentum.[4] Usually the target atom M_2 is considered to be initially at rest while the projectile M_1 has an initial energy E_0. The energy E of the projectile after being scattered into an angle ϑ_1 in the laboratory system (see Fig. 1) is

$$\frac{E}{E_0} = \frac{1}{1 + (M_2/M_1)^2} \left\{ \cos \vartheta_1 \pm \left[\left(\frac{M_2}{M_1} \right)^2 - \sin^2 \vartheta_1 \right]^{1/2} \right\}^2 \quad (1)$$

Thus E/E_0 is a function of the mass ratio $A = M_2/M_1$ and the scattering angles ϑ_1 [i.e., $E = E_0 f(\vartheta A)$]. For $M_2 > M_1$ the positive sign holds, for $M_2 < M_1$ both signs apply, i.e., in the latter case there are two final energies for each scattering angle ϑ_1. On the other hand, there is only a limited regime of scattering angles accessible for heavier projectiles, the limit being given by $\sin \vartheta_1 = M_2/M_1$. The function $f(\vartheta, A)$ is plotted in Fig. 2 showing the strong variation of the angular dependence with the

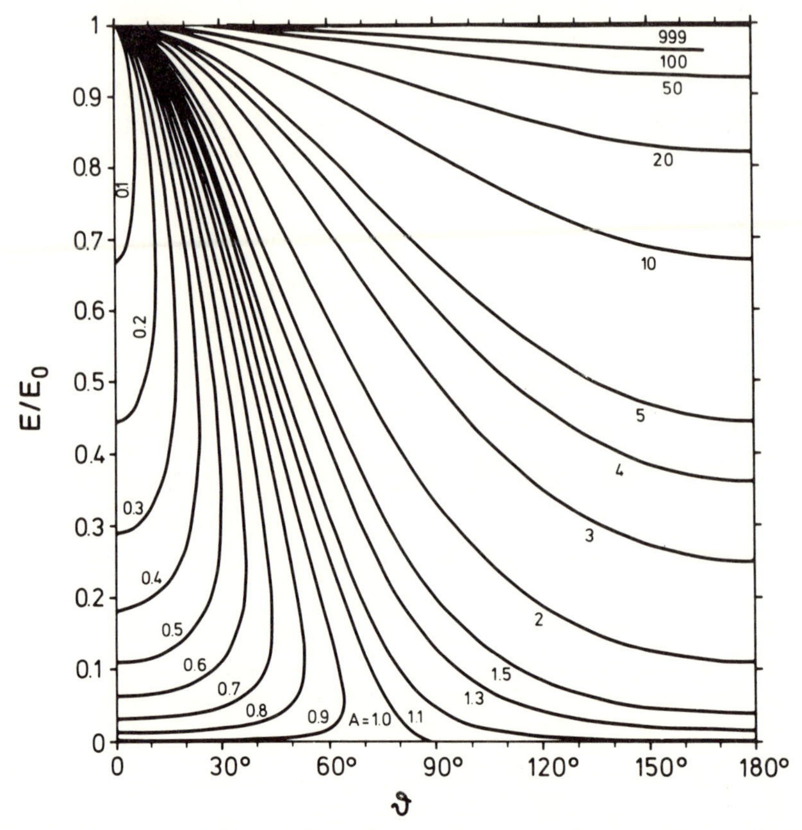

Figure 2. The function $E/E_0 = f(\vartheta)$; see Eq. (1). Parameter is the mass ratio $A = M_2/M_1$.

mass ratio A. Similar masses give the strongest functional variation with scattering angle, and correspondingly, for a given scattering angle the maximum energy loss, i.e. the widest separation in the energy spectrum is obtained for scattering from masses that are close to the projectile mass.

Equation (1) becomes particularly simple for a scattering angle of $\vartheta_1 = 90°$:

$$E/E_0 = (M_2 - M_1)/(M_2 + M_1) \tag{1a}$$

Examples of measured energy spectra are shown in Fig. 3 that demonstrate that the positions of the energy peaks in ISS can be correctly calculated from Eq. (1).

Similarly, the energy of the recoiling target atom can be calculated to be

$$E_2/E_0 = [4M_1M_2/(M_1 + M_2)^2]\cos^2 \vartheta_2 \tag{2}$$

and consequently also peaks due to recoiling ions can be observed in the energy spectra as shown in Fig. 4. It follows from the derivation of Eq. (2) that direct recoil detection requires scattering angles below 90°.

Only a two-body interaction has been considered and is sufficient to describe the collision kinematics. That is, for the scattering events, the

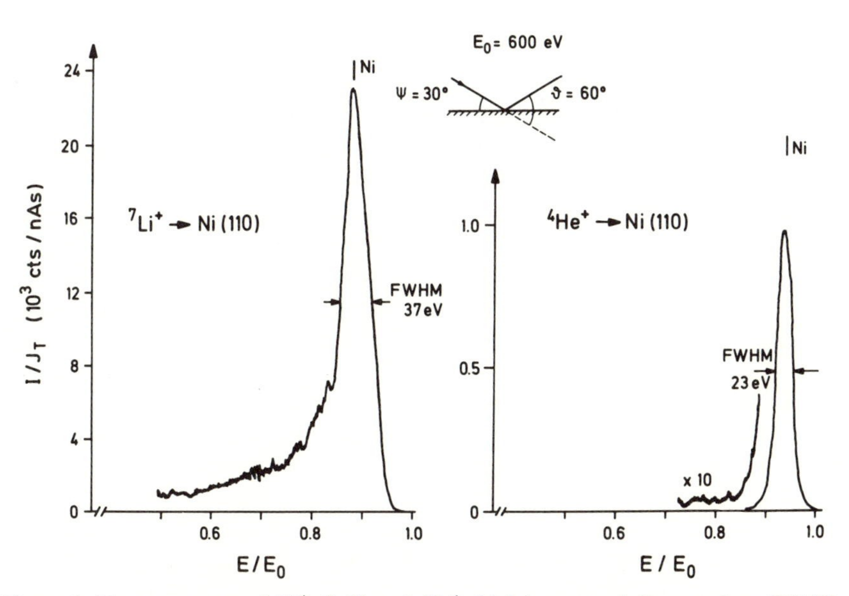

Figure 3. Energy spectra of Li^+ (left) and He^+ (right) scattered from a clean Ni(110) surface. The indicated positions of the Ni peaks are calculated using Eq. (1). Note the differences in ion yield (given relative to the target current) and in full width at half-maximum (FWHM).[5]

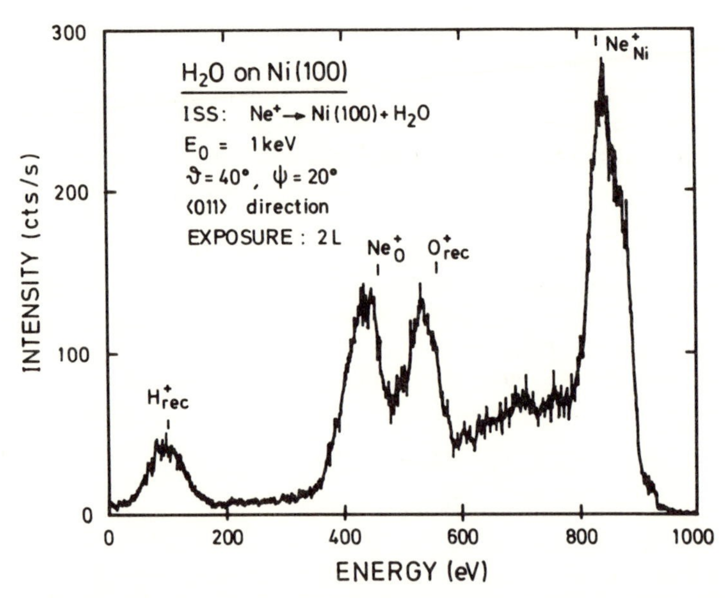

Figure 4. Ion energy spectrum for Ne^+ bombardment of a Ni(100) surface with H_2O adsorbed at 104 K. Ne^+ scattering and O^+ and H^+ recoil peaks are indicated according to Eqs. (1) and (2), respectively.[6]

surface appears to consist of single isolated atoms; if we have conditions as in Figs. 3 and 4, for one impinging ion the surface consists only of one atom. In the case of multiple scattering events that are considered below, a *series* of single collisions describes the situation. It has in fact been shown for He^+ scattering from Pb that the same energy spectra are obtained for scattering from a solid surface and from Pb atoms in the vapor phase.[7]

The question of a low-energy limit for the validity of the binary collision model was addressed by Veksler[8] in 1963 who claimed scattering of alkali ions from several surface atoms at the same time, having an effective mass M_2 corresponding to the sum of these masses. This effect was observed for primary energies below 50 eV. For He^+ and Ne^+ scattering from clean surfaces no such effect could be found down to 20 eV.[9] For Li^+ it was observed[10] that for primary energies below about 15 eV the attractive image force had to be taken into account to obtain reasonable agreement between experimental and calculated peak positions. Equation (1) is then modified by a potential step of height ϵ:

$$(E + \epsilon)/(E - \epsilon) = f(\vartheta, A) \tag{3}$$

These limiting cases are, however, not significant for the practical purposes of surface analysis.

Detection of very light surface atoms, e.g., hydrogen isotopes, is not very easy as can be deduced from Eq. (1) and Fig. 2. Although He^+ scattering from deuterium at very small scattering angles has been reported,[11] it is generally more effective to detect recoiling particles. In particular, negative recoil ions have the advantage that there is no background from scattered particles in the negative spectrum.[12,13] The kinematics for light target atoms is illustrated in Fig. 5 (using the representation of Ref. 14) showing that for recoils also much larger detection angles are possible.

The mass resolution in the energy spectra, i.e., the separation of peaks as far as collision kinematics are concerned, can also be derived

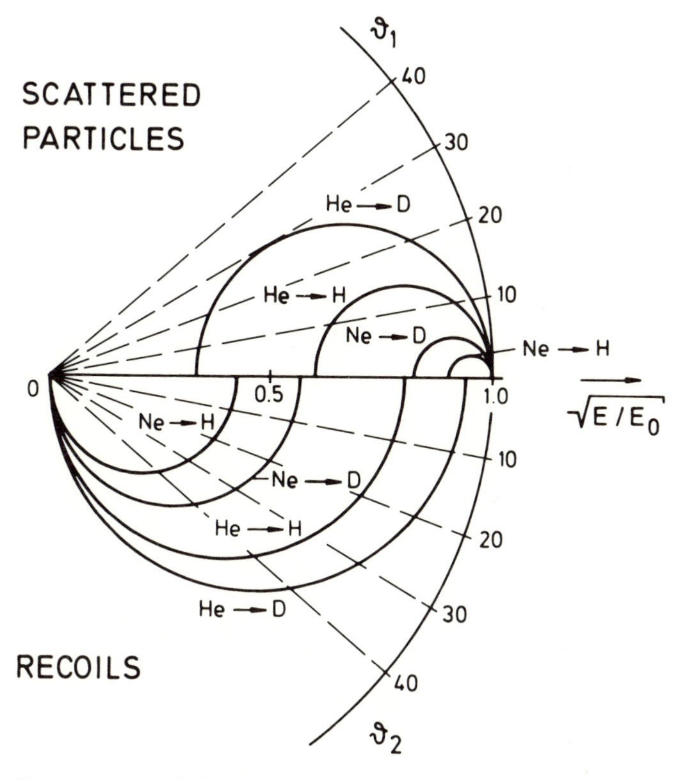

Figure 5. Representation of the kinematics of scattered and recoil particles (He and Ne projectiles, H and D target atoms).

from Eq. (1) and the slope of the curves in Fig. 2:

$$\frac{M_2}{\Delta M_2} = \frac{E}{\Delta E} \cdot \frac{2A}{A+1} \frac{A + \sin^2 \vartheta_1 - \cos \vartheta_1 (A^2 - \sin^2 \vartheta_1)^{1/2}}{A^2 - \sin^2 \vartheta_1 + \cos \vartheta_1 (A^2 - \sin^2 \vartheta_1)^{1/2}} \tag{4}$$

For the special case of $\vartheta_1 = 90°$ the mass resolution is

$$\frac{M_2}{\Delta M_2} = \frac{E}{\Delta E} \cdot \frac{2A}{A^2 - 1} \tag{4a}$$

In Fig. 6 data from Eq. (4a) are plotted for an electrostatic analyzer with a resolution of $\Delta E/E = 3\%$.

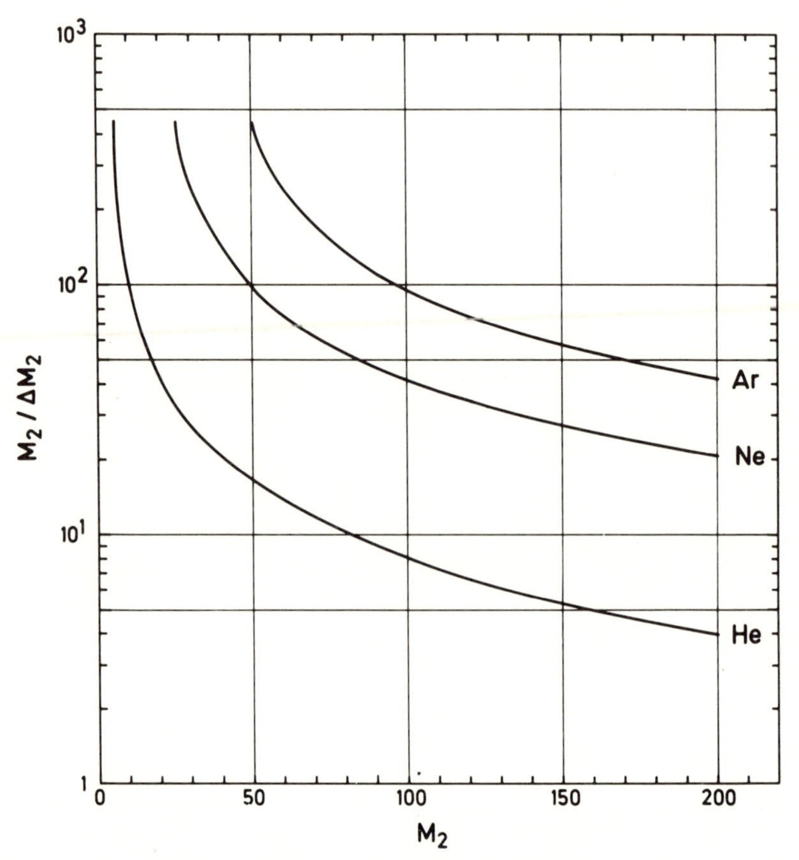

Figure 6. Mass resolution $M_2/\Delta M_2$ as a function of the target atom mass M_2 for three different projectile masses, according to Eq. (4a).

2.2.2. Interaction Potentials and Cross Sections

The energy of a particle scattered into a certain angle is determined by the conservation laws as explained in the previous section. The probability for this scattering process, i.e., the scattered particle intensity, is determined by the potential acting between the ion and the target atom (we do not consider electronic charge exchange processes at this stage). This probability is usually described by a cross section σ, and according to the experimental situation the double differential scattering cross section $d^2\sigma/d\Omega\, dE$ with respect to the intervals of energy, dE, and solid angle, $d\Omega$, is relevant. Since for a given pair of collision partners the energy for scattering into dE is determined by Eq. (1), the differential cross section $d\sigma/d\Omega$ has to be found. In the center-of-mass system a flux of particles through the infinitesimal area $d\sigma$ with an impact parameter p is scattered into an angle θ (see Fig. 1); thus $d\sigma = 2\pi p\, dp$. For scattering angles, θ will be consistently used for a center-of-mass system and ϑ for a laboratory system. For a spherically symmetric potential the connection between θ and the impact parameter p is given by the scattering integral (see, e.g., Ref. 4)

$$\theta = \pi - 2\int_{r_0}^{\infty} \frac{p\, dr}{r^2\left[1 - \frac{p^2}{r^2} + \frac{V(r)}{E_r}\right]^{1/2}} \tag{5}$$

where r_0 is the distance of closest approach of the collision partners, E_r is the energy of their relative motion, and $V(r)$ is the interaction potential.

The scattering angles ϑ in the laboratory system and θ in the center-of-mass system are connected via the mass ratio:

$$\tan \vartheta_1 = \sin \theta/[(M_1/M_2) + \cos \theta], \qquad \vartheta_2 = \tfrac{1}{2}(\pi - \theta) \tag{6}$$

Using these formulas $d\sigma/d\Omega$ can be found if the appropriate interaction potential is known. In the energy range considered here only the repulsive interaction between two atoms is relevant, which is determined by their nuclear charge and the screening of this charge by the electron clouds.[3,15] Therefore screened Coulomb potentials are usually considered that have the general form,

$$V(r) = \frac{Z_1 Z_2 e^2}{r}\, \phi\left(\frac{r}{a}\right) \tag{7}$$

where r is again the internuclear separation and a the so-called screening parameter in the screening function ϕ. For the energy range of interest here, it is assumed that the interaction potential of singly charged ions is the same as for neutral atoms;[3] i.e., for cross-section calculations there is usually no distinction made between He and He^+ or Ne and Ne^+. For the screening parameter an expression according to Firsov[16] can be used

$$a_F = \frac{0.8853a_0}{(Z_1^{1/2} + Z_2^{1/2})^{2/3}} \tag{8}$$

in which Z_1 and Z_2 are the nuclear charge numbers of the colliding atoms, and a_0 is the Bohr radius of 0.529 Å [the numerical factor is $(9\pi^2/128)^{1/3}$].

A value of the order of $0.8a_F$ has been reported to give better agreement with experimental results.[17–19] A similar expression for the screening length was developed by Lindhard.[20]

There are several analytical approximations given for the screening function ϕ. The Molière[21] approximation to the Thomas–Fermi function has become most widely used in ISS, it is given by a sum of three exponentials:

$$\phi(\chi) = 0.35e^{-0.3\chi} + 0.55e^{-1.2\chi} + 0.10e^{-6\chi} \tag{9}$$

with

$$\chi = r/a$$

The corresponding potential $V(r)$ is plotted in Fig. 7, together with two

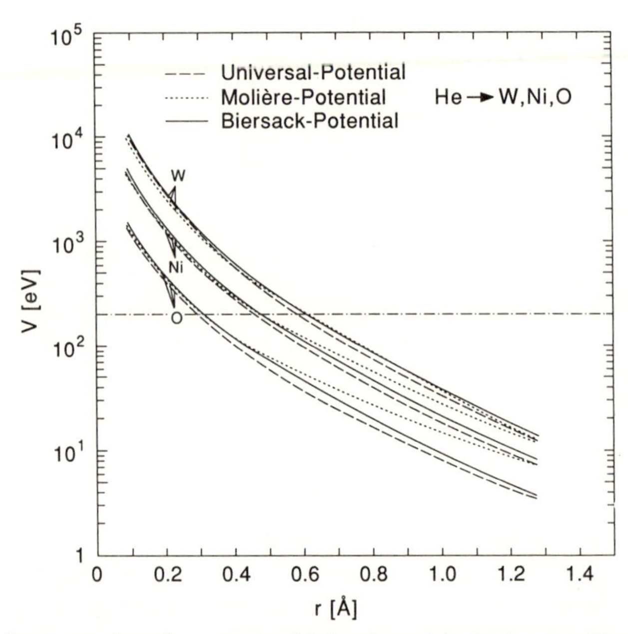

Figure 7. Representation of some potential functions given in the text. The dash–dotted line at 200 eV is the typical lowest energy used for practical applications of ISS.

other recently proposed potential functions, namely, the "universal potential" by Ziegler *et al.*[22] and an improved version by Biersack *et al.*[23] The universal potential which has the form

$$\phi_u(\chi) = 0.1818e^{-3.2\chi} + 0.5099e^{-0.9423\chi} + 0.2802e^{-0.4029\chi} + 0.02817e^{-0.2016\chi} \quad (10)$$

results from empirical fitting of nuclear stopping power data. All

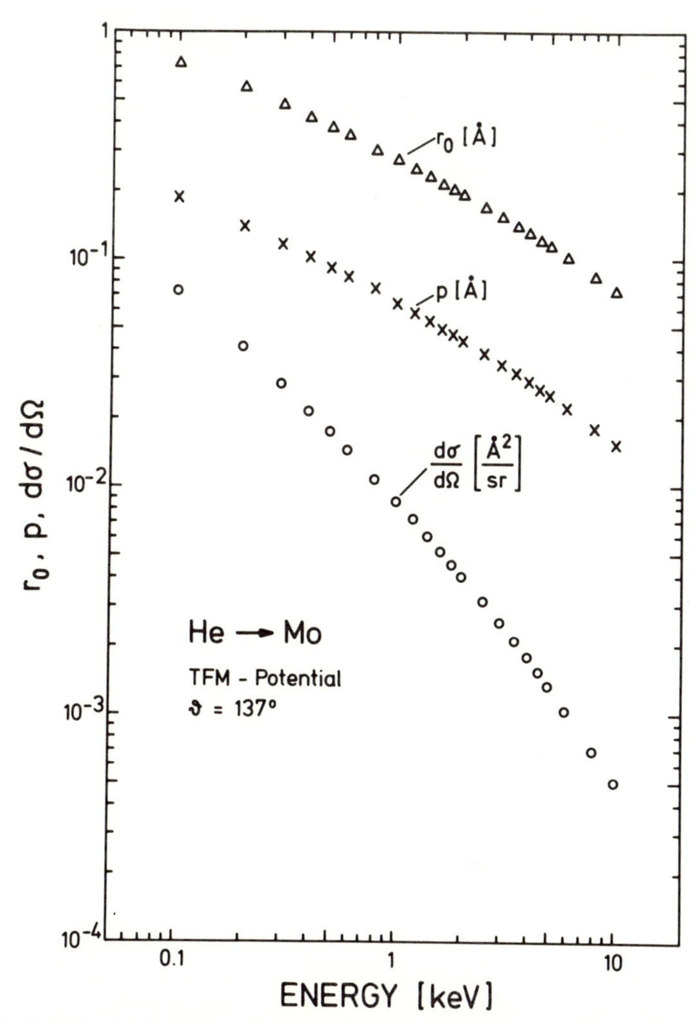

Figure 8. Calculated values of the distance of closest approach r_0, impact parameter p, and scattering cross section $d\sigma/d\Omega$ as a function of primary energy of He scattered from Mo using a Thomas–Fermi–Molière potential.

potentials agree quite well up to separations of about 0.4 Å; the deviations become more pronounced at lower target atomic masses. A potential that was frequently used in early ion scattering work is the two parameter Born–Mayer potential[15]

$$V(r) = Ce^{-r/a} \tag{11}$$

which was deduced to describe the closed-shell repulsion in ionic crystals. It should only be used for small overlap of the closed shells, i.e., $E_0 < C/2$. In the following, a few examples are given for calculations using the Thomas–Fermi–Molière potentials (TFM) with $a = a_F$ (from Eq. (8)].

Figures 8 and 9 show calculated values of the distance of closest approach r_0, the impact parameter p, and the differential scattering cross section $d\sigma/d\Omega$ as a function of energy and laboratory scattering angle, respectively. The variation of $d\sigma/d\Omega$ with the nuclear charge number Z_2 for 1-keV He is given in Fig. 10 for two different scattering angles that are frequently used in analytical work. Examples for the dependence of the scattering cross section on the primary energy of the helium projectile, E_0, are plotted in Fig. 11. These numbers are useful for quantitative evaluation of ion scattering data. It has been shown that the

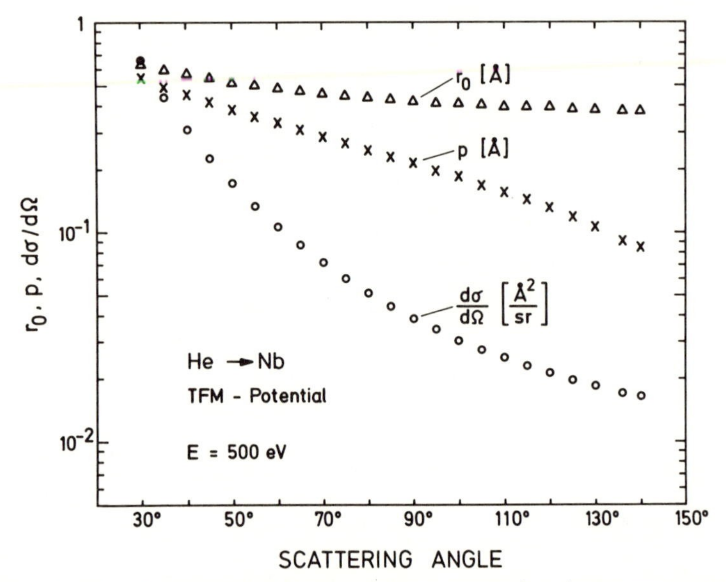

Figure 9. Calculated values of the distance of closest approach r_0, impact parameter p, and scattering cross section $d\sigma/d\Omega$ as a function of the scattering angle for He scattering from Nb.

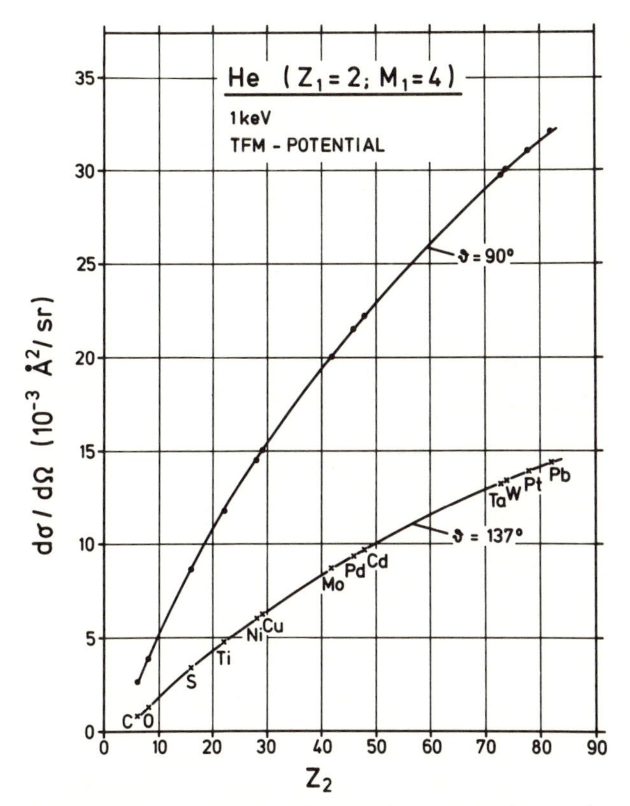

Figure 10. Calculated dependence of the He scattering cross section on the target atomic number for two different scattering angles and 1 keV primary energy.

cross sections calculated with the TFM potential agree well with experimental data.[18,19] For quantitative analysis, however, the neutralization of incident ions is extremely important (see Section 2.4.).

2.3. Multiple Scattering

The objective in ISS is to determine the masses of atoms that are present on the surface. For this purpose the single scattering model, i.e. the collision of a projectile ion with an 'isolated' single surface atom, is applied and turns out to be adequate. The influence of other atoms around the target atom is, however, not completely negligible. In a certain parameter range this is relevant for the scattering kinematics as well as for neutralization effects. As far as surface analysis is concerned, multiple scattering can be used for structure determination, and applica-

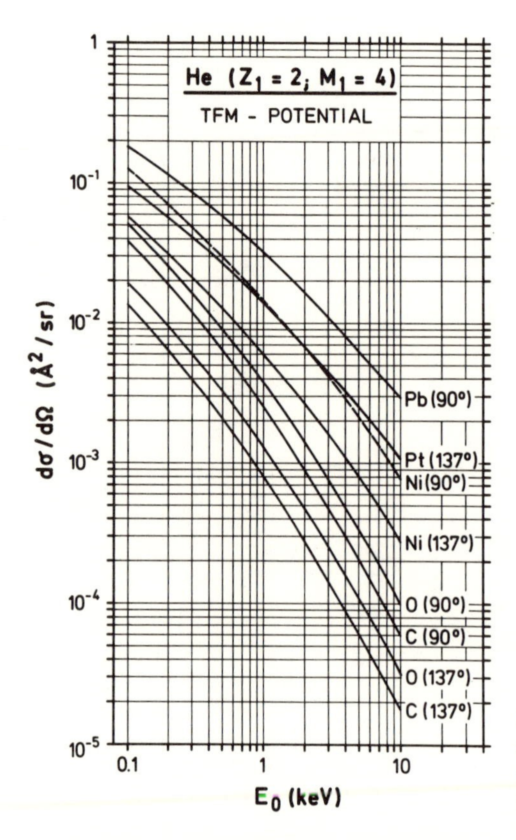

Figure 11. Calculated He scattering cross sections as a function of primary energy for various target atoms and scattering angles.

tions of this will be given in Section 6. Here a few characteristics of multiple scattering kinematics will be briefly outlined (for a detailed discussion see Ref. 15).

For a scattering experiment with Ne^+, Ar^+, or heavier ions and grazing incidence ($\psi < 30°$), the energy spectra obtained are not as simple as those shown in Fig. 3. Additional scattering peaks appear that can be attributed to multiple scattering events. In the simplest case, the ion collides with a single second atom after the initial collision. Owing to the nonlinear dependence of E on ϑ [see Eq. (1) and Fig. 2], the final energy after two consecutive collisions with a scattering angle of $\vartheta/2$ each is higher than a single scattering at an angle ϑ. This has been experimentally verified.[24,25] Repeated application of Eq. (1) yields for the special case of equal scattering angles ϑ/n and scattering from equal masses for the final energy

$$E_n = E_0 f^n(\vartheta/n, A) \tag{12}$$

where f is the scattering function of Eq. (1) and n is the number of collision events. Scattering from different masses and with a symmetric scattering angle was recently analyzed by Aono.[26] An important feature in double scattering can be seen by considering the impact parameters involved. The impact parameter for the second collision is given by

$$p_2 = p_1 - d \sin(\psi - \vartheta) \tag{13}$$

Thus the impact parameter p_2 and therefore the final energy depend on the scattering geometry, i.e., the angle of incidence and on the distance d of neighboring atoms. This demonstrates that for the case of multiple scattering the position of the peaks in the energy spectrum depends on the surface structure, which in turn can be analyzed by using this effect.

Multiple scattering, in particular the quasitriple peak (QT), was used to determine the displacement of atoms due to thermal vibrations.[27] Using the Debye model, thermal vibrations can be characterized by a Debye temperature that generally yields lower values for surfaces than for the bulk of the crystal. This will be further discussed in Section 6.

2.4. Neutralization

The neutralization of the primary ions in the scattering process is a very important phenomenon in ISS. While high neutralization probabilities make ISS extremely surface selective, they also make it more difficult to perform quantitative analysis. For noble gas ions only a few percent of the primary projectiles are generally backscattered as ions from the topmost layer of target atoms; ions penetrating to deeper layers of the target are almost 100% neutralized. Some typical values for "ion survival probabilities" are given in Table 2. The use of alkali ions for which the neutralization probability is much smaller, or the detection of backscattered neutral particles can make the quantification easier, but the energy spectra quite often contain a large amount of multiple scattering contributions (see also Figs. 3 and 20) and therefore their interpretation

Table 2. Typical Values of Ion Survival Probabilities for Scattering from Metal Surfaces ($E_0 = 1\,\text{keV}$, $\vartheta = 60°$)

Scattering from	He^+	Ne^+	Li^+	Na^+
First layer	<0.1	Single < 0.04 Double $< 4 \times 10^{-3}$	0.9–1	0.8
Deeper layers	$<3 \times 10^{-3}$	$<4 \times 10^{-3}$	0.9–1	0.8

is not as simple as in the case of noble gas ion scattering. The situation is tolerable for large backscattering angles (ICISS, Section 3) and computer simulations are useful for the understanding of the spectra.

In general variations in ion escape probabilities of noble gas ions for different elements are within one order of magnitude and there are only minor matrix effects. Therefore quantification is still much easier than in the case of SIMS. Figure 12 shows the backscattering intensities for a number of clean metal surfaces.[28] Neutralization effects appear as deviations from a smooth dependence of the scattering intensity on the target atomic number.

Quantitative analysis is possible in many practical cases with proper calibration. For example, surface composition analysis of metal alloys can be done by using pure elemental standards.[29–31] A linear dependence on coverage was also demonstrated for several adsorbates on metal surfaces.[32,33]

The physical model describing the neutralization of ions close to a

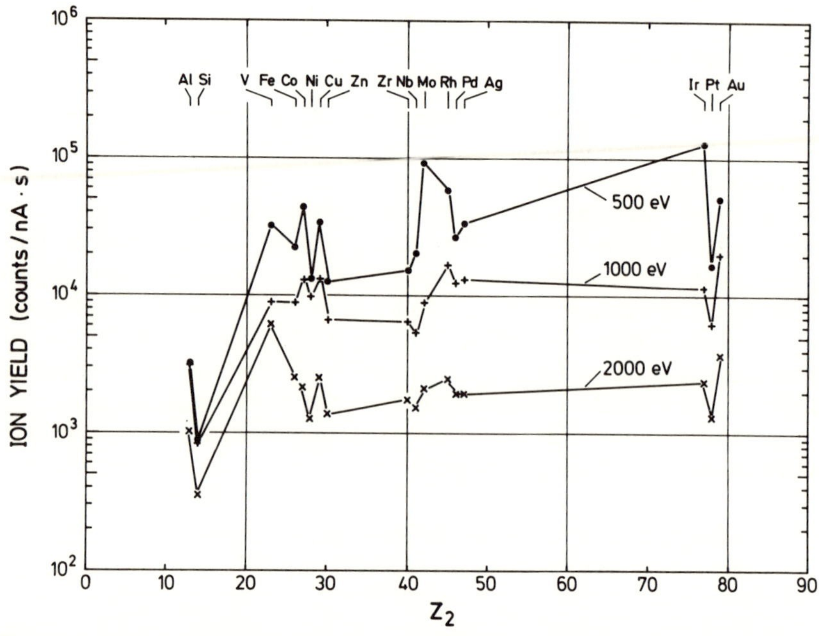

Figure 12. Yield (counts/nA s) of He^+ ions scattered from various elemental surfaces, given in relation to the target current. The data were taken with a cylindrical mirror analyzer ($\vartheta = 137°$) at three different primary energies.[28]

surface was developed by Hagstrum.[34] Unoccupied electronic states of the ion can be filled by an Auger transition if the atomic levels are well below the conduction band states of the solid (see Fig. 8 of Chapter 1). If the ionic states are close to the Fermi level of the solid, resonance neutralization and ionization can occur.[35,36] Auger transitions are mainly responsible for neutralization of noble gas ions; resonance transitions are most important for alkali ions. The basic concept is an exponential decrease of the neutralization probability with distance s from the surface $P_n \sim e^{-as}$, the constant a being of the order of $1\,\text{Å}^{-1}$. As a result the neutralization probability depends on the time the ion spends within a certain distance from the surface, this time being given by the perpendicular velocity component. Therefore the probability that the ion escapes neutralization can be generally described by an expression like $P \sim e^{-v_0/v_\perp}$, the neutralization probability being $1 - P$. v_0 is connected to the electron transition rate constant $A \approx 3 \times 10^{15}\,\text{s}^{-1}$ through the parameter a given above, $v_0 = A/a$, and is of the order of $3 \times 10^7\,\text{cm s}^{-1}$. Such a dependence has been verified experimentally in several cases.[5,37] In modified concepts a combination of a local contribution from the closest approach to the colliding target atom and a second contribution corresponding to the path inside or close to the surface was considered.[38–40] This results in an accordingly extended exponential that is able to describe the angular or energy dependence of scattered ion intensities, at least for the cases selected. For backscattering into large scattering angles (ICISS geometry, see Section 3.2) such concepts have been successfully applied to structure investigations.[41]

In some cases, such as oxygen adsorption on metals, trajectory-dependent ion escape probabilities have been reported[42–44] if the ions are scattered along different azimuthal directions from single crystalline surfaces at grazing angles. While this effect can be analyzed in detail, the analysis is complicated and tedious. Trajectory-dependent ion escape probabilities can be minimized or avoided by using both large incidence and scattering angles.

For special ion target combinations, quasiresonant neutralization has been reported which results in oscillatory ion yields as a function of ion velocity.[45] This occurs when an electronic energy level of a metal (e.g., $3d$ or $5d$) is energetically very close to the energy level of the He $1s$ ground state, a good example is the He–Pb couple.[45,46] Since the effect is element specific and is also influenced by chemical bonds to surrounding atomic species, exploitation for chemical information has been suggested,[47] but only limited results have been reported in the literature so far.

3. Experimental Techniques

3.1. Apparatus

Since ISS probes the outermost atomic layer of a surface, vacuum conditions are required under which the sample surface can be cleaned or prepared and maintained in a defined state. Therefore the scattering chamber must be an UHV system with a base pressure of reactive gases (CO, H_2, H_2O) below 10^{-9} mbar. For example, if a clean surface with a contamination below 10^{-2} monolayers is to be maintained for an hour, a pressure below 10^{-11} mbar is required for a gas with a sticking coefficient of unity. Values of that order are not uncommon for reactive gases on clean metal surfaces.[48] The partial pressure of noble gas (He, Ne) from the ion source reaches pressures between 10^{-7} and 10^{-6} mbar in the scattering chamber of many systems. Since the sticking probability of these thermal rare gas atoms is virtually zero on all surfaces at room temperatures,[48,49] these partial pressures are all tolerable for the practical purposes of ISS surface analysis.

The essential components of an ion scattering apparatus are the ion source and the energy analysis and detector system. Examples of typical arrangements are shown in Figs. 13 and 14. For noble gas ions electron impact ion sources are most convenient.[50] For ion energies around 1 keV they easily yield an ion current of 10–30 nA that is very constant in time, provided the neutral gas pressure is stable. Currents of that order are sufficient for surface analysis and cause negligible surface erosion due to sputtering. For a beam spot diameter of about 1–2 mm a total fluence of the order of 10^{13} He^+/cm^2 is required to record an energy spectrum. With a sputtering yield of the order of 10^{-1} atoms/ion which is typical for low-energy helium ions on solids,[51] this fluence is tolerable, since only about 10^{-3} of a monolayer is removed from the surface for one spectrum.

However, much higher yields are possible for adsorbates and therefore ion impact desorption during ISS measurements can occur.[52] For example, for CO yields up to 10 molecules/ion have been reported. To obtain good ISS measurements the adsorbed layer must be restored after short bombardment intervals and it is also useful to restrict the measurement to those parts of the energy spectrum that are of particular interest. The situation is better for alkali ions or neutral particle detection, as discussed below. If erosion of the surface by ion bombardment is desirable for cleaning purposes or in order to obtain near-surface depth profiles, then higher yields are useful and also higher current densities can generally be applied.[53]

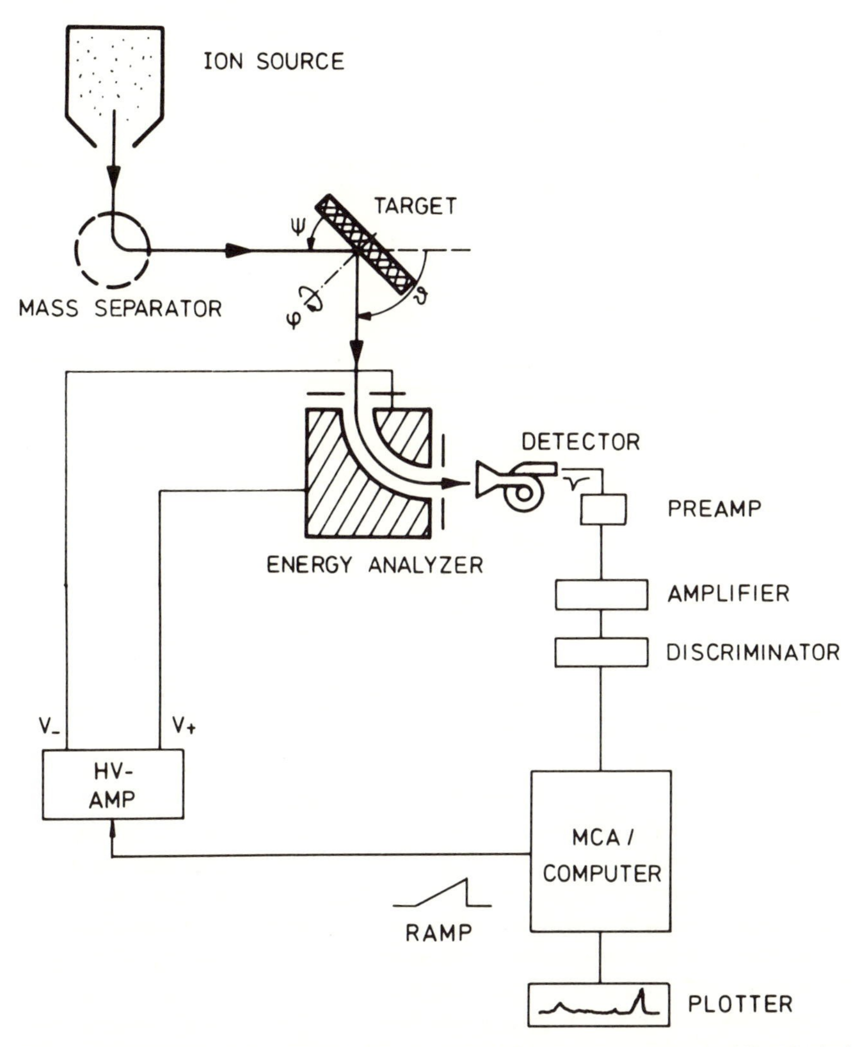

Figure 13. Schematic of a typical experimental ISS apparatus using a 90° spherical electrostatic energy analyzer.[28]

For electron impact noble gas ion sources, mass separation of the extracted ions is generally not necessary, but with plasma ion sources and solid dispenser sources for alkali ions, mass separation is recommended in order to avoid contamination problems, because a variety of ionic species and also neutral gas particles are emitted from these sources.

Sources for alkali ions are commercially available.[54] They contain appropriately doped minerals (e.g., eucryptite) that release Li^+, Na^+, or

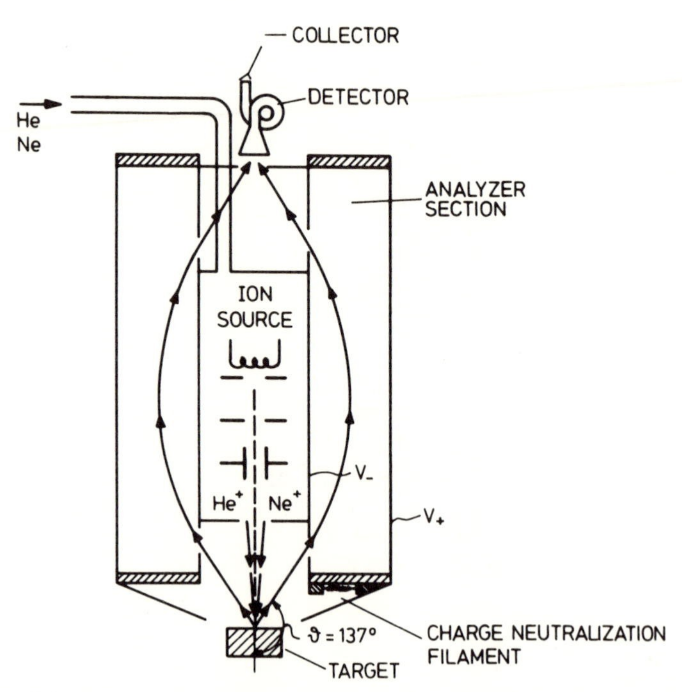

Figure 14. Schematic of an ISS apparatus using a cylindrical mirror analyzer with integrated ion gun (modified from 3M model 515).

K^+ ions at elevated temperatures. Since the neutralization probability for alkali ions is very low (see Table 2), ion currents of 20–50 pA are sufficient to give a reasonable scattering signal[5,55] and the total ion fluence per spectrum is greatly reduced. Moreover, this high ion yield makes quantitative ISS measurements more reliable. The disadvantage of the high ion escape probability, however, is the substantial contribution of multiply scattered ions to the energy spectrum. These also include ions that are backscattered from deeper layers in the sample. That is, the advantage of high ion yield is obtained at the expense of a more complicated ion energy spectrum whose interpretation is not as easy as in the case of noble gas ion scattering, for which the signal arises only from the outermost layer of the target. For the interpretation of alkali ion spectra therefore computer simulations are usually necessary.[55–57]

For the energy analysis of the scattered ions electrostatic sector fields are very convenient and most commonly applied. Their relative energy resolution $\Delta E/E$ is given by the width of the aperture and the radius of the mean path, r_0, i.e., $\Delta E/E = s/r_0$. From this relation, it follows that the energy window of an electrostatic analyzer increases in proportion to the energy, and this has to be corrected for absolute measurements. For ISS purposes a resolution of 1%–2% is sufficient. Spherical sector analyzers (with a sector angle of 90°; see Fig. 13) can be mounted on a UHV manipulator system such that the scattering angle is variable from 0° to large scattering angles of 160° or more. This is very useful for experimentally determining the direction, primary energy, and the energy and angular spread of the incident beam. Variation of the scattering angle is often useful for peak assignment and large scattering angles have become important for structure determination, as will be discussed below. Cylindrical mirror analyzers (CMA; see Fig. 14) have the advantage of higher scattered ion intensities due to the large acceptance solid angle. In practical cases this amounts to about a factor of 30 compared to spherical sectors.[28] This high intensity and the relatively good mass resolution due the large scattering angle (137°) make CMAs very useful for standard surface compositional analysis. Since the scattering geomtry is fixed, they are generally not suited for structure analysis.

For very high detection efficiency special systems were developed with annular[58] detectors or a spiral shaped position sensitive detector for simultaneous angle and energy distribution recording.[59] An energy dispersive toroidal prism has also been successfully used to measure simultaneously energy and angular distributions in a multichannel mode.[60]

An alternative and very successful means for analyzing scattered particle distributions is the time-of-flight (TOF) method.[61–67] In this case the time distribution of a chopped ion beam is measured after scattering and this distribution is to be transformed into an energy spectrum through the equation

$$E = \tfrac{1}{2}M_1L^2/t^2 \tag{14}$$

where L is the length of the flight path. The counts per constant time increment $\Delta N(t)$ are converted into constant energy increments via the expression $\Delta N(E) = (t^3/M_1L^2)\Delta N(t)$. For paths of about 1 m in length, flight times of microseconds are obtained and the corresponding electronics have to chop the beam with rise times of some ten nanoseconds in order to obtain an energy resolution of about 1% as in the case of electrostatic analyzers. This, however, is not always achieved.

Particle detection can be accomplished by using channeltrons or open multipliers, just as in the case of electrostatic analyzers. In some cases,[63] a stripping cell is also used for postionization of scattered particles. If the particle energies are sufficiently high, secondary electron multipliers respond equally well to neutral particles as to ions. Therefore

TOF systems have the advantage of being able to record neutral particle energy spectra simultaneously with ion energy spectra, the latter can be separated by applying an additional acceleration voltage.[64] In order to operate in a regime in which the detector response is independent (or only weakly dependent) on the neutral particle energy, these energies

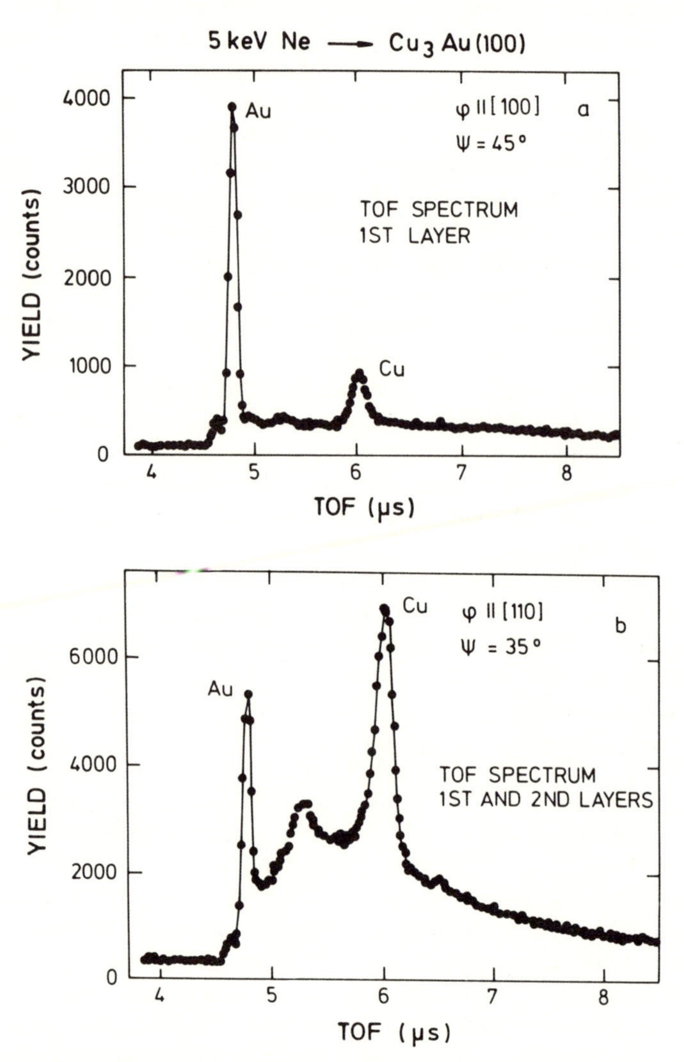

Figure 15. Time-of-flight spectra for Ne scattered from $Cu_3Au(100)$. (a) For scattering parallel to the [100] azimuth and with an incidence angle $\psi = 45°$ single scattering is restricted to the first layer. (b) Scattering parallel to the [110] azimuth and $\psi = 35°$ shows single scattering peaks from first and second layers.[69]

should be above about 1 keV.[68] Consequently TOF experiments are usually carried out with primary energies of 2 keV or more, up to 10 keV, and that reduces the surface specificity of the technique to some extent. Generally, the neutral particle spectra exhibit similar features as alkali ion spectra: quantification is easier in principle, but large contributions from multiple scattering can make the interpretation of spectra more complicated and computer simulations are occasionally necessary. Since the neutral particle fraction is usually more than 90% and due to the simultaneous recording of the entire spectrum, primary ion fluences for one spectrum are two orders of magnitude or more lower than for noble gas ISS, i.e., 10^{11} ions/cm^2 or less. Therefore, the surface damage is correspondingly lower and thus this method is quasi nondestructive. As an example, Fig. 15 shows TOF spectra for 5-keV neon scattering from Cu_3Au (neutral particle detection).[69] By changing the angles of incidence φ and ψ the scattering from Cu atoms in the second layer is strongly enhanced and thus the ordered phase of the Cu_3Au crystal is analyzed.

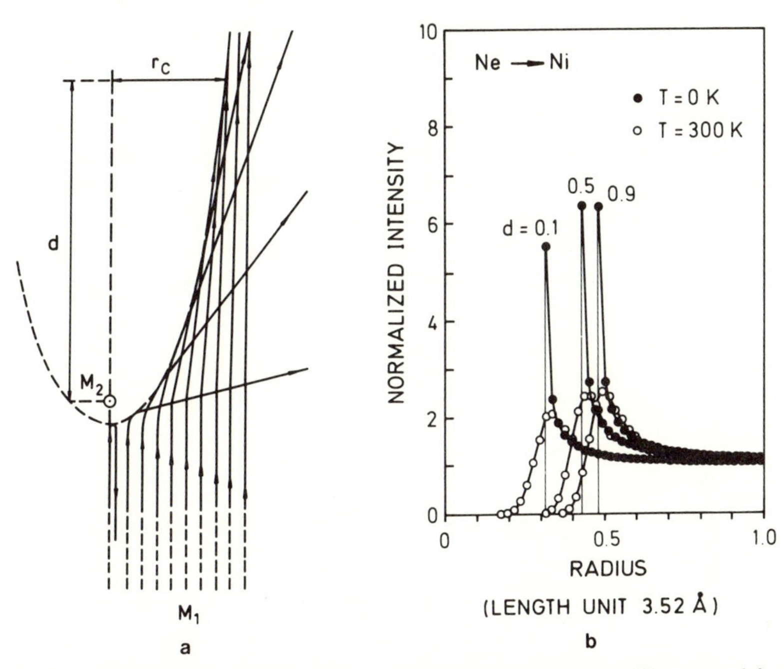

Figure 16. (a) Shadow cone formed by the trajectories of projectiles M_1 scattered from a target atom M_2. (b) Intensity distributions in the shadow cone of fixed and thermally vibrating Ni atoms. Parameter is the distance d from the scattering center that is located at the origin.[19]

3.2. Shadow Cones and Backscattering

A very useful concept for the consideration of scattering from neighboring atoms is the shadow cone.[70,71] This is the cone-shaped space downstream from a scattering atom that is formed by the trajectories of a uniform parallel flux of incoming ions (see Fig. 16a). The primary ion flux is deflected in such a way that there is no flux immediately behind the scattering atoms, the trajectories are focused to the edge of the cone, and the flux decreases again to its average value at large distance from the cone axis; see Fig. 16b. For a parallel primary ion flux and the scattering atom at rest the flux density has a singularity at the cone edge (surface rainbow); for atoms in thermal vibration the intensity distribution is broadened. If a neighboring atom is now inside, at the edge, or outside the shadow cone, its corresponding scattering contribution is zero, maximum, or average. If the radius of the shadow cone as a function of distance from the scatterer, $r(d)$, is known, the position of the neighboring atom can be measured. The shadow cone radius can be obtained experimentally by scattering from an ordered surface of known structure. Thus the method is self-calibrating because the interaction potentials need not be explicitly known. Calculations can also be made with sufficient accuracy (see Section 4).

The shadow cone concept is very successfully applied with large scattering angles, a technique that was called "Impact Collision ISS" (ICISS).[72] Its principle is shown in Fig. 17. At low angles of incidence ψ scattering into large angles ϑ is not possible from a chain of atoms due to shadowing. At the critical angle ψ_c the scattering reaches a maximum and decreases subsequently. A corresponding intensity distribution is dem-

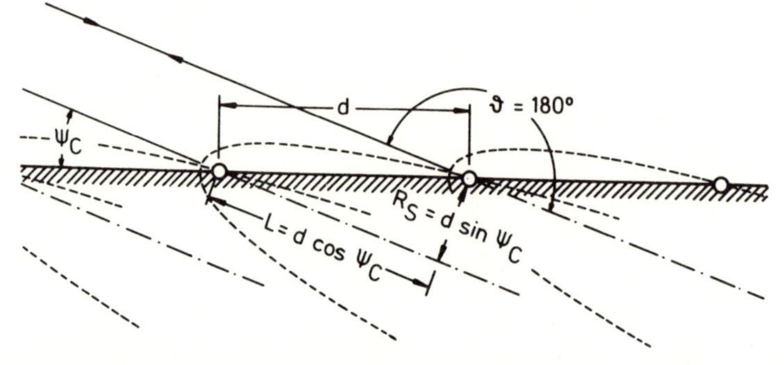

Figure 17. Shadow cones and scattering geometry for ICISS. The relations between the shadow cone radius R_s, the interatomic distance d, and the angle of incidence ψ_c are given, and similarly for the distance L along the shadow cone axis. (After Aono.[72])

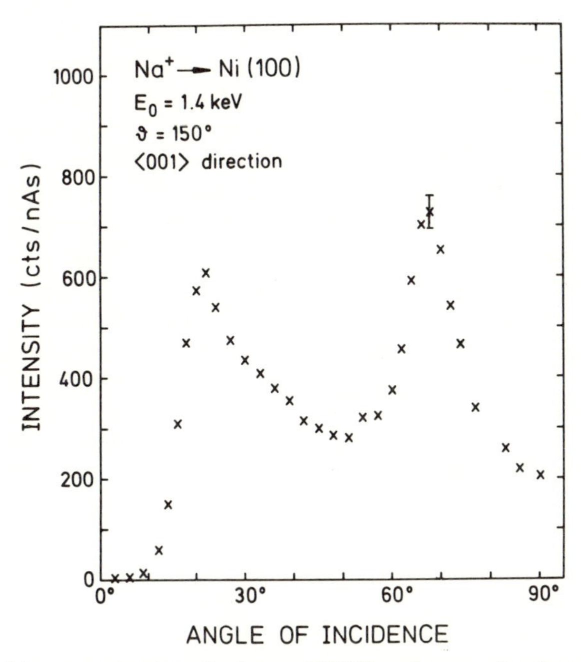

Figure 18. Na^+ ion scattering intensity from a Ni(100) surface as a function of the angle of incidence. The maximum at 20° corresponds to the critical angle for the interception of the shadow cone edge with nearest-neighbor atoms in the ⟨001⟩ direction. The second maximum results from the second layer of atoms in the face-centered position.[19]

onstrated in Fig. 18. Ideally a scattering angle of 180° should be used, because in that case the impact parameter of the atom at the edge of the cone is zero and thus it can be exactly located. However, good ICISS results can be obtained for scattering angles above about 145°.[73,19] Because of thermal vibrations and finite angular resolution, the slope in $I(\psi)$ curves, Fig. 18, has a certain width and this gives some arbitrariness in determining the critical angle ψ_c from such distributions. Values of 50%,[72,18] 70%,[73] and 100%[19] of the maximum were chosen. From Fig. 16b the choice of the intensity maximum does not seem to be unreasonable. Apparently this choice is not too critical if the same procedure is used for calibration and application of the shadow cone.

3.3. Direct Recoil Detection

A technique that is closely related to ion scattering consists in the detection of particles that are ejected from the surface in direct collisions as recoiling atoms or ions. In contrast to the particles sputtered through a collision cascade, recoil particles have a distinct final energy and therefore can be identified according to Eq. (2). The technique has been called Ion Desorption Spectroscopy (IDS)[12] or Direct Recoil Spectroscopy (DRS).[13] An example is shown in Fig. 19. DRS is particularly useful for light particles and for detection of negative ions, because of the low background from negative species. DRS is favorably applied to light particles on the surface, such as hydrogen isotopes, scattering from which is very restricted. It has been shown[11] that He^+ ion scattering from adsorbed deuterium is possible, but owing to the scattering kinematics it is limited to small scattering angles.[14,74,75] The situation is depicted in Fig. 5, showing that scattering of He from H is restricted to angles below

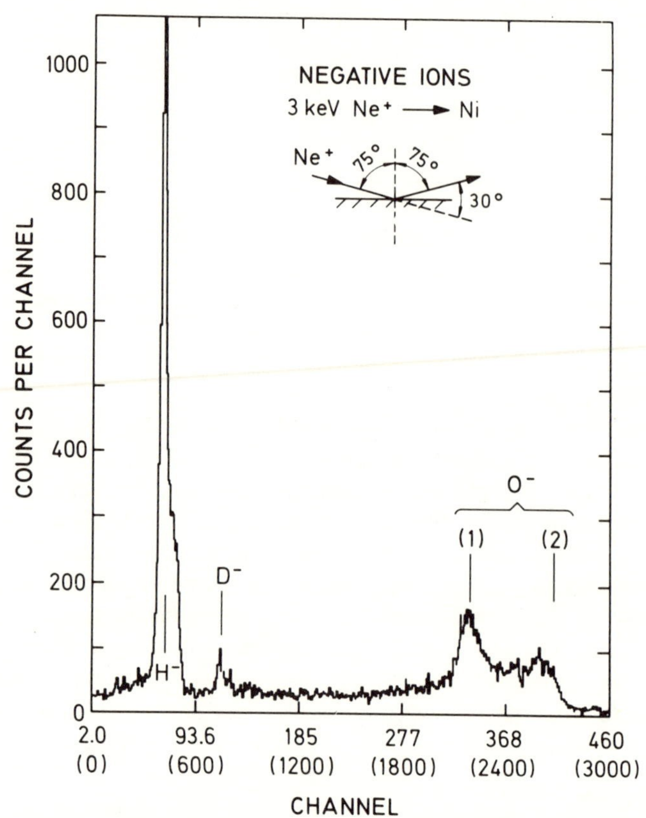

Figure 19. Energy spectrum of negative recoil ions for the experimental situation indicated in the insert. The bracketed numbers on the abscissa give the recoil energy in eV. The oxygen peak (1) corresponds to a direct recoil from a Ne–O two-body collision. For the oxygen peak (2) the neon projectile is first scattered from the Ni surface before causing the O atom recoil. Note the low background intensities.[12]

20° whereas recoils can be detected up to 90°. The larger angles are useful for structure or position determination. The corresponding cross sections are comparable in value to scattering cross sections.[75]

4. Calculations

4.1. General Considerations

For the basic application of ISS, namely, the determination of elemental masses on the surface, the peaks in the energy spectra can be calculated using the fundamental relations discussed in Section 2. Depending on the material, the energy of the scattering peaks is sufficiently well given by Eq. (1). Small deviations are observed, particularly for light elements, e.g., oxygen, but they rarely prohibit positive peak assignment. The peak intensities depend on both the scattering cross sections and the ion survival probabilities, as discussed above. In this kind of analysis the surface consists only of single isolated atoms.

For surface structure determination, the analysis is more involved. Since the relative positions of neighboring atoms are to be determined, in principle three-dimensional calculations including neutralization phenomena are required. There are, however, certain approximations possible which still lead to sufficient agreement with experimental results. In diffraction methods, the distribution of atoms in space, which is given by a pair correlation function $G(\mathbf{r})$, can be obtained as the Fourier transform of the experimental scattering function $S(\mathbf{k})$; no such general method is available in ion scattering. The usual procedure for deducing the corresponding distribution of atoms in the scattering plane, $G(r)$, from a measured distribution $I(\psi)$ (see Fig. 18) consists of testing various "reasonable" structure models. By this means it is frequently possible to select the model that agrees best with the experimental data and to exclude others. In this respect the procedure is not too different from LEED analysis methods, but ISS analysis is generally much more straightforward (at the expense of less detailed information).

4.2. Numerical Codes

All model calculations that are presently used do not include neutralization phenomena. Therefore they can only give qualitative agreement with noble gas ion scattering, but with alkali ions or neutrals quantitative agreement can be obtained. There are several three-dimensional numerical codes available in the literature, such as MARLOWE[76] or ARGUS.[77] They make use of the binary collision approximation, i.e., the projectile trajectories are determined by a sequence of two-body interactions. Thermal vibrations and inelastic energy losses can generally be included. A detailed trajectory analysis is also possible that can be used to analyze the various collision sequences contributing to one peak in the energy spectrum (see Fig. 20).

These Monte Carlo calculations yield all the information needed for data interpretation, but they require a large amount of computer time even on big machines and therefore they generally cannot be used routinely for data analysis. Many-body interactions have been taken into

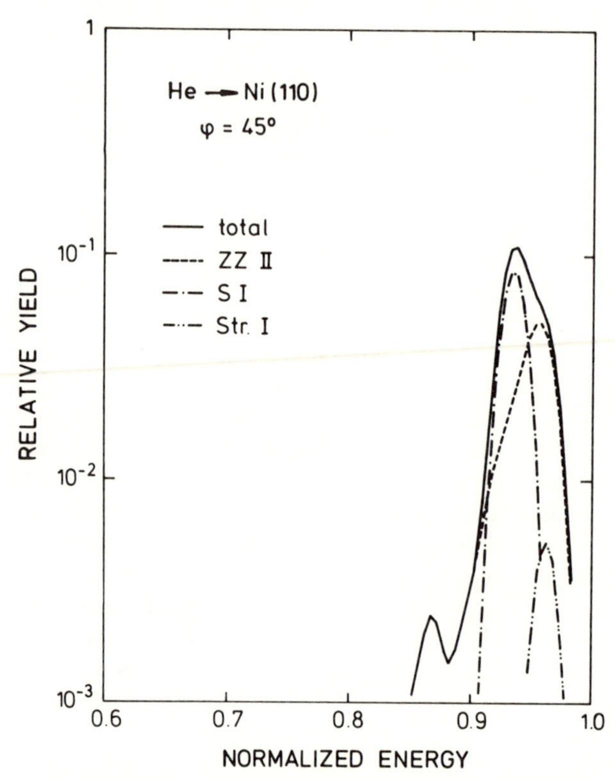

Figure 20. Energy spectrum for scattering of He from Ni(110) as shown in Fig. 3. The spectrum was calculated using the computer code ARGUS and demonstrates the contribution of various scattering classes but does not take neutralization into account (i.e., He and Li give similar results). I and II refer to scattering from the first and second atomic layer, respectively, S indicates single scattering, Str. scattering from a string of atoms, and ZZ zig-zag collisions.

account in addition to calculations using the binary collision model, and distinctive differences have been reported,[78] but in general the binary collision model appears to be appropriate.

4.3. Shadow Cones

A simple and very useful description of the shadow cone was given by Oen,[79] who calculated the radius r_c of the shadow cone in the momentum approximation using a Thomas–Fermi–Molière (TFM) screened Coulomb potential. The cone radius at a distance d from the scattering center is given by the expression

$$r = 2(Z_1 Z_2 e^2 d/E_0)^{1/2}(1 + 0.12\alpha + 0.01\alpha^2) \tag{15}$$

with $\alpha = (2/a)(Z_1 Z_2 e^2 d/E_0)^{1/2}$ being between 0 and 4.5 and a similar expression for larger values of α. Figure 21 demonstrates that these

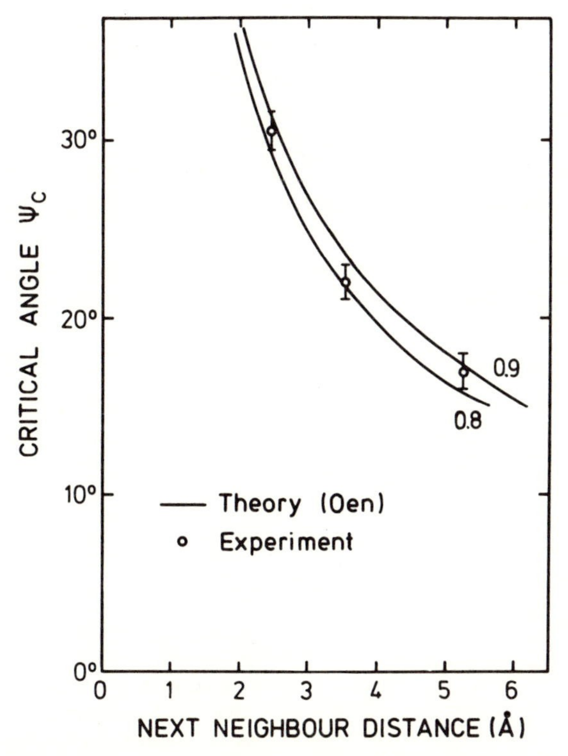

Figure 21. Critical angle ψ_c as a function of next neighbor distance for 1.4-keV Na on Ni. Calculations are made using Eq. (15) with a screening parameter of $0.8a_F$ and $0.9a_F$, respectively.[19]

calculated radii agree quite well with experimental results for a screening length a between $0.8a_F$ and $0.9a_F$.

Since the quantity directly available from the experiment is the critical angle ψ_c rather than the shadow cone radius, this was calculated according to Oen's procedure.[18] For a TFM potential a numerical fit yields

$$\ln \psi_c = 4.6239 + \ln(d/a)(-0.0403 \ln A - 0.6730) + \ln A(-0.0158 \ln A + 0.4647) \tag{16}$$

with $A = Z_1 Z_2 e^2/(E_0 a 4\pi\varepsilon_0)$.

For a given target–projectile combination this reduces to an expression like $\psi_c \sim d^{-\gamma}$, with the exponent γ between 0.7 and 0.8,[18] in agreement with the results in Fig. 21.

4.4. "Hitting Probability" Model

A numerical model calculation that yields $I(\psi)$ distributions for ICISS measurements was used by Daly *et al.*[80] on the basis of the so-called "hitting probability" concept developed by Tromp *et al.*[81] Because of its simplifications this model needs much less computer time compared to a three-dimensional Monte Carlo calculation and it yields more information than simple shadow cone considerations. The two-atom shadowing geometry used is given in Fig. 22. Atom 1 is displaced by the rms value of the motions of both atoms $\rho^2 = \rho_1^2 + \rho_2^2$; atom 2 is at rest. The ions are impinging along the z axis and are scattered into an angle ϕ.

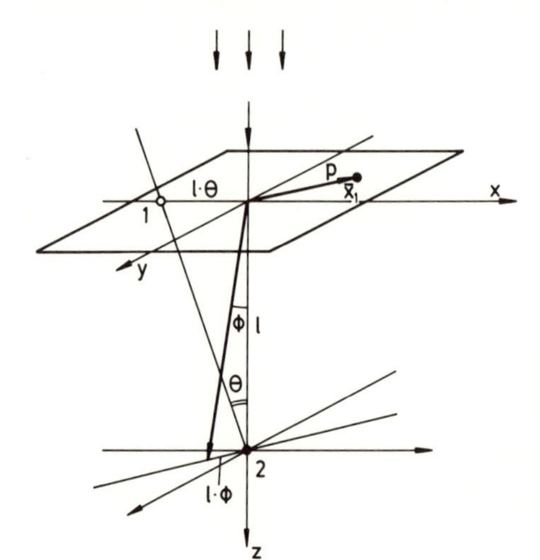

Figure 22. Geometry of a two-atom shadowing model. The equilibrium position of atom 1 is indicated by the open circle, its actual position x_1 by the solid circle. Atom 2 is fixed. The ion entering along the z axis with an impact parameter p relative to atom 1 is deflected by the angle ϕ. (After Ref. 81.)

The probability that atom 2 at a distance l from the x–y plane is hit by an incoming ion is given by the expression

$$P_2 = 2\int_0^\infty \frac{p\,dp}{\rho^2}\exp[-(a^2+b^2)/\rho^2]I_0\left(\frac{2ab}{\rho^2}\right) \tag{17}$$

which integrates over all possible impact parameters p. $a = s + l\tan\phi(p)$, and $b = l\tan\theta$. I_0 is the modified Bessel function of the first kind. The scattering interaction is contained in the term $\phi(p)$ and can be given in the form of tables in the program. The impact parameter with atom 2 is considered to be negligibly small. For a given ρ (see, e.g., Ref. 82) and a given angle of incidence θ, the hitting probability for atom 2 with shadowing by atom 1 can be calculated. A similar expression is obtained for blocking of the scattered ion beam. Using this model, the experimental $I(\psi)$ distributions for 5-keV Li^+ scattered from Cu(110) could be well reproduced.[80]

5. Analysis of Surface Composition

5.1. Compounds and Alloys

One of the main advantages of ISS is the possibility to obtain a mass specific signal from the outermost atomic layer of a surface. Therefore the composition of the terminating layers of a solid can be uniquely analyzed. This was demonstrated very early in ISS research by investigating the surfaces of polar crystals, such as CdS or ZnTe,[83,84] which clearly showed that one of the two elements was abundant on opposite surfaces. Important applications of ISS lie therefore in the field of preferential sputtering,[85] segregation,[31,86,87] the investigation of adsorption layers, and catalysts.[88]

For the analysis the ion scattering signal has to be proportional to the surface density N_i of a certain elemental species and can be written as

$$I_i^+ = I_0^+ K N_i \frac{d\sigma_i}{d\Omega}\Delta\Omega P_i \tag{18}$$

where I_0^+ is the primary ion current, K is a constant determined by the transmission and response of the apparatus, σ_i is the scattering cross section, $\Delta\Omega$ is the solid angle accepted by the analyzer, and P_i is the probability that primary ions will escape as ions after colliding with species i. Such a linear dependence between I_i and N_i was proven in several cases,[29–33] for which a proper calibration was possible. A complication can arise from neutralization-reionization effects. It was observed[89–91,18] that noble gas ions can be neutralized approaching the target surface and reionized during a binary collision, depending on the energy of the projectile. Since the ionization energy of the projectile or target atom has to be taken from the kinetic energy of the projectile,[9] satellite peaks can occasionally be observed.

In the case of binary alloys a calibration by using the pure elements as standards is generally possible and then the surface concentration X_i of one species is given by

$$X_i = \left(1 + \frac{I_j N_j I_i^\infty}{I_i N_i I_j^\infty}\right)^{-1} \tag{19}$$

where I_i^∞ is the scattering intensity of the elemental standard. In the case of a series of Pd/Pt alloys, the validity of this expression could be proven by analyzing samples that were prepared by milling in UHV, a treatment that is expected to produce the bulk composition on the surface.[31] Such studies also show that other techniques, such as AES, with small but finite information depth, give signals that are the average over a few atomic layers and therefore often different from ISS results.

Thus the monolayer sensitivity of ISS makes it the method of choice for studying surface segregation. This sensitivity is necessary, since in Gibbsian segregation the composition of the topmost atomic layer of an alloy is generally different from the composition of deeper layers.[92] ISS has been very successfully applied to study equilibrium surface compositions of a number of binary alloys such as CuNi,[29,84] AuPd,[30] NiAu, Cu_3Au, FeSn,[87] etc.; for Cu_3Au also the order–disorder phase transition could be analyzed using ISS. The kinetics of surface segregation has also been studied for several alloys, particularly in connection with preferential sputtering and radiation enhanced segregation.[29–31,86] For these studies it is useful to combine the monolayer sensitivity of ISS with the several layer sensitivity of AES to obtain compositional depth profiles of the near-surface region of the alloy. For example, this technique has been applied for PdPt,[31] CuLi,[93] and CuTi.[94]

5.2. Adsorption Layers

The surface specificity of ISS makes it particularly sensitive to adsorption layers. The increase of the signal from the adsorbed species A is accompanied by a decrease of the substrate signal I_S:[95]

$$I_S^+ = K I_0^+ (N_S - \alpha N_A)(d\sigma_s/d\Omega)\Delta\Omega P_S \tag{20}$$

The screening factor α arises from the shadowing effect of the adsorbate, and it was pointed out[44,96] that it can also contain neutralization effects. An example of the variation of the adsorbate–substrate signals during the adsorption of CO on Ni(100) is shown in Fig. 23.[33] Comparison with the work function change shows a linear increase of the oxygen signal with coverage up to one monolayer. A similar linearity had been found earlier for S on Ni.[32] It can also be seen in Fig. 23 that the Ni signal decreases drastically, giving an initial value of 4 for α in Eq. (20). The linearity between the ISS signal and coverage is not influenced by the large change in work function of 0.9 eV. This is probably due to the fact that neutralization occurs by electron capture into the helium ground state and this process yields about 20 eV in energy for the Auger electron and the corresponding transition probability is not strongly influenced by the work function change. ISS is therefore very well suited for studying adsorption layers, their structure and physical properties.

Ion impact desorption of adsorbed layers is a special field of fundamental interest for investigating sputtering processes[97,98] and is of practical interest for surface cleaning,[99] for plasma-surface effects in fusion research, etc. For such studies the ion beam can be used simultaneously for the analysis and for the sputtering action. The signal then decreases as a function of time t or, more generally, fluence (it) like

$$I/I_0 = \exp(-\sigma_D it) \tag{21}$$

where i is the primary current density and σ_D is the desorption or sputtering cross section. These cross sections have already been determined for a large number of adsorbate–substrate–projectile combinations.[100,101]

5.3. Catalysts

Successful applications of ISS have been made in the field of heterogeneous catalysis, and more specifically in the composition of near-surface layers of supported catalysts.[102–107] Here the surface sensitivity of ISS is used together with the sputtering action of the probing beam to obtain monolayer "depth profiles" (Chapter 1, this volume). Supported catalysts have to have a large surface area (of the order of 100 m^2/g) and therefore are prepared from powders with grain sizes of micrometers. Samples are made by pressing the powdered material into wafers. The support material generally consists of oxides, such as Al_2O_3, TiO_2, or SiO_2. It is usually impregnated in a liquid solution with the active component of the catalyst, e.g., molybdate,

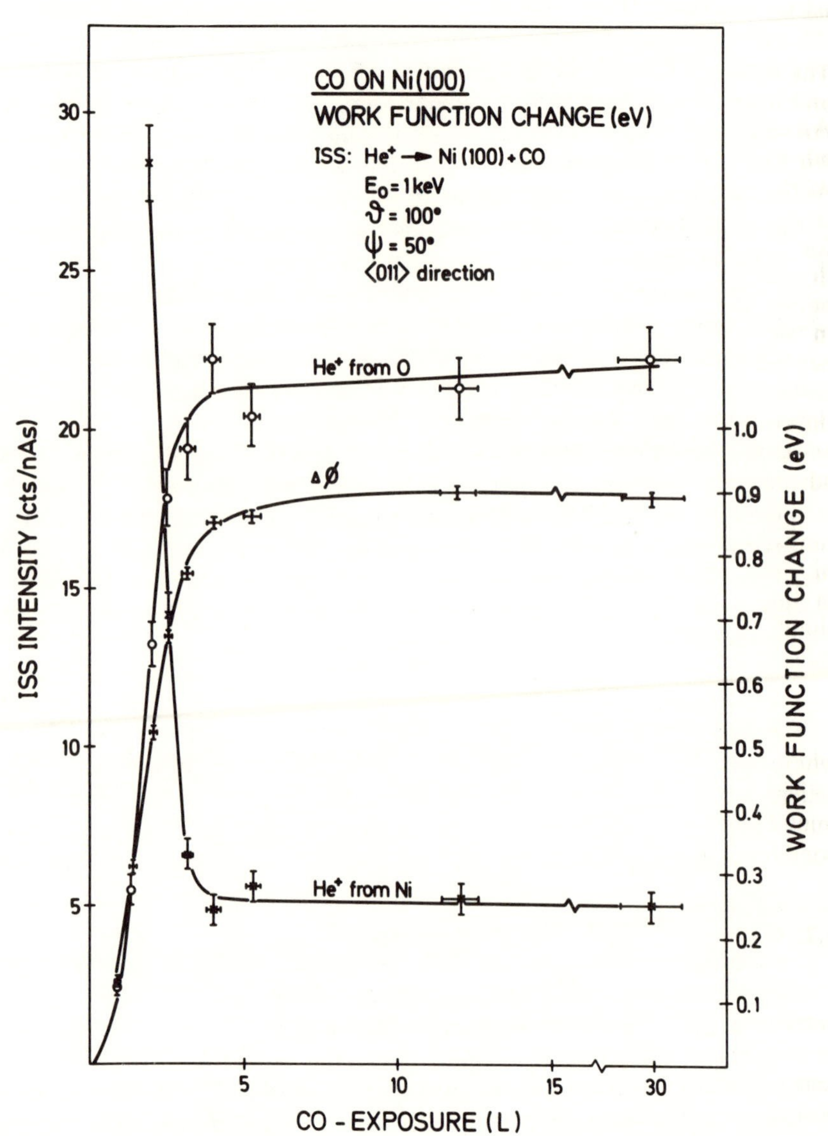

Figure 23. He^+ ion scattering intensities and work function change for the adsorption of CO on Ni(100) at 300 K.[33]

vanadate, or tungstate, in an amount close to, but below, the monolayer capacity of the support. It cannot be expected that accurate monolayer coverage takes place for such support materials and this preparation method, nor can it be expected that a layer-by-layer removal results from ion sputtering. Nevertheless ISS yields useful information about the outermost (i.e., active) species and the near-surface depth distribution. In the case of double impregnation, for example, ISS easily showed which species (e.g., molybdena) was first deposited on the support (Al_2O_3) and which species was subsequently adsorbed (nickel compound); see Fig. 24.

Since the support material is highly insulating, charging effects, can pose serious problems. Electron spectroscopies, such as AES, can therefore generally not be applied to such systems. For ISS, neutralizing of the samples with thermal electrons from a flood gun (see Fig. 14) can largely overcome this difficulty.

A good dispersion of the active phase on the support is very essential for the efficiency of a catalyst. As an example for the applicability of ISS for studying supported catalysts, investigations on the spreading behavior of oxides can be taken. The spreading of oxides used for catalysts, such as MoO_3 on Al_2O_3, has been studied for the usual way of preparation by impregnation[103] and more recently for the interesting case of the formation of monolayers by solid–solid wetting.[108] A similar example is shown for molybdena on titania in Fig. 25 by the Mo/Ti intensity ratio as a function of He^+ ion fluence. It is clearly seen that heat treatment ("calcination") in an oxygen gas stream with or without water vapor (curves b and c) leads to a steep increase of the Mo/Ti intensity ratio close to the surface, i.e., a spreading of the active Mo-compound on the TiO_2 support. Similar results were found for WO_3 and V_2O_5 as active components and Al_2O_3 as support.[109] On SiO_2 no spreading is observed. It could further be shown that the formation of surface polymolybdate, which is considered a necessary precursor state for active catalysts,[110] is a process that has a quite different dependence on calcination time compared to spreading (see Fig. 26); and that spreading occurs independently of the polymolybdate formation. For a calcination temperature of 723 K spreading is completed after about 5 h, whereas the polymolybdate formation requires about 30 h to be completed as shown by laser Raman spectroscopy. Therefore it can be concluded that spreading and polymolybdate formation occur in separate processes. Moreover, polymolybdate formation requires the presence of water vapor, in contrast to the ISS results, which are similar for H_2O-saturated and dry streams of oxygen gas.

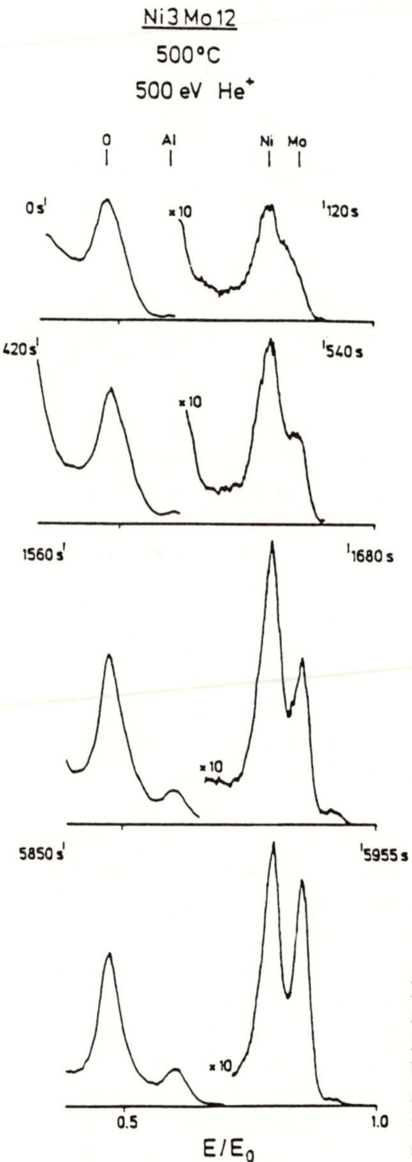

Figure 24. Sequence of energy spectra from an alumina-supported Ni–Mo catalyst calcined at 500°C. The time marks indicate the total bombardment time (see text). The Ni was impregnated after the Mo impregnation and therefore the initial Ni signal is higher than the Mo signal.[88]

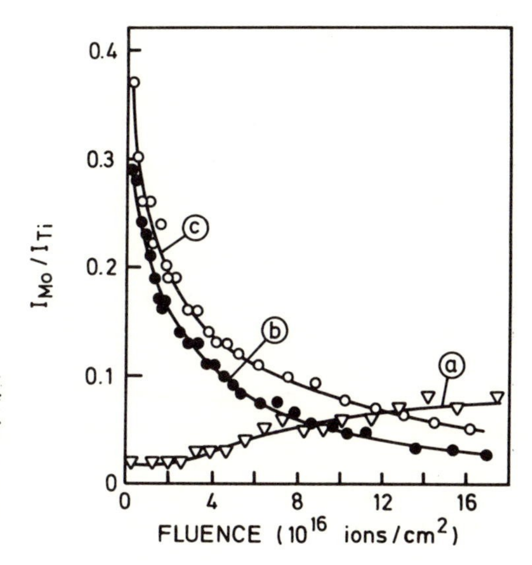

Figure 25. Scattered He^+ ion intensity ratios Mo/Ti as a function of He^+ fluence yield "depth profiles" of MoO_3/TiO_2: (a) physical mixture, (b) after calcination at 720 K for 24 h in dry O_2, and (c) in H_2O-saturated O_2.[109]

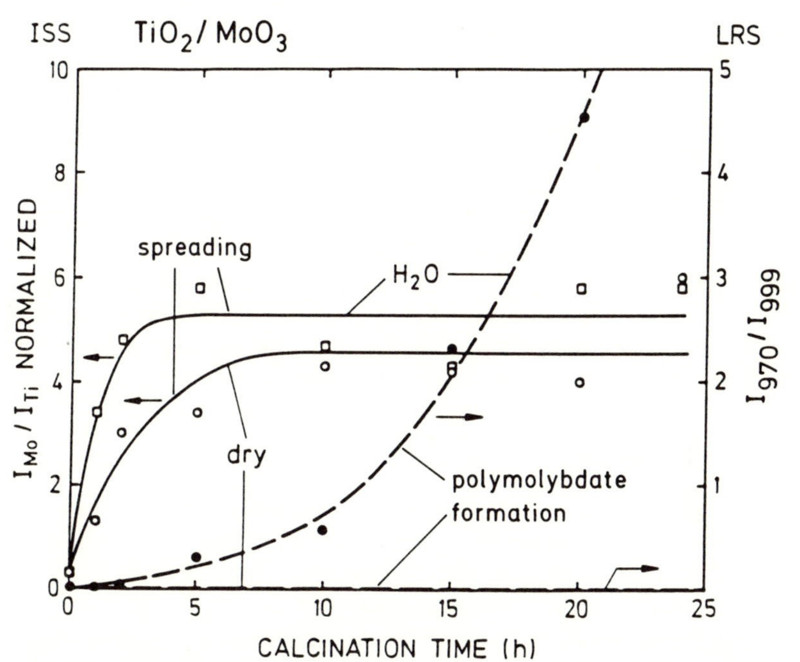

Figure 26. Dependence of the normalized Mo/Ti He^+ ion scattering intensity on calcination time at 723 K in H_2O-saturated and dry streams of oxygen. The ratios of the initial to final intensity values are taken from Fig. 25 to obtain a normalized measure for the degree of spreading. The Laser Raman signal ratio for the 970-cm^{-1} and 999-cm^{-1} bands, indicative of polymolybdate formation, is also shown.[19]

5.4. Surface Roughness

The studies of real catalyst systems, which have rather heterogeneous rough surfaces, pose the question of the influence of surface roughness on ISS studies. Rough surfaces can be expected to give lower scattering intensities compared to atomically smooth surfaces because microscopically there is a broad variation of the angles of incidence and also the exit angles relative to the average scattering surfaces. Therefore the angular distribution of backscattered ions is expected to be different for rough and smooth surfaces resulting in an intensity decrease for the former case. Parts of the surface that are struck by the impinging beam can also be shadowed by other surface structures in a given scattering geometry. Such a case was modeled by Nelson,[111] who found an intensity decrease of a factor of 1.7 for a moderate surface roughness. Studies of scattered ion intensities from Al_2O_3 samples with a large variation of surface roughness showed more than a factor of 5 difference in the Al peaks; see Fig. 27a.[112] In contrast to the relatively large differences in the absolute scattering intensities, the intensity ratios of oxygen to aluminum were relatively similar, showing initial values between 4 and 6 and steady state ratios close to 4; Fig. 27b. (The porous film probably contained SiO_4^{2-} ions from the preparation, which are located in the pores of the film and influence the oxygen values.) These results obviously show good agreement in the intensity ratios for different surface roughness. It was therefore concluded that ISS data for supported catalysts in the form of intensity ratios give valid information about the surface composition also for rough surfaces and the obtained sputter depth profiles bear relevance for relative elemental distributions near the surface. Moreover, it appears justified in these cases to apply such results from plane model catalysts to rough real catalyst systems.

5.5. Isotopic Labeling

The mass sensitivity of ISS permits mechanistic studies of isotopically labeled surfaces or gas phase species. The allowable isotopic pairs are restricted by mass resolution, but ^{1}H–^{2}H, ^{14}N–^{15}N, ^{16}O–^{18}O, and ^{32}S–^{34}S are all resolvable by He^+ ion scattering.

Czanderna and co-workers have reported several ISS results involving the use of ^{16}O–^{18}O.[113–115] By using copper oxidized in oxygen-36,[113] it was shown that the initial step in the copper-catalyzed oxidative degradation of polypropylene involves reduction of the copper oxide by the polymer. As the degradation in oxygen-32 proceeded, the copper oxide eventually is all labeled, with ^{16}O via continual oxidation–reduction

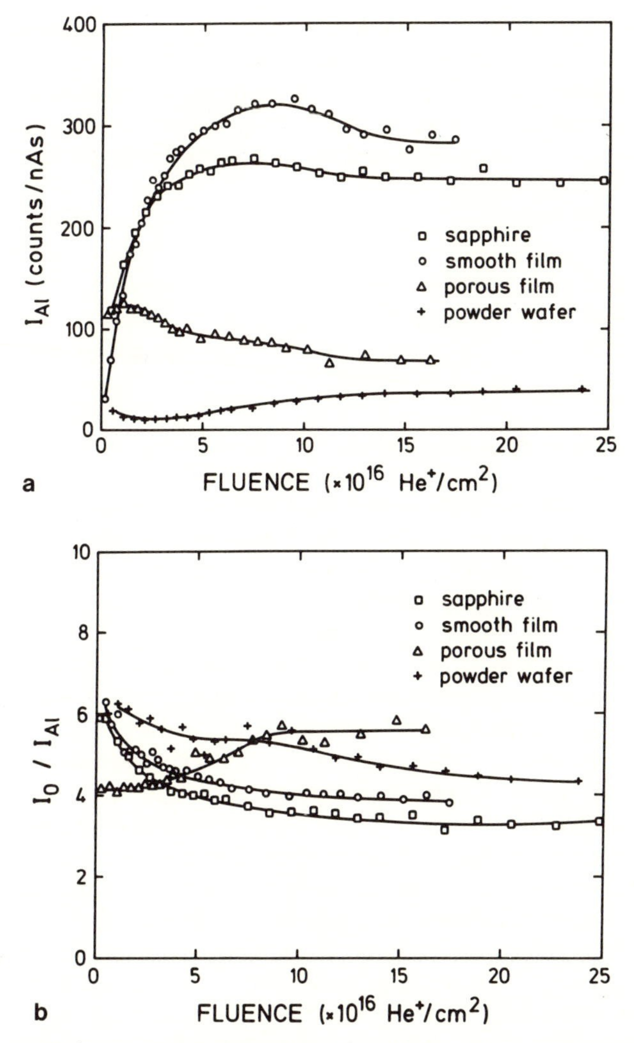

Figure 27. (a) He^+ ion scattering intentieis from Al for Al_2O_3 samples with different surface roughness as a function of He^+ ion fluence. (b) Oxygen to aluminum intensity ratios as a function of fluence for alumina samples with different surface roughness.(112)

cycles. In later work,(114) the mechanism of electron beam enhanced oxidation of aluminum in CO_2 at room temperature was established. The oxide product was traced to ^{18}O in the CO_2; the absence of carbon buildup on the surface (confirmed with XPS) proved the scission of CO_2 was to O and CO, where CO was evolved during the reaction. Finally, silica surfaces (with oxygen-16), which were bombarded with 1-keV $^3He^+$

in $C^{18}O_2$ showed an increase in oxygen-18.[115] Silica surfaces reduced by ion or electron bombardment to SiO_x, where x is less than 2, did not chemisorb oxygen from $^{18}O_2$ or $C^{18}O_2$ in the absence of a bombarding beam. The possibility of detecting hydrogen isotopes by scattering or recoil techniques is discussed in Section 2.2 and is further treated in detail in Section 6.2.

6. Structure of Crystalline Surfaces

"Physics is ruled by geometry"
B. Mandelbrot

6.1. Reconstructed surfaces, ICISS

The structure of single crystalline surfaces is of fundamental interest in surface physics. It is therefore quite natural that besides other techniques (see chapters in this book) low-energy ion scattering is applied for structure analysis, particularly in view of its surface sensitivity and by making use of its mass dependence in the case of multicomponent material. Consequently, in recent years numerous ISS structure studies have been published, mostly on metal surfaces and also some results of semiconductor surfaces. Most investigations are based on the shadow cone concept and in particular the ICISS technique as described in Section 3. By these studies the structure of a number of reconstructed surfaces has been determined. These reconstructions take place on clean surfaces of noble metals and quite often under the influence of a light adsorbate, such as hydrogen or oxygen. The procedure of structure determination with ISS has been usually to consider various reasonable models, which have been deduced from diffraction experiments (LEED), and then compare scattering from the model structures with the experimental results. This procedure then generally leads to the exclusion of all but one model structure. Different procedures can be followed to obtain the scattering distribution from the theoretical models. If a three-dimensional scattering program such as MARLOWE is available, the corresponding scattering intensities can be calculated, but virtually only for heavy projectiles and forward scattering, because of the cost of the computer time. This method was used, for example, to determine the Au(110) surface reconstruction with K^+ scattering at a scattering angle of 70°.[116] Extensive MARLOWE calculations[117] showed that only a missing row structure with layer contraction of 0.25 Å and a small lateral shift of second-layer atoms gave good agreement with the experimental results. This model had been proposed earlier from LEED investigations.[118]

A generally simpler but equally successful way is to apply the ICISS method and to determine the critical angles ψ_c for various interatomic spacings occurring in the structure considered. The required shadow cone radii can be determined either experimentally or by calculations, as was pointed out in Section 3. An example is given in Fig. 28, showing the

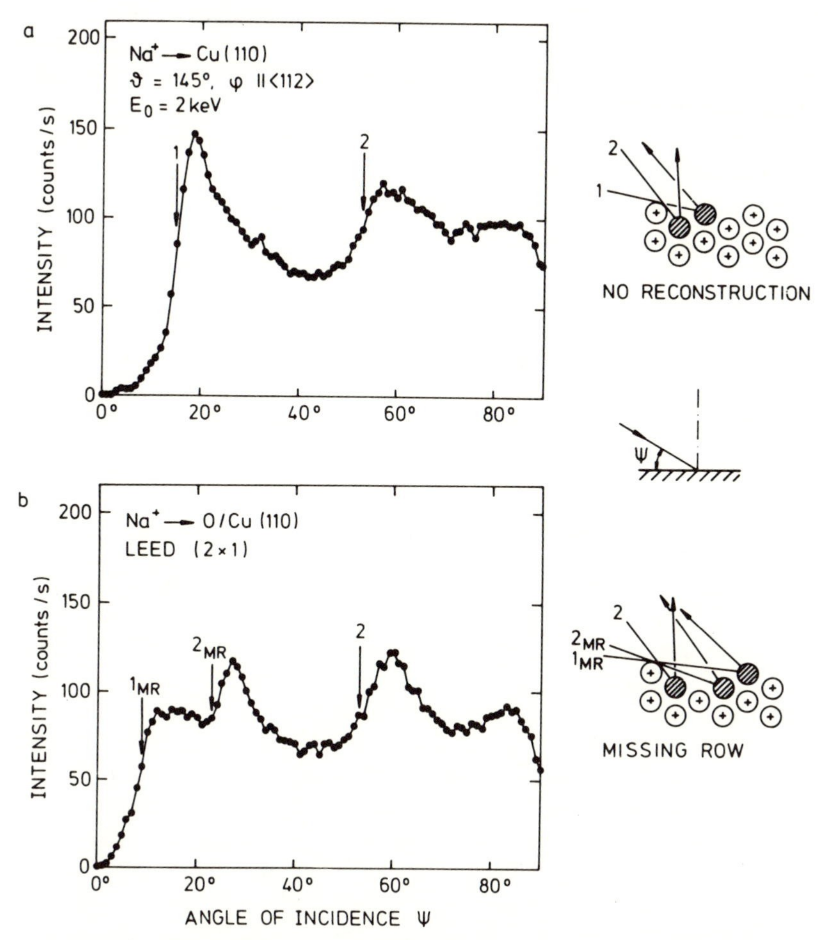

Figure 28. Na^+ scattering spectra from ICISS measurements of a Cu(110) crystal surface before and after reconstruction due to oxygen adsorption. The intensity as a function of the angle of incidence exhibits steep increases if a distinct neighboring atom gets outside the shadow cone as indicated on the right-hand side. For the reconstructed surface (lower part), the measured onsets can only be explained by the missing row (MR) model.[119]

reconstruction of the Cu(110) surface due to oxygen adsorption.[119] The measured intensity distributions can only be explained by the missing row model and not by others, e.g., the sawtooth model. A similar result was obtained for the O–Ni (110) – (2 × 1) surface.[120]

A detailed study of the reconstructed (2 × 1) structure of Si(001) led to the determination of an asymmetric dimer structure that can exactly account for the azimuthal variation of 1-keV He^+ scattering intensities at 164° scattering angle.[121] Even the long-time reconstruction evergreen Si(111) – (7 × 7) could be satisfactorily analyzed by ICISS measurements.[122] It was, however, pointed out[123] that several models could be supported by the data, showing the limitations of the ICISS method with structures of large unit cells.

Some very exciting investigations have been reported with ordered alloy surfaces. TOF measurements on Cu_3Au (001) with 5- and 10-keV Ne^+ could be used to study the surface structure, the order–disorder phase transition, and surface segregation[87] (see also Figs. 15a and 15b). For NiAl alloy surfaces ICISS measurements showed that domains of Ni and Al rich areas exist on the surface.[124]

Some of the surface structure studies are listed in Table 3 (which is intended to give important examples but not a complete listing). It can be concluded from the presently available results that ICISS is a very powerful method for surface structure investigation that is able to give unambiguous answers to specific questions.

6.3. Adsorption Layers, Recoil Detection

Structure investigations of ordered adsorption layers started very early in ISS research. They were obtained with forward scattering, which is very sensitive to overlayers, and included frequently studied systems such as O(2 × 1)–NI(110), O(2 × 1)–Ag(110), and O–W(110) and the position and orientation of CO on metal surfaces (see also Table 3 and references therein). The large O and negligible C signal from adsorbed CO gives direct evidence for upright adsorption of the CO molecule with the oxygen pointing away from the surface. Only after dissociation is a C signal visible.[130] Neutralization can pose problems in this context because of trajectory-dependent effects.[42] Therefore, Li^+ ion scattering was also applied[43] and used for a good qualitative description of C overlayers on W(100).[129,131] Usually the symmetry of the ordered adsorption layer is known from electron diffraction, and ion scattering can contribute by giving the adsorbate position relative to the substrate lattice and the bond lengths involved. Again the ICISS method was very useful in structure analysis, although the signal from light adsorbates is

Table 3. Surface Structures Studied by ISS

Surface	Structure model	ISS Method	Ref.
Reconstruction of clean surfaces			
Pt (110)—(1 × 2)	Missing row	Na^+—ICISS	120
Au (110)—(1 × 2)	Missing row	K^+—ISS	116
		MARLOWE	117
Si (001)—(2 × 1)	Dimer formation	He^+—ICISS	121
Si (111)—(7 × 7)	Reconstruction	He^+—ICISS	122
Adsorption induced reconstruction			
H/Pd (110)—(1 × 2)	Pairing row	Ne—NICISS	125
O/Cu (110)—(2 × 1)	Missing row	Na^+—ICISS	119
		Li^+—ICISS	138
		Model calc.	80
		Ne^+—ISS	139
O/Ni (110)—(2 × 1)	Missing row	Na^+—ICISS	120
Adsorbate structure, adsorption sites			
O on Ni(110)—(2 × 1)		He^+—ISS	126
		Ne^+—ISS	140
		Ne^+—DRS	141
O on Ag(110)—(2 × 1)		He^+—ISS	127
O on W (110)		He^+—ISS	128
O on Pt (111)		He^+—ISS	133
O on W (211)		Ar^+—ISS	136
		—DRS	
		—ICISS	
O, C on Mo(100)		Li^+, K^+—ISS	44
C on W (001)		Li^+—ICISS	131
S on Ni (001)—c(2 × 2)		Ne^+—ICISS	132
(110)—c(2 × 2)			
(111) (2 × 2)			
D on W(211)		$^3He^+$—ISS	146
D on Pd(110)		He^+—ISS	11
H on Ru(001)		Ne^+—DRS	137
Ag on Si(111)—($\sqrt{3} \times \sqrt{3}$)		He^+—ICISS	41, 143
		Li^+—ICISS	142
Au on Si(111)—($\sqrt{3} \times \sqrt{3}$)		Na^+—ICISS	144
		Li^+—ICISS	145
CO on Ni(111)—($\sqrt{7}/2 \times \sqrt{7}/2$)		$^3He^+$—ISS	129
CO on W(100) α, β		He^+—ISS	130
CO on NiAl(110)		He^+—,Ne^+ISS, ICISS	147
CO on Ni(100)		Na^+, He^+—ICISS	33

generally very weak in the backscattering geometry. It should be noted in this context that in the adsorbate-induced surface reconstruction studies presented in the preceding section, the signal from the adsorbate itself is generally not recorded. A favorable case in that respect is S on Ni, and the positions of adsorbed S atoms on the low index nickel surfaces could in fact be determined by 5-keV Ne^+ ion scattering.[132] The result for $c(2 \times 2)$S on Ni(110) and Ni(100) and (2×2)S on Ni(111) was that S adsorbs in the hollow sites with S–Ni bond lengths of 2.2 Å, in agreement with LEED data. Another interesting observation is subsurface adsorption, which was reported for O on Pd(111),[133] and strong evidence was also found for N on W(110).[134]

As mentioned before, scattering from light adsorbates is generally weak. Nevertheless 1-keV He^+ ion scattering from deuterium adsorbed on Pd(110) at angles less than 30° could be successfully used to determine D adsorption in the troughs between atom rows.[11] In general, this restriction to small scattering angles may be a problem for structure determination. In this context a promising variety of ion techniques consists in recoil detection (cf. Section 2). This was used for analyzing adsorbate positions of H on Pt[135] and O on W.[136] To illustrate the technique we discuss the case of H adsorbed on Ru(001).[137] Ne^+–ICISS measurements yield a critical angle $\psi_c = 23°$ for scattering in the [100] azimuthal direction; Fig. 29a. After adsorption of hydrogen at 138 K a quite different distribution is found for the H^+ recoil intensity; Fig. 29b (at 68 eV for $E_{Ne} = 1.5$ keV and $\vartheta_2 = 60°$), with the critical angle at 9.5°. From these measurements a position of H atoms at 1.01 ± 0.07 Å above the plane of Ru atoms was determined. The variations with the azimuthal angle of the scattering plane are shown in Fig. 30 and also demonstrate the differences for adsorbate and substrate lattices. The H^+-recoil peaks at $\pm 7.5°$ from the [100] direction give clear evidence for H adsorption in a three-fold coordinated site. No such lateral ordering was found for adsorbed hydrogen at 300 K. These examples demonstrate that recoil detection can be a very powerful tool for locating adsorbed atoms, particularly for light adsorbates. For exact structure determination an accompanying computer simulation seems to be necessary.[135,137]

6.3. Defects, Thermal Displacements

It has been shown in the preceding section that for a rigid lattice with fixed interatomic distances, scattering is restricted to well-defined angular regions. That is, for a fixed angle of incidence of a parallel ion beam, there are "forbidden zones" into which scattering is not possible from a rigid lattice, e.g., if $\psi < \psi_c$. From this it is obvious that deviations from

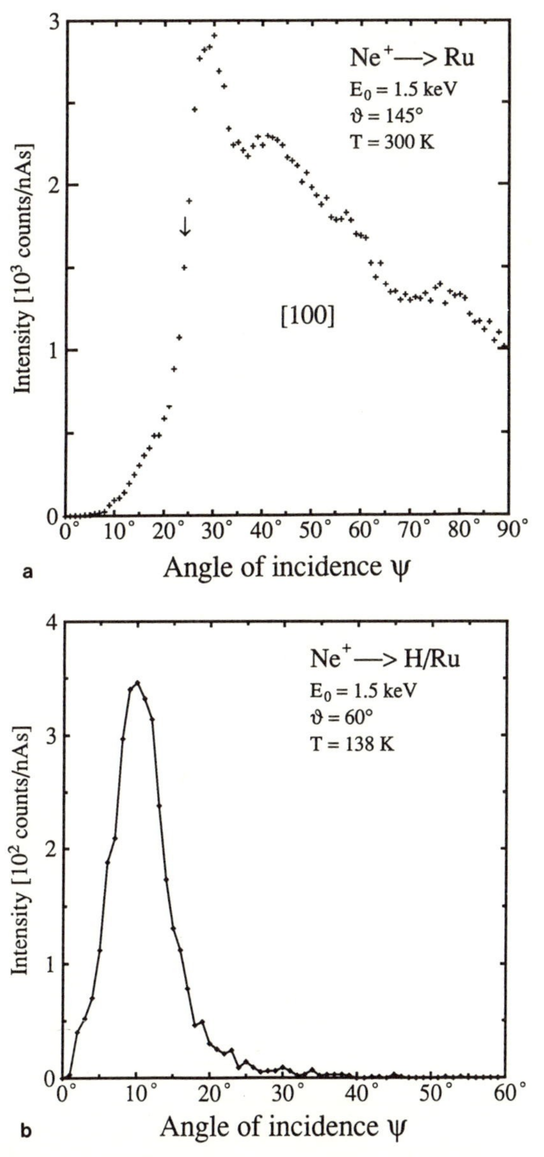

Figure 29. (a) ICISS yield of Ne^+ from clean Ru(001). (b) H^+ recoil ICISS yield after adsorption of a saturation layer of hydrogen on Ru(001) at 138 K.[137]

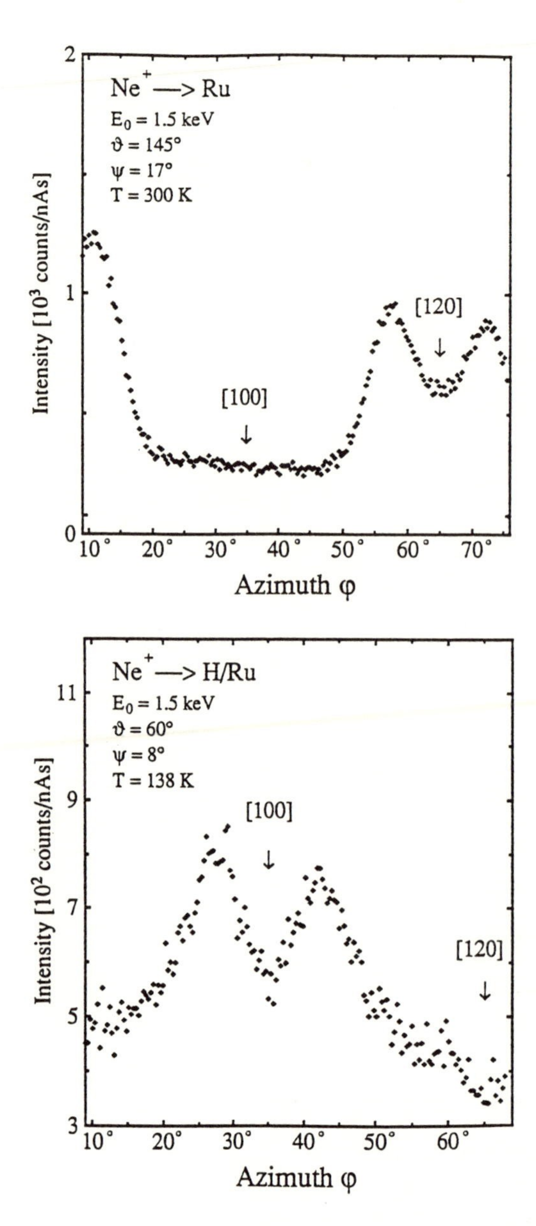

Figure 30. Variation of the Ne^+ scattering intensity (top) and H^+ recoil intensity (bottom) with the azimuthal angle.[137]

a fixed lattice arrangement should be detectable by measuring the scattering intensity into those forbidden regions (see also Fig. 16). Such irregularities can be defects, adatoms, or steps on the surface, or they can also be due to thermal vibrations of the surface atoms. Owing to the short scattering times of 10^{-15} s, lattice vibrations are only sensed by the passing projectile as static displacements of surface atoms. Both defects and thermal displacements have in fact been observed by ISS, although the technique has probably not yet been exploited to its full potential in this respect.

For example, an increase of the scattering intensity into a shadowed region was observed on a Cu(110) surface for ion bombardment at temperatures below 200 K.[40] From the fluence dependence of the scattering intensity at 100 K, it could be concluded that single atomic defects are created and not two-dimensional craters. A characteristic fluence of about 10^{14} ions/cm^2 is necessary to reach steady state in surface damage with 1-keV Ne^+ ions. A detailed study of surface defects on TiC(001) was done using He^+ ICISS.[26] Special kinds of defects in the lattice can thus be identified by determining the related critical angle. Similarly, the creation and annealing of defect structures on MgO(110) were studied by ICISS; single vacancies and various kinds of vacancy clusters could be detected.[148] More work along these lines will be necessary to verify the potential of ISS for defect analysis, their distribution along lattice directions, and particularly for quantification of defect concentrations.

As mentioned above, thermal vibrations represent a quasistatic surface disordering for the probing ion. Therefore attempts have been made to detect vibrational amplitudes of surface atoms by ISS. Poelsema *et al.*[149] investigated the temperature dependence of the quasitriple scattering peak resulting from three-atom thermal pits for 10-keV Kr^+ on Cu(100) and deduced a surface Debye temperature from these measurements. They were further interpreted by Martin *et al.*,[150] who considered the effect of correlated vibrations on quasidouble (QD), and quasitriple collisions. The temperature dependence was found to be rather weak; for $T \gg \theta_D$ one obtains

$$I_{QD} = \text{const}(mk)^{1/2}[\hbar(1 - \rho)^{-1/2}]\theta_D/T^{1/2} \quad (22)$$

where ρ is the correlation coefficient, θ_D the Debye temperature, and other symbols have the usual meaning. However, for correlations no conclusive results could be obtained from the experimental data available. The ICISS technique was also used to determine mean square displacements of Ti atoms on TiC surfaces, using a three-parameter fit procedure to reproduce the experimental $I(\psi)$ curves.[151]

The temperature dependence of the (quasi) single collision scattering intensity found experimentally for Ne^+ scattering from Cu(110) could be interpreted with a simple two-atom model using the shadow cone concept,[152] the results of which were confirmed by three-dimensional computer calculations. This model nicely illustrates the concept and is therefore demonstrated in Fig. 31. Depending on the angle of incidence ψ, the second atom can have its mean position s outside ($\psi > \psi_c$, $s > 0$), inside ($\psi < \psi_c$, $s < 0$), or right at the edge of the shadow cone. In the harmonic approximation and neglecting correlations, the probability for this second atom to be outside the shadow cone due to thermal vibrations is

$$W(\overline{\Delta s^2}) = \tfrac{1}{2} \pm \tfrac{1}{2}\phi((s^2/\overline{\Delta s^2})^{1/2}) \tag{23}$$

where ϕ is the error function and the mean square displacement $\overline{s^2}$ is related to the Debye temperature in the high-temperature approximation of the Debye model by

$$\overline{\Delta s^2} = \frac{6\hbar T}{Mk\theta_D^2} \tag{24}$$

if we consider displacements perpendicular to the surface. Thus, depending on the mean position, the probability for an atom to be outside the shadow cone, and therefore to contribute to the scattering intensity, can be an increasing or decreasing function with temperature. This can be directly studied in the experiment by varying ψ accordingly and by choosing a scattering angle large enough so that—due to the chain effect—no scattered ions are detected unless target atoms are displaced from their ideal mean position in the chain. Figure 32 shows corresponding results for 1-keV Ne^+ on Cu(110). The agreement between experi-

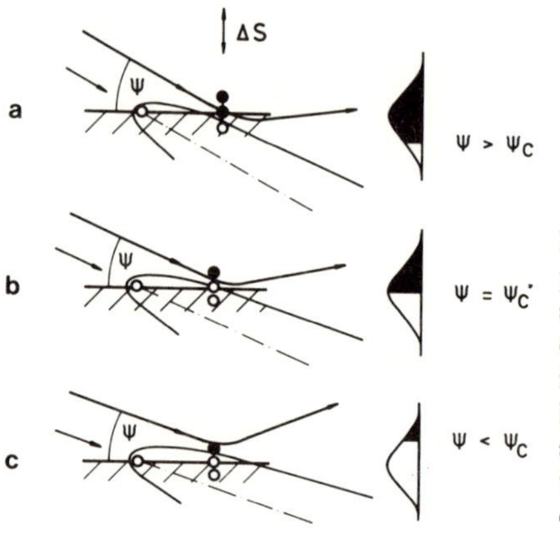

Figure 31. Two-atom model indicating the probability of a second atom being in the shadow of the preceding scattering atom for different angles of incidence. The dark part in the distributions plotted on the right-hand side represents the probability that the second atom is outside the shadow cone due to thermal vibrations and therefore can contribute to the scattering.[152]

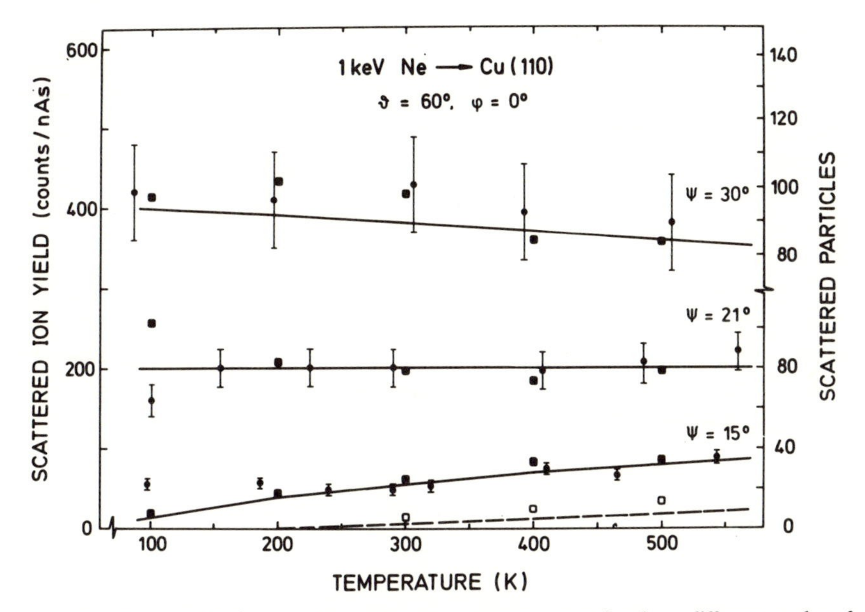

Figure 32. Scattered Ne^+ ion yield as a function of temperature for three different angles of incidence according to Fig. 31. Dots with error bars are experimental values, dark squares are three dimensional numerical calculations, and solid lines represent calculations applying the two-atom model, both using a surface Debye temperature of 150 K. The dashed line and the open squares correspond to calculations for $\psi = 15°$ with the bulk value of $\Theta_D = 325$ K.(152)

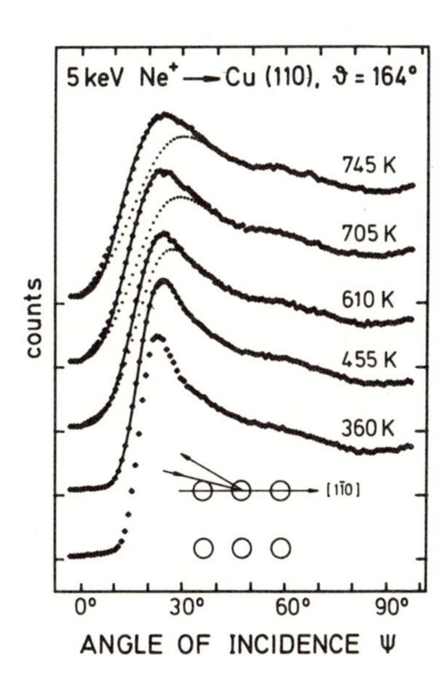

Figure 33. Number of 5-keV Ne^+ ions scattered through 164° along the [110] rows of atoms on a Cu(110) surface as a function of the angle of incidence ψ. Parameter is the sample temperature. The inset illustrates the trajectory at the critical angle ψ_c. The dotted lines for higher temperatures give the extrapolation from the data for 360 K using the Debye model. The solid lines represent a fit taking next-nearest neighbors into account.(153)

ment and calculations is good if a surface Debye temperature of 150 K is chosen; no agreement can be obtained with the bulk value of 325 K. These results were further supported by experiments in the ICISS geometry,[153] as shown in Fig. 33, from which similar values of surface Debye temperatures for various surface lattice directions were obtained.

Lattice dynamics of surfaces can be better analyzed by diffraction of thermal helium atom beams, which is used to measure surface phonon dispersion curves.[154] Such measurements are not within the scope of ISS, as can be deduced from the considerations in Section 1. But the determination of thermal displacements may still be useful, particularly in view of possible anharmonicities[155] and phase transitions such as roughening and melting.[156] This aspect of ISS has only been marginally touched so far and leaves ample opportunity for future development.

References

1. D. P. Smith, *J. Appl. Phys.* **18,** 340 (1967).
2. N. Bohr, *Mat.-Fys. Medd. Kgl. Dan. Vid. Selsk.* **18**(8) (1948).
3. I. M. Torrens, *Interatomic Potentials,* Academic, New York (1972).
4. See, e.g., H. Goldstein, *Classical Mechanics,* Addison-Wesley, Reading, Massachusetts (1965).
5. E. Taglauer, W. Englert, W. Heiland, and D. P. Jackson, *Phys. Rev. Lett.* **45,** 740 (1980).
6. G. van Wyk, W. Englert, and E. Taglauer, *Nucl. Instrum. Methods* **B35,** 504 (1988).
7. A. Zartner, W. Heiland, and E. Taglauer, *Phys. Rev. Lett.* **40,** 1259 (1978).
8. V. I. Veksler, *Sov. Phys. JETP* **17,** 9 (1963).
9. W. Heiland and E. Taglauer, *Nucl. Instrum. Methods* **132,** 535 (1976).
10. E. Hulpke, *Surf. Sci.* **52,** 615 (1975).
11. R. Bastasz, T. E. Felter, and W. P. Ellis, *Phys. Rev. Lett.* **63,** 558 (1989).
12. P. J. Schneider, W. Eckstein, and H. Verbeek, *Nucl. Instrum. Methods* **218,** 713 (1983).
13. J. W. Rabalais, J. A. Schultz, and R. Kurnov, *Nucl. Instrum. Methods Phys. Res.* **218,** 719 (1983).
14. W. Eckstein and R. Bastasz, *Nucl. Instrum. Methods Phys. Res.* **B29,** 603 (1988).
15. E. S. Mashkova and V. A. Molchanov, *Medium Energy Ion Reflection from Solids,* North-Holland, Amsterdam (1985).
16. O. B. Firsov, *Sov. Phys. JETP* **6,** 534 (1958).
17. W. Heiland, E. Taglauer, and M. T. Robinson, *Nucl. Instrum. Methods* **132,** 655 (1976).
18. Th. Fauster, *Vacuum* **38,** 129 (1988).
19. E. Taglauer, M. Beckschulte, R. Magraf, and D. Mehl, *Nucl. Instrum. Methods Phys. Res.* **B35,** 404 (1988).
20. J. Lindhard, V. Nielsen, and M. Scharff, *Mat. Fys. Medd. Dan. Vid. Selsk.* **36**(10) (196)8.
21. G. Molière, *Z. Naturforsch.* **2A,** 133 (1947).

22. J. F. Ziegler, J. P. Biersack, and U. Littmark, *Stopping Powers and Ranges of Ions in Matter,* Vol. 1 (J. F. Ziegler, ed.), Pergamon, New York (1985).
23. J. P. Biersack and J. F. Ziegler, *Nucl. Instrum. Methods* **194,** 93 (1982).
24. V. M. Kivilis, E. S. Parilis, and N. Yu Turaev, *Sov. Phys.-Dokl.* **12,** 328 (1967).
25. E. Taglauer and W. Heiland, *Surf. Sci.* **33,** 27 (1972).
26. M. Aono, *Nucl. Instrum. Methods Phys. Res.* **B2,** 374 (1984).
27. B. Poelsema, L. K. Verheij, and A. L. Boers, *Surf. Sci.* **133,** 344 (1983).
28. E. Taglauer, *Appl. Phys.* **A3,** 161 (1985).
29. D. G. Swartzfager, S. B. Ziemecki, and M. J. Kelley, *J. Vac. Sci. Technol.* **19,** 185 (1981).
30. P. Varga and G. Hetzendorf, *Surf. Sci.* **162,** 544 (1985); G. Hetzendorf and P. Varga, *Nucl. Instrum. Methods Phys. Res.* **B18,** 501 (1987).
31. J. du Plessis, G. N. van Wyk, and E. Taglauer, *Surf. Sci.* **220,** 381 (1989).
32. E. Taglauer and W. Heiland, *Appl. Phys. Lett.* **24,** 437 (1974).
33. M. Beckschulte, D. Mehl, and E. Taglauer, *Vacuum* **41,** 67 (1990).
34. H. D. Hagstrum, *Phys. Rev.* **96,** 336 (1954); **104,** 672 (1956).
35. E. G. Overbosch, B. Rasser, A. D. Tenner, and J. Los, *Surf. Sci.* **92,** 310 (1980).
36. A. L. Boers, *Nucl. Instrum. Methods Phys. Res.* **B2,** 353 (1984).
37. W. Heiland and E. Taglauer, in: *Applied Atomic Collision Physics,* Vol. 4, p. 237 (S. Datz, ed.), Academic, New York (1983).
38. R. J. MacDonald and D. J. O'Connor, *Surf. Sci.* **124,** 423, (1983).
R. J. MacDonald and P. J. Martin, *Surf. Sci.* **111,** L739 (1981).
D. J. O'Connor, Y. G. Shen, J. M. Wilson, and R. J. MacDonald, *Surf. Sci.* **197,** 277 (1988).
39. A. Richard and H. Eschenbacher, *Nucl. Instrum. Methods Phys. Res.* **B2,** 444 (1984).
40. G. Engelmann, E. Taglauer, and D. P. Jackson, *Nucl. Instrum. Methods Phys. Res.* **B13,** 240 (1986).
41. C. S. Chang, T. L. Porter, and I. S. T. Tsong, *Vacuum* **39,** 1195 (1989).
42. D. J. Godfrey and D. P. Woodruff, *Surf. Sci.* **105,** 438 (1981).
43. W. Englert, E. Taglauer, W. Heiland, and D. P. Jackson, *Phys. Scr.* **T6,** 38 (1983).
44. S. H. Overbury, B. M. Dekoven, and P. C. Stair, *Nucl. Instrum. Methods Phys. Res.* **B2,** 384 (1984).
45. R. L. Erickson and D. P. Smith, *Phys. Rev. Lett.* **34,** 297 (1975).
46. A. Zartner, E. Taglauer, and W. Heiland, *Phys. Rev. Lett.* **40,** 1259 (1978).
47. T. W. Rusch and R. L. Erickson, in: *Inelastic Ion Surface Collisions,* N. H. Tolk, J. C. Tully, W. Heiland, and C. W. White, eds., p. 73, Academic, New York (1977).
48. P. A. Redhead, J. P. Hobson, and E. V. Kornelsen, *The Physical Basis of Ultrahigh Vacuum,* Chapman and Hall, London (1968).
49. H. J. Kreuzer and Z. W. Gortel, *Physorption Kinetics,* Springer, Berlin (1986).
50. W. Heiland and E. Taglauer, in: *Methods of Experimental Physics,* Vol. 22, p. 99, R. L. Park and M. G. Lagally, eds., Academic, Orlando, Florida (1985).
51. H. H. Andersen and H. L. Bay, in: *Sputtering by Particle Bombardment I,* p. 145, R. Behrisch, ed., Springer, Berlin (1981).
52. E. Taglauer, W. Heiland, and J. Onsgaard, *Nucl. Instrum. Methods* **168,** 571 (1980).
53. E. Taglauer, *Appl. Phys. A* **51,** 238 (1990).
54. Spectra-Mat. Inc., Watsonville, California.
55. S. H. Overbury, *Nucl. Instrum. Methods Phys. Res.* **B27,** 65 (1987).
56. A. J. Algra, S. B. Luitjens, E. P. Th. M. Suurmeijer, and A. L. Boers, *Nucl. Instrum. Methods* **203** (1982) 515.
57. D. P. Jackson, W. Heiland, and E. Taglauer, *Phys. Rev. B* **24,** 4118 (1981).

58. H. H. Brongersma, N. Hazewindus, J. M. van Nieuwland, A. M. M. Otten, and A. J. Smets, *Rev. Sci. Instrum.* **49,** 707 (1978).
59. P. A. J. Ackermans, P. F. H. M. van der Meulen, H. Ottevanger, and H. H. Brongersma, *Proceedings of the Symposium Surface Science,* Kaprun, Austria, P. Varga and G. Betz, eds., Institut für Allgemeine Physik, Tu Wien, p. 165 (1988).
60. H. A. Engelhardt, W. Bäck, D. Menzel, and H. Liebl, *Rev. Sci. Instrum.* **52,** 835 (1981).
H. A. Engelhardt, A. Zartner, and D. Menzel, *Rev. Sci. Instrum.* **52,** 1161 (1981).
61. T. M. Buck, G. H. Wheatley, G. L. Miller, D. A. H. Robinson, and Y.-S. Chen, *Nucl. Instrum. Methods* **149,** 591 (1978).
62. Y.-S. Chen, G. L. Miller, D. A. H. Robinson, G. H. Wheatley, and T. M. Buck, *Surf. Sci.* **62,** 133 (1977).
63. S. B. Luitjens, A. J. Algra, E. P. Th. M. Suurmeijer, and A. L. Boers, *Appl. Phys.* **21,** 205 (1980).
64. H. Niehus and G. Comsa, *Nucl. Instrum. Methods Phys. Res.* **B15,** 122 (1986).
65. J. W. Rabalais, J. A. Schultz, and R. Kumar, *Nucl. Instrum. Methods Phys. Res.* **218,** 719 (1983).
66. D. Rathmann, N. Exeler, and B. Willerding, *J. Phys. E* **18,** 17 (1985).
67. R. Aratari, *Nucl. Instrum. Methods Phys. Res.* **B34,** 493 (1988).
68. H. Verbeek, W. Eckstein, and F. E. P. Matschke, *J. Phys. E* **10,** 944 (1977).
69. T. M. Buck, G. H. Wheatley, and L. Marchut, *Phys. Rev. Lett.* **51,** 43 (1983).
70. E. Bøgh, in: *Channeling,* D. M. Morgan, ed., Wiley, London (1973).
71. A. G. J. de Wit, R. P. N. Bronckers, and J. M. Fluit, *Surf. Sci.* **82,** 177 (1979).
72. M. Aono, C. Oshima, S. Zaima, S. Otani, and Y. Ishizawa, *Japn. J. Appl. Phys.* **20,** L829 (1981).
73. H. Niehus and G. Comsa, *Surf. Sci.* **140,** 18 (1984).
74. B. J. J. Koeleman, S. T. de Zwart, A. L. de Boers, B. Poelsema, and L. K. Verheij, *Nucl. Instrum. Methods* **218,** 225 (1983).
75. W. Eckstein, *Nucl. Instrum. Methods Phys. Res.* **B27,** 78 (1987).
76. M. T. Robinson and I. M. Torrens, *Phys. Rev. B* **9,** 5008 (1974).
77. D. P. Jackson, W. Heiland, and E. Taglauer, *Phys. Rev. B* **24,** 4189 (1981).
78. B. J. Garrison, N. Winograd, C. T. Reimann, and D. E. Harrison, Jr., *Phys. Rev. B* **36,** 3516 (1987).
79. O. S. Oen, *Surf. Sci.* **131,** L407 (1983).
80. R. S. Daley, J. H. Huang, and R. S. Williams, *Surf. Sci.* **215,** 281 (1989).
81. R. M. Tromp and J. F. van der Veen, *Surf. Sci.* **133,** 159 (1983).
82. D. S. Gemmel, *Rev. Mod. Phys.* **46,** 129 (1974).
83. W. H. Strehlow and D. P. Smith, *Appl. Phys. Lett.* **13,** 34 (1968).
84. H. H. Brongersma and P. M. Mul, *Chem. Phys. Lett.* **14,** 380 (1972).
85. E. Taglauer, *Appl. Surf. Sci.* **13,** 80 (1982).
R. Shimizu, *Nucl. Instrum. Methods Phys. Res.* **B18,** 486 (1987).
86. N. Q. Lam and H. Wiedersich, *Nucl. Instrum. Methods Phys. Res.* **B18,** 471 (1987).
87. T. M. Buck, in: *Chemistry and Physics of Solid Surfaces* (R. Vanselow and R. Howe, eds.), p. 435, Springer, Berlin (1982).
88. H. Jeziorowski, H. Knözinger, E. Taglauer, and C. Vogdt, *J. Catal.* **80,** 286 (1983).
89. R. C. McCune, J. E. Chelgren, and M. A. Z. Wheeler, *Surf. Sci.* **84,** L515 (1979).
90. T. M. Thomas, H. Neumann, A. W. Czanderna, and J. R. Pitts, *Surf. Sci.* **175,** L737 (1986).
91. M. Aono and R. Souda, *Nucl. Instrum. Methods Phys. Res.* **B27,** 55 (1987).

92. R. Kelly, *Surf. Interface Anal.* **7,** 1 (1985).
93. B. Baretzky, E. Taglauer, W. Möller, and R. P. Schorn, *J. Nucl. Mater.* **162–164,** 920 (1989).
94. G. N. van Wyk, J. du Plessis, and E. Taglauer, *Surf. Sci.* (1991).
95. E. Taglauer and W. Heiland, *Surf. Sci.* **47,** 234 (1975).
96. E. Taglauer and W. Heiland, *Appl. Phys.* **9,** 261 (1976).
97. E. Taglauer, W. Heiland and J. Onsgaard, *Nucl. Instrum. Methods* **168,** 571 (1980).
98. H. F. Winters and P. Sigmund, *J. Appl. Phys.* **45,** 4760 (1974).
99. E. Taglauer, *Appl. Phys.* **A51,** 238 (1990).
100. E. Taglauer, *Nucl. Fusion* Special Issue (1984).
101. A. Koma, *IPPJ-AM* **22** (1982).
102. M. Shelef, M. A. Z. Wheeler, and H. C. Yao, *Surf. Sci.* **47,** 697 (1975).
103. H. Knözinger, H. Jeziorowski, and E. Taglauer, in: *Proceedings, 7th International Congress on Catalysis,* p. 604, Elsevier-Kodansha, Amsterdam/Tokyo (1981).
104. R. L. Chin and D. M. Hercules, *J. Phys. Chem.* **86,** 360 (1982).
105. B. A. Horrell and D. L. Cocke, *Catal. Rev.-Sci. Eng.* **29,** 447 (1987).
106. F. Delannay, E. N. Haeussler, and B. Delmon, *J. Catal.* **66,** 469 (1980).
107. P. Bertrand, J.-M. Beuken, and M. Delvaux, *Nucl. Instrum. Methods* **218,** 249 (1983).
108. R. Margraf, J. Leyrer, H. Knözinger, and E. Taglauer, *Surf. Sci.* **189/190,** 842 (1987).
109. J. Leyrer, R. Margraf, E. Taglauer, and H. Knözinger, *Surf. Sci.* **201,** 603 (1988).
110. H. Jeziorowski and H. Knözinger, *J. Phys. Chem.* **83,** 1166 (1979).
111. G. C. Nelson, *J. Appl. Phys.* **47,** 1253 (1976).
112. R. Margraf, H. Knözinger, and E. Taglauer, *Surf. Sci.* **211/212,** 1083 (1989).
113. A. C. Miller, A. W. Czanderna, H. H. G. Jellinek, and H. Kachi, *J. Colloid Interface Sci.* **85,** 244 (1982).
114. J. R. Pitts, S. D. Bischke, J. L. Falconer, and A. W. Czanderna, *J. Vac. Sci. Technol.* **A2,** 1000 (1984).
115. J. R. Pitts and A. W. Czanderna, *Nucl. Instrum. Methods Phys. Res.* **B13,** 245 (1986).
116. H. Derks, H. Hemme, W. Heiland, and S. H. Overbury, *Nucl. Instrum. Methods Phys. Res.* **B23,** 374 (1987).
117. H. Hemme and W. Heiland, *Nucl. Instrum. Methods* **B9,** 41 (1985).
118. W. Moritz and D. Wolf, *Surf. Sci.* **77,** L29 (1979).
W. Moritz and D. Wolf, *Surf. Sci.* **163,** L655 (1985).
119. N. Niehus and G. Comsa, *Surf. Sci.* **140,** 18 (1984).
120. H. Niehus, *J. Vac. Sci. Technol.* **A5,** 751 (1987).
121. M. Aono, Y. Hou, C. Oshima, and Y. Ishizawa, *Phys. Rev. Lett.* **49,** 567 (1982).
122. M. Aono, R. Souda, C. Oshima, and Y. Ishizawa, *Phys. Rev. Lett.* **51,** 801 (1983).
123. R. M. Tromp and E. J. van Loenen, *Surf. Sci.* **155,** 441 (1985).
124. H. Niehus, *Nucl. Instrum. Methods* **B33,** 876 (1988).
D. R. Mullins and S. H. Overbury, *Surf. Sci.* **199,** 141 (1988).
125. H. Niehus, Ch. Hiller and G. Comsa, *Surf. Sci.* **173,** L599 (1986).
126. W. Heiland and E. Taglauer, *J. Vac. Sci. Technol.* **9,** 620 (1972).
127. W. Heiland, F. Iberl, E. Taglauer, and D. Menzel, *Surf. Sci.* **53,** 383 (1975).
128. H. Niehus and E. Bauer, *Surf. Sci.* **47,** 222 (1975).
129. W. Englert, W. Heiland, E. Taglauer, and D. Menzel, *Surf. Sci.* **83,** 243 (1979).
130. W. Heiland, W. Englert, and E. Taglauer, *J. Vac. Sci. Technol.* **15,** 419 (1978).
131. D. R. Mullins and S. H. Overbury, *Surf. Sci.* **193,** 455 (1988).
132. Th. Fauster, H. Dürr, and D. Hartwig, *Surf. Sci.* **178,** 657 (1986).
133. H. Niehus and G. Comsa, *Surf. Sci.* **93,** L147 (1980); **102,** L14 (1981).

134. C. Somerton and D. A. King, *Surf. Sci.* **89,** 391 (1979).
135. B. J. J. Koeleman, S. T. de Zwart, A. L. Boers, B. Poelsema, and L. K. Verheij, *Phys. Rev. Lett.* **56,** 1152 (1986).
136. J. W. Rabalais, O. Grizzi, M. Shi, and H. Bu, *Phys. Rev. Lett.* **63,** 51 (1989).
137. J. Schulz, Diploma Thesis, Technical University of Munich (1990), unpublished; J. Schulz, E. Taglauer, P. Feulner, and D. Menzel, *Verhandl. DPG (VI)* **25,** 1166 (1990).
138. J. A. Yarmoff, D. M. Cyr, J. H. Huang, S. Kim, and R. S. Williams, *Phys. Rev. B* **33,** 3856 (1986).
139. E. van de Riet, J. B. J. Smeets, J. M. Fluit, and A. Niehaus, *Surf. Sci.* **214,** 111 (1989).
140. J. A. van den Berg, L. K. Verheij, and D. G. Armour, *Surf. Sci.* **91,** 218 (1980).
141. D. J. O'Connor, *Surf. Sci.* **173,** 593 (1986).
142. M. Aono, R. Souda, C. Oshima, and Y. Ishizawa, *Surf. Sci.* **168,** 713 (1986).
143. K. Sumitomo, K. Tanaka, Y. Izawa, I. Katayama, F. Shoji, K. Oura, and T. Hanawa, *Appl. Surf. Sci.* **41/42,** 112 (1989).
144. K. Oura, M. Katayama, F. Shoji, and T. Hanawa, *Phys. Rev. Lett.* **55,** 1486 (1985).
145. J. H. Huang and R. S. Williams, *J. Vac. Sci. Technol.* **A6,** 689 (1988).
146. W. P. Ellis and R. R. Rye, *Surf. Sci.* **161,** 278 (1985).
147. C. H. Patterson and T. M. Buck, *Surf. Sci.* **218,** 431 (1989).
148. H. Nakamatsu, A. Sudo, and S. Kawai, *Surf. Sci.* **223,** 193 (1989).
149. B. Poelsema, L. K. Verheij, and A. L. Boers, *Nucl. Instrum. Methods* **132,** 623 (1976).
150. D. J. Martin, *Surf. Sci.* **97,** 586 (1980).
R. P. Walker and D. J. Martin, *Surf. Sci.* **118,** 659 (1982).
151. R. Souda, M. Aono, C. Oshima, S. Otani, and Y. Ishizawa, *Surf. Sci.* **128,** L236 (1983).
152. G. Engelmann, E. Taglauer, and D. P. Jackson, *Surf. Sci.* **162,** 921 (1985).
153. Th. Fauster, R. Schneider, H. Dürr, G. Engelmann, and E. Taglauer, *Surf. Sci.* **189/190,** 610 (1987).
154. J. P. Toennies, *J. Vac. Sci. Technol.* **A5,** 440 (1987).
155. P. Zeppenfeld, K. Kern, R. David, and G. Comsa, *Phys. Rev. Lett.* **62,** 63 (1989).
156. J. W. M. Frenken and J. F. van der Veen, *Phys. Rev. Lett.* **54,** 134 (1985).

7

Comparisons of SIMS, SNMS, ISS, RBS, AES, and XPS Methods for Surface Compositional Analysis

C. J. Powell, D. M. Hercules, and A. W. Czanderna

1. Purpose

The purpose of this chapter is to summarize and compare available information about the six major techniques for surface compositional analysis (SIMS, SNMS, ISS, RBS, AES, and XPS) in tabular form and in brief narratives. It is assumed that the reader is familiar with the principles of these techniques. Detailed information on the ion spectroscopies (SIMS, SNMS, ISS, and RBS) is presented in earlier chapters of this volume; information on the electron spectroscopies can be found in the references given below and in a forthcoming volume of this series. Acronyms and abbreviations used in this chapter are defined in Chapter 1 of this volume.

C. J. Powell • Surface Science Division, National Institute of Standards and Technology, Gaithersburg, Maryland 20899 **D. M. Hercules** • Department of Chemistry, University of Pittsburgh, Pittsburgh, Pennsylvania 15260 **A. W. Czanderna** • Photovoltaics Measurements and Performance Branch, Solar Energy Research Institute, Golden, Colorado 80401.

Ion Spectroscopies for Surface Analysis (Methods of Surface Characterization series, Volume 2), edited by Alvin W. Czanderna and David M. Hercules. Plenum Press, New York, 1991.

2. Introduction

Our comparison of the six methods of surface analysis is based on a number of criteria and categories of instrument use. These criteria and categories include the physical principles of each method, principal information output, sample damage from the excitation beam, the spatial resolution, rate of collection of data, quantification, detectability, versatility, ease of use, and availability of data bases. The nature of the specimen also has an important influence on the results obtainable. Each method has particular advantages and limitations that are evident in the following comparisons.

Our summaries are intended only as a semiquantitative guide. The information pertains to commercially available surface analysis systems, in their normal configurations, as opposed to custom-made instruments that optimize a particular parameter. In many cases, tradeoffs can and often must be made among the various parameters, e.g., spatial resolution, accuracy and precision of analysis, sensitivity, sample damage, and cost. The authors offer judgments for some of the categories, and these will be obvious from our comments. We refer readers to other chapters in this book series where appropriate. More details are provided on the ion spectroscopies (SIMS, SNMS, ISS, RBS) in this volume and on the electron spectroscopies (AES, XPS) in a forthcoming volume in this series. We include information for SALI under the generic acronym SNMS. For definitions of terms used in this chapter, we refer readers to the Appendix (ASTM Standard E-673).

3. Comparison Categories or Criteria

3.1. Input/Output Particles, Sample Damage, Measured Quantity, Principal Information Output, and Sampled Depth

Table 1 provides overview information about the input/output particles (stimulation/emission), sample damage, measured quantity, and sampled depth for SIMS, SNMS, ISS, RBS, AES, and XPS. All methods provide the elemental composition of the surface region, for the sampled depth, as the principal information output.

Specimens can be damaged by the incident beams. Such damage can take the form of removal of specimen material (e.g., by sputtering or desorption), of changes in specimen composition (e.g., decomposition or polymerization), or of atomic displacements. XPS and RBS are relatively nondestructive, but some materials may be damaged during a prolonged surface analysis. Decomposition is not common in XPS, but prolonged

Table 1. Summary of Features of SIMS, SNMS, ISS, RBS, AES, and XPS

Category	SIMS	SNMS	ISS	RBS	AES	XPS
Input "particle"	0.5–20-keV ions	0.5–20-keV ions	0.5–3.0-keV ions	0.5–3-MeV ions	2–20-keV electrons	Photons (x-rays)
Damage to sample by input particle	Minimal to extensive depending on beam parameters	Minimal to extensive	Minimal for defocused beam to moderate	Minimal except for organics and polymers	Minimal to extensive for focused *e*-beam	Minimal (x-radiation)
Output particles	Secondary ions	Secondary neutrals	Input ion	Input ions	Electrons	Electrons
Measured quantity of output particle	Mass/charge of sputtered surface ions	Mass/charge of postionized surface species	Energy of scattered ions after binary collisions	Energy of ion backscattered from elastic collisions	Energy of ejected Auger electrons	Energy of ejected core-level and Auger electrons
Sampled depth	1–2 monolayers ("static")	1–2 monolayers ("static")	Surface monolayer	Many μm into the solid	2–20 monolayers	2–20 monolayers

x-ray exposures may damage adsorbates, organic compounds, and some ionic solids.[1,2] AES is essentially nondestructive for metals, alloys, and many semiconductors, but electron beams will rapidly damage organic materials, adsorbates, glasses, and some compounds.[2,3] Electron beam effects are generally proportional to the electron beam dose (primary current density times bombardment time). An electron-beam-induced temperature increase is most detrimental for thin films on substrates with low thermal conductivity. Electron-stimulated desorption is often encountered in studies of ionic compounds. In extreme cases, e.g., for alkali halides, the number of desorbed atoms or molecules per incident electron is greater than unity.[3] Charging of insulators is sometimes difficult to avoid and can even lead to induced electromigration. In general, electron-beam effects on fragile specimens can be minimized with a low primary electron beam current density, short measurement times, or scanning the incident beam to different regions of the surface.

SIMS, SNMS, and ISS are all intrinsically destructive; that is, ion bombardment removes specimen material from the exposed surface and it is not possible to remeasure the composition of a previously sputtered surface. A trade-off must then be made between operating in the "static" mode for true surface analysis and the data acquisition time. The "static" mode requires very low erosion rates,[4] and thus low current densities, which also may require optimization between the current density and detection limits (i.e., signal-to-noise ratio). A low incident current density for AES, SIMS, SNMS, and ISS also means that the lateral resolution will be degraded.

Electron-beam-induced, x-ray-induced, ion-beam effects will be reviewed by Pantano, Thomas, and Czanderna in chapters in a forthcoming companion volume of this series. These chapters provide in-depth support for the qualitative summaries made in Table 1 and in the text. Ion beams are also used for compositional depth profiling (CDP) and damage specimens in many ways (e.g., surface roughening, knock-on effects, atomic-scale cascade mixing, preferential sputtering, amorphization and structural changes, decomposition, implantation, chemical state changes, enhanced diffusion, enhanced adsorption, and redeposition).

The sampled depth (or information depth) for SIMS and SNMS is listed in Table 1 for the "static" mode; in the dynamic mode (high erosion rates),[4] as needed for thin-film analysis, the sampled depth may range from 1 to 5 nm or more depending on the erosion rate, incident beam angle, and beam energy used. In RBS, surface monolayers can be studied on single crystals using channeling techniques. The sampled depth depends on the escape depth of the ejected electrons in XPS and AES, on neutralization probabilities in ISS, and erosion rates in SIMS and SNMS. The ASTM definition for information depth (see Appendix) gives the depth from the exposed surface from which a specified percentage of the detected signal originates. This depth should be distinguished from two other terms, the detection limit (or detectability) and the sensitivity. The detection limit for a particular technique is the minimum quantity of material that can be detected under specified operating conditions and is often expressed as a fraction of an atomic layer; for example, it is often assumed that the detected material is present as an identifiable phase on a substrate of another material. The sensitivity is the slope of a calibration curve in which the measured signal strength for a given element under specified conditions is plotted versus elemental concentration.

3.2. Data Collection

Factors important in data collection include vacuum requirements, data acquisition rates, signal-to-noise (S/N) ratio, turnaround time, and depth profiling compatibility. Measurements must be made with a pressure in the vacuum chamber of 10^{-6} Torr or better so that collisions between electrons or ions and residual gas molecules are minimized. Residual gases may react with specimens being analyzed, and pressures of 10^{-9} Torr or lower are generally required to minimize reaction (see Section 2.2, Chapter 1). Another factor is electron-stimulated adsorption, which can lead to oxidation or carburization.[3] Table 2 provides an overview of other factors in data collection. Average data acquisition times are given in the table, although the actual times may range from seconds to hours depending on the information desired. The S/N ratio governs the limit of detection; the ranges given should be compared with the detection limits in Table 3.

For CDP, XPS is clearly inferior because the ion beam etching and surface analysis must be done alternately as a result of the relatively long data acquisition time and the relatively large x-ray beam size, even for current "small-spot" XPS systems. Otherwise, a depth profile is obtained with an inferior depth resolution. However, XPS (as well as AES) can be used nondestructively to obtain depth profiles for shallow depths ($\leq$50 Å) by detection of ejected electrons at different takeoff angles. This technique makes use of the variation of electron escape depth with takeoff angle; it is especially powerful when there are electrons with widely different energies and thus escape depths for the same element. In custom-designed RBS systems, near-surface nondestructive CDP may be obtained using channeling and grazing-exit detector techniques. SIMS and AES combined with ion erosion are the two most widely used CDP techniques.

In recent years, the analog instrumentation for recording spectral intensities has been replaced by digital data systems. For example, the

Table 2. Factors for Data Collection in Surface Analysis by SIMS, SNMS, ISS, RBS, AES, and XPS

Factor	SIMS	SNMS	ISS	RBS	AES	XPS
Vacuum (Torr)[a]	10^{-6}–10^{-10}	10^{-6}–10^{-10}	10^{-6}–10^{-10}	$<10^{-6}$	10^{-6}–10^{-10}	10^{-6}–10^{-10}
Acquisition time	Seconds	Seconds	Minutes	Minutes	Seconds	Minutes
Signal/noise ratio	Up to 10^9	10^6	10^2–10^4	10–10^3	10^2–10^3	10^2–10^3
Turnaround time	Fast[b]	Fast[b]	Fast[b]	Fast	Fast[b]	Fast[b]
CDP[c]	Most natural for SIMS/SNMS because sample erosion is required		Easily done with bombarding beam	Automatically done and nondestructively	Easily done with an auxiliary ion gun	Laboriously done with an auxiliary ion gun
Other depth profiling methods	NA	NA	NA	Yes[d]	Yes[d]	Yes[d]

[a] Although data can be collected at pressures as high as 10^{-4} Torr, contamination of the specimen by the residual gases is a serious concern depending on the reactivity of the surface (see Chapter 1, Section 2.2).
[b] Fast with instruments equipped with rapid-sample-introduction or load-lock facility. Turnaround is slow if samples are loaded onto a carousel and a flange has to be rebolted to the vacuum system.
[c] Does not include artifacts in the sample resulting from ion bombardment, which are the same for all analytical techniques except RBS.
[d] See text.

common lock-in amplifier with high-frequency modulation that yields directly the derivate spectrum in AES has been replaced by a voltage-to-frequency converter to obtain a digital output. For lower currents, individual pulses from the detector are counted to give energy distributions ("direct" spectra) in AES, XPS, and ISS, or mass spectra in SIMS and SNMS. The digital data are stored in a microcomputer memory, can be displayed in real time on a monitor, and can be subsequently processed (e.g., in operations such as peak identification, smoothing, differentiation, background subtraction, peak fitting, and quantitative analyses). In modern equipment, all analyzer and excitation source functions are controlled by the computer software which has become an increasingly important part of the instrumentation.

3.3. Features of the Analytical Methods

Important features of SIMS, SNMS, ISS, RBS, AES, and XPS are summarized in Table 3. Each technique is able to identify directly elements with atomic numbers of 3 and higher. The elements H and He can be detected by SIMS and SNMS, by using forward scattering in ISS, or by using nuclear reaction analysis (NRA) with RBS equipment. Hydrogen has been detected in some materials by AES and XPS from a detailed analysis of the spectral line shapes of other elements. The mass-sensitive techniques can detect isotopes, but this capability is limited in ISS and RBS by poor peak resolution for mid- to high-Z elements. SIMS and SNMS have the lowest detection limits, but SIMS suffers from wide sensitivity variations with Z.

None of the ion spectroscopies provides the detailed chemical information obtained with the electron spectroscopies, especially XPS. Small shifts (chemical shifts) of elemental lines in XPS can be used to identify different chemical states of an element; such identifications are extremely valuable (e.g., in identifying surface reactions or processes). Auger lines also appear in XPS data and typically also show chemical shifts. The difference in XPS and AES peak positions for an element, the so-called Auger parameter, often has a chemical shift. Not only is the latter shift a useful diagnostic but it may be the easiest to measure reliably on nonconducting specimens. AES, as practiced with electron excitation, has usually been done with analyzers of low energy resolution. Thus, the chemical information (shifts of elemental lines and changes of peak shapes) obtained from the technique has been limited. In some cases chemical species can be identified by AES data alone, but the supporting data are much less than for XPS. Static SIMS represents a largely untapped resource for obtaining chemical information (see

Table 3. Summary of Analytical Features of SIMS, SNMS, ISS, RBS, AES, and XPS

Features	SIMS	SNMS	ISS	RBS	AES	XPS
Elements not directly detected	None	None	H, He	H, He	H, He	H, He
Detection of H, He	Yes	Yes	$\theta < 19°$ (H)[a] <42° (D)	With NRA	Line shapes of other elements in a compound	
Isotopes	Yes	Yes	Low Z	Low Z	No	No
Detection limit (atomic fraction)	10^{-6}–10^{-9}	10^{-6}	10^{-2}–10^{-4}	10^{-1}–10^{-4}	10^{-2}–10^{-3}	10^{-2}–10^{-3}
Variation of detection limit with Z	10^{4}–10^{5}	10 to 100	$\sim 10^{2}$ (Li to U)	$\sim 10^{3}$ (Li to U)	~20	~20
Chemical information and value/usefulness	Yes Some	Yes Some	Yes, but only for eight elements	No NA	Yes Considerable	Yes Outstanding
Other information	Polymers and organics, m/Z to 10,000 and beyond for SIMS; spectra are matrix dependent, but much less so for SNMS		Shadowing of adsorbate overlayers on single crystals	Surfaces of epitaxial films	Plasmon loss structure, chemical shifts, line shape changes	Chemical shifts, oxidation states, valence band structure
Information on compounds	Yes	Yes	Rarely	No	Frequently	Yes

Organic samples	Yes	Yes	No	No	Damage likely	Yes
Surface structure	Some	Some	Adsorbates on single-crystal substrates	Epitaxial films	Adsorbates on single-crystal substrates, epitaxial films	
Peak interference[b]	Rare	Some	None	None[b]	Occasional	Occasional
Elemental specificity[c]	Excellent	Very good	Good—low Z; Poor—high Z	Good—low Z; Poor—high Z	Very good	Very good
Lateral $(x-y)$ resolution	0.1 μm	0.2–5 μm	0.1 mm	1 mm	20–100 nm	20–100 μm
Lateral imaging and value	Yes Very good	No NA	Yes Poor	No NA	Yes Outstanding	Yes Good
Charging problems	Yes	Yes	Yes	No	Yes	Yes
Quantification sensitivity factors	Vary by up to 10^5 in different matrices	Yes for given instrument	Yes but small data base	Complete	Yes	Yes
Standard deviation (for repeated measurements)	5–100%	5%–10%	5%	2%	1%–5%	1%

[a] θ is ion scattering angle.
[b] Extent to which there may be accidental overlaps of spectral peaks due to two or more elements. For RBS, signals at energies corresponding to certain elements can overlap signals due to other elements at different depths.
[c] Extent to which two or more elements can be distinguished due to finite instrumental resolution.

Winograd and Garrison, Chapter 2, this volume). The mass spectrum derived from SIMS contains considerable chemical information. The problem is to extract this information from the signals due to sputtered cluster ions and molecular fragments. Recent experiments using SIMS with high mass resolution have been particularly valuable in identifying organic materials on surfaces. It is possible now to measure femtomoles of small organic molecules (molecular weight ~300) and to obtain structural and molecular weight information for polymers. One can detect small molecules on polymer surfaces or one polymer on another.

Each technique can provide different additional information about a specimen material. For example, information about the electronic structure can be obtained from XPS and AES. Structural information for adsorbed molecules or thin epitaxial films can be obtained from the angular distributions of photoelectrons or Auger electrons. The energy-loss structures in the vicinity of the elastic peak due to excitations of valence and core electrons in electron-excited AES can be analyzed to give electronic and local structural information.

Peak interferences can occur in AES and XPS but are usually only an annoyance (e.g., for compounds with a large number of elements). An advantage of XPS is that accidental overlaps of photoelectron and Auger peaks can be identified and overcome by changing the x-ray energy. High-resolution mass spectrometers can separate peaks with nearly identical mass/charge ratios in SIMS, but the broader peaks obtained with many SNMS instruments can lead to some ambiguity in peak identification for more complex specimens. Both ISS and RBS are severely limited in elemental identification when analyzing samples with several high-Z elements.

The highest lateral resolution is attained at present with AES. The indicated AES resolution in Table 3 is achieved, however, only with robust specimens which are not damaged by the high bombarding current density. In addition, only major constituents of the surface region (with atomic concentrations $\gtrsim 10\%$) can be detected with such lateral resolutions.[4] Very good lateral resolution, about 0.1 μm, is achieved in SIMS although then with relatively high erosion rates of the surface (i.e., "dynamic" SIMS). Considerable effort has been made in recent years to improve the spatial resolution of XPS. Although the lateral resolution with XPS is now poorer than that in AES or SIMS, it should be remembered that x-ray excitation is generally much less damaging to a surface than electron or ion bombardment and that the chemical-state information in XPS can be very useful.

High spatial resolution is clearly important if the specimen is inhomogeneous. Practical materials can have different types of in-

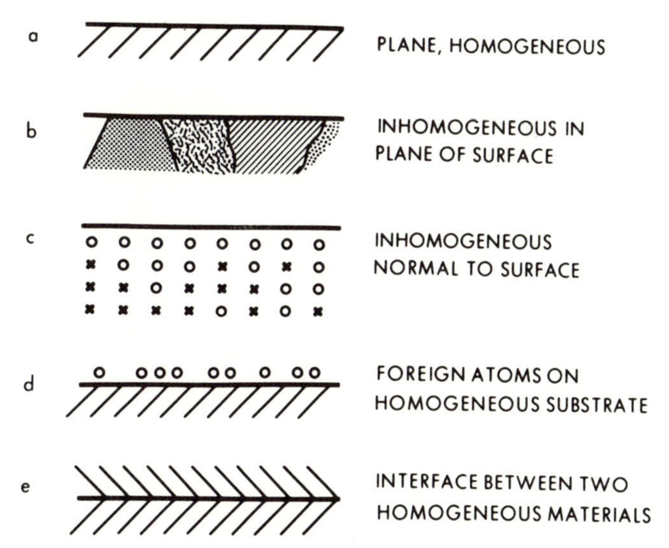

Figure 1. Idealized sketches of different types of surface morphologies to illustrate different types of inhomogeneities and surface phases: (a) A plane homogeneous surface; (b) a surface with lateral inhomogeneities consisting of several different surface phases; (c) a surface with depth inhomogeneities (the circles and the crosses represent different types of atoms); (d) a surface phase consisting of a submonolayer of foreign atoms on an otherwise homogeneous surface; and (e) an interface between two homogeneous phases (e.g., as in a thin-film system). Actual surfaces will have various combinations of these inhomogeneities, which may occur on different length scales together with finite roughness and defects. (Reference 5.)

homogeneities (some simplified limiting cases are sketched in Fig. 1) as well as finite roughness and perhaps complex morphologies. High spatial resolution is then desirable for determining whether multiple surface phases exist and their spatial arrangement. Even if a detailed compositional map of the surface is not needed, multiple high-resolution analyses can be required at different surface regions to ensure that representative and statistically adequate information is acquired.[4]

While most surface analyses performed are qualitative, there is increasing interest in quantitative measurements, particularly with the availability of data systems for processing acquired spectra. Physical models have been developed for relating observed spectral intensities to elemental concentrations, but only in the case of RBS is the model well established and the needed parameters readily available.[4,5] Most quantitative analyses with AES and XPS are made with the use of elemental sensitivity factors, but matrix effects, mainly the variation of attenuation lengths for Auger electrons or photoelectrons from a given element from one matrix to another, can cause errors of about 20–50%.[6] Matrix effects in SIMS can be substantial (up to a factor of 10^5) but useful quantitative analyses can be obtained with appropriate standards.[4,7] Matrix effects in SNMS are much reduced compared to SIMS, but more studies are needed of matrix effects on the energy and angular distributions of sputtered neutrals. In ISS, variations of elemental signals in different matrices of up to an order of magnitude can occur; these variations are due in large part to changing ion neutralization probabilities.

Standard materials, with compositions approximating those of the specimens to be analyzed, provide a relatively straightforward means of minimizing matrix and instrumental uncertainties. It is not generally easy to prepare standards with similar properties (distribution of surface phases, roughness, etc.) to the specimens, but ion implantation has been shown to be a useful method for generating a reference signal in SIMS.[7] The high surface sensitivity of ISS is such that care has to be exercised in avoiding surface impurities such as adsorbed gases.

3.4. Versatility, Ease of Use, and Supporting Data

Table 4 shows qualitative comparisons of selected features of the analytical techniques. The symbols give a rough guide to the various features to supplement the information given in Tables 1–3.

Successful use of a particular technique depends on the availability of needed reference data. Table 4 gives an indication of the extent to which needed data are available.[7–9]

Information is also given in Table 4 on the relative use of the six techniques in recent years and the effective take-off year, that is, the year in which publications started to grow following the introduction of commercial equipment.[8]

3.5. Specimen and Vacuum Requirements

One or more specimens are generally mounted on a manipulator which allows *x*, *y*, and *z* movements and sometimes rotation. These motions are needed for locating the specimen with respect to the analyzer and for selecting a particular region for analysis.

Modern instruments are equipped with a specimen introduction stage which allows transfer of the specimen from the atmosphere through an intermediate chamber to the main analysis chamber. The intermediate chamber can be evacuated to a pressure of $\leq 10^{-3}$ Torr before the sample is transferred to the main chamber; in this way, samples can be

Table 4. Comparisons of Selected Features of SIMS, SNMS, ISS, RBS, AES, and XPS[a]

Feature	SIMS	SNMS	ISS	RBS	AES	XPS
Versatility	+ + + +	+ + +	+ + +	+ + +	+ + + +	+ + + +
Ease of use	+ + + (dynamic) + + (static)	+ + +	+ + + +	+ + + +	+ + + + +	+ + + + +
Thin film CDP	+ + + + +	+ + + + +	+ + +	+ + + + +	+ + + + +	+ +
Simplicity of data interpretation	+ + + +	+ + + + +	+ + + + +	+ + + + +	+ + + +	+ + + +
Accuracy of quantitative analysis	+	+ + + +	+ +	+ + + + +	+ + + +	+ + + +
Availability of reference data	+ + + + (dynamic) + + (static)	+	+ + +	+ + +	+ + + + +	+ + + + +
Approximate percent of use 1985–1990	24	2	3	5	33	33
Effective take-off year	1972	1984	1973	1971	1969	1969

[a] The number of plus symbols indicates a value from poor (+) to very good (+ + + + +)

transferred rapidly into and out of the main chamber without degrading the ultrahigh vacuum environment (or requiring lengthy pumping times or bakeouts). The specimen may also be transferred to other chambers for processing (e.g., heating, cooling, fracture, deposition of films, reactions, etc.), again without significantly degrading the pressure in the analysis chamber.

Since an ultrahigh vacuum is required for all of the techniques (except for RBS when used for thin-film analysis), it is important that the specimens be stable while in this environment. The specimen's vapor pressure should also be low enough so that evaporation of potentially harmful vapors does not cause deterioration of analyzer and multiplier surfaces. ASTM Standard E-1078 in the Appendix gives guidance on specimen handling, mounting, and treatment for surface analysis. This guide also gives suggestions for reducing the charging of nonconductive specimens during analysis and for handling particles, pellets, and fibers.

3.6. Summary of Advantages and Limitations

We summarize particular advantages and limitations of SIMS, SNMS, ISS, RBS, AES, and XPS in Tables 5–10. These tables together with the information in Tables 1–4 give an overview of these techniques for surface analysis. More details concerning instrumental capabilities can be found in earlier chapters of this book and in other review articles.[4,8,10–17]

Table 5. Major Advantages and Limitations of SIMS for Surface Analysis and Compositional Depth Profiling (CDP)

Advantages
Is sensitive to outermost 1–2 monolayers (static mode)
Can detect 10 ppm or less
Acquires CDP data as surface is eroded during analysis
Can detect isotopes
Can detect H and D
Can acquire data rapidly
Has lateral imaging capability
Limitations
Requires destruction of sample for the analysis
Is quantitative with difficulty, at best
Has varying elemental sensitivity
Has complex spectra
May cause chemical state changes from ion bombardment

Table 6. Major Advantages and Limitations of SNMS[a] for Surface Analysis and Compositional Depth Profiling

Advantages
- Is sensitive to outermost 1–2 monolayers (static mode)
- Can detect 10 ppm or less
- Acquires CDP data as surface is eroded during analysis
- Can detect isotopes
- Can detect H and D
- Is quantitative with modest use of standards (a major improvement over SIMS)
- Can acquire data rapidly

Limitations
- Requires sample destruction
- Has poor lateral resolution
- May cause chemical state changes from ion bombardment

[a] SNMS, SALI.

Table 7. Major Advantages and Limitations of ISS for Surface Analysis and Compositional Depth Profiling

Advantages
- Is sensitive to first monolayer
- Can detect 10^{-2}–10^{-4} monolayer
- Is good for CDP
- Is useful for studying ordered surfaces and adsorbates
- Is excellent for segregation and/or altered layer studies
- Can detect isotopes (^{16}O–^{18}O, ^{14}N–^{15}N, ^{32}S–^{34}S)
- Has simple spectra
- Provides chemical information for a few surface compounds
- Can be quantitative with standards

Limitations
- Has poor lateral resolution
- Has poor specificity for high-Z elements[a]
- Is destructive although erosion rate can be less than a monolayer/h
- Requires better understanding of ion neutralization
- Generally provides no chemical information

[a] For example, with incident ^{4}He ions and a scattering angle of 138°, the following pairs of elements can be resolved: Na and Mg, Fe and Cu, and Ba and Ta. Better resolution can be obtained for high-Z elements with incident neon or argon ions.

Table 8. Major Advantages and Limitations of RBS for Surface Analysis and Compositional Depth Profiling

Advantages
- Is quantitative without standards
- Can detect atomic fractions of 10^{-1}–10^{-4} depending on Z
- Directly measures depth distributions below the surface without sample destruction
- Has rapid data acquisition
- Is especially sensitive to high-Z elements
- Has modest vacuum requirements except for surface studies
- Can be customized to study surface monolayers
- Can resolve some low-Z isotopes
- Can be combined with NRA for detection of H, He, and other low-Z elements

Limitations
- Has poor lateral resolution
- Provides no chemical information
- Cannot resolve different elements of the same mass
- Is limited for detecting low-Z elements
- Requires lateral uniformity of sample
- Is limited by overlapping signals due to mass and depth
- Is limited to specimens with only a few elements

Table 9. Major Advantages and Limitations of AES for Surface Analysis and Compositional Depth Profiling

Advantages
- Is sensitive to 2–20 monolayers
- Can detect ca. 10^{-3} atomic fraction
- Is outstanding for CDP
- Has a sensitivity range within a factor of 20
- Has superb lateral resolution
- Gives chemical information for some elements
- Can acquire data rapidly

Limitations
- May alter surface composition
- May have severe charging problems
- Is of limited value for organic materials
- Has a slow rate of element mapping

Table 10. Major Advantages and Limitations of XPS for Surface Analysis and Compositional Depth Profiling

Advantages
- Is sensitive to 2–20 monolayers
- Can detect ca. 10^{-3} atomic fraction
- Is especially useful for chemical shifts from the same element in different compounds
- Is least destructive of all techniques
- Has a sensitivity range within a factor of 20
- Has minimal sample charging

Limitations
- Has moderate lateral resolution
- Is slower for CDP than other methods

3.7. Selection of a Technique

Surface analysis has been successful in solving a wide range of scientific and technological problems.[8,12–20] Table 11 gives a summary of applications in basic and applied research. The main types of applications in which surface analysis is used are listed in Table 12.

Many factors are involved in considering whether or not to use a particular surface analysis technique.[4] Table 13 shows issues and questions that should be kept in mind when selecting a technique or a

Table 11. Applications of Surface Analysis in Basic and Applied Research

Surfaces	Interfaces	Thin films
Adsorption/desorption	Adhesion	Epitaxial growth
Catalysis	Composites	Evaporated layers
Condensation	Contamination	Growth
Contamination	Corrosion	Interdiffusion
Corrosion	Delamination	Ion implantation
Diffusion	Diffusion	Ion (electro) migration
Epitaxy	Embrittlement	Microelectronic devices
Erosion/abrasion	Grain boundaries	Passivation layers
Friction/wear	Mechanical stability	Protective coatings
Oxidation	Oxidation	Solar-energy multilayer stacks
Reaction	Photovoltaic materials	Solid-state reactions
Reconstruction	Precipitation	
Reduction	Segregation	
Relaxation	Sintering	
Subsurface layers	Vapor-deposited films	
Surface states		

Table 12. Major Uses of Surface Analysis

1. A–B comparisons (e.g., failure analysis)
2. Correlations of surface composition with other properties (e.g., process and device development)
3. Characterization of model systems that approximate complex systems (e.g., catalysts, interfaces in electronic materials)
4. Depth profiling (e.g., surface segregation, thin-film devices)
5. Identification of surface contaminants (e.g., process control)
6. Microscopic view of surface
   - Lateral and depth inhomogeneities
   - Understanding of properties and phenomena
   - Design of improved products and processes

Table 13. Issues and Questions to Consider in Selecting a Surface Analysis Instrument and/or Technique

Issue	Question
1. Cost versus return on investment	
2. Nature of specimens	Detection limit
• Range of materials	Complexity
• Number of elements	Interferences
3. Presence of multiple surface phases	Spatial resolution
4. Analytical question(s)	Trace elements
	Impurities
	Chemical state
	Segregation/diffusion
5. Qualitative analysis	Combination of techniques
• Chemical state	Specificity
6. Quantitative analysis	Feasibility
	Accuracy
7. *In situ* processing	Heating/cooling
	Fracture
	Evaporation/dosing/treatment
	Sputtering
8. Other properties to be measured concurrently	Compatibility
	Space
9. Automation	Data manipulation/reduction
	Data presentation
10. Vacuum requirements	High vapor pressure materials
	Specimen reaction chamber

particular instrument. If a potential user is not sure whether a particular technique or instrument will be useful for a given range of problems, it is recommended that test samples be sent for analysis to a service laboratory or to an instrument manufacturer.

4. The Surface Analysis Community

While the principles of the various surface analysis techniques are well established, there are many complexities in the operation of the instruments and the interpretation of the observations. There are, in addition, many artifacts to be avoided. As a result, users new to the field may not operate their instruments or process their data as efficiently and as effectively as possible. This book series is, in fact, designed to assist such users as well as more experienced scientists. We give here additional information on other sources of guidance.

Investigations in which surface analysis has been employed are reported at the conferences and in the journals of many professional societies and other groups. In the United States, relevant organizations include the American Vacuum Society (AVS), American Physical Society, American Chemical Society, Materials Research Society (MRS), and the Pittsburgh Conference on Analytical Chemistry and Applied Spectroscopy. The AVS and MRS also offer short courses on surface analysis topics in conjunction with their major meetings, and the AVS sponsors a biennial Topical Conference on Quantitative Surface Analysis. In Europe, the biennial European Conferences on Surface and Interface Analysis (ECASIA) and International Conferences on Quantitative Surface Analysis are held in alternate years. The Surface Science Society of Japan holds regular meetings and publishes a journal. The International Conference on Secondary Ion Mass Spectrometry is held every two years. In addition to the publications of the societies mentioned above, the journals *Surface and Interface Analysis, Surface Science, Applied Surface Science,* and *Journal of Electron Spectroscopy* contain many articles relevant to surface analysis.

As the use of surface analysis has grown, the need for reference procedures, reference data, and reference material has also increased. Three groups are now active in developing such standards. The American Society for Testing and Materials (ASTM) established Committee E-42 on Surface Analysis in 1976 to advance the field of surface analysis and the quality of surface analyses through the development of appropriate standards, standard procedures, standard materials, round robins, symposia, workshops, and publications. This Committee and its Subcommittees for AES, XPS, ISS, SIMS, Energetic Ion Analysis, Ion Beam Sputtering, Reference Materials, and Reference Data meet twice each year; one of these meetings is held in conjunction with the annual National Symposium of the AVS. This Committee has developed documentary standards for surface analyses, and four of these are presented in the Appendix. Standards of the ASTM E-42 Committee are published in Volume 3.06 of the *Annual Book of ASTM Standards* and are republished in the journal *Surface and Interface Analysis.* Status reports of the E-42 Committee's activities are published at intervals.[(21)]

The Versailles Project on Advanced Materials and Standards (VAMAS) is a cooperative agreement among the Governments of Canada, the Federal Republic of Germany, France, Italy, Japan, the United Kingdom, and the United States of America and with the Commission of the European Communities. VAMAS established a Surface Chemical Analysis Technical Working Party in 1984 to produce the reference procedures, reference data, and reference materials necessary to establish standards for surface analysis. There are now over 30 multinational VAMAS projects, and summary reports of these are published.[(21,22)]

The International Union of Pure and Applied Chemistry has established a Subcommittee on Surface Analysis of Commission V.2 on Microchemical Techniques and Trace Analysis. This Subcommittee considers terminology, symbols, and units for surface analysis, provides data collections relevant to surface analysis, makes critical evaluations of methods and techniques for surface analysis, and develops recommended procedures. Reports from this group have been published in the journal *Pure and Applied Chemistry* and status reports of Subcommittee activities are available.[(21,23)]

In a number of countries there are active groups of users of surface analysis equipment that meet regularly to discuss progress and common problems. Information on these groups can be obtained from the instrument manufacturers.

References

1. R. G. Copperthwaite, *Surf. Interface Anal.* **2,** 17 (1980).
2. J. Cazau, *Appl. Surf. Sci.* **20,** 457 (1985).
3. C. G. Pantano and T. E. Madey, *Appl. Surf. Sci.* **7,** 115 (1981); T. E. Madey, in: *Analytical Electron Microscopy—1987* (D. C. Joy, ed.), p. 345, San Francisco Press, San Francisco (1987).
4. H. W. Werner and R. P. H. Garten, *Rep. Prog. Phys.* **47,** 221 (1984).
5. C. J. Powell and M. P. Seah, *J. Vac. Sci. Technol. A* **8,** 735 (1990).

6. A. I. Zagorenko and V. I. Zaporozchenko, *Surf. Interface Anal.* **14,** 438 (1989).
7. J. T. Grant, P. Williams, J. Fine, and C. J. Powell, *Surf. Interface Anal.* **13,** 46 (1988).
8. S. Hofmann, *Surf. Interface Anal.* **9,** 3 (1986).
9. M. P. Seah, *Surface. Interface Anal.* **9,** 85 (1986).
10. A. W. Czanderna, *Solar Energy Mater.* **5,** 349 (1981).
11. C. J. Powell, in: *Applied Electron Spectroscopy for Surface Analysis* (H. Windawi and F. Ho, eds.), pp. 19–36, Wiley, New York (1982).
12. M. P. Seah and D. Briggs, in: *Practical Surface Analysis, second edition, Vol. 1, Auger and X-Ray Photoelectron Spectroscopy* (D. Briggs and M. P. Seah, eds.), p. 5, Wiley, New York (1990).
13. H. W. Werner, in: *Thin Film and Depth Profile Analysis* (H. Oechsner, ed.), p. 5, Springer-Verlag, Berlin (1984).
14. A. Benninghoven, F. G. Rüdenauer, and H. W. Werner, *Secondary Ion Mass Spectrometry,* pp. 1022–1047, Wiley, New York (1987).
15. J. R. Bird and J. S. Williams, in: *Ion Beams for Materials Analysis* (J. R. Bird and J. S. Williams, eds.), pp. 515–537, Academic Press, New York (1989).
16. I. V. Bletso, D. M. Hercules, D. van Leyen, and A. Benninghoven, *Macromolecules* **20,** 407 (1987).
17. H. Hantsche, *Scanning* **11,** 257 (1989).
18. C. J. Powell, *Appl. Surf. Sci.* **1,** 143 (1978).
19. C. J. Powell, *Aust. J. Phys.* **35,** 769 (1982).
20. D. Briggs, *Surf. Sci.* **189/190,** 801 (1987).
21. C. J. Powell, *Surf. Interface Anal.* **11,** 103 (1988).
22. M. Seah, *Surf. Interface Anal.* **14,** 407 (1989); *Surf. Interface Anal.* **16,** 135 (1990).
23. W. H. Gries, *Fresenius Z. Anal. Chem.* **333,** 596 (1989).

Appendix

Standard Terminology Relating to Surface Analysis[1]

[ε1] NOTE—Editorial corrections were made throughout in March 1990.

1. Scope

1.1 These definitions are related to the various disciplines involved in surface analysis.

1.2 The definitions listed apply to (*a*) Auger electron spectroscopy (AES), (*b*) X-ray photoelectron spectroscopy (XPS), (*c*) ion-scattering spectroscopy (ISS), (*d*) secondary ion mass spectrometry (SIMS), and (*e*) energetic ion analysis (EIA).

2. Abbreviations

2.1 Abbreviations commonly used in surface analysis are as follows:

AES	Auger electron spectroscopy
BS	backscattering spectroscopy
CHA	concentric hemispherical analyzer
CMA	cylindrical mirror analyzer
EIA	energetic ion analysis
eV	electron-volts
ESCA	electron spectroscopy for chemical analysis
FABMS	fast atom bombardment mass spectrometry
FWHM	full width at half maximum peak height
ISS	ion spectroscopy scattering
pp	peak-to-peak
RBS	Rutherford backscattering spectroscopy
RFA	retarding field analyzer
SAM	scanning Auger microprobe
SIMS	secondary ion mass spectrometry
SNMS	sputtered neutral mass spectrometry
XPS	X-ray photoelectron spectroscopy

3. Definitions

adventitious carbon referencing—*XPS,* a method of determining the charging potential of a particular specimen by comparing the experimentally determined binding energy of the *C* 1s peak maximum from contaminating hydrocarbon or hydrocarbon groups on the specimen to a standard binding energy value.

analysis:

area—a two-dimensional region of a specimen surface, usually measured in the plane of the specimen surface, from which a specified percentage of the signal is detected.

volume—a three dimensional region of a specimen from which a specified percentage of the measured signal is generated. Also see **information depth.**

analyzer transmission—see **spectrometer transmission.**

angle:

collection—SIMS, the angle between the normal to the original specimen surface and the axis of the secondary ion collection optics.

of detector—EIA, SIMS, the angle between the incident beam direction and the direction pointing from the beam spot to the center of the detector.

of emission—AES, XPS, the angle of emission or ejection of electrons from a solid measured relative to the normal to the surface.

of incidence—the angle between the incident beam and the normal to the surface.

of scattering—EIA, the angle between the incident beam direction and the direction in which a particle is traveling after it is scattered. If the particle is incident on the detector, this angle will be the same as **angle** *of detector.*

solid, of detector—EIA, the solid angle intercepted by the detector, with the radius originating at the beam spot.

takeoff—AES, XPS the angle at which particles leave a specimen measured relative to the plane of the specimen surface. (see **angle** *of emission*).

angular distribution of secondary ions—see **secondary ions.**

Auger:

analysis volume—see *volume* under **analysis.**

chemical effects—AES, see **chemical.**

chemical shift—AES, see **chemical.**

current—the electron current due to the emission of Auger electrons.

electron—an electron emitted as the result of an Auger process.

electron yield—the probability that an atom with a vacancy in a particular inner shell will relax by an Auger process.

line scan—a plot of Auger signal strength as a function of displacement along a designated line on the specimen surface. Normally, the abscissa is the line along which the signal is measured and the ordinate is directly proportional to signal strength.

line shape—the energy distribution in an Auger spectrum for a particular Auger transition.

map—two dimensional image of the specimen surface showing the location of emission of Auger electrons from a particular element. A map is normally produced

[1] This terminology is under the jurisdiction of ASTM Committee E-42 on Surface Analysis and are the direct responsibility of Subcommittee E42.02 on Terminology.

Current edition approved April 28, 1989. Published July 1989. Originally published as E 673 - 79a. Last previous edition E 673 - 89.

by rastering the incident electron beam over the specimen surface and simultaneously recording the Auger signal strength for a particular transition as a function of position.

matrix effects—see **matrix effects,** *Auger.*

parameter—XPS, the kinetic energy of the sharpest Auger peak in the spectrum minus the kinetic energy of the most intense photoelectron peak from the same element; the energy of the ionizing photons must be specified.

peak energy for dN(E)/dE, N(E)—the designation of the energy of the Auger electron distribution. In *dN/dE* spectra, peak energies should be measured at the most negative excursions of the Auger features. In *N(E)* spectra, peak energies are measured at peak maxima. (Peak energies in *dN/dE* spectra are not the same as those in *N(E)* spectra.)

process—the relaxation, by electron emission, of an atom with a vacancy in an inner electron shell.

signal strengths—AES, XPS, in *dN/dE* spectra, signal strengths are measured as the peak-to-peak heights of the Auger features. In *N(E)* spectra, signal strengths are measured as the heights of the Auger peaks above background. In *I(E),* signal strengths are measured as the areas under the electron energy distribution, *N(E).*

spectrum, dN(E)/dE, N(E), I(E)—AES, the display of Auger signal strength as a function of electron energy. Auger spectra from solids may be measured as the first derivative of the electron energy distribution and may be designated by *dN/dE.* The Auger electron energy distribution may be designated as *N(E).* With certain type analyzers (for example, the CMA) the displays are *dEN(E)/*dE and *EN(E).* The area under Auger peaks may be designated as *I(E)* with background substration method, and integration limits specified.

transition—transitions involved in electron emission by an Auger process are designated by indicating the electron shells. The first letter designates the shell containing the initial vacancy and the last two letters designate the shells containing electron vacancies created by Auger emission (for example, KLL, and LMN). When a bonding electron is involved the letter V is used (for example, LMV and KVV). When a particular subshell involved is known this can also be indicated (for example, KL_1L_2). Coupling terms may also be added where known ($L_3M_{4,5}M_{4,5};{}^1D$). More complicated Auger processes (such as, multiple initial ionizations and additional electronic excitations) can be designated by separating the initial and final states by a dash (for example, LL-VV and K-VVV). When an Auger process involves an electron from the same principal shell as the initial vacancy (for example, L_1L_2M) it is referred to as a Coster-Kronig transition. If both electrons are from the same principal shell as the initial vacancy (for example, $M_1M_2M_3$) it is called a super Coster-Kronig transition.

transition rate—the probability per unit time for two bound electrons to undergo energy state transitions such that one will fill an initial core hole vacancy and the other will go to a final state in the positive energy continuum.

background:

inelastic—ISS, the response of the energy filtering and detection system to probe ions that have undergone inelastic scattering events at the specimen surface.

instrumental—ISS, the response of the energy filtering and detection system to events other than those induced by bombardment of the specimen surface by a beam of probe ions.

secondary ion—ISS, the response of the energy filtering and detection system to secondary ions produced by bombardment of the target material with probe ions.

signal—for a specific measurement, any signal present at a particular position due to processes or sources other than those of primary interest.

backscattered electrons—*AES,* electrons originating in the incident beam which are re-emitted after interaction with the target. By convention, electrons with energies greater than 50 eV are considered as backscattered electrons.

backscattering:

energy—EIA, energy of a particle from the analyzing beam after it has undergone a backscattering collision and escaped the specimen.

factor—AES, the fractional increase in the Auger current due to backscattered electrons.

spectrum—EIA, a plot of backscattering yield (ordinate) versus backscattering energy (abscissa).

yield—EIA, the number of particles detected (counts) per unit backscattering energy per incident ion.

beam:

analyzing—same as *incident.*

current—the total current incident on the specimen by the primary particle source.

current density—the current incident on the specimen per unit area.

diameter—in surface analysis, the full width of the incident beam at half maximum intensity measured in a plane normal to the beam direction. This plane must be specified and is often taken at the intersection of the beam center with the specimen.

divergence, convergence—angles spanned by the directions of all particles of the incident beam.

energy—the energy of the particles incident on the specimen surface, expressed in electron volts (eV).

energy, primary—the kinetic energy of the primary beam, usually expressed in kiloelectronvolts (keV).

incident—the energetic particles incident on the specimen.

particle—atomic or molecular species contained in the incident beam, regardless of state of ionization.

primary beam—a directed flux of particles (ions or neutrals) incident on the specimen.

profile, primary ion—the spatial distribution of the primary ion current in a plane perpendicular to the primary ion beam axis.

size—the full width at half-maximum of the beam at a given point in space that must be defined.

spot—the area on the specimen surface illuminated by the incident beam.

binary elastic scattering event—*ISS,* the collision between an incident probe ion and a single surface atom in which the total kinetic energy and momentum are conserved.

binary elastic scattering peak—*ISS*, an increase in the spectrometer detection system response above the background level which can be attributed to binary elastic scattering of the probe ion from a surface atom of a particular mass.

binding energy—the work that must be expended in removing an electron from a given electronic level to a reference level, such as the vacuum level or the Fermi level.

blocking geometry—*EIA*, experimental situation wherein the atom rows or planes of a single crystal target are aligned parallel to a vector from the specimen to the detector.

Bragg's rule—an empirical rule formulated by W. H. Bragg and R. Kleeman that states that the stopping cross section of a compound specimen is equal to the sum of the products of the elemental stopping cross sections for each constituent and its atomic fraction, that is,

$$\epsilon(A_xB_y) = x\epsilon_A + y\epsilon_B$$

where:
$\epsilon(A_xB_y)$ = the stopping cross section of the compound, and A_xB_y and ϵ_A and ϵ_B = the stopping cross section of elements A and B respectively.

bremsstrahlung—*XPS*, photon radiation, continuously distributed in energy up to the energy of the incident electrons, emitted from an anode due to deceleration of incident electrons within the anode. The bremsstrahlung from a conventional X-ray source contributes to the background and the Auger signal strengths in an XPS spectrum.

cascade mixing—the rearrangement of the constituents of a solid, within the penetration depth of an incident particle, caused by collisions between the incident particles and the atoms of the solid.

channel—*EIA*, an interval of the measured energy of backscattered particles defined by adjacent energy thresholds in the analog-to-digital converter used for spectrum production.

channeling—motion of energetic particles along certain axial or planar directions of a crystalline solid as the particles penetrate the specimen. The potentials of the individual atoms of the solid combine to reduce scattering with those atoms.

channeling—*SIMS*, the process by which particles preferentially penetrate crystalline specimens in certain crystallographic directions because of the relatively open arrangement of atoms presented to the impinging particle beam.

characteristic electron loss phenomena—*AES*, the inelastic scattering of electrons in solids that produces a discrete energy loss determined by the characteristics of the material. The most probable form is due to excitation of valence electrons. For some solids (for example, nontransition metals), inelastic scattering is dominated by plasmon excitations (a collective excitation of valence electrons). For other solids, the inelastic scattering may be due to a combination of plasmon excitation and single valence electron excitations. Inelastic scattering can also occur through the excitation of core level electrons when this is energetically possible.

characteristic X-rays—photons emitted by ionized atoms and having a particular distribution in energy and intensity characteristic of the atomic number and chemical environment of the atom; in XPS, the term is ordinarily used in reference to the X-ray source of the spectrometer.

charge:

charge modification—any method used to alter the amount or the distribution of charge on a specimen surface.

charge neutralization—*ISS, SIMS*, a technique in which a surface under ion bombardment is maintained at a known potential by compensating for the accumulated charge.

charge referencing—any method used to adjust the energy scale calibration of a spectrometer to accommodate the effects of steady-state charging of a specimen surface.

charging potential—*in surface analysis*, the electrical potential of the surface of an insulating specimen caused by irradiation. If the specimen is heterogeneous, there may be different charging potentials on different areas of the surface.

chemical:

effects—*AES*, any change in the shape of an Auger spectrum or in the Auger peak energy for an element which is due to chemical bonding.

shift—*AES, XPS*, a change in peak energy because of a change in the chemical environment of the atom.

collection angle—See **angle.**

collision cascade—a sequential energy transfer between atoms in a solid as a result of bombardment by an energetic species.

compositional depth profile—the chemical composition and the atomic concentration measured as a function of distance from the surface.

constant energy resolution—*AES, XPS*, a mode of operation in which the instrumental resolution is constant over the spectrum. See **fixed analyzer transmission.**

Coster-Kronig transition—*AES, XPS* Auger process involving an electron from the same principal shell as the initial vacancy (for example, $L_1 L_2M$).

counts—*EIA*, events recorded by the detector and registered in a channel of a backscattering spectrum. Counts constitute the ordinate of a BS spectrum.

integrated—sum of all counts registered in a given set of channels or their corresponding energy ranges.

per channel—number of counts in a single channel; unit of the ordinate of a backscattering spectrum.

crater edge effect—*SIMS*, a signal caused by secondary ions that originate from depths shallower than the maximum depth of the crater formed by primary bombardment.

cross section:

enhanced elastic—*EIA*, cross section for elastic scattering that is larger than that predicted by *Rutherford* due to partial penetration of a nucleus in the specimen by the incident particle.

nuclear reaction—*EIA*, the probability of a particular nuclear reaction as a function of energy and the emission direction of the detected product. Usually expressed as an area in units of barns = 10^{-28} m^2.

Rutherford—*EIA*, nuclear reaction cross section for the particular case of elastic scattering as calculated from classical mechanics. First evaluated by *Rutherford.*

E 673

stopping—EIA, the energy loss of a particle incident on the specimen per unit area density of specimen atoms. Usually expressed in units of $eV \cdot cm^2$/atom.

current integration—the measurement of total electric charge deposited into a specimen by the incident beam.

curve resolving—the construction of the individual peaks of a spectrum that consists of overlapping peaks, also called curve fitting or peak fitting.

deconvolution—*AES, XPS,* a mathematical procedure to (*1*) remove the contribution to a peak of one of the factors contributing to its line width, for example, X-ray linewidth, analyzer broadening; or

(*2*) remove the energy loss background by deconvoluting the spectrum with an electron energy loss spectrum.

depth resolution—*AES, XPS, SIMS,* the depth range over which a signal increases (or decreases) by a specified amount when profiling an ideally sharp interface between two media. By convention, the depth resolution corresponds to the distance over which a 16 % to 84 % (or 84 % to 16 %) change in signal is measured.

depth scale—*EIA,* a relationship between energy loss and target depth that allows a direct correlation between multichannel analyzer channel number and depth in the specimen.

detection limit—the smallest concentration of an element or compound that can be measured for specific analysis conditions and data collection periods.

DISCUSSION—By convention, the detection limit is often taken to occur when the total signal minus the background signal is two or three times the standard deviation of the background signal above the background signal. This convention may not be applicable to all measurements.

detector:

angle—EIA, see **angle** *of detector:*

efficiency—EIA, fraction of particles incident on the detector that actually generate a detectable signal.

foil—EIA, a thin sheet, usually metal or plastic, placed over a detector to absorb low energy or high mass products, or both, from nuclear reactions, while transmitting other reaction products.

solid angle—see **angle,** *solid, of detector.*

dose—number of beam particles per unit area that impinge on the specimen. Alternatively, the dose may be defined as the charge per unit area that impinges on the specimen.

rate—number of beam particles per unit area per second that impinge on the specimen. Alternatively, the dose rate may be defined as the current per unit area that impinges on the specimen.

dynamic SIMS—SIMS analysis at a primary ion current density and dose such that more than one monolayer of material is removed during the analysis.

electron:

electron attenuation length—the average distance that an electron with a given energy travels between successive inelastic collisions as derived from a particular model in which elastic scattering is assumed to be insignificant. With this model, electrons are assumed to be scattered only inelastically and predominantly in the forward direction although, in reality, elastic scattering will modify the electron trajectories. For example, Auger electrons or photoelectrons reaching the surface in surface analysis experiments are thus assumed to have travelled along straight-line paths from their point of generation.

electron escape depth—the distance (in nanometres) normal to the surface at which probability of an electron escaping without significant energy loss due to inelastic scattering processes drops to e^{-1} (36.8 %) of its original value.

electron flooding—in surface analysis, irradiation of a specimen with low-energy electrons in order to change (generally to reduce) or stabilize the charging potential.

electron retardation—AES, XPS, a method of measuring the kinetic energy distribution by retarding the emitted electrons before or within the analyzer.

electron spectrometer—see **electron energy analyzer.**

inelastic mean free path—the average distance (in nanometres) that an electron with a given energy travels between successive inelastic collisions.

electron energy analyzer—*AES, XPS,* a device for measuring the number of electrons as a function of kinetic energy. (See also **spectrometer**)

pass energy—XPS, AES, the mean kinetic energy of electrons in the energy dispersive portion of an electron energy analyzer that will allow them to traverse the analyzer and be counted.

fixed analyzer transmission—AES, XPS, a mode of analyzer operation that varies the electron retardation but keeps the pass energy constant in the final analyzer stage.

fixed retarding ratio—AES, XPS, a mode of operation in which the electron kinetic energy is analyzed by varying the retarding potential on lens elements preceding the analyzer and the analyzer pass energy so that the analyzer pass energy is a constant fraction of the kinetic energy.

electron energy loss spectrum—*XPS,* the energy spectrum of electrons from a monoenergetic electron source after interaction with the specimen, exhibiting peaks due to inelastic loss processes. The spectrum obtained using an incident electron beam of about the same energy as an XPS peak approximates the loss spectrum associated with that XPS peak. Also see **characteristic electron loss phenomena.**

electron flooding—see **electron.**

electron retardation—see **electron.**

electron spectrometer—see **electron energy analyzer.**

energy:

per channel—EIA, energy differences between two successive channels.

edge—EIA, values of the backscattering energy in a BS spectrum for an element (or isotope) that is located at the surface of the specimen.

loss—EIA, energy dissipated by the particles of the incident beam as they penetrate through the specimen.

of incident beam—average energy of analyzing particle in the incident beam at the moment of impact.

pass—(See *pass energy* under **electron energy analyzer.**)

surface approximation—EIA, see **surface energy approximation.**

equilibrium surface composition—see **sputtering.**

ESCA—acronym for "electron spectroscopy for chemical analysis," a term historically used to describe a technique whereby one generates electron spectra by irradiating a specimen with narrow band characteristic X-rays.

escape depth—see **electron escape depth.**

extra-atomic relaxation energy—*XPS,* see **screening energy.**

fast atom bombardment mass spectrometry (FABMS)—the technique in which a mass spectrometer is used to measure the mass-to-charge ratio and abundance of secondary ions emitted from a target as a result of the bombardment of fast neutral atoms.

Fermi energy (level)—*for metals,* the energy of the top-most filled electron level at zero Kelvin. *For insulators and semiconductors,* the Fermi level is usually between the valence and conduction bands.

Fermi level referencing—*XPS,* a method of establishing the binding energy scale for a particular specimen by assigning the kinetic energy corresponding to the Fermi level, as determined by analysis of the specimen's XPS or UPS spectrum, as the point of zero binding energy. See also **Fermi energy (level).**

fixed analyzer transmission—see **electron energy analyzer.**

fixed retarding ratio—see **electron energy analyzer.**

fluence—*EIA,* same as **dose.**

flux—*EIA,* same as **dose** *rate.*

fractional ion yield—*SIMS,* the ratio of the number of secondary ions of a particular species to the total number of secondary ions emitted by a specimen.

glancing exit—*AES, EIA, XPS,* geometrical arrangement in which the scattered (or emitted) particles are near 90° from the normal to the specimen surface. This results in improved depth resolution.

glancing incidence—*AES, EIA,* geometrical arrangement in which the incident particles are near 90° from the normal to the specimen surface. This results in improved depth resolution.

gold decoration—*XPS,* a method whereby a very thin coat of evaporated gold on an insulator is used as a charge reference; the gold should be deposited as unconnected islands covering the area analyzed.

grazing exit (incidence)—*EIA,* same as **glancing exit (incidence).**

image depth profile—*AES, XPS, SIMS,* a three-dimensional representation of the spatial distribution of a particular elemental or molecular species (as indicated by emitted secondary ions or electrons) as a function depth or material removed by sputtering.

incident particle energy—the effective energy of the primary particles incident on the specimen surface, usually expressed in kiloelectronvolts (keV) per atomic particle.

inelastic:

inelastic mean free path— see **electron.**

scattering correction to background—*XPS,* a method of correcting background for contributions of inelastic scattering processes, most often approximated by simulating the background through a peak by assuming that the rise in background is proportional to the peak area at higher kinetic energy. A more accurate correction is done by deconvolving the energy loss spectrum itself.

scattering cross-section—*AES, XPS,* a measure of the probability that an electron traversing a material will undergo an inelastic scattering process, expressed as an area per unit event.

scattering event—*ISS,* a collision process in which a fraction of the kinetic energy imparted by the probe ion contributes to an increase in the internal energy of the target material, and is not recovered as kinetic energy of the scattered probe ion or target atom recoil.

information depth—in surface analysis, the distance, normal to the surface, from which a specified percentage of the directed signal is generated. For example, in *AES* or *XPS,* if the percentage of electrons detected varies exponentially with distance from the surface, then 63.2, 86.5, 95.0, 98.2, and 99.3 % of the detected signal from a homogeneous material originates from within a depth of 1, 2, 3, 4, and 5 times the electron escape depth respectively. (See **electron,** *inelastic mean free path.*)

instrumental detection efficiency—*SIMS,* the ratio of ions for a particular species detected to ions produced.

interatomic Auger process—*AES, XPS,* an Auger transition in which final electron vacancies are in valence levels or molecular orbitals, some of which may be predominantly orbitals of a neighboring bonded atom.

interference signal—*SIMS,* signal measured at the mass position of interest due to another, undesired species.

interface width, observed—*AES, XPS, SIMS,* the distance over which a 16 % to 84 % (or 84 % to 16 %) change in signal is measured at the junction of two dissimilar matrices.

internal carbon referencing—*XPS,* a method of determining the charging potential of a specimen by comparing the experimentally determined binding energy of the C 1s peak maximum from a specific carbon group within the specimen to a standard binding energy value for that carbon group. A hydrocarbon group within the specimen is often used for this purpose.

intrinsic linewidth, of specimen—*AES, XPS,* the linewidth contribution arising from the specimen. The measured linewidth is a convolution of this function and broadening contributions of the instrument (for example, X-ray source radiation linewidth, spectrometer energy resolution).

ion beam—a directed flux of charged atoms or molecules.

current—the measured rate of flow of charged atoms or molecules incident upon the specimen per unit time, usually expressed in amperes (A).

current density—the ion beam current incident on the specimen per unit cross-sectional area, usually expressed in amperes per square centimetre (A/cm^2).

energy—in surface analysis, the mean kinetic energy of the ions in the beam (see **beam,** *energy*).

ion image—*SIMS,* a two-dimensional representation of the spatial distribution of a particular secondary ion emitted from a specific area of the specimen.

ion implantation—the injection of ions into a specimen.

ionization cross-section—the probability that an incident particle traversing a gas or solid will produce an ionizing collision. The total ionization cross-section includes all electron vacancies produced by a primary collision and subsequent Coster-Kronig or Auger decay process. The partial ionization cross-section results from one particular process such as a primary collision to produce an initial innershell vacancy in a particular shell, a Coster-Kronig process, or an Auger ejection process to produce particular distributions of electron vacancies.

ion lifetime—the average time that an ion exists in a particular electronic configuration, for example, a vacancy in a particular shell in an atom.

ion neutralization—*ISS, SIMS,* the charge exchange processes in which a probe is neutralized by the material surface or gas phase species with which it interacts.

ion-scattering:

spectrometer—*ISS,* an instrument capable of generating a beam of principally monoenergetic, singly charged, low-energy ions and determining the energy distribution of the probe ions that have been scattered from the solid surface through a known angle.

spectrometry—*ISS,* a technique to elucidate composition and structure of the outermost atomic layers of a solid material, in which principally monoenergetic, singly charged, low-energy (less than 10 keV) probe ions are scattered from the surface and are subsequently detected and recorded as a function of the energy or scattering angle, or both.

spectrum—*ISS,* a representation in which the scattered ion intensity is presented as a function of the ratio of the scattered ion energy to the incident ion energy.

ion species—type and charge of ion such as Ar^+, O^-, and H_2^+. If an isotope is used, it should be specified.

knock-on—the movement of a constituent of the specimen deeper into the specimen matrix as a result of collisions with the primary particle.

Koopmans energy—a calculated energy of an electron in an orbital, on the assumption that its removal to infinity is unaccompanied by electronic relaxation.

K-value—*EIA,* a kinematic factor (between 0 and 1) that relates the backscattered energy to the incident energy.

mass:

analyzer—a device for dispersing ions as a function of their mass-to-charge ratio.

resolution—the ratio $M/\Delta M$ where ΔM is the full width at half-maximum peak height for the ion peak of mass M.

resolving power—the peak-to-valley ratio between adjacent, equal-sized peaks, separated by one mass unit.

spectrum—a plot of the measured ion signal as a function of mass-to-charge ratio.

matrix effects:

Auger—any change of an Auger spectrum (for example, shape or signal strength) due to the physical environment (for example, amorphous/crystalline, thin layer/thick layer, or rough/smooth surface) of the emitting element and not due to chemical bonding or changes in concentration.

SIMS—any change in the secondary ion yields which are caused by changes in the chemical composition or structure of a particular specimen.

AES—see **Auger.**

mean free path—See **electron,** *inelastic mean free path.*

modified Auger parameter—the Auger parameter plus the photon energy, which equals the kinetic energy of the sharpest Auger peak plus the binding energy of the most intense photoelectron peak.

modulation—*AES,* the periodic waveform added to the spectrometer pass energy to obtain the desired Auger spectrum display. The modulation should be given as eV peak-to-peak, thereby including the geometrical factor of the spectrometer, rather than volts peak-to-peak. The frequency and waveform should also be given.

molecular SIMS—the SIMS technique when applied to molecular or polyatomic secondary ions.

multiple scattering event—*ISS,* a collision process that may be described as a sequence of binary scattering events which may or may not be elastic.

multiplet or exchange splitting—*XPS,* splitting of a photoelectron line caused by the interaction of the unpaired electron created by photoemission with other unpaired electrons in the atom.

natural linewidth—See **intrinsic linewidth.**

noise—*in surface analysis,* the random fluctuation of the measured intensity at a particular location in a spectrum, usually expressed as an RMS (root-mean-square), standard deviation, or a peak-to-peak value.

orbital energy—*XPS,* Koopmans energy corrected for intra-atomic relaxation.

organic (inorganic) SIMS—the SIMS technique when applied to organic (inorganic) specimens or organic (inorganic) molecules placed on a solid.

peak width, FWHM—full width at half-maximum peak height above background.

photoelectric cross-section—the probability that an incident photon traversing a material will produce a photoelectron from a given subshell, expressed as an area unit per event.

photoelectric effect—a dipole interaction involving the interaction of photons with bound electrons in atoms, molecules, and solids, resulting in production of photoelectrons and excited ions.

photoelectron satellite peaks—See **photoelectron X-ray satellite peaks,** and **shakeup lines or shakeup satellites.**

photoelectron X-ray satellite peaks—photoelectron peaks in a spectrum resulting from photoemission induced by characteristic minor X-ray lines associated with the X-ray spectrum of the anode material.

photoelectron X-ray satellite subtraction—the removal of photoelectron X-ray satellite peaks from a spectrum.

photoemission—the emission of electrons from atoms or molecules caused by photoelectric effects.

pileup—*EIA,* counts in a backscattering spectrum arising from two separate events that occur so closely in time that the signals are not resolved by the detection system and cause counts to be recorded in erroneous channels.

plasmon loss lines—peaks in an electron spectrum that are due to certain characteristic energy losses of electrons emitted from the specimen. These losses occur as a result of the excitation of collective oscillations among the valence band electrons.

polarizability—the ability of a neutral atom or group to stabilize a nearby ion by coulombic attraction, without electron transfer.

polarization energy—see **screening energy.**

polyatomic ion—a charged multi-atom species.

positive (negative) ion yield—the total number of positive (negative) secondary ions sputtered from the specimen per incident primary particle.

E 673

preferential sputtering—See **sputtering.**

probe ion—an ionic species intentionally produced by an ion source and directed onto the specimen surface at a known incident angle with a known energy.

quantitative analysis—the determination of the concentration and distribution of elements within the Auger analysis volume.

DISCUSSION—For elements uniformly distributed over the analyzed volume, the concentrations should be reported as weight or atomic percentages. For uniform distributions in a plane, but nonuniform distributions normal to this plane, concentrations should be reported as weight or atoms per unit area. For nonuniform distributions in a plane, but uniform distributions normal to the plane (over the detected depth), the concentration should be reported in percent area covered.

radiation induced (or enhanced) diffusion—atom movement in the solid, well beyond the typical penetration depth of an incident particle, due to particle beam damage or bombardment induced defects.

raster—*SIMS*, the two-dimensional pattern swept out by the deflection of a primary ion beam.

recoil implantation or knock-on—the injection, due to collisions caused by incident particles, of surface or near surface atoms into the bulk along the path of the incident beam.

relaxation energy—*XPS*, the energy associated with intra-atomic or extra-atomic electronic readjustment to the removal of an atomic electron, so as to minimize the energy of the final state of the system.

resolution:

depth—*EIA*, energy resolution translated into an equivalent resolution of depth in the specimen.

energy—*EIA*, the full width at half-maximum (FWHM) of the measured energy distribution when the energy distribution of the backscattered particles is monoenergetic.

lateral—*EIA*, the distance measured on the surface of a specimen over which changes in composition can be established with confidence by BS. This resolution is generally determined by the size of the beam spot.

system—*EIA*, the energy or depth resolution measured in a BS spectrum for a monoenergetic incident beam.

resonance reaction—*EIA*, a nuclear reaction that has a narrow peak in the nuclear reaction cross section, which is so much larger than the nuclear reaction cross sections at adjacent energies both above and below the peak that essentially all the particles detected from the reaction are due to the peak.

satellite peaks—See **photoelectron X-ray satellite peaks** and **shake-up lines or shake-up satellites.**

scattered ion:

energy—*ISS*, *for a binary elastic collision*, the kinetic energy of the probe ion following a binary elastic collision, E_s, is given by:

$$E_s = E_0[M_0/(M_0 + M_1)]^2(\cos\theta + [(M_1/M_0)^2 - \sin^2\theta]^{1/2})^2$$

where:

E_s = kinetic energy for the scattered probe ion,

E_0 = energy of the incident probe ion prior to collision, determined from the product of ionic charge and accelerating potential,

M_0 = mass of the probe ion,

M_1 = mass of the target atom, and

θ = angle between the initial and final velocity vectors for the probe ion, as determined from a common origin in the laboratory coordinate system, expressed as a value between 0 and 180°.

energy ratio—*ISS*, the value E_S/E_0 which may be used as the abscissa of an ion-scattering spectrum. For definition of E_s and E_0, see **scattered ion** *energy.*

intensity, experimental—*ISS*, the measured response of the energy filtering and detection system as a consequence of bombarding the specimen material with a beam of probe ions, usually presented as the ordinate of an ion-scattering spectrum.

intensity, theoretical—*ISS*, defined by an equation of the form:

$$I_i(\theta) = I_0 N_i P_i \alpha_i (d\sigma_i/d\Omega)\theta \Delta\Omega T$$

where:

$I_i(\theta)$ = scattered ion intensity from atoms of species, i, at a given scattering angle, θ, ions s^{-1},

I_0 = intensity of incident probe ions, ions s^{-1},

N_i = number of scattering centers of species i per unit area of surface, or per unit volume accessible to the incident beam, atoms $metre^{-2}$,

P_i = probability that the probe ion remains ionized after interacting with an atom of species i,

α_i = geometric or shadowing factor for species i in the given environment and geometry,

$(d\sigma_i/d\Omega)\theta$ = differential scattering cross section per unit solid angle, for species i, taken at the angle for which scattering is measured; that is, the angular distribution of scattered ion intensity per unit flux of incident ions, per atom of species i, $metre^2$ $atom^{-1}$ $steradian^{-1}$,

$\Delta\Omega$ = solid angle of acceptance determined by the entrance aperture of the filtering and detection system, steradians, and

T = fractional transmission of the analyzing and detection system.

screening energy—the diminished energy of an ion due to coulombic attraction of electrons in the immediate environment.

secondary electrons—*AES*, electrons leaving a surface, produced through various mechanisms of energy transfer from the incident beam. By convention, electrons with energies ≤50 eV are considered as secondary electrons.

secondary ion—ions ejected from a specimen surface as a result of energy transfer from a primary beam.

angular distribution—*SIMS*, the secondary ion yield as a function of emission angle.

energy distribution—the number of secondary ions as a function of the energy at a specified collection angle.

mass spectrometry—the technique in which a mass spectrometer is used to measure the mass-to-charge ratio of secondary ions emitted from a target as a result of particle (ion or neutral) bombardment.

signal gating—the process of accepting secondary ion signal from only a portion of the sputtered area of the specimen to minimize crater edge effects.

E 673

yield—the total number of ions sputtered from the specimen per incident ion of given mass, energy, charge, and angle of incidence.

selected area aperture—*SIMS*, the mechanical equivalent of electronic signal gating, commonly used in stigmatic mass spectrometers.

sensitivity factor:

elemental—*XPS*, intensities of peaks relative to those of a standard, for example, F 1s, for atoms in typical homogeneous environments. Division of peak height or peak area intensities by the appropriate sensitivity factors gives the relative number of atoms detected, on the assumption of sample homogeneity.

relative Auger elemental—*AES*, the ratio of the Auger signal strength of a specified Auger transition from a single element to that from a selected standard element (for example, silver), as measured under identical conditions.

SIMS—the factor used to convert the net counts per unit time, for a particular species, matrix and experimental conditions, to concentration.

shakeoff process—*XPS*, a multi-electron photoelectric process in which two or more electrons are emitted, partitioning between them the excess kinetic energy.

shakeup lines or shakeup satellites—*XPS*, photoelectrons originating from photoelectric processes in which the final ion is left in an excited state, so that the photoelectron has a characteristic energy slightly less than that of the normal photoelectron.

signal height—*EIA*, the number of counts in the channels of a backscattering spectrum due to a specific element in the target.

signal-to-background ratio—*AES*, the ratio of signal (above background) to that of the nearby background on the high kinetic energy side of the elastically scattered Auger electrons.

signal to background ratio—*SIMS, XPS, for a spectral peak*, the ratio of the maximum counts in the peak above the background to the magnitude of the background.

signal-to-noise ratio—the ratio of the signal intensity (above background) to that of noise in determining that signal.

DISCUSSION—The method of noise measurement must be specified, such as, rms (root mean square), or peak-to-peak.

smoothing—*XPS*, a mathematical treatment of the data to reduce the noise.

specimen charging—the accumulation of electrical charge on the specimen caused by particle bombardment.

spectrometer: (See also **electron energy analyzer**)

dispersion—*AES*, the change in electron image position at the exit of a spectrometer per unit change in electron energy.

energy resolution—*AES, XPS*, the ratio of the full width at half-maximum intensity of the response curve for monoenergetic electrons at a given energy to the energy of the electrons.

transmission—*AES, XPS*, the ratio of the number of electrons at a given energy transmitted through the spectrometer to the number entering the entrance aperture of the spectrometer at that energy.

spectrum:

aligned incidence—*EIA*, a backscattering spectrum recorded with the analyzing beam aligned with crystallographic axes or planes of the specimen that produce channeling.

random (incidence)—*EIA*, a backscattering spectrum recorded with the analyzing beam incident on the specimen in a direction such as to produce no channeling.

spin orbit splitting—the splitting of p-, d-, or f-levels arising from coupling of the spin and orbital angular momentum.

sputter depth profile—*AES*, the compositional depth profile obtained when material is removed by sputtering as a result of ion bombardment.

sputter depth profile—*SIMS*, the compositional depth profile obtained when material is removed by sputtering as a result of primary bombardment.

sputtered neutral mass spectrometry (SNMS)—a technique in which a mass spectrometer is used to measure the mass-to-charge ratio and abundance of secondary neutral species emitted from a target as the result of particle bombardment. These neutral species are analyzed by using plasma, electron, or photon ionization methods.

sputtering—the phenomenon which occurs when atoms and ions are ejected from the specimen as a result of particle bombardment.

equilibrium surface composition—the steady-state surface composition produced by sputter-etching a homogeneous specimen under nonvarying conditions for the ambient vacuum and the primary beam.

fractional yield—*SIMS*, the ratio of the number of atoms and ions of a particular element to the total number of atoms and ions ejected from the specimen.

preferential—the phenomenon which may occur when the sputtering of multicomponent specimens causes a change in the equilibrium surface composition of the specimen.

rate—the amount of specimen material removed per unit time as a result of particle bombardment.

yield—the number of atoms and ions ejected from the specimen per incident ion.

static SIMS—SIMS analysis at a primary ion current density and dose such that during the analysis each sputtering event occurs from a previously unsputtered region.

statistical noise—*XPS*, the noise in the spectrum due solely to the statistics of randomly detected single events; the root mean square of the deviations in neighboring channels is equal to the square root of the average counts per channel.

stopping:

cross section—*EIA*, see **cross section,** *stopping*.

cross section factor—*EIA*, the stopping factor expressed per unit volume density of the constituent molecular entity of the specimen at the appropriate depth.

factor—*EIA*, the ratio between a (differential) energy interval in a backscattering spectrum and the corresponding (differential) depth interval in the specimen.

power—*EIA*, same as **stopping,** *cross section*.

surface energy approximation—*EIA*, a simplification of calculations involving the energy of an ion passing through a solid specimen. The energy of the ion at the surface is used in place of a properly averaged energy. This approximation is used to determine the energy at which scattering or stopping cross sections, or both, are evaluated.

surface roughness—*AES*, the deviation of the topography of an actual surface from an ideal atomically smooth and planar surface. The rms deviation from the center line average is a measure of surface roughness.

surface segregation—a diffusion controlled process (as opposed to evaporation, preferential sputtering, or other processes) that causes the surface composition of a homogeneous solid to differ from the bulk composition.

synchrotron radiation—*XPS*, a continuous radiation created by the acceleration of high energy electrons, as in a synchrotron or storage ring. Monochromatized, it is a practical variable energy source of photons for photoelectron spectroscopy.

target—specimen under investigation.

thick—*EIA*, specimen whose thickness produces backscattered particles whose energies, for each constitutive element, vary greatly with respect to the system resolution.

thin—*EIA*, specimen whose thickness is sufficiently small that the variations in energies of particles backscattered from atoms of each constitutive element is small with respect to the system resolution.

tilt, of target—same as **angle** *of incidence*.

time constant—*AES*, the time required for a signal to change by $1 - (1/e)$ (63.2 %) of its final value in response to a step function input.

UPS—acronym for "ultra-violet photoelectron spectroscopy," similar to ESCA but employing characteristic ultra-violet radiation.

useful ion yield—*SIMS*, the ratio of the number of ions of a particular isotope detected to the total number of atoms and ions of the same element sputtered.

vacuum level—the potential of the vacuum space at a sufficiently large distance outside the specimen such that electric fields caused by different work functions of different parts of the surface are zero or extremely small.

vacuum level referencing—*XPS*, a method of establishing the binding energy scale for a particular specimen by assigning the kinetic energy corresponding to the vacuum level as the point of zero binding energy. See also **vacuum level**.

valence band spectrum—*XPS*, photoelectron energy distribution arising from the less tightly bound electrons involved in the chemical bonds of the specimen material.

work function—the potential barrier that must be overcome to remove an electron from the Fermi level of a specimen to the vacuum level.

X-ray ghost line—*XPS*, lines in a spectrum due to presence of contaminating X-ray photons from an impurity in the X-ray anode, from the X-ray window, or from certain elements present in the specimen.

X-ray linewidth—the energy width of the principal characteristic X-ray; in XPS it usually refers to that of the X-ray source. The X-ray linewidth contributes to the photoelectron peak widths.

X-ray monochromator—a device used to eliminate photons of energies other than those in a narrow band.

zone of mixing—the layer of the specimen surface within which the primary beam causes atomic mixing.

Designation: E 684 – 88

Standard Practice for Approximate Determination of Current Density of Large-Diameter Ion Beams for Sputter Depth Profiling of Solid Surfaces[1]

1. Scope

1.1 This practice describes a simple and approximate method for determining the shape and current density of ion beams. The practice is limited to ion beams of diameter greater than 0.5 mm of the type used for sputtering of solid surfaces to obtain sputter depth profiles. It is assumed that the ion-beam current density is symmetrical about the beam axis.

1.2 *This standard may involve hazardous materials, operations, and equipment. This standard does not purport to address all of the safety problems associated with its use. It is the responsibility of the user of this standard to establish appropriate safety and health practices and determine the applicability of regulatory limitations prior to use.*

2. Referenced Documents

2.1 *ASTM Standards:*

E 673 Definitions of Terms Relating to Surface Analysis[2]

E 1127 Guide for Depth Profiling in Auger Electron Spectroscopy[2]

3. Terminology

3.1 Terms used in Auger electron spectroscopy are defined in Definitions E 673.

4. Significance and Use

4.1 Sputter depth profiling is used in conjunction with Auger electron spectroscopy to determine the chemical composition and atomic concentration as a function of distance from the surface of a specimen. See Guide E 1127.

4.2 The diameter of the ion beam used for sputtering must be specified and this practice is a relatively quick method of measuring the shape (that is, current density distribution) of the ion beam if a suitable Faraday cup is not available.

[1] This practice is under the jurisdiction of ASTM Committee E-42 on Surface Analysis and is the direct responsibility of Subcommittee E42.03 on Auger Electron Spectroscopy.

Current edition approved May 27, 1988. Published August 1988. Originally published as E 684 – 79. Last previous edition E 684 – 83.

[2] *Annual Book of ASTM Standards*, Vol 03.06.

5. Procedure

5.1 Measure the total ion current in the beam by allowing the total beam to strike the carousel (or other specimen holder). Apply a +30 V d-c bias to the carousel to return low-energy secondary electrons created by the ion beam.

5.2 Attach a straight wire to the carousel extending over the edge such that the ion beam will strike the wire but not the carousel. The wire may be tungsten or other suitable material with a diameter of about 25 μm. The wire diameter should be sufficient to intercept measurable ion current but small with respect to the ion beam diameter to minimize distortion. The carousel may then be translated or rotated (with rotation converted to arc length) to determine the ion beam shape.

6. Interpretation of Results

6.1 In general, a Gaussian current distribution will be observed and the full width at half-maximum peak height can be determined. The maximum ion beam current density may then be determined as follows:

$$J_{max} = 0.88 \frac{i^+}{(FWHM)^2}$$

where:

J_{max} = maximum ion beam current density,

i^+ = total ion current, and

FWHM = full width at half-maximum peak height.

6.2 This method is simple, but it suffers from the collection of stray secondary electrons during biased ion current measurements. This situation can be alleviated somewhat if the ion detection wire is electrically isolated from the carousel and both are biased independently. However, the stray electron problem still exists. Also, the application of a bias can distort the beam profile to be measured. Thus, this practice should be used only as an approximate method for determining the shape and current density of the ion beam. The errors of measurement have not been investigated but can be expected to depend on the wire material, the electric field in the neighborhood of the wire, and the ion species and energy. A Faraday-cup detector should be used to obtain more accurate measurements of ion current.

Designation: E 1078 – 90

Standard Guide for Specimen Handling In Auger Electron Spectroscopy, X-Ray Photoelectron Spectroscopy, and Secondary Ion Mass Spectrometry[1]

1. Scope

1.1 This guide covers specimen handling prior to, during, and following surface analysis.

1.2 This guide applies to the following surface analysis disciplines:

1.2.1 Auger electron spectroscopy (AES),

1.2.2 X-ray photoelectron spectroscopy (XPS or ESCA), and

1.2.3 Secondary ion mass spectrometry, SIMS.

1.2.4 Although primarily written for AES, XPS, and SIMS, methods will also apply to many surface sensitive analysis methods such as ion scattering spectrometry, low energy electron diffraction, and electron energy loss spectroscopy, where specimen handling can influence surface sensitive measurements.

1.3 *This standard does not purport to address all of the safety problems associated with its use. It is the responsibility of the user of this standard to establish appropriate safety and health practices and determine the applicability of regulatory limitations prior to use.*

2. Referenced Documents

2.1 *ASTM Standards:*

E 673 Definitions of Terms Relating to Surface Analysis[2]

E 983 Guide for Electron Beam Effects in Auger Electron Spectroscopy[2]

3. Terminology

3.1 *Definitions*—For definitions of surface analysis terms used in this guide, see Definitions E 673.

4. Significance and Use

4.1 Proper handling of specimens is particularly critical for surface analysis. Improper handling of specimens can result in erroneous data.

4.2 Auger electron spectroscopy (AES), X-ray photoelectron spectroscopy (XPS or ESCA), and secondary ion mass spectrometry (SIMS) are sensitive to surface layers that are typically a few nanometres (nm) in thickness. Such thin layers can be subject to severe perturbations due to specimen handling (**1**).[3]

4.3 This guide describes methods to minimize the effects of specimen handling on the results obtained using surface sensitive analytical techniques. Combinations of these techniques are often used.

5. General Requirements

5.1 Although the handling techniques for AES, XPS, and SIMS are basically similar, there are some differences. In general, handling of specimens for AES and SIMS requires more attention because of potential problems with electron or ion beam damage or charging, or both. This guide will note when specimen handling is significantly different among the three techniques.

5.2 The degree of cleanliness required by surface sensitive analytical techniques is often much greater than for other forms of analysis. Analysts new to AES, XPS, and SIMS often need to be educated regarding these more stringent requirements.

5.3 *Contact*—Any handling of the surface to be analyzed should be eliminated or minimized whenever possible.

5.4 *Visual Inspection:*

5.4.1 One should make a visual inspection, possibly using a light microscope, prior to analysis.

5.4.2 Not all features that are visually apparent will be observable with the system's usual imaging method. When such a situation occurs, it may be necessary to mark the specimen with scratches while examining it visually so that the correct location for analysis can be found.

5.4.3 Following analysis, visual examination of the specimen is recommended to look for possible effects of sputtering, electron beam exposure, x-ray exposure, or vacuum. Changes that may have occurred during analysis may influence data interpretation.

6. Specimen Influences

6.1 *History*—The history of a specimen can influence the handling of that specimen. For example, if a specimen has previously been exposed to a contaminating environment, the need for exceptional care in handling might be less than for a specimen that came from a very clean environment.

6.1.1 If a specimen is known to be contaminated, precleaning may be warranted in order to reduce the risk of vacuum system contamination.

---

[1] This guide is under the jurisdiction of ASTM Committee E-42 on Surface Analysis and is the direct responsibility of Subcommittees E42.03 on Auger Electron Spectroscopy, E42.04 on X-Ray Photoelectron Spectroscopy, and E42.06 on Secondary Ion Mass Spectrometry.

Current edition approved Aug. 31, 1990. Published October 1990.

[2] *Annual Book of ASTM Standards*, Vol 03.06.

[3] The boldface numbers in parentheses refer to the list of references at the end of this standard.

6.1.2 Special caution should be exercised with specimens containing potential toxins.

6.2 *Information Sought*—The information sought can influence the handling of a specimen. If the information sought lies beneath an overlayer that must be sputtered away in the analytical chamber, or can be exposed by in-situ fracture, cleaving, or other means, then more handling may be allowed than if the information sought comes from the exterior surface of a specimen.

6.3 *Information Available From Other Analytical Techniques*—Information available from other analytical techniques can influence handling of a specimen. Specimens that have been previously analyzed may have also been contaminated on their surfaces. In general, it is best to perform surface analysis before applying other techniques.

7. Sources of Specimen Contamination

7.1 *Tools, Gloves, Etc.:*

7.1.1 When the handling of specimens is necessary, it should only be done with clean tools. Use of clean tools ensures that the specimen surface is not altered prior to analysis and that the best possible vacuum conditions exist in the analytical chamber. Tools used to handle specimens should be made of materials that will not transfer to the specimen, and these tools should be cleaned in high purity solvents prior to use. Tools should also be demagnetized.

7.1.2 Although gloves and wiping materials are sometimes used to handle specimens, it is likely that their use will result in some contamination. Care should be taken to avoid contamination by talc, silicone compounds, and other materials that are often found on gloves. "Clean-room" quality gloves have no talc and may be better suited. The surface to be analyzed should never be touched by the glove or other tools unless necessary.

7.1.3 Specimens must never be handled by hand, even though the skin does not touch the surface of interest. Fingerprint materials contain mobile species which may contaminate the surface of interest. Skin oils and other skin material are not suitable for high vacuum.

7.2 *Particulate Debris*—Compressed gases from cans or from air lines used to blow particles from the surface of a specimen must be considered a source of possible contamination. In-line particle filters can reduce particles from these sources. Blowing one's breath on the specimen is also likely to cause contamination. Certainly, particles are removed from specimens in both methods but caution is advised in critical cases. A gas stream can produce static charge in many specimens, and this could result in attraction of more particulate debris. Use of an ionizing nozzle on the gas stream may eliminate this problem.

7.3 *Vacuum Conditions and Time:*

7.3.1 Because AES, XPS, and SIMS are sensitive to even the first atomic layer of contamination, the vacuum conditions in the analytical chamber can have an important influence on the data obtained.

7.3.2 Assuming a worst-case situation, that every gas atom/molecule striking the surface sticks, then about one atomic layer can adsorb in one second at a chamber pressure of 1×10^{-4} Pa (133 Pa = 1 Torr). The exact time required for adsorption of one atomic layer will depend on several factors including chamber pressure, gas species, chemical reactivity of the surface, and surface temperature. Reactive gas species such as oxygen, water vapor, carbon dioxide, carbon monoxide, hydrogen, and methane tend to have high sticking coefficients. Their partial pressures are, therefore, of importance. Nearby hot filaments can increase the sticking coefficient of less reactive species even on inactive surfaces. Less volatile species can also be deposited on a specimen from warm surfaces, such as the x-ray anode housing.

7.3.3 Specimens that are sputtered, fractured, cleaved, or scribed in the analytical chamber have surfaces that are generally very chemically active. In such cases special attention must be paid to vacuum conditions and exposure time.

7.4 *Effects of the Incident Electrons/Photons/Ions:*

7.4.1 The incident electron flux in AES, ion flux in SIMS, and to a lesser extent the incident photon flux in XPS, can cause changes in the specimen being analyzed (**2**). Such a flux may cause enhanced reactions between the surface of a specimen and the residual gases in the analytical chamber. The incident flux may also degrade the specimen resulting in a possible rise in chamber pressure and in contamination of the analytical chamber.

7.4.2 One can test for the effects of incident electron or photon beams by monitoring signals from the specimen as a function of time. This could be done by setting up the system for a sputter depth profile and then not turning on the ion gun. If changes occur with time, then the incident beam or residual gases may be altering the surface.

7.4.3 The incident ion beams used during SIMS, AES, and XPS depth profiles not only erode the surface of interest but can also affect surfaces nearby. This can be caused by poor focusing of the primary ion beam and impact of neutrals from the primary beam. These adjacent areas may not be suitable for subsequent analysis by surface analysis methods.

7.5 *Analytical Chamber Contamination:*

7.5.1 The analyst should be alert to materials that will lead to contamination of the vacuum chamber as well as other specimens in the chamber. High vapor pressure elements such as Hg, Te, Cs, K, Na, As, I, Zn, Se, P, S, etc. should be analyzed with caution.

7.5.2 Even if an unperturbed specimen meets the vacuum requirements of the analytical chamber, the probing beam required for analysis may degrade the specimen and result in serious contamination, as discussed in 7.4. If there are questions regarding possible contamination, tests should be done before the specimen is admitted to the analytical chamber, during insertion for the case of rapid insertion probes, or using low intensity beams for initial analyses.

7.5.3 Contamination by surface diffusion can be a problem, especially with silicone compounds (**3**). It is possible to have excellent vacuum conditions in the analytical chamber and still have contamination by surface diffusion.

7.5.4 In SIMS, atoms sputtered onto the secondary ion extraction lens can be resputtered back onto the surface of the specimen. This effect can be reduced by not having the secondary ion extraction lens close to the specimen.

7.5.5 The order of incidence of probing beams can be important, especially when dealing with organic material or other materials as discussed in 9.9.4.

ASTM E 1078

8. Specimen Storage and Transfer

8.1 *Storage:*

8.1.1 *Time*—The longer a specimen is in storage, the more care must be taken to ensure that the surface to be analyzed will not be contaminated. Even in clean laboratory environments, surfaces can quickly become contaminated to the depth analyzed by AES, XPS, SIMS, and other surface sensitive analytical techniques.

8.1.2 *Containers:*

8.1.2.1 Containers suitable for storage should not transfer contaminants to the specimen via particles, liquids, gases, or surface diffusion. Preferably, the surface to be analyzed should not contact the container or any other object.

8.1.2.2 Containers such as glove boxes, vacuum chambers, and desiccators may be excellent choices for storage of specimens. Keep in mind that volatile species such as plasticizers may leave a surface in such containers, further contaminating the surface. Cross contamination between specimens may also occur in such cases.

8.1.3 *Temperature*—Possible temperature effects should be considered when storing or shipping specimens. Most detrimental effects result from elevated temperatures. Additionally, low specimen temperatures can lead to moisture condensation on the surface.

8.2 *Transfer:*

8.2.1 *Chambers*—Chambers that allow transfer of specimens from a controlled environment to an analytical chamber have been reported **(4, 5, 6)**. Controlled environments could be other vacuum chambers, glove boxes (dry boxes), glove bags, reaction chambers, etc. Other vacuum chambers, glove boxes, and reaction chambers can be attached directly to an analytical chamber with the transfer made through a permanent valve. Glove bags can be temporarily attached to an analytical chamber with transfer of a specimen done by removal and then replacement of a flange on the analytical chamber.

8.2.2 *Coatings*—Coatings can sometimes be applied to specimens allowing transfer in atmosphere. The coating is then removed by heating or vacuum pumping in either the analytical chamber or its introduction chamber. This concept has been successfully applied to the transfer of GaAs **(7)**. Surfaces to be analyzed by SIMS or AES can be covered with a uniform layer, such as polysilicon for silicon-based technology **(8)**. In this case, the coating is removed during analysis.

9. Techniques for Removal of Overlayers

9.1 *General Considerations:*

9.1.1 Often the surface or interface of interest lies beneath a layer of contaminants or other constituents. The problem is then to remove the overlayer without perturbing the surface or interface of interest **(9)**.

9.1.2 For electronic devices, information regarding preparation of specimens can be found in Ref **(10)**.

9.1.3 Organic layers can be removed from smooth metal substrates by rubbing with cork **(11)**.

9.2 *Mechanical Separation:*

9.2.1 Sometimes it is possible to mechanically separate layers and expose the surface of interest. Except for possible reactions with the atmosphere, a surface exposed in this way is generally excellent for analysis.

9.2.2 The inside surfaces of blister-like structures are often investigated in this way. Sputter depth profiling is generally not a good method to use on blister-like structures. At the point when the outer skin is penetrated by the ion beam, the data become dominated by artifacts.

9.3 *Thinning versus Removal*—Complete removal of an overlayer may not be possible, or desirable. It may be sufficient to thin the overlayer and continue using sputter depth profiling as discussed in 9.9.

9.4 *Removing the Substrate*—In some specimens it may be easier to approach the interface of interest by removing the substrate rather than the overlayer. This could be the case when the composition of the substrate is not of interest, and the composition of the overlayer material is unknown. Chemical etches may be used more effectively and perhaps selectively when the composition of the material to be etched is known. In SIMS, if the overlayers are characterized by non-uniform sputtering, substrate removal may provide good depth resolution **(12)**. As discussed in 9.3, complete removal of the substrate may not be necessary.

9.5 *Sectioning Techniques:*

9.5.1 *General*—Sectioning is most often applied to metals, but it can often be applied to other materials equally well. When using sectioning techniques, it is important to section such that minimum alteration occurs to the region of the specimen that will be analyzed. After sectioning it is usually necessary to clean the specimen by sputtering in the analytical chamber prior to analysis.

9.5.2 *Methods of Sectioning*—Cutting can be accomplished with an abrasive wheel, sawing, or shearing. The extent of damage is generally increased as cutting speed is increased. Chemical changes can be extensive if local heating occurs. Coarse grinding is usually done with abrasive belts or disks. Fine grinding is usually done with silicon carbide, emery, or aluminum oxide abrasives. Lubricating oils from cutting tools and grinding materials may contaminate the surface and should be removed. If possible, sectioning (cutting) should be done without lubricants.

9.5.3 *Mechanical Polishing*—Polishing is often the most crucial step in the sequence of preparing a lapped or polished specimen. The abrasives used may be aluminum oxide, chromium oxide, magnesium oxide, cerium oxide, silicon dioxide, or diamond. Choice of suspension medium (normally oil or water) and polishing cloth must be carefully considered.

9.5.4 *Chemical or Electrochemical Polishing*—Chemical or electrochemical polishing is sometimes applied after the final mechanical polishing. In chemical polishing the specimen is immersed in a polishing solution without external potentials being applied. In electrochemical polishing, a constant current or voltage is applied to the specimen. The solution and temperature selected will depend upon the specimen. These polishing methods usually prevent surface damage introduced by mechanical polishing. Polishing may alter the chemistry of the surface.

9.5.5 *Mounting Materials*—Compression and thermosetting materials are normally used for mounting specimens for sectioning. These mounting block materials are often detrimental to the vacuum environment of the analytical chamber, and the specimens are normally removed from the mounting blocks prior to analysis.

9.5.6 *Angle Lapping*—Angle lapping is a technique used to expand the area available for analyzing a thin layer at some depth into a specimen **(13)**. The diameter of the probing electron beam for AES must be small relative to the expanded dimensions of the layer to be analyzed. The same considerations and techniques applicable to sectioning described in 9.5 would also be applicable to lapping.

9.5.7 *Ball Cratering*—Ball cratering is similar to angle lapping **(14)**. Ball cratering is applicable when the radius of curvature of the spherical surface is large relative to the thickness of the films being analyzed.

9.5.8 *Crater Edge Profiling*—Crater edge profiling is similar to angle lapping. The craters left by fixed or rastered ion beams have a slightly slanting sidewall. The electron beam can be deflected across the crater wall to obtain composition versus depth information **(15)**.

9.5.9 *Location of the Interface*—For some angle lapped specimens, the location of the interface of interest may not be apparent from either visual inspection or secondary electron images. If there is a difference in the presence or concentration of an element across the interface, then either an Auger electron line scan or map may be used to locate the interface.

9.5.10 *Combining Sputter Depth Profiles With Angle Lapping*—Angle lapping can be used to reduce the amount of overlayer that must be sputtered away to expose the interface of interest. The sputter depth profile would be started at a location on the lapped surface adjacent to the interface where the overlayer material would be a minimum.

9.6 *Growth of Overlayers*—The interface between some overlayer material and a substrate can be analyzed by AES and XPS if the overlayer can be grown slowly or in discrete steps (for example, amounts of about one monatomic layer thickness). AES and XPS can thus be used to probe interface properties and possible reactions as the interface is grown. The composition at the interface measured in this way, however, may not always be identical with that for a thicker overlayer film. Many gas-metal and metal-metal interactions can be studied in this fashion.

9.7 *Solvents:*

9.7.1 High purity solvents can be used to remove soluble contaminants and overlayers. Ethanol and acetone are the most commonly used solvents. These solvents are often used in conjunction with ultrasonic agitation. A residue from the solvent may, however, be left on the specimen; in addition, use of acetone could temporarily reduce emission from LaB_6 cathodes if used in AES equipment.

9.7.2 Wiping a specimen with a tissue or other material that has been soaked with solvent can result in transfer of contaminants from the tissue to the specimen.

9.8 *Chemical Etching*—Chemical etches can be used to remove or thin an overlayer. In some cases an etch will be selective and etch down to, but not through, an interface. Specific etches can be found for many types of overlayers **(16)**.

9.9 *Sputtering:*

9.9.1 *General Conditions*—Sputtering (ion etching) is often used to expose subsurface layers or, combined with analysis, to produce sputter depth profiles. One typically uses noble gas ions at 1 to 5 kilo electron volts incident energy for sputtering. The effects of sputtering in surface analysis can be quite complex, and reviews of sputtering can be found in Refs **(17 and 18)**. Some of the more important aspects are discussed in the following paragraphs:

9.9.2 *Mixed Layer*—Ion bombardment will normally mix the top layers of a specimen to a depth that is comparable with the depth of analysis for AES and XPS **(19)**. The extent of mixing will depend upon the composition of the specimen, the incident ion species, and the energy of the incident ions. Lower incident energies, and more grazing angle of incidence will reduce the depth of the mixed layer.

9.9.3 *Preferential Sputtering*—The constituents of a specimen may not sputter at uniform rates. This means that within the mixed layer the species that sputters most rapidly will be depleted relative to the bulk composition of the material. This may be an important consideration in quantitative studies, especially when dealing with metal alloys **(20)**.

9.9.4 *Chemical Changes*—The energetic ion beam used for sputtering can cause chemical changes in the specimen. The composition of the specimen will be dominant in determining if this will occur. For example, nitrates, phosphates, and carbonates can be converted to oxides under bombardment by 1 to 3 keV argon ions **(21)**.

9.9.5 *Sputtering with Hydrogen*—Sputtering with hydrogen might remove contaminants in some cases with minimum alteration of the surface of interest **(22)**.

9.9.6 *Changes in Surface Topography*—Unidirectional ion bombardment often produces changes in surface topography. This can seriously reduce the chances of properly exposing or determining a subsurface interface. The depth resolution is usually 3 to 15 % of the sputtered depth **(23)**. Use of two ion guns incident at different angles can reduce sputter induced topographical features **(24)**. Specimen rotation during sputtering improves depth resolution **(25)**. Lower incident energies can also improve depth resolution **(26)**. Both smaller **(27)** and higher **(26)** angles of incidence have been shown to improve depth resolution for certain specimens.

9.9.7 *Sputtering and Heating*—Sputtering and heating (either simultaneously or sequentially) can be used to remove bulk impurities from metal foils or crystals when impurities segregate to the surface during heating. With single crystals, heating should be the final step to remove lattice damage.

9.9.8 *Sputter Enhanced Diffusion*—Sputtering can result in enhanced diffusion away from or toward the surface layer, producing distorted depth profiles. This can be a particular problem in SIMS **(28)**.

9.10 *Plasma Etching*—Plasma etching using a reactive species such as oxygen has been used to etch specimens when directional ion beams would produce artifacts in the data.

9.11 *Heating:*

9.11.1 Heating is not often used to clean specimens because only a small number of materials can withstand the high temperatures required to drive off many contaminants. The technique should be considered for refractory metals and, possibly, ceramics. Heating can cause many changes in a specimen, so this technique should be used with discretion. Heating is also useful for the outgassing of specimens, the removal of implanted rare gas ions, and annealing out lattice damage caused by ion bombardment of single crystals.

Methods of heating include resistive, electron bombardment, quartz lamp, laser, and indirect heating by conduction.

9.11.2 A variation of the heating technique is to combine lower temperatures with a reactive environment such as oxygen or hydrogen. This may result in the transformation of contaminants to volatile species that can be pumped away. This approach would normally be used in a chamber separate from the analysis chamber.

9.12 *Vacuum Pumping*—When the overlayers to be removed consist of materials with higher vapor pressures than the surface of interest, then the overlayers may be pumped away in an auxiliary vacuum chamber. As discussed in 9.11.2, vacuum pumping may be used in conjunction with heating. This approach may require several days and is generally applicable to organic overlayers on inorganic substrates.

9.13 *Ultraviolet Radiation*—Exposure of a specimen to ultraviolet radiation in air can remove organic contaminants, including photoresist residues, from the surfaces of specimens **(29)**. Note that some specimens may decompose under ultraviolet radiation.

10. In-Situ Exposure and Chemical Reaction Techniques—Fracture, Cleaving, and Scribing

10.1 *Vacuum Conditions, Time, and Contamination*—See 7.3.

10.2 *Fracture:*

10.2.1 *General Conditions*—Although in-situ fracture has been applied most extensively to metal specimens, it could be applied equally well to a broad range of materials. In-situ fracture has also found considerable use with composite materials, glasses, and ceramics.

10.2.2 *Impact or Tensile Fracture*—Impact fracture is utilized more than tensile fracture, possible because such devices are simpler and readily available, and multiple specimens can be analyzed without breaking vacuum. Devices for tensile fracture have been reported in Ref **(30)**, and have recently become commercially available. Such devices are usually limited to single specimens per pump-down of the vacuum chamber. Specimens can be intergranularly fractured at proper strain rate by tensile devices.

10.2.2.1 *Pretest*—It is possible to pretest specimens for impact fracture by mounting the specimen in a vise, and hitting it with a hammer or other methods to simulate the fracture stage or method. If an intergranular surface is exposed in this fashion, then it is likely that an intergranular failure will occur using the impact fracture mechanism in an ultra high vacuum chamber. Pretesting is also suggested for hydrogen charged specimens.

10.2.3 *Preparation of Specimens:*

10.2.3.1 *Geometry, Location of Fracture*—Impact fracture devices generally have a preferred geometry for the specimen to be fractured. The specimens are usually notched in an attempt to control the location of the fracture. Tensile fracture devices generally require a particular geometry for the specimen.

10.2.3.2 *Nonideal Geometries*—Specimens with nonideal geometries for impact fracture can still be fractured in the impact device by using additional pieces to allow the nonideal shape to approximate the ideal shape or by using special mounting in the fracture devices. When the geometry of a specimen does not fit the mounting mechanism well, or if the specimen is brittle, then it is advisable to wrap the end of the specimen held in the mount with a foil such as aluminum or indium. This should help prevent premature and poorly located fractures.

10.2.3.3 *Hydrogen Charging*—Many metal specimens can be charged with hydrogen to increase the probability of intergranular fracture. The time and temperature required for charging will depend upon the specimen. Specimens that have been charged with hydrogen will usually lose the hydrogen if they are allowed to remain at room temperature for a relatively short time. Such specimens can be shipped in dry ice via overnight express and stored in liquid nitrogen for many days without serious degradation of the charging. Specimens charged with hydrogen may need to be stressed or slowly strained in order for hydrogen embrittlement and in-situ fractures to occur.

10.2.3.4 *Coatings on Electrical Insulators*—When electrical insulators such as ceramic materials are fractured, problems with electrical charging may develop during analysis. To reduce these problems, it may be helpful to coat the outer surface of the insulator with a conducting material such as gold, prior to fracture.

10.3 *Cleaving*—Cleaving a single crystal specimen in an analytical chamber requires a special mechanism. Descriptions of such mechanisms can be found in Refs **(31 and 32)**.

10.4 *Scribing*—Scribing in-situ can be done to expose bulk material simply by scraping the specimen with a hard, sharp point. Caution should be observed regarding possible smearing of the constituents. The scribe mark should be large enough to contain the probing beam. A variation of this concept is to use a wire brush within a load-lock chamber.

10.5 *Reaction Chambers*—Specialized UHV chambers for controlled exposure of specimens to unique environments are available. Generally these chambers are separated from the analytical chamber by UHV valves and a suitable specimen transfer mechanism is available. Such chambers allow for specimen modifications by chemical or thermal means and minimize possible contamination to the analytical chamber.

11. Mounting of Specimens

11.1 *Methods of Reducing Charging:*

11.1.1 *General Considerations*—Specimen charging can be a serious problem with poorly conducting specimens. For many specimens, charging problems are usually more severe with incident electron or ion beams.

11.1.2 *Conductive: Mask, Wrap, or Coating*—Any conductive mask, grid, wrap, or coating should be connected to ground. A mask or grid of a conducting material can be used to cover insulating specimens and make contact as close as possible to the surface that will be analyzed. Wraps using metal foils have been used for the same purpose. In many cases, it is important to cover insulating parts of the specimen that are not in the immediate area of analysis because they can charge enough to deflect the electron beam to and from the specimen and will perturb the analysis accordingly. Whenever sputtering is used in conjunction with a mask, grid, or wrap, care should be taken to ensure that material is not sputtered from the mask, grid, or wrap to the surface of the specimen. Materials such as colloidal silver or colloidal graphite can be used to provide a conducting path from near the point of analysis to ground; beware that outgassing of the solvent may cause a problem. Coating a specimen with a thin conducting layer and subsequently removing the coating by sputtering may be useful, but information regarding the topmost layer of the specimen will generally be lost. This approach can be useful for sputter depth profiling. Combinations of coatings and masks or wraps may be used.

11.1.3 *Flood Gun*—Low-energy electrons from a nearby filament can be useful for reducing charging of specimens. Relative location of electron and ion optics in SIMS analysis of insulators can influence charging phenomena **(33, 34)**.

11.1.4 *Incident Electron and Ion Beams:*

11.1.4.1 *Angle of Incidence*—The secondary electron emission coefficient and the incident beam current density are functions of the angle of incidence of the primary electron beam. Grazing angles of incidence increase the secondary electron emission coefficient and are, therefore, generally better for reduction of charging.

11.1.4.2 *Energy*—The secondary electron emission coefficient is also a function of the energy of the incident electron beam. Generally, incident energies where the secondary electron emission coefficient is greater than unity, are better for reducing specimen charging. For some specimens that are layered, it might be possible to achieve reduced specimen charging by increasing the energy of the incident electron beam such that penetration is made to a conducting layer beneath the layer being analyzed. This will result in charge neutralization through the insulating layer to the conducting layer if the conducting layer is suitably grounded. In SIMS, the energy of the incident ion affects specimen charging **(33)**.

11.1.4.3 *Current Density*—Specimen charging can sometimes be reduced by decreasing the current density of the incident electron or ion beam. Reduction of the beam density can be achieved by reducing the total current, defocusing the beam, rastering the beam over a part of the specimen surface, or by increasing the angle of incidence.

11.2 *Methods of Reducing Thermal Damage*—To reduce thermal damage, specimens can be mounted on a cold probe or stage with liquid nitrogen or other cold liquids or gases flowing through it. Some specimens such as powders could benefit from being compacted to pellets, thereby increasing conducting heat dissipation. Good thermal contact between the specimen and the mounting system should be considered. Wrapping a specimen in a metal foil may be of value in some cases. Reducing the energy input during analysis would also be beneficial as discussed in 11.1.4.2 and 11.1.4.3, but this may result in longer data acquisition times.

11.3 *Wires and Wire-like Specimens*—Wire-like specimens may be of such size that it is not possible for the probing beam to remain on the specimen only, and background artifacts may result. In such instances it may be possible to mount the specimen such that the background is sufficiently out of focus that it does not contribute to the signal. Many wires can also be placed side-by-side to fill the field of view.

11.4 *Powders and Particles:*

11.4.1 *Substrates*—Powders and particles are often easier to analyze if they can be placed on a conducting substrate. Indium foil has been used because it is soft at room temperatures, and powders or particles will imbed partly into the foil. A problem with indium foil is that it splatters if sputtering is attempted. Aluminum, copper, and other metal foils can be used, though only a small percentage of the powder particles may adhere to them. For XPS, powders can be placed on adhesive tape. The metallized kind is usually best and can meet the vacuum requirements of most XPS systems. The adhesive tape used should be pretested for vacuum compatibility.

11.4.2 *Pellets*—Many powders can be formed into pellets. Forming pellets can be an excellent approach for XPS but often leads to specimen charging in AES and SIMS. Note that pressure and temperature-induced changes may occur.

11.4.3 *Transfer of Particles*—Particles can sometimes be transferred to a suitable substrate by using a very sharp needle and by working under a microscope. Particles that are not soluble can sometimes be floated on solvents and picked up on conducting filters.

11.4.4 *Fibers and Filaments*—Fibers and filaments can often be handled in much the same ways as particles. Bundles of fibers can be mounted on sticky tape, under masks, etc. If possible, individual fibers or filaments can be supported or hung as discussed in 11.3.

11.4.5 *Pedestal Mounting*—For some analytical systems, it is possible to mount a specimen on a pedestal so that only the specimen will be seen by the analyzer. This approach may allow analysis of specimens that are smaller than the probing beam.

12. Special Handling Techniques

12.1 *Prepumping of Gassy Specimens*—Some specimens will emit gases and cannot be analyzed due to problems with the vacuum environment in the analytical chamber. These specimens can be prepumped in an auxiliary vacuum chamber and quickly transferred to the analytical chamber without appreciable pickup of gases during the transfer. Perhaps the easiest method for prepumping is in the introduction chamber of a fast insertion probe. Cross contamination between specimens may occur in such cases.

12.2 *Material Transfer*—Material transfer can be of value when the specimen is too large to be inserted into an analytical chamber. The film or particles to be analyzed must transfer from the specimen to the replicating compound. The replicating compound should be conductive **(35)**.

12.3 *Preservation of Salt Films Formed by Electrochemistry*—When forming specimens by electrochemical means, salt films of interest may be present only when a potential is applied and will dissolve once the potential is removed. Such films may be preserved by withdrawing the specimen from the electrolyte solution while maintaining the applied potential. The specimen may be withdrawn into a nonreacting atmosphere, or in some cases, into a liquid layer in which the salt film is insoluble.

12.4 *Viscous Liquids*—Viscous liquids can be analyzed by XPS by placing a thick layer on a smooth substrate material and wiping away most of the liquid. Often the remaining specimen layer is of such thickness that no signal from the substrate is detected, yet the vacuum requirements of the analytical chamber are met.

ASTM E 1078

REFERENCES

(1) Rivière, J. C., "Instrumentation," *Practical Surface Analysis by Auger and X-ray Photoelectron Spectroscopy*, Briggs, D. and Seah, M. P., eds., John Wiley & Sons, Chichester, 1983, pp. 17–85.

(2) Pantano, C. G. and Madey, T., "Electron Beam Damage in Auger Electron Spectroscopy," *Applications of Surface Science*, Vol 7, 1981, pp. 115–141.

(3) Fote, A. A., Slade, R. A., and Feurstein, S., "Thermally Induced Migration of Hydrocarbon Oil," *Journal of Lubrication Technology*, Vol 99, 1977, pp. 158–162.

(4) Hobson, J. P. and Kornelsen, E. V., "A Target Transfer System at Ultrahigh Vacuum," *Proceedings of the 7th International Vacuum Congress and the 3rd International Conference on Solid Surfaces*, Vienna, Austria, Vol 3, September 12–16, 1977, pp. 2663–2666.

(5) Hobson, J. P., "First Intercontinental Test of UHV Transfer Device," *Journal of Vacuum Science and Technology*, Vol 15, No. 4, 1978, pp. 1609–1611.

(6) Fleisch, T., Shepherd, A. T., Ridley, T. Y., Vaughn, W. E., Winograd, N., Baitinger, W. E., Ott, G. L., and Delgass, W. N. "System for Transferring Samples Between Chambers in UHV," *Journal of Vacuum Science and Technology*, Vol 15, No. 5, 1978, pp. 1756–1760.

(7) Kowalczyk, S. P., Miller, D. L., Waldrop, J. R., Newman, P. G., and Grant, R. W., "Protection of Molecular Beam Epitaxy Grown Al-Ga-As Epilayers During Ambient Transfer," *Journal of Vacuum Science and Technology*, Vol 19, No. 2, 1981, pp. 255–256.

(8) Williams, P. and Baker, J. E., "Quantitative Analysis of Interfacial Impurities Using Secondary Ion Mass Spectrometry," Applied Physics Letters, Vol. 36, 1980, pp. 840–845.

(9) Musket, R. G., McLean, W., Colmenares, C. A., Makowiecki, D. M., and Siekhaus, W. J., "Preparation of Atomically Clean Surfaces of Selected Elements: A Review," *Applications of Surface Science*, Vol 10, 1982, pp. 143–207.

(10) Holloway, P. H. and McGuire, G. E., "Characterization of Electronic Devices and Materials by Surface-Sensitive Analytical Techniques," *Applications of Surface Science*, Vol 4, 1980, pp. 410–444.

(11) Lindfors, P. A., "Using a Cork to Remove Organic Materials from the Surface of Metals," The PHI Interface Vol. 8, No. 1, 1985, p. 4 (published by Perkin-Elmer, Physical Electronics Division, Eden Prairie, MN, USA).

(12) Lareau, R. T., *Secondary Ion Mass Spectrometry, SIMS VI*, eds. A. Benninghoven, A. M. Huber, and H. W. Werner, J. Wiley, New York, 1988, pp. 437–440.

(13) Tarng, M. L. and Fisher, D. G., "Auger Depth Profiling of Thick Insulating Films by Angle Lapping," *Journal of Vacuum Science and Technology*, Vol 15, No. 1, 1978, pp. 50–53.

(14) Walls, J. M., Brown, I. K., and Hall, D. D., "The Application of Taper-sectioning Techniques for Depth Profiling Using Auger Electron Spectroscopy," *Applications of Surface Science*, Vol 15, No. 1-4, 1983, pp. 93–107.

(15) Taylor, N. J., Johannessen, J. S., and Spicer, W. E., "Crater-edge Profiling in Interface Analysis Employing Ion-beam Etching and AES," *Applied Physics Letters*, Vol 29, No. 8, 15 October 1976, pp. 497–499.

(16) Beadle, W. E., Tsai, J. C. C., and Plummer, R. D., eds., *Quick Reference Manual for Silicon Integrated Circuit Technology*, Chapter 5, J. Wiley, New York, 1985.

(17) Wehner, G. K., "The Aspects of Sputtering in Surface Analysis Methods," *Methods of Surface Analysis*, A. W. Czanderna, ed. Elsevier, Amsterdam, 1975, pp. 5–37.

(18) Holloway, P. H. and Hofmeister, S. K., "Correction Factors and Sputtering Effects in Quantitative Auger Electron Spectroscopy," *Surface and Interface Analysis*, Vol 4, 1982, pp. 181–184.

(19) Sanz, J. M. and Hofmann, S., "Quantitative Evaluation of AES Depth Profiles of Thin Anodic Oxide Films (Ta_2O_5/Ta, Nb_2O_5/Nb)," *Surface and Interface Analysis*, Vol 5, No. 5, 1983, pp. 210–216.

(20) Shimizu, R. and Saeki, N., "Study of Preferential Sputtering on Binary Alloy by In-situ Auger Measurement of Sputtered and Sputter Deposited Surfaces," *Surface Science*, Vol 62, 1977, pp. 751–755.

(21) Christie, A. B., Lee, J., Sutherland, I., and Walls, J. M., "An XPS Study of Ion-Induced Compositional Changes With Group II and Group IV Compounds," *Applications of Surface Science*, Vol 15, 1983, pp. 224–237.

(22) Bouwman, R., van Mechelen, J. B., and Holsher, A. A., "Surface Cleaning by Low Temperature Bombardment with Hydrogen Particles: An AES Investigation on Copper and Fe-Cr-Ni Steel Surfaces," *Journal of Vacuum Science Technology*, Vol 15, No. 1, 1978, pp. 91–94.

(23) Seah, M. P. and Hunt, C. P., "The Depth Dependence of the Depth Resolution in Composition-Depth Profiling with Auger Electron Spectroscopy," *Surface and Interface Analysis*, Vol 5, No. 1, 1983, pp. 33–37.

(24) Sykes, D. E., Hall, D. D., Thurstans, R. E., and Walls, J. M., "Improved Sputter-Depth Profiles Using Two Ion Guns," *Applications of Surface Science*, Vol 5, 1980, pp. 103–106.

(25) Zalar, A., "Improved Depth Resolution by Sample Rotation During Auger Electron Spectroscopy Depth Profiling," Thin Solid Films, Vol. 124, 1985, pp. 223–230.

(26) Fine, J., Lindfors, P. A., Gorman, M. E., Gerlach, R. L., Navinsek, B. D., Mitchell, F., and Chambers, G. P., "Interface Depth Resolution of Auger Sputter Profiled Ni/Cr Interfaces: Dependence on Ion Bombardment Parameters," Journal of Vacuum Science and Technology, Vol. A(3), 1985, pp. 1413–1417.

(27) Seah, M. P. and Lea, C., "Depth Resolution in Composition Profiles by Ion Sputtering and Surface Analysis for Single-Layer and Multilayer Structures on Real Surfaces," Thin Solid Films, Vol. 81, 1981, pp. 257–270.

(28) Deline, V. R., Reuter, W., and Kelly, R., *Secondary Ion Mass Spectrometry, SIMS V*, eds. A. Benninghoven, R. J. Colton, D. S. Simons, and H. W. Werner, eds., Springer-Verlag, Berlin, 1986, pp. 299–302.

(29) Holloway, P. H. and Bushmire, D. W., "Detection by Auger Electron Spectroscopy and Removal by Ozonization of Photoresist Residues," *Proceedings of the 12th Annual Reliability Physics Symposium*, Las Vegas, IEEE Electron Device and Reliability Groups, NY, 1974, p. 180.

(30) Seah, M. P. and Hondros, E. D., "Grain Boundary Segregation," *Proc. Royal Society London*, Vol A335, 1973, pp. 191–212.

(31) Lander, J. J., Gobeli, G. W., and Morrison, J., "Structural Properties of Cleaved Silicon and Germanium Surfaces," *Journal of Applied Physics*, Vol 34, No. 8, 1963, pp. 2298–2306.

(32) Grant, J. T. and Haas, T. W., "Auger Electron Spectroscopy of Si," *Surface Science*, Vol 23, 1970, pp. 347–362.

(33) Stevie, F. A., Rana, V. V. S., A. Harris, S., Briggs, T. H., and Skeath, P., "High Sputter Rate Secondary Ion Mass Spectrometry Analysis of Insulators Used in Microelectronics and Lightwave Applications," Journal of Vacuum Science and Technology, Vol. A6, 1988, pp. 2082–2084.

(34) Wilson, R. G., Stevie, F. A., and Magee, C. W., *Secondary Ion Mass Spectrometry: A Practical Handbook for Depth Profiling and Bulk Impurity Analysis*, J. Wiley, New York, 1989, Section 4.3.

(35) DeGroot, P. B. and Scott, R. H., "Extending Replication Methods to Auger Electron Spectroscopy by Using Conductive Replicas," *Microbeam Analysis 1979*; Newbury, D. E., ed., San Francisco Press, Inc., 547 Howard Street, San Francisco, CA 94105, pp. 321–323.

Designation: E 1162 – 87

Standard Practice for Reporting Sputter Depth Profile Data in Secondary Ion Mass Spectrometry (SIMS)[1]

1. Scope

1.1 This practice covers the information needed to describe and report instrumentation, specimen parameters, experimental conditions, and data reduction procedures. SIMS sputter depth profiles can be obtained using a wide variety of primary beam excitation conditions, mass analysis, data acquisition, and processing techniques **(1–4)**.[2]

1.2 *Limitations*—This practice is limited to conventional sputter depth profiles in which information is averaged over the analyzed area in the plane of the specimen. Ion microprobe or microscope techniques permitting lateral spatial resolution of secondary ions within the analyzed area, for example, image depth profiling, are excluded.

2. Referenced Document

2.1 *ASTM Standard:*

E 673 Definitions of Terms Relating to Surface Analysis[3]

3. Terminology

3.1 For definitions of terms used in this practice, see Definitions E 673.

4. Summary of Practice

4.1 Experimental conditions and variables that affect SIMS sputter depth profiles **(1–4)** and tabulated raw data (where feasible) are reported to facilitate comparisons to other laboratories or specimens, and to results of other analytical techniques.

5. Significance and Use

5.1 This practice is used for reporting the experimental conditions as specified in Section 6 in the "Methods" or "Experimental" sections of other publications (subject to editorial restrictions).

5.2 The report would include specific conditions for each data set, particularly, if any parameters are changed for different sputter depth profile data sets in a publication. For example, footnotes of tables or figure captions would be used to specify differing conditions.

6. Information to Be Reported

6.1 *Instrumentation:*

6.1.1 If a standard commercial SIMS system is used, specify the manufacturer and instrument model number. Specify, the model numbers and manufacturer of any accessory or auxiliary equipment relevant to the depth profiling study (for example, special specimen stage, primary mass filter, electron flood gun, vacuum pumps, data acquisition system, and source of software, etc.).

6.1.2 If a nonstandard commercial SIMS system is used, specify the manufacturer and model numbers of components (for example, primary ion source, mass analyzer, data system, and accessory equipment).

6.2 *Specimen:*

6.2.1 Describe the specimen as completely as possible. For example, specify its bulk composition, preanalysis history, physical dimensions. If the specimen contains dopants, for example, semiconductors, report the dopant type and concentration. For multicomponent specimens, state the degree of specimen homogeneity.

6.2.2 State the method of mounting and positioning the specimen for analysis. Specify any physical treatment of the specimen mounted in the SIMS analysis chamber (for example, heated, cooled, electron bombarded, etc.). Note the specimen potential relative to ground. Describe the method of specimen charge compensation used (if any), for example, conductive coatings or grid, electron flooding, etc.

6.3 *Experimental Conditions:*

6.3.1 *Primary Ion Source*—Give the following parameters whenever possible: Composition (if mass filtered, give the specific ion and isotope, for example, $^{16}O^-$); angle of incidence (relative to the surface normal); ion beam energy; current (including the method used for measurement, for example, Faraday cup); beam diameter (including the method used for measurement); size and shape of sputtered area; primary beam current density for a stationary beam (A/m^2); beam raster size and rate (if used); primary ion dose rate averaged over the sputtered area ($ions/m^2 \cdot s$).

6.3.2 *Secondary Ion Mass Spectrometer*—Give the following parameters whenever possible: analyzed area versus total sputtered area (for example, image filed/selected area aperture size for stigmatic ion microscopes; raster/electronic signal gating for ion microprobes, etc.); collection angle (angle between surface normal and secondary ion collection optics); the nature of secondary ion energy distributions and the spectrometer energy acceptance/bandpass within the energy distribution used during depth profiles (particularly important if energy discrimination is used to remove polyatomic ion interferences); mass resolution ($M/\Delta M$ where ΔM is the full width at half maximum intensity for an ion peak of mass M); the nature of specimen charge compensation if any (for example, changes in sample poten-

[1] This practice is under the jurisdiction of ASTM Committee E-42 on Surface Analysis and is the direct reponsibility of Subcommittee E42.06 on SIMS.

Current edition approved April 24, 1987. Published June 1987.

[2] The boldface numbers in parentheses refer to the references at the end of this standard.

[3] *Annual Book of ASTM Standards*, Vol 03.06.

tial biasing during depth profile): method used to perform selected ion monitoring during sputtering (for example, electrostatic or magnetic peak switching procedures for double focusing instruments).

6.3.3 *Secondary Ion Intensity Measurement*—Specify the type of detector (for example, electron multiplier, Faraday cup) and detector bias used including the counting (integration) time used for each measurement of each ion of interest. For analog detection, give the detector system time constant. For pulse counting detection, give the pulse pair resolution including dead time corrections. For rapidly rastered primary beams, correct intensities (counts/second) to instantaneous values by multiplying by the ratio of total sputtered area to the analyzed area (important procedure to help assess possible detector saturation limitations).

6.3.4 *Vacuum*—Specify pressures in the primary column, specimen chamber, mass spectrometer prior to and during sputter depth profiling, including the type of vacuum pumping. Also give the composition of the residual gas, if available. If flooding of the sample surface region or backfilling of the analysis chamber with reactive gases (for example, oxygen) is used give the details of the procedure including the partial pressure of the reactive gas.

6.4 *Quantification by Data Reduction:*

6.4.1 *Concentrations*—If any elemental concentrations are presented, state clearly the methodology used for quantification (**5, 6**). In addition, specify the nature of any external or internal standards used including methods for normalization in comparing ion intensities in standards to ion intensities in specimen depth profiles. Specify standards made by ion implantation according to ion species, dose, energy, matrix, and reference data used to calculate peak concentration of the implant in the standard. Report analytical precisions for multiple determinations of concentrations.

6.4.2 *Depth Scales*—Specify the methods used (if any) to relate elapsed sputter time to a depth sputtered (that is, depth scale calibration). Possible techniques include measurements of: times to remove standard films of known thickness, ion implant standards with peak concentrations occurring at calculated depths (for example, LSS model), or crater depths via various stylus, profilometry or interferometry techniques. Report any nonuniform sputtering of the specimen, if observed.

6.5 *Display of SIMS Sputter Depth Profile Figures:*

6.5.1 *Raw Ion Intensity Versus Sputtering Time Profiles*—The left hand vertical axis gives ion intensities measured in arbitrary units (analog detection), or in instantaneous counts per second (pulse counting, see 6.3.3). The intensity axis can be either linear or logarithmic depending upon suitability relative to the dynamic range of the profile. The scale selected should be clearly indicated. The bottom horizontal axis will be the sputtering time reported in time units. If the primary ion parameters are changed during the profile in a manner that affects the sputter rate, the time axis must be adjusted accordingly.

6.5.2 *Quantified Depth Profiles*—If elemental concentrations or depth scales are quantified as described in 6.4.1 and 6.4.2, use the following procedure. The right hand verticle axis can be reported in units of atomic percent, weight percent, or atoms per cubic metre, whichever is most convenient or appropriate. The top horizontal axis can be indicated in units of depth (typically nanometres or micrometres). An example of the format is shown in Fig. 1 for a ^{11}B implant profile in silicon.

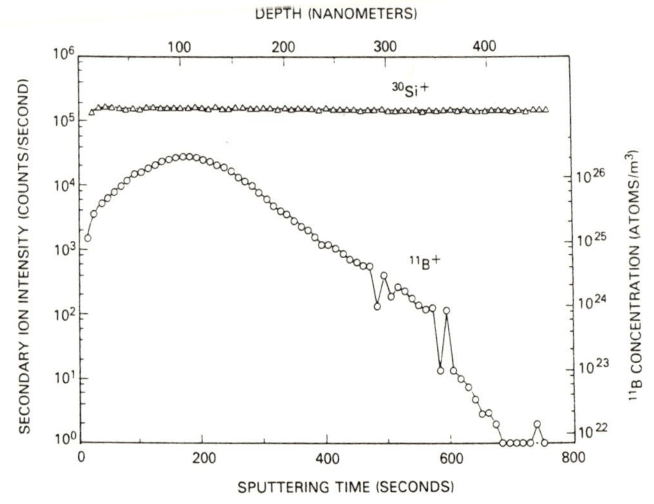

FIG. 1 SIMS Sputter Depth Profile of Boron in Silicon

ASTM E 1162

REFERENCES

(1) Hofmann, S., "Quantitative Depth Profiling in Surface Analysis," *Surface and Interface Analysis*, Vol 2, 1980, p. 148.
(2) Zinner, E., "Depth Profiling by Secondary Ion Mass Spectrometry," *Scanning*, Vol 3, 1980, p. 57.
(3) Wittmaack, K., "Depth Profiling by Means of SIMS: Recent Progress and Current Problems," *Radiation Effects*, Vol 63, 1982, p. 205.
(4) Williams, P., "Secondary Ion Mass Spectrometry," *Applied Atomic Collision Physics*, Vol 4, 1983, p. 327.
(5) Werner, H. W., "Quantitative Secondary Ion Mass Spectrometry: A Review," *Surface and Interface Analysis*, Vol 2, 1980, p. 56.
(6) Wittmaack, K., "Aspects of Quantitative Secondary Ion Mass Spectrometry," *Nuclear Instruments and Methods*, Vol 168, 1980, p. 343.

Index